Monitoring the Comprehensive Nuclear-Test-Ban Treaty: Sourse Location

Edited by
Frode Ringdal
Brian L. N. Kennett

Springer Basel AG

Reprint from Pure and Applied Geophysics
(PAGEOPH), Volume 158 (2001), No. 1/2

Editors:

Frode Ringdal
NORSAR
P.O. Box 51
N-2027 Kjeller
Norway
e-mail: frode.ringdal@norsa.no

Brian L. N. Kennett
Research School of Earth Sciences
Australian National University
Canberra ACT 0200
Australia
e-mail: brian@rses.anu.edu.au

A CIP catalogue record for this book is available from the Library of Congress,
Washington D.C., USA

Deutsche Bibliothek Cataloging-in-Publication Data

Monitoring the comprehensive nuclear test ban treaty. - Basel ; Boston ; Berlin : Birkhäuser
(Pageoph topical volumes)

Sourse location / ed. by Frode Ringdal ; Brian L. N. Kennett. - 2001
ISBN 978-3-7643-6534-9 ISBN 978-3-0348-8250-7 (eBook)
DOI 10.1007/978-3-0348-8250-7

Originally published by Birkhäuser Verlag, Basel - Boston - Berlin in 2001
Printed on acid-free paper produced from chlorine-free pulp

9 8 7 6 5 4 3 2 1

Contents

Pure appl. geophys. 158 (2001) 1–2
0033–4553/01/020001–2 $ 1.50 + 0.20/0

Pure and Applied Geophysics

Monitoring the Comprehensive Nuclear-Test-Ban Treaty

Preface

The first nuclear bomb was detonated in 1945, thus ushering in the nuclear age. A few political leaders quickly saw a need to limit nuclear weapons through international cooperation and the first proposals to do so were made later that same year. The issue of nuclear testing, however, was not formally addressed until 1958 when the United States, the United Kingdom, and the Soviet Union, initiated talks intended to establish a total ban on that testing (a Comprehensive Test-Ban Treaty or CTBT). Those talks ended unsuccessfully, ostensibly because the participants could not agree on the issue of on-site verification.

Less comprehensive treaties did, however, place some constraints on nuclear testing. The United States, the United Kingdom, and the Soviet Union, in 1963, negotiated the Limited Test-Ban Treaty (LTBT) which prohibited nuclear explosions in the atmosphere, outer space and under water. The Threshold Test-Ban Treaty (TTBT), signed by the United States and the Soviet Union in 1974, limited the size, or yield, of explosions permitted in nuclear tests to 150 kilotons.

Seismological observations played an important role in monitoring compliance with those treaties. Many of the world's seismologists set aside other research projects and contributed to that effort. They devised new techniques and made important discoveries about the Earth's properties that affect our ability to detect nuclear events, to determine their yield, and to distinguish them from earthquakes. Seismologists are rightfully proud of their success in developing methods for monitoring compliance with the LTBT and TTBT.

Although seismologists have also worked for many years on research related to CTBT monitoring, events of recent years have caused them to redouble their efforts in that area. Between 1992 and 1996 Russia, France and the United States all placed moratoria on their nuclear testing, though France did carry out a few tests at the end of that period. In addition, the United States decided to use means other than testing to ensure the safety and reliability of its nuclear arsenal and all three countries, as well as the United Kingdom, agreed to continue moratoria as long as no other country tested. Those developments, as well as diplomatic efforts by many nations, led to the renewal of multilateral talks on a CTBT that began in January 1994. The talks led to the Comprehensive Nuclear-Test-Ban Treaty. It was adopted by the United Nations General Assembly on 10 September, 1996 and, as of November 2000, has been signed by 160 nations. Entry of the treaty into force, however, is still uncertain since it requires ratification by all 44 nations that have some nuclear capability and, as of November 2000, only 30 of those nations have done so.

Although entry of the CTBT into force is still uncertain, seismologists and scientists in related fields, such as radionuclides, have proceeded with new research on issues relevant to monitoring compliance with it. Results of that research may be used by the International Monitoring System, headquartered in Vienna, and several national centers and individual institutions to monitor compliance with the CTBT. New issues associated with CTBT monitoring in the 21st century have presented them with many new challenges. They must now be able to effectively monitor compliance by several countries that have not previously been nuclear powers. Effective monitoring requires that we be able to detect and locate considerably smaller nuclear events than ever before and to distinguish them from small earthquakes and other types of explosions. We must have those capabilities in regions that are seismically active and geologically complex, and where seismic waves might not propagate efficiently.

Major research issues that have emerged for monitoring a CTBT are the precise location of events, and discrimination between nuclear explosions, earthquakes, and chemical explosions, even when those events are relatively small. These issues further require that we understand how seismic waves propagate in the solid Earth, the oceans and atmosphere, especially in regions that are structurally complex, where waves undergo scattering and, perhaps, a high degree of absorption. In addition, we must understand how processes occurring at the sources of explosions and earthquakes manifest themselves in recordings of ground motion.

Monitoring a CTBT has required, and will continue to require, the best efforts of some the world's best seismologists. They, with few exceptions, believe that methods and facilities that are currently in place will provide an effective means for monitoring a CTBT. Moreover, they expect that continuing improvements in those methods and facilities will make verification even more effective in the future. This topical series on several aspects of CTBT monitoring is intended to inform readers of the breadth of the CTBT research program, and of the significant progress that has been made toward effectively monitoring compliance with the CTBT.

The following set of papers, edited by Drs. Frode Ringdal and Brian Kennett, presents research results on methods used to locate seismic sources in complex media. It is the first of eight topics addressed by this important series on *Monitoring the Comprehensive Nuclear-Test-Ban Treaty*. Topics to appear in later issues are Hydroacoustics, Surface Waves, Source Processes and Explosion Yield Estimation. Regional Wave Propagation and Crustal Structure, Infrasound, Source Discrimination, and Data Processing.

Brian J. Mitchell
Saint Louis University
Series Editor

Pure appl. geophys. 158 (2001) 3–5
0033–4553/01/020003–3 $ 1.50 + 0.20/0

Pure and Applied Geophysics

Introduction

FRODE RINGDAL[1] and BRIAN L. N. KENNETT[2]

With the adoption of the Comprehensive Nuclear-Test-Ban Treaty (CTBT), accurate location of small seismic events has become more important than ever before in seismic monitoring programs. The seismic component of the International Monitoring System (IMS) includes 50 primary stations with continuous data transmission to the International Data Centre (IDC) in Vienna, Austria and 120 auxiliary stations which will provide data to the IDC on request. This high-quality and carefully operated network will enable significantly more accurate location of earthquakes and other seismic sources worldwide as compared to traditional practice. Regional location calibration of various areas of the world and improved data processing algorithms will be needed to achieve these improvements.

The first paper in this volume (by Kennett and Ringdal) gives an overview of the technical background and the requirements for seismic event location in the CTBT context. It outlines the procedures currently employed in locating seismic events using global or regional networks of stations, with emphasis on the problems associated with low-magnitude events. The authors address the issue of using global versus regional travel-time models, and illustrate the respective uncertainties in location estimates by analyzing a small mining explosion recorded by a sparse array network in Fennoscandia.

The next two contributions describe various aspects of the operation of the prototype IDC in Arlington, Virginia, USA. Bondár *et al.* give an introduction to the location calibration database at the prototype IDC and briefly describe techniques used in the calibration procedure. A paper by Yang *et al.* describes in detail how source-specific station corrections have been employed at the prototype IDC to obtain significantly improved location accuracy of seismic events in Fennoscandia and adjacent regions. This is seen as a model for future developments in other regions of the world.

One of the major cooperative efforts in source location in recent years is the joint United States/Russian Federation project for calibrating North America and

[1] NORSAR, PO Box 51, Kjeller, N-2007, Norway.
[2] Research School of Earth Sciences, Australian National University, Canberra ACT 0200, Australia.

Northern Eurasia. Two papers from this project are presented, one by the joint working group describing their approach to the problem and initial results on deriving regional travel-time curves for these areas, and the second one by Ryaboy *et al.* which discusses the application of three-dimensional modelling techniques for deriving accurate travel-time corrections for North America.

Several other contributions also address the development and application of regional models to provide improved location estimates and to validate these improvements by considering events with known location. Finn *et al.* show that significant improvement in location accuracy can be achieved for small mining events in Wyoming using regional calibration. Kremenetskaya *et al.* develop a regional velocity model for the European Arctic, and demonstrate by examples that significant location improvements can be obtained when using this model compared to the use of a global velocity model. Hicks *et al.* compare the results from using local and sparse regional networks to locate small earthquakes in Norway, and point out that the current procedures at the prototype IDC do not fully exploit the potential to detect and locate low-magnitude seismic events in this region.

Seismic events in Kazakhstan have been studied in three papers. This region is of particular interest because the main nuclear test site for the former Soviet Union was located near Semipalatinsk in Kazakhstan, and because several recent large chemical calibration explosions have been conducted there. Khalturin *et al.* summarize the history of nuclear testing at Semipalatinsk and provide important new information regarding some of the smaller historic nuclear tests. They also discuss the detectability of seismic events by the global network in the perspective of this new information. Phillips *et al.* use a master-event technique to obtain very accurate location estimates of some of the calibration explosions at the Semipalatinsk test site, using IMS data (or in some cases surrogate IMS stations). Belyashova *et al.* apply data from the calibration explosions to develop improved regional travel-time curves for this area.

A paper on location of seismic events in China is presented by Steck *et al.* They develop correction surfaces based upon a special interpolation technique, and demonstrate that such corrections improve the precision and accuracy of regional event location at the Lop Nor test site in China, and that they can also be applied to improve location estimates for events at the Balapan test site near Semipalatinsk in Kazakhstan.

The last group of papers deals with various methodological aspects of seismic event location. Sambridge and Kennett present a recently developed method of inversion, known as the neighborhood algorithm, and apply it to the location problem. They demonstrate that the method provides robust location estimates with a significantly reduced computational load compared to other nonlinear inversion techniques. Uhrhammer *et al.* introduce a new location algorithm which makes use of data recorded by a network of three-component broadband seismic stations and applies an adaptive migrating grid search technique. A paper by Schweitzer introduces several enhancements to the conventional seismic event location method, including the option to use travel-time differences between different seismic phases observed at the same station.

Antolik *et al.* investigate the ability of current three-dimensional models of mantle *P*-wave velocity to provide corrections that enable improved accuracy in locating seismic events at teleseismic distances. Piromallo and Morrelli present an alternative to three-dimensional structural models, demonstrating that about the same level of effectiveness in improving teleseismic location estimates can be obtained by using an approach based on empirical heterogeneity corrections.

The importance of accurate slowness and azimuth estimates from individual three-component stations and arrays is implicit in many of the contributions, especially when associating phases to define seismic events and for locating low-magnitude events. This topic is explicitly addressed in two contributions: Claassen describes a method for robust and accurate azimuth estimation using three-component data, and Schweitzer discusses the use of slowness and azimuth corrections at seismic arrays for obtaining improvements in the IDC processing.

The final paper in the volume, written by Bartal *et al.*, discusses a somewhat different approach to the location problem. It proposes and compares several techniques to define "exclusion zones," i.e. areas where a given seismic event could not possibly have occurred. This approach would be particularly relevant in the context of a possible on-site inspection under the CTBT.

In conclusion, this volume presents recent advances both in the methodology of locating seismic events and in the large-scale international regional calibration effort that is now underway. The editors hope that this volume will stimulate increased interest and further advances in this important field.

Pure appl. geophys. 158 (2001) 7–18
0033-4553/01/020007-12 $ 1.50 + 0.20/0

Pure and Applied Geophysics

Locating Seismic Events in the CTBT Context

B. L. N. KENNETT[1] and F. RINGDAL[2]

Abstract—The verification of the Comprehensive Nuclear-Test-Ban Treaty (CTBT) requires the determination of accurate location of seismic events from a fixed network of seismic stations across the globe. The requirements of possible on-site inspections mean that the goal is to place the location estimate in a zone smaller than 1000 km^2 that includes the true location. Because a defined set of stations will be used, corrections can be refined to represent the influence of departures from the global reference model IASPEI91. The primary stations in the International Monitoring Scheme (IMS) are mostly seismic arrays and therefore the present location scheme is based on minimisation of a misfit function built from arrival time, azimuth and array slowness residuals. The effective network will change markedly with the magnitude of the event and as a result regional information has to be integrated into the location process.

Key words: Comprehensive Nuclear-Test-Ban Treaty (CTBT), event location, regional corrections.

1. Introduction

The Comprehensive Nuclear-Test-Ban Treaty (CTBT) establishes an International Monitoring System with a seismological component of 50 primary stations and 120 auxiliary stations designed to provide as uniform coverage of the globe as possible. Data from the primary stations, which are mostly seismic arrays, will be transmitted to the International Data Centre in Vienna in real time, whereas data will be requested from the auxiliary stations as required.

The location of seismic events for treaty purposes is thus based on a relatively sparse global network. The on-site inspection protocols of the treaty specify:

> The one-site inspection shall be carried out in the area where the event that triggered the on-site inspection request occurred.
>
> The area of an on-site inspection shall be continuous and its size shall not exceed 1,000 square kilometres. There shall be no linear distance greater than 50 kilometres in any direction.

Thus the goal of the location process for the CTBT is to determine the position of events sufficiently well that, for any event where the nature of the source is in question, the true location lies within the designated zone of area 1000 km^2 or less.

---

[1] Research School of Earth Sciences, Australian National University, Canberra ACT 0200, Australia.
[2] NORSAR, PO Boks 51, Kjeller, N-2007, Norway.

The major part of the information used in the location process is a set of arrival times of seismic phases which have been measured from displays of seismic waveforms by experienced analysts. For seismic arrays it is possible to provide additional information such as the apparent azimuth of the arriving energy and its velocity across the array. For each seismic station the object is to provide arrival times (and array information) for a number of different seismic phases which have followed different propagation paths through the earth, together with an indication of the likely character of the arrival. Particular effort is made to try to extract information which is sensitive to the depth of the event e.g., by recognising the phases *pP*, *sP* which are reflected back from the surface above the source.

The initial identifications of phase character may need to be modified when information from many stations are combined. One of the most difficult tasks has then to be faced. From the assemblage of arrival time picks, azimuth and apparent velocity, those readings from different stations which should be grouped together, have to be recognised and a preliminary location assigned. This location will then be refined by matching the patterns of observed and predicted arrival times.

Around the globe, many different seismic events can occur in a given time period so there is no guarantee that all readings correspond to the same event. An aid to event location is provided by the continuous monitoring scheme of RINGDAL and KVAERNA (1989) in which a net of points across the globe are monitored to provide an indication of the most likely source which can match the pattern of waveform change. This approach has been progressively extended and can now be used across a full global network, but is particularly valuable for scanning for events in specified regions. It can therefore be used as a means of keeping track of the vast majority of seismic events globally, which occur associated with subduction zones, particularly around the Pacific.

Once an initial estimated location is produced, an attempt is made to improve this location using a form of analysis which assumes that the true location is close to this starting point. As we shall see, such an iterative scheme is quite sensitive to the starting estimate and it may be necessary to consider a grid of different starting locations to achieve a satisfactory result. A variety of different schemes have been suggested which do not require a progressive linearisation and in the future may well displace the conventional approach. For example SAMBRIDGE and KENNETT (2000) demonstrate that a fully nonlinear approach based on the use of a "neighbourhood algorithm" is able to achieve good convergence in a very large initial search area.

All location schemes for seismic events depend on having a suitable model of the propagation characteristics of the seismic phases which are observed. In particular, information on the depth and origin time of the seismic energy cannot be extracted without invoking a model of the earth (which in turn is dependent on knowledge of the position of events).

The history of seismology has therefore included a long cycle of improvements to earth models based on improved observations and timing, and better location estimates. The reference model for seismic wavespeeds currently used at the International Data Centre of the CTBT is the IASPEI91 model of KENNETT and ENGDAHL (1991). This model, which depends only on radius, was derived as a representation of the travel times of seismic phases (notably P and S) across the whole globe. The times of passage for different types of seismic phases can be found by tracing rays through the model and then these are tabulated for use in location.

The outer regions of the earth differ significantly in seismic structure and therefore a single wavespeed model such as IASPEI91 will not match the variations in different regions. The global reference model enables the rapid determination of a reasonable hypocentre estimate, which then needs refinement to take account of the specific propagation environment for the region in which the source is situated and the propagation characteristics to those stations which have been used in the location process. The generation of suitable regional models and correction schemes is currently a topic of major interest and many different approaches are being employed as illustrated in the papers in this special volume.

2. *Procedure for Event Location*

The aim of the event location is to determine the point in space and time at which seismic energy was initiated, which is specified by latitude, longitude, depth from the surface and origin time; these four parameters represent the *hypocentre* for the event. The four hypocentral parameters have to be estimated from the phase information collected at a limited number of seismic stations. At first sight this does not appear to be too difficult a problem, but in reality it is hard to secure high accuracy in location with just a few stations.

The minimal information needed is 4 sets of arrival times for a single phase such as P. The use of multiple phases or additional classes of information can reduce the number of stations required. In many circumstances, the geometry of the effective network will have a major influence on the accuracy of the hypocentral estimates. Locations will be most accurate when there is a good azimuthal distribution of recording stations around the source.

The specification of the location of the event requires the determination of four parameters:

t_h — the origin time of the seismic disturbance,

$\mathbf{x}_h = (x_h, y_h, z_h)$ — the spatial coordinates of the hypocentre,

which together define the 4-vector $\hat{\mathbf{h}}$ of hypocentral parameters. The effectiveness of a location estimate $\hat{\mathbf{h}}$ is judged by comparison of the observed phase information (arrival times, slowness, azimuth) for the different phases with the corresponding quantities calculated for a trial hypocentre using a particular earth model, which may include corrections for heterogeneity.

Consider an event grouping for which we have N_T observations t_i of the arrival times of identified phases at a set of seismic receivers; these may be associated with different ray paths for a single wave type or different wave types. In addition, since the primary network for the CTBT is largely comprised of seismic arrays we can expect to have measurements of the azimuth of arrival and apparent velocity (slowness) of a seismic phase across the array. Depending on the nature of the arrival we may not have both measurements for each phase so we will consider N_A azimuth observations and N_S slowness observations.

Figures 1–3 illustrate the process of picking arrivals and calculating array information for a regional event near Scandinavia. This event, which was a small mining explosion in the Kola Peninsula, was detected by the three Fennoscandian arrays Noress, Arcess and Finess (Fig. 1). Individual seismograms and phase arrival picks for the *Pn* and *Sn* phase of each array are illustrated in Figure 2. Note that the signal-to-noise ratio is much higher for Arcess than for Finess and Noress (Arcess

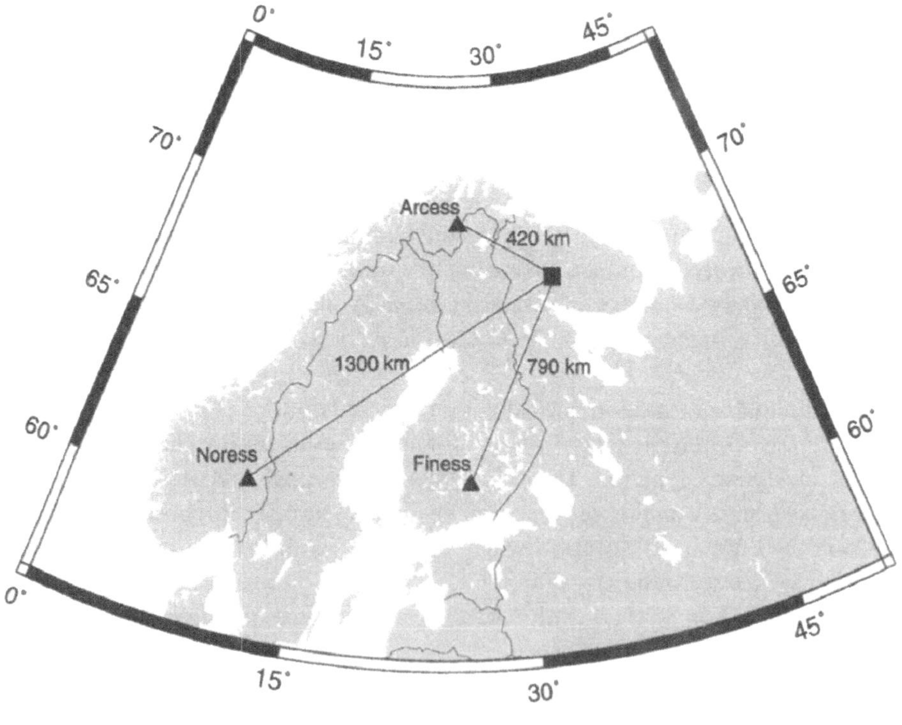

Figure 1
Map of northern Europe showing the known location of a small (m_b 2.8) mining explosion in the Kola Peninsula together with the three Fennoscandian seismic arrays Noress, Finess, Arcess. The distance from each array to the mine is indicated.

being the closest station) and consequently the Arcess time picks can be made with much better accuracy. Figure 3 shows a display of the *P*-phase azimuth and slowness at the Arcess and Noress arrays using a frequency-wavenumber estimation routine. The circle marked as representing 6 km/s is commonly used in array processing in

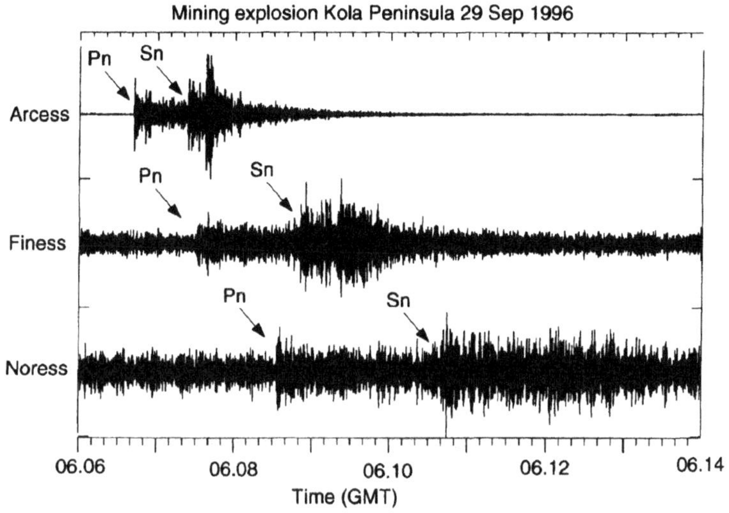

Figure 2
Illustration of arrival time picking for the event shown in Figure 1.

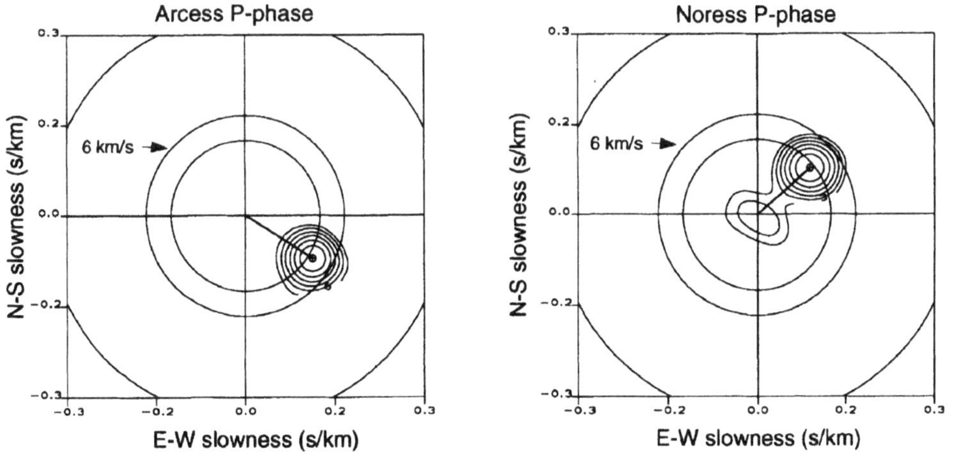

Figure 3
Illustration of frequency-wavenumber array analysis for the *P*-phase of the Arcess and Noress arrays. The estimated azimuth and slowness can be used to supplement the arrival times in the location procedure, as discussed in the text.

Fennoscandia to separate between *P*-type and *S*-type phases, thus providing automatic phase identification.

For the *i*-th seismic phase we calculate the travel time $t_{ri}(\mathbf{x}_s)$ for passage between a source at $\mathbf{x}_s = (x_s, y_s, z_s)$ and the requisite receiver by ray-tracing in a particular earth model. From the travel times we construct estimates t_{ci} of the arrival times of the phases for an assumed origin time t_s as

$$t_{ci}(\mathbf{x}_s, t_s) = t_s + t_{ri}(\mathbf{x}_s), \tag{1}$$

and these arrival times have to be compared with the observations t_i. We note the clear separation between the dependence of the estimated arrival time t_{ci} on the spatial and temporal components of the estimated location.

A measure of the misfit between observed and calculated arrival times is derived from the time residuals,

$$C_T = \frac{1}{N_T}\sum_{i=1}^{N_T}\frac{(t_i - t_{ci})^2}{\sigma_{ti}^2}, \tag{2}$$

where σ_{ti}^2 is the estimate of the variance of the *i*-th arrival time observation.

In a similar way, for each piece of array information we compare the theoretical predictions for the current source location, a_{cj} for azimuth and s_{ck} for slowness, with the observations a_j or s_k. We construct a misfit contribution for the array information with terms corresponding to azimuth and slowness residuals,

$$C_R = C_A + C_S = \frac{1}{N_A}\sum_{j=1}^{N_A}\frac{(a_j - a_{cj})^2}{\sigma_{aj}^2} + \frac{1}{N_S}\sum_{k=1}^{N_S}\frac{(s_k - s_{ck})^2}{\sigma_{sk}^2}, \tag{3}$$

where σ_{aj} is the error in the *j*-th azimuth observation, and σ_{sk} is the error in the *k*-th slowness observation.

We then combine the misfit functions for travel time and array information to provide a composite measure of the quality of a hypocentral estimate

$$\mathcal{F} = C_T + C_R = C_T + C_A + C_S. \tag{4}$$

2.1. Location Algorithm

The misfit functions introduced above depend on the square of the residuals between observed and calculated quantities. The process of refinement of location estimates which builds on the work of GEIGER (1910) is based on an iterative search for a minimum of the misfit function $\mathcal{F}$.

We assume that the current estimate $\mathbf{h}_s = (\mathbf{x}_s, t_s)$ for the hypocentral parameters lies close to the true values to that we can make a quadratic expansion of the misfit function $\mathcal{F}$ about this location

$$\mathcal{F}(\mathbf{h}_s + \Delta\mathbf{h}) = \mathcal{F}(\mathbf{h}_s) + \nabla_h\mathcal{F}\cdot\Delta\mathbf{h} + \tfrac{1}{2}\Delta\mathbf{h}\cdot\nabla_h\nabla_h\mathcal{F}\cdot\Delta\mathbf{h} + O(\Delta\mathbf{h}^3), \tag{5}$$

where ∇_h is the gradient with respect to the hypocentral parameters. The gradient $\nabla_h \mathscr{F}$ is given by

$$\Theta = \nabla_h \mathscr{F} = -2\sum_i \frac{1}{\sigma_{ti}^2}(t_i - t_{ci})\nabla_h t_{ci} - 2\sum_j \frac{1}{\sigma_{aj}^2}(a_j - a_{cj})\nabla_h a_{cj} - 2\sum_k \frac{1}{\sigma_{sk}^2}(s_k - s_{ck})\nabla_h s_{ck} \tag{6}$$

and the Hessian

$$\begin{aligned}\mathbf{H} &= \nabla_h \nabla_h \mathscr{F} \\ &= 2\sum_i \frac{1}{\sigma_{ti}^2}\{\nabla_h t_{ci}\nabla_h t_{ci} - (t_i - t_{ci})\nabla_h \nabla_h t_{ci}\} \\ &\quad + 2\sum_j \frac{1}{\sigma_{aj}^2}\{\nabla_h a_{cj}\nabla_h a_{cj} - (a_j - a_{cj})\nabla_h \nabla_h a_{cj}\} \\ &\quad + 2\sum_k \frac{1}{\sigma_{sk}^2}\{\nabla_h s_{ck}\nabla_h s_{ck} - (s_k - s_{ck})\nabla_h \nabla_h s_{ck}\}.\end{aligned} \tag{7}$$

We now look for a perturbation $\Delta\mathbf{h}$ in the hypocentral parameters such that $\mathscr{F}$ is minimised, so that we require the gradient of $\mathscr{F}$ with respect to $\Delta\mathbf{h}$ to vanish. Using the quadratic approximation (5) we obtain

$$\Delta\mathbf{h} = -\mathbf{H}^{-}\Theta, \tag{8}$$

where $\mathbf{H}$-represents a generalised inverse of the Hessian determined by using singular value decomposition, to avoid any problems with ill-conditioning. From the expressions (6) and (7) we see that the shift in hypocentral parameters is determined from the arrival time residuals $(t_i - t_{ci})$, the slowness residuals $(\mathbf{s}_j - \mathbf{s}_{cj})$ and their derivatives with respect to the hypocentral parameters. Frequently the terms involving the second derivative of the travel times and slowness are neglected, which gives a slightly simpler formulation. The calculated shift in the hypocentre is then used to update the estimate $\mathbf{h}_s$ of the source location, and the process is iterated to generate improved hypocentre estimates.

Some care needs to be taken in numerical implementation because there can be significant differences in the sizes of the derivatives of the arrival times with respect to the hypocentral parameters. We note that $\partial t_{ci}/\partial t_s = 1$, whilst the spatial derivatives of travel times are often much smaller. Convergence to a suitable hypocentre can generally be obtained if the errors in the arrival times are small and the assumed earth model gives a good representation of the travel times for the paths employed. However, there is a strong trade-off between the depth estimate z_{he} and the estimated origin time t_{he} that is best resolved if observations of the different wave types P and S can be made. BULAND (1976) provides a detailed analysis of the properties of the location estimates using just travel-time information. The extension to include array information is discussed by BRATT and BACHE (1988).

If the starting location is suitable the iterative procedure will converge towards a minimum of the misfit function, even though the convergence can be so slow on occasions that premature termination may be applied resulting in a systematic error. However, if the initial hypocentral estimate is too far from the true location the method may either not converge or, more insidiously, converge to a local minimum in the misfit which may well have little relation to the true location. Frequently, the iterative procedure is applied with a range of starting hypocentres and the hypocentral estimate with least misfit $\mathscr{F}$ is selected.

2.2. Location Assessment

In the immediate neighbourhood of the hypocentral estimate $\hat{\mathbf{h}}$ which minimises the misfit measure $\mathscr{F}$ the misfit surface is quadratic in hypocentral space

$$\mathscr{F}(\mathbf{h}_e) - \mathscr{F}(\hat{\mathbf{h}}_s) = \frac{1}{2}(\mathbf{h}_e - \hat{\mathbf{h}}_s) \cdot \mathsf{H}(\hat{\mathbf{h}}_s) \cdot (\mathbf{h}_e - \hat{\mathbf{h}}_s). \tag{9}$$

The points $\mathbf{h}_e$ on the α percent confidence ellipsoid in hypocentral space are then determined by

$$(\mathbf{h}_e - \hat{\mathbf{h}}_s) \cdot \mathsf{H}(\hat{\mathbf{h}}_s) \cdot (\mathbf{h}_e - \hat{\mathbf{h}}_s) = \kappa_\alpha^2. \tag{10}$$

As shown by JORDAN and SVERDRUP (1981), the confidence coefficient κ_α is determined from the α percent confidence level for the F-statistic on 4 and $N - 4$ degrees of freedom, where N is the total number of observations $N_T + N_S$; the factor 4 arises from the four hypocentral parameters. Thus

$$\kappa_\alpha^2 = 4\hat{s}^2 F_\alpha[4,\ N - 4]. \tag{11}$$

The quantity $\hat{s}^2$ is the variance scale factor which can be thought of as the mean ratio between the actual and assumed data variances. An *a posteriori* estimate is provided by

$$s_s^2 = \frac{N_T C_T(\hat{\mathbf{h}}) + N_A C_A(\hat{\mathbf{h}}) + N_S C_S(\hat{\mathbf{h}})}{N_T + N_A + N_S - 4}. \tag{12}$$

This estimate is very effective when the number of observations is large but is not so effective for only a smaller number of data. With the sparse global network for the CTBT we can expect that for many events there will only be a limited amount of data and so is preferable to adopt the Bayesian scheme proposed by JORDAN and SVERDRUP (1981) which includes both *a priori* and *a posteriori* information for which the variance scale factor $\hat{s}^2$ is estimated by

$$s_e^2 = \frac{K s_K^2 + N_T C_T(\hat{\mathbf{h}}) + N_A C_A(\hat{\mathbf{h}}) + N_S C_S(\hat{\mathbf{h}})}{K + N_T + N_A + N_S - 4}. \tag{13}$$

The quantity s_K^2 is an *a priori* estimate for $\hat{s}^2$ and can be found from the average of a set of previous estimates of s_s^2 for a region

$$s_K^2 = \frac{1}{L}\sum_{l=1}^{L}(s_s^2)l. \tag{14}$$

The quantity K is inversely related to the standard deviation in $1/s_s$ (JORDAN and SVERDRUP, 1981). Area specific information can be used to estimate K, but in the absence of other information $K = 8$ is a reasonable starting point. For a fixed network the accumulation of information over time will allow improved estimates of s_K and K leading to smaller confidence regions.

As an illustration of event location using a limited amount information we display in Figure 4, the event location and confidence ellipses estimated for the regional event for which seismograms were displayed in Figure 2. We show two different solutions, one obtained by using the IASPEI91 travel-time model in the location procedure and the other obtained by using a regional travel-time model developed specifically for Fennoscandia and adjacent regions (the Barents model, KREMENETSKAYA *et al.*, 2000). Not surprisingly, the location estimate provided by

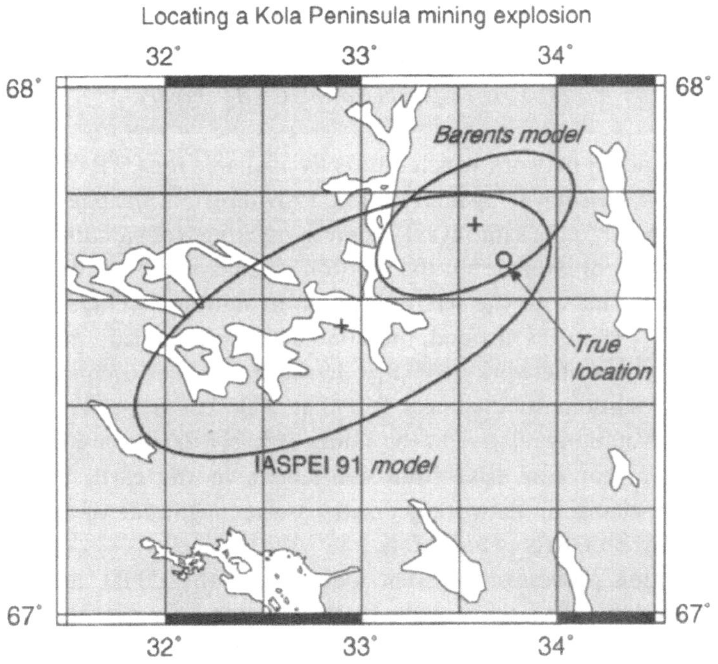

Figure 4
Location of the regional event described in Figures 1–3 using the network of the three Fennoscandian arrays. The estimated location and the associated error ellipses are shown both when using the IASPEI91 model and the Barents model in the location procedure (see text for details).

the regional model is closer to the true (known) location of this mining explosion, and the error ellipse is correspondingly smaller than for IASPEI91. This illustrates precisely the point made earlier in this paper, that the IASPEI91 global model will provide locations that are generally of high quality in all regions of the globe, whereas regional calibration can improve the locations even further in areas where sufficient calibration data is available.

We should note that the ellipsoidal shape of the confidence region is a direct consequence of using a quadratic approximation about the estimated location and is commonly an oversimplification of the shape of the true confidence region. Further, the estimate obtained is only a self-consistent value with the location procedure and does not provide any information about possible systematic errors arising from network geometry or the influence of heterogeneity within the earth.

Also, as pointed out by BILLINGS *et al.* (1994) the projection of the four-dimensional hypocentral ellipsoid into e.g., the horizontal plane will not necessarily coincide with the domain which might be mapped out by looking at the variation of the misfit function in that plane. For critical events, a direct search of the misfit function about the proposed hypocentre will allow a more reliable mapping of the true confidence region including non-linear effects such as e.g., the network geometry.

3. Issues Specific to the CTBT Treaty

The sparse global network which is directly available for CTBT purposes forces a strong reliance on seismic array information. Providing that the phases from different stations correspond to the same event, azimuth tracking from a number of arrays can provide a good initial estimate for the location scheme. Large events will be recorded at many stations and with the relatively uniform station coverage we can expect a broad range of azimuths around the source to be sampled, this will minimise systematic error from network geometry. However, smaller events will be recorded by only a few stations at distances less that 1500 km from the source and the geometry of the stations relative to the source is likely to be non-ideal. Some of the classes of error that can arise from deficiencies in the earth model, and also the influence of change in network geometry as the magnitude of an event changes, are discussed by BILLINGS *et al.* (1994).

The correction processes for event location for the CTBT need to take into account not only the deviations of the earth from the reference model but also the way in which the available network varies with magnitude. The influence of the network can be reduced if suitable regional corrections are introduced but will always pose a limitation on the ultimate accuracy which can be achieved. In particular, in the absence of close stations, resolution of depth will be dependent on observing depth sensitive phases such as the surface reflections *pP*, *sP* or core reflections such as

PcP. Otherwise the depth estimates will depend on rather subtle differences in the propagation characteristics of regional phases. Waveform analysis techniques offer the hope for aiding the definitions of the depth of the event but have yet to be brought into operational use.

A large-scale international effort is currently underway to develop regional corrections to the IASPEI91 travel-time table for all areas of the world. If this effort is successful, it will mean that routine location of seismic events on a global scale will be significantly improved in the future as compared to current practice. Among the prerequisites for this development is the systematic collection of well-located seismic events for calibration purposes, the compilation of regional travel-time models and the conduct of specific chemical calibration explosions in various parts of the world.

The global network that is being developed for CTBT monitoring will in practical operation be seen in conjunction with supplemental data from national and regional networks established for other purposes. The use of data from such networks will be important in the continued calibration of the International Monitoring System, and will in addition provide useful data for improving the location estimate and source characterisation of seismic events of monitoring interest. This synergy between the international system and national resources will thus be important both in context of CTBT verification and for future developments in seismology.

Acknowledgement

The development of travel-time tables and improved methods for locating seismic events has involved many members of the seismological community. We would like to acknowledge the broad range of contributions beyond those explicitly referenced.

References

BILLINGS, S., SAMBRIDGE, M. S., and KENNETT, B. L. N. (1994), *Errors in Hypocentre Location: Picking, Model and Magnitude Dependence*, Bull. Seismol. Soc. Am. *84*, 1978–1990.

BRATT, S. R., and BACHE, T. C. (1988), *Locating Events with a Sparse Network of Regional Arrays*, Bull. Seismol. Soc. Am. *78*, 780–798.

BULAND, R. (1976), *The Mechanics of Locating Earthquakes*, Bull. Seismol. Soc. Am. *66*, 173–187.

JORDAN, T. H., GEIGER, (1910) and SVERDRUP, K. A. (1981), *Teleseismic Location Techniques and their Applications to Earthquake Clusters in the South-Central Pacific*, Bull. Seismol. Soc. Am. *71*, 1105–1130.

KENNETT, B. L. N., and ENGDAHL, E. R. (1991), *Travel Times for Global Earthquake Location and Phase Identification*, Geophys. J. Int. *105*, 429–465.

KREMENETSKAYA, E. O., ASMING, V. E., and RINGDAL, F. (2000), *Seismic Location Calibration of the European Arctic*, Pure appl. geophys., this issue.

KVAERNA, T., and DOORNBOS, D. J. (1986), *An Integrated Approach to Slowness Analysis with Arrays and Three-component Stations*, Seminannual Technical Summary, 1 October 1985–31 March 1986, NORSAR Sci. Report No. 2-85/86, Kjeller, Norway.

RINGDAL, F., and KVAERNA, T. (1989), *A Multi-channel Processing Approach to Real-time Network Detection, Phase Association and Threshold Monitoring*, Bull. Seismol. Soc. Am. *79*, 1927–1940.
SAMBRIDGE, M. S., and KENNETT, B. L. N. (2000), *Seismic Event Location: Nonlinear Inversion Using a Neighbourhood Algorithm*, Pure appl. geophys., this issue.

Pure appl. geophys. 158 (2001) 19–34
0033–4553/01/020019–16 $ 1.50 + 0.20/0

Pure and Applied Geophysics

Location Calibration Data for CTBT Monitoring at the Prototype International Data Center

ISTVÁN BONDÁR, XIAOPING YANG, ROBERT G. NORTH and CARL ROMNEY

Abstract—Ground-truth information is essential for location calibration of the International Monitoring System network being developed under the Comprehensive Nuclear-Test-Ban Treaty. The objective of the calibration effort is to improve the accuracy of seismic event locations and to reduce the size of the error ellipse, both in automatic and in human analyst-reviewed bulletins, in order to meet the On-Site Inspection requirement for the size of the inspection area. Several databases were compiled and are maintained at the Prototype International Data Center (PIDC) to support calibration efforts. The Nuclear Explosion Database contains most readily accessible information about all nuclear explosions worldwide. The events in the Calibration Event Bulletin (CEB) carefully selected well located events from the PIDC Reviewed Event Bulletin and relocated using additional arrivals from regional networks requested from various National Data Centers. The Ground-Truth Database contains carefully selected events with known or well estimated location accuracies from the Nuclear Explosion Database, CEB, as well as from bulletins of U.S. National Earthquake Information Center and International Seismic Centre. It also contains data on chemical explosions and quarry blasts when confirmed by local or national authorities. Ground-truth events are subdivided into various ground-truth categories according to their location accuracy. The databases have been used in various calibration studies to derive and test corrections to improve event locations. Several location calibration techniques are briefly described. The validation test for any proposed operational change requires that the results meet the location calibration metrics developed and implemented at the PIDC.

Key words: Nuclear explosion database, calibration event bulletin, ground-truth database, location calibration, error ellipse, PIDC.

1. Introduction

During the final phase of the negotiations of the Comprehensive Nuclear-Test-Ban Treaty (CTBT) the Group of Scientific Experts (GSE) in 1995 launched its third experiment, GSETT-3, to test the feasibility of real-time data communication and processing of seismic, hydroacoustic, infrasonic and radionuclide data from a global station network. With the signing of the CTBT in 1996 the GSETT-3 station network has been converted or replaced by existing or substitute stations to meet the

---

Science Applications International Corporation, Center for Monitoring Research, 1300 N 17th Street Suite 1450 Arlington, VA 22209, U.S.A. E-mail: xiaoping@cmr.gov requirements of the International Monitoring System (IMS) specified by the CTBT. The final IMS network will consist of 50 primary seismic stations sending data continuously and 120 auxiliary seismic stations; 6 hydrophone stations and 5 seismic T-phase stations; 60 infrasonic stations; and 80 radionuclide stations. The data are routinely processed at the Prototype International Data Center (PIDC) which has been established at the Center for Monitoring Research (CMR) in Arlington, Virginia, USA. The PIDC has delivered the products described in the Treaty since 1 January 1995, including one of the most important ones, the Reviewed Event Bulletin (REB).

The On-Site Inspection provision of the Comprehensive Nuclear-Test-Ban Treaty limits the inspection area to 1000 km^2. Thus, epicenters must be accurately located, and the regions of uncertainty should be well defined so that the inspection can focus on the most probable area. The underlying earth model used at the PIDC for calculating seismic event locations is the IASPEI91 model (KENNETT, 1991). In the location procedure, not only arrival times but also slowness and azimuth measurements are used (MYKKELTVEIT and BUNGUM, 1984; BACHE and BRATT, 1988). Automatic phase identification is also partly based on slowness values. Frequency-wavenumber (f-k) analysis is used to estimate slowness, azimuth and associated errors for signals detected at seismic arrays. For signals detected at 3-component stations, slowness and azimuth are derived from polarization analysis using the method of JURKEVICS (1988).

The uncertainties in the location are represented by an error ellipse scaled to 90% confidence probability. In calculating this ellipse, it is assumed that *a priori* standard errors in time, azimuth, and other parameters that enter into the epicentral calculations are exactly known, so that the coverage ellipse follows a χ^2 distribution equivalent to an F-distribution with infinite number of degrees of freedom (JORDAN and SVERDRUP, 1981). The confidence error ellipse is itself an approximation of a more complicated misfit space since it relies heavily on the assumption of linearity. Furthermore, the *a priori* errors are never exactly known, and thus their estimates play an important role in the reliability of the error ellipses. If the *a priori* errors are under-or overestimated, the 90% confidence level could only be taken as a nominal value. In addition to these statistically described errors, systematic effects biasing locations are known to exist. Thus the true (known) location may lie outside of the error ellipses.

ISRAELSSON *et al.* (1997) have revisited the estimation procedure of *a priori* errors and partitioned the errors into measurement and modeling errors. Measurement errors represent the uncertainties in the signal characterization (arrival-time picking error) and are calculated as a linear function of the signal-to-noise ratio only, although they are also phase-dependent. Larger errors are assigned to the automatic processing picks than to the human analyst phase picks. Modeling errors characterize the uncertainties in the model itself due to the differences between the model and the real earth. These depend on both phase and distance. A revised representation of the error budget has been in operational use at the PIDC since September 1997. Any proposed calibration technique is expected to further decrease the modeling errors, thus leading to more reliable error ellipse estimates.

The size of the error ellipse depends not only on the *a priori* errors but also on network geometry (SWEENEY, 1998). Figure 1 shows the dependency of the area of the error ellipse of PIDC REB events between October 1998 and October 1999 on azimuthal gap (top) and on the number of stations contributing to the location (bottom). The mean ellipse size is shown by a solid line while the mean ellipse size in 1997 is drawn by a dashed line. The dependency on azimuthal gap is almost linear. Although the new error estimates reduced the size of the error ellipse, especially for events recorded by just a few stations and with a large azimuthal gap, the ellipses are still too large. The figures indicate that to achieve the 1000 km^2 upper limit imposed

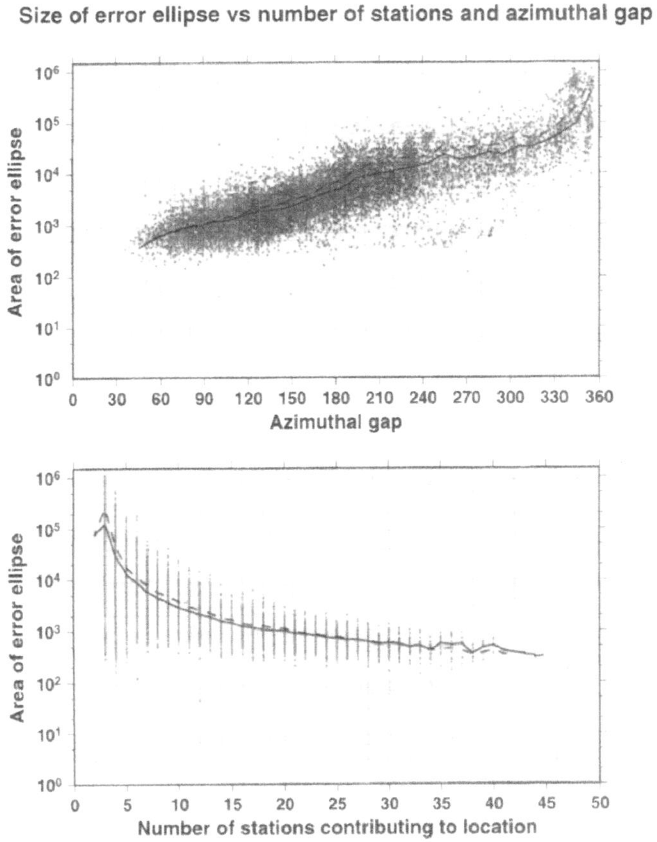

Figure 1
Area of PIDC REB error ellipses between October 1, 1998 and October 1, 1999 as the function of azimuthal gap (top) and number of stations contributing to location (bottom). The mean ellipse size is shown as a solid line. The dashed line shows the mean ellipse size in 1997.

by the On-Site Inspection limitation, an event must be recorded by at least 20 stations or with an azimuthal gap less than 90 degrees. With a basically sparse, teleseismic network like the IMS seismic stations, these conditions are met only for the higher magnitude events. To achieve satisfactory results for less well recorded events, both measurement and modeling errors must be reduced. Recognizing these needs, the Group of Scientific Experts (GSE) concluded in 1996 that a concentrated location calibration effort would be required to address the modeling errors.

The GSE recommended the collection of ground-truth locations (events with known locations) to be used to generate and test location corrections. The collection of well-located events began immediately at the PIDC and the locations are stored in the Calibration Event Bulletin (CEB). The calibration effort has relied on international cooperation from the very beginning; until now about 30 countries have provided data pertaining to calibration events or submitted information regarding ground-truth events.

Besides the continually growing Calibration Event Bulletin (YANG and MCLAUGHLIN, 1999), a Nuclear Explosion Database (YANG *et al.*, 1999) has been compiled that contains most readily accessible information concerning nuclear explosions conducted to date. The Ground Truth (GT) Database (YANG and ROMNEY, 1999) contains carefully selected events divided into several categories of known location accuracy. A more complete description of these databases follows in Sections 2, 3, and 4. All the databases can be publicly accessed through the PIDC web page, www.pidc.org.

Some countries have also voluntarily provided event locations from their own network. These are stored in the so-called "Gamma" Database. Although the quality of the Gamma Database varies from region to region, when properly checked, the Gamma Database provides an invaluable source of ground-truth information.

The location calibration effort also includes the determination and testing of station and/or path-dependent corrections, derived either empirically from GSETT-3 data, or using model-based calculations. Prior to implementation the corrections must be thoroughly tested to validate that they actually improve locations. The performance of the proposed corrections is measured by calculating various metrics.

2. *Nuclear Explosion Database*

The PIDC has compiled a Nuclear Explosion Database for worldwide nuclear explosions conducted from 1945–1998 to include most readily accessible information relevant to seismology (e.g., origin time, location, yield, seismic magnitude, burial depth, etc.) and, for many, seismic waveform data. The data are drawn from a wide variety of official and unofficial sources, including the official listings of all the US and USSR nuclear explosions, so that the data quality is heterogeneous. For many US and USSR tests the accuracy of the origin information (latitude, longitude, time) should be well within ±1 km in location and ±1 s in time. Data sources with location accuracies better than 5 km are identified. This database is updated as new information and digital data are collected.

As of October 1999, the Nuclear Explosion Database includes descriptions of 2041 nuclear explosions worldwide from 1945–1998, identified by responsible countries (Table 1). For most of the events there are multiple origins from different data sources. Among them, the available official lists provide accurate information on the nuclear explosions in terms of event confirmation and yield, although usually they do not include information on origin time or event coordinates. We identified the preferred data source for any given event from data supplied by a set of preferred authors according to a hierarchy which represents our judgement as to the source of the most accurate information on the coordinates and time of each explosion. In the absence of data from these preferred authors, preferred origins were selected from data supplied by other authors weighted according to the completeness of the information reported.

3. Calibration Event Bulletin

Calibration events are selected daily from the REB on the basis of quality, location and magnitude of the events. The selection of CEB events is a two-stage procedure that consists of an automatic and a manual selection. In the automatic selection procedure the earth is divided into geographic regions following the Flinn-Engdahl scheme, and each region is further divided into three depth sets (shallow: <35 km, intermediate, deep: >130 km). To prevent oversampling of high seismicity areas, the number of events in a set is limited to 10. Each set has its self-adjusting acceptance threshold that depends on the number of defining phases. This ensures that only the 'best' events are selected in active regions, and still allows the selection of somewhat lower quality but rare events in seismically inactive regions. In

Table 1

Explosions in the Nuclear Explosion Database

Country	Number of events	Number of origins	Official listing
United States	1032	6350	yes
Soviet Union	715	5107	yes
France	198	1385	partially
China	45	229	no
Joint United States-Great Britain	24	180	yes
Great Britain	21	93	yes
India	3	15	no
Pakistan	2	12	no
Unknown	1	2	no
Total	2041	13373	

general only events with mb between 4 and 6 are selected. The selection of larger events is questionable for calibration purposes as the effect of extended sources may no longer be negligible. Small events are poorly recorded by the IMS network and their location accuracy can be questionable. The automatic selection is performed each time a REB is released.

From the automatic daily selection one or two events are picked manually as calibration events. The manual selection provides further quality control. Waveforms for the selected CEB events are requested from all auxiliary IMS stations. The waveforms associated with the event are scheduled for careful reanalysis, which may lead to a better location.

If an event occurred within regional distances from a national network, the network's location as well as the station readings are requested from the corresponding National Data Center (NDC). Finally, the event is relocated using both the NDC station readings and those from the reanalyzed IMS waveforms. This location is considered the preferred CEB location of the event. Calibration events as of 21 October 1999 are shown in Figure 2.

4. Ground-Truth Database

The location accuracy of calibration events varies from event to event and many of them do not meet the location accuracy required for calibration studies. Therefore

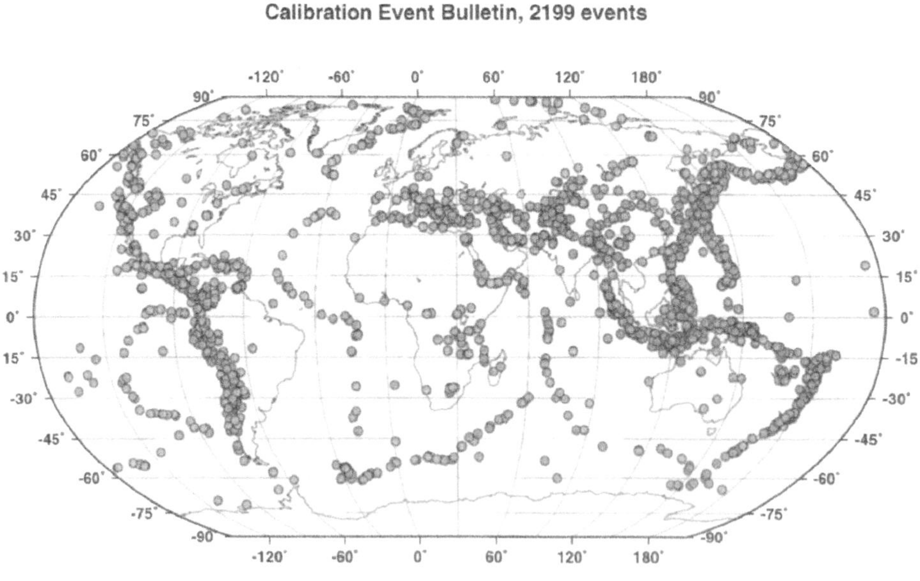

Figure 2
Preferred origins from the calibration event bulletin (CEB) as of 21 October, 1999. The CEB contains the locations and station readings from both IMS and national networks for well-located events.

a separate database has been constructed that contains ground-truth events with known location accuracy. Although origin time and depth are also important parameters, at this time we have focused primarily on location.

Ground-truth data are collected from independent sources (announced chemical explosions, confirmed quarry blasts, etc.), the Nuclear Explosion Database, and from seismic bulletins such as the REB, CEB, and the bulletins of the U.S. National Earthquake Information Center (NEIC) and the International Seismic Centre (ISC). As part of the calibration effort, many countries (Australia, Canada, Israel, Japan, Kazakhstan, Russian Federation, South Africa, Switzerland, USA) have reported ground-truth information. Canada, Finland and Great Britain have also provided the locations of major mines in their territory.

Although there are many historical events that may meet GT standards, for direct calibration of the IMS network recent events are the most useful. The use of historical ground-truth events for CTBT purposes is limited to those that were recorded by existing IMS stations or by stations that can be considered IMS substitute stations (ENGDAHL, 1998), i.e., colocated, or stations close to IMS stations.

The definition of each ground-truth category is given below. In order to maintain a high quality database, ground-truth information has to be validated. The validation can be based on confirmation from national authorities or mining companies, as well as accompanying documentation and publications. Events located by seismic networks must meet well-defined acceptance criteria.

4.1. GT0, GT1, GT2 (0–2 km Location Accuracy from Independent Sources)

These categories consist of events with known, accurately surveyed locations and often with accurately measured origin times. Locations are typically known to better than 1 km (<0.5 km for GT0) or can be tied to a mine or quarry with dimensions of about 2 km. Except for the nuclear explosions, these are usually calibration shots, announced chemical explosions, underwater explosions, quarry blasts or mine explosions. The origin time may be erroneous in the case of industrial explosions. Apart from the mostly historical nuclear explosions, few of these GT0-GT2 events were recorded teleseismically. Figure 3 shows the geographic distribution of ground-truth events belonging to categories GT0 (circles), GT1 (triangles) and GT2 (inverted triangles). Confirmation from local, national or industrial authorities is required to label an event as GT0-GT2.

4.2. GT5, GT10 (5–10 km Location Accuracy from Seismic Network Locations)

GT5 and GT10 events are seismic network locations where both the location and origin time can be biased. Reference events from Joint Hypocentre Determination (DOUGLAS, 1967; DEWEY, 1991) of large aftershock sequences are the best candidates for GT5. GT10 events are selected from CEB, NEIC and ISC bulletins,

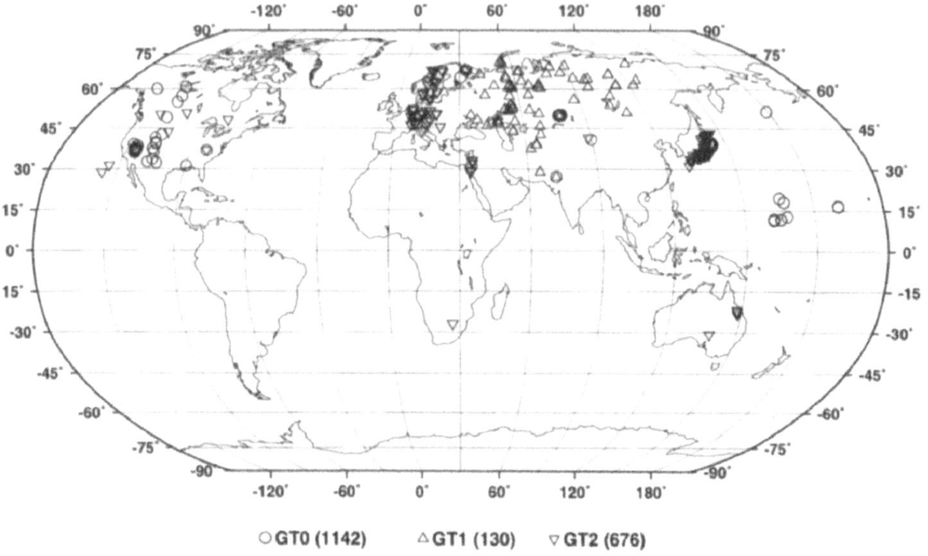

Figure 3
Ground truth from independent sources with 0 (circles), 1 (triangles) and 2 (inverted triangles) km location accuracy as of October, 1999. GT0-GT2 events are mostly nuclear explosions as well as calibration shots, underwater explosions, quarry blasts and mine explosions.

provided that at least three IMS stations recorded the event. The acceptance criteria for a GT10 event require that it be recorded by at least 5 stations within 3 degrees distance, and that the largest azimuthal gap between the recording stations within 5 degrees distance cannot exceed 180 degrees. GT10 events are well-constrained regional network solutions, however only a few such networks can satisfy the above requirements. Figure 4 delineates the geographic distribution of GT5 (circles) and GT10 (triangles) events.

To better represent the earth's seismicity, a GT25 category is devised requiring that the number of defining phases (i.e., arrivals contributing to the location) be at least 50, and the largest azimuthal gap be less than 90 degrees. Because of their fairly large location errors, their use for calibration purposes is rather limited. We doubt the usefulness of GT25 events in regional calibration studies. However, they may be suitable for deriving teleseismic corrections.

5. *Applications of Calibration Data*

The objective of the calibration effort at the PIDC is to develop and explore various calibration techniques that can be used to improve the locations provided by

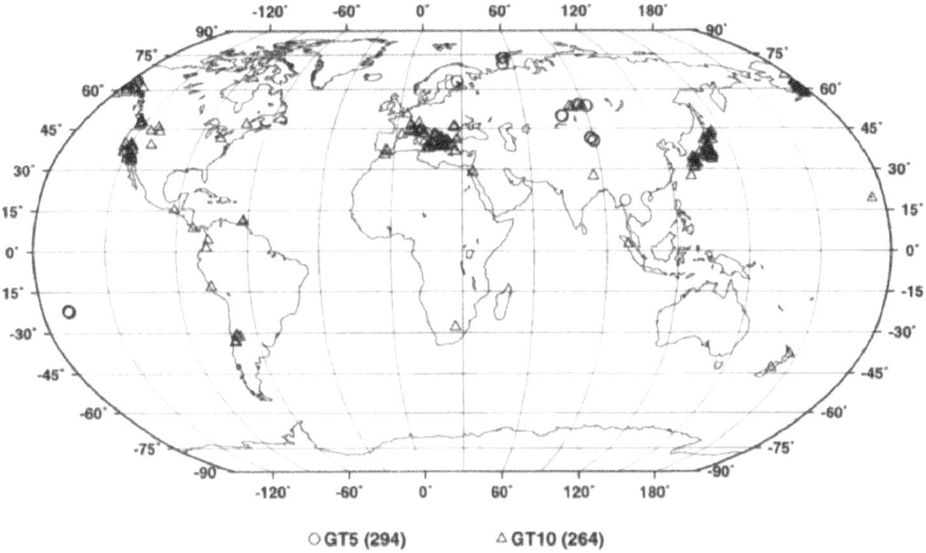

Figure 4
Ground truth from seismic network locations with 5 (circles) and 10 (triangles) km location accuracy as of October, 1999. GT5-GT10 events are reference events from joint hypocenter determination of large aftershock sequences and locations from very dense networks.

the IMS network (BONDÁR and NORTH, 1999). These techniques include the development of empirical or model-based corrections, application of regional travel-time curves and improved representation of the error budget. In most cases calibration events were used in deriving corrections. The calibration techniques are validated using ground-truth events. REB events, of which the CEB events are a subset with better quality, are often employed in evaluating the impact of corrections to the system.

The azimuth and slowness estimates can also be biased due to effects of local geology (orientation and polarization errors should be corrected at instrument level). To remove systematic biases from teleseismic azimuth and slowness estimates, BONDÁR *et al.* (1999) have empirically derived slowness-azimuth station corrections (SASC) for most of the IMS stations that provide data to the PIDC. SASCs have been applied in routine operations since February 1998. The corrections have improved signal association and event location and increased the ability of the automatic processing system to locate events recorded by only few stations. Owing to the better estimates of slowness and azimuth error, the areas of error ellipses have also been reduced. The constant increase in data provided by the IMS network will allow us to define SASCs for regional phases as well.

The IASPEI91 travel-time tables, being global averages, generally do not accurately represent the true velocity distributions, especially at regional distances. The tables are known to be too slow for shield and platform areas and too fast for some tectonic regions. The regional travel times can deviate from the tables by as much as 5 seconds beyond 12 degrees (e.g., YANG *et al.*, 2001). Thus, the modeling errors for regional phases in the IASPEI91 travel-time curves must be large in order to incorporate the observed time residuals. This results in enlarged error ellipses. To overcome this problem the IASPEI91 travel-time tables for regional phases can be replaced by appropriate 1-D regional travel-time curves.

At Fennoscandian stations regional travel-time curves have been applied in routine operations since September 1997 (NORTH *et al.*, 1997). Figure 5 shows a mislocation scatter plot of GT2 events relocated with and without applying the Baltic Shield model travel times in Fennoscandia. Circles above the diagonal indicate improvement in location. Besides the overall improvement in location, owing to the better modeling error estimates, the area of the error ellipses has been reduced from 3800 km^2 to 1100 km^2 on average.

With sufficient number of ground-truth events it is possible to measure the actual performance of the error ellipses. The coverage parameter, which is defined by the equation:

$$\vartheta = \left(\frac{x}{a}\right)^2 + \left(\frac{y}{b}\right)^2$$

where a and b are the estimated semi-axes of the error ellipse and x and y are differences between coordinates of the true and estimated epicenters, follows a χ^2 distribution with 2 degrees of freedom. This can be tested against a 90 percentile theoretical χ^2 distribution with 2 degrees of freedom. The coverage parameter is less than one if the true location is covered by the error ellipse, and larger than one if the true location falls outside the error ellipse. Thus, the intersection of the cumulative probability distribution of the coverage parameter with a vertical line drawn where the coverage parameter is equal to one gives the actual coverage value.

Figure 6 shows the cumulative probability distributions of the coverage parameter with (triangles) and without (circles) applying the Baltic Shield model. The theoretical 90 percentile χ^2 distribution with 2 degrees of freedom is plotted with dotted lines. As the figure indicates, the *a priori* errors were overestimated for Fennoscandia before introducing the regional travel-time curves for the Baltic Shield. This resulted not only in very large error ellipses but also 93% coverage level. The better modeling error estimates of the Baltic Shield model reduced the average size of the error ellipses by a factor of 3.5 while still maintaining more than 80% coverage level.

However, there are few large homogeneous regions in the world that can adequately be described by a single 1-D velocity model. Using the Baltic Shield model for Fennoscandian stations, locations became poorer outside of the Baltic

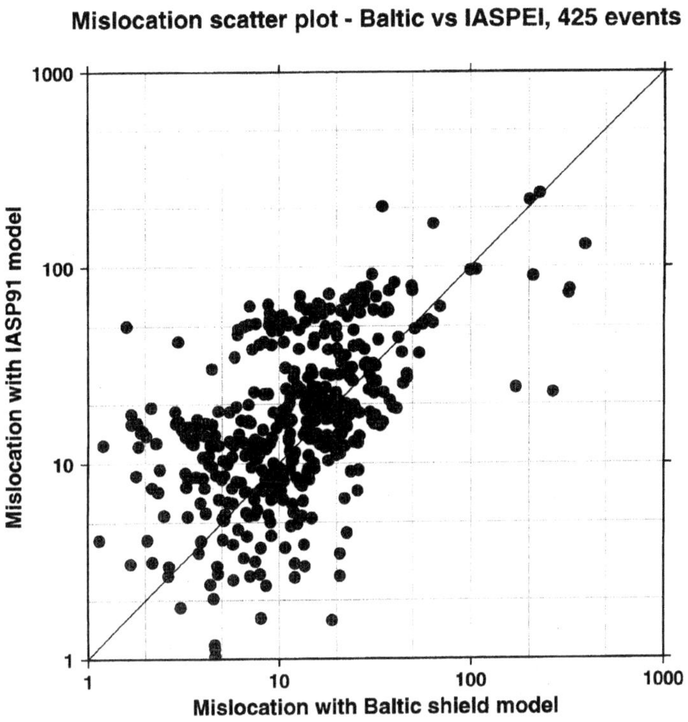

Figure 5
Mislocation scatter plot of GT2 events in Fennoscandia located with and without applying the travel-time curves from the Baltic Shield model.

Shield (North Atlantic ridge, South Europe). Therefore, we are focusing on developing source specific station corrections (SSSC) for regional phases in several regions of the world (BONDÁR *et al.*, 1998). Path-dependent corrections are able to take into consideration heterogeneities on a regional scale.

SSSCs for regional phases at Fennoscandian stations have been shown to improve locations both within and outside the Baltic Shield, and yielded similar or smaller error ellipses to those provided by the single 1-D velocity model (YANG *et al.*, 2001). The Fennoscandian SSSCs were implemented in routine operations at the PIDC in April 1999 and the single 1-D velocity model approach was replaced for REB analysis.

The location improvements with SSSCs for events inside Fennoscandia are almost identical to those of the Baltic Shield model, consequently we demonstrate improvements in locating events outside Fennoscandia using only stations in Fennoscandia. Figure 7 illustrates the location improvements for North Atlantic ridge events when SSSCs are applied. The cumulative probability distributions of the coverage parameter with SSSCs (triangles) and with the Baltic Shield model (circles) are shown in Figure 8. While for events within Fennoscandia the Baltic Shield model

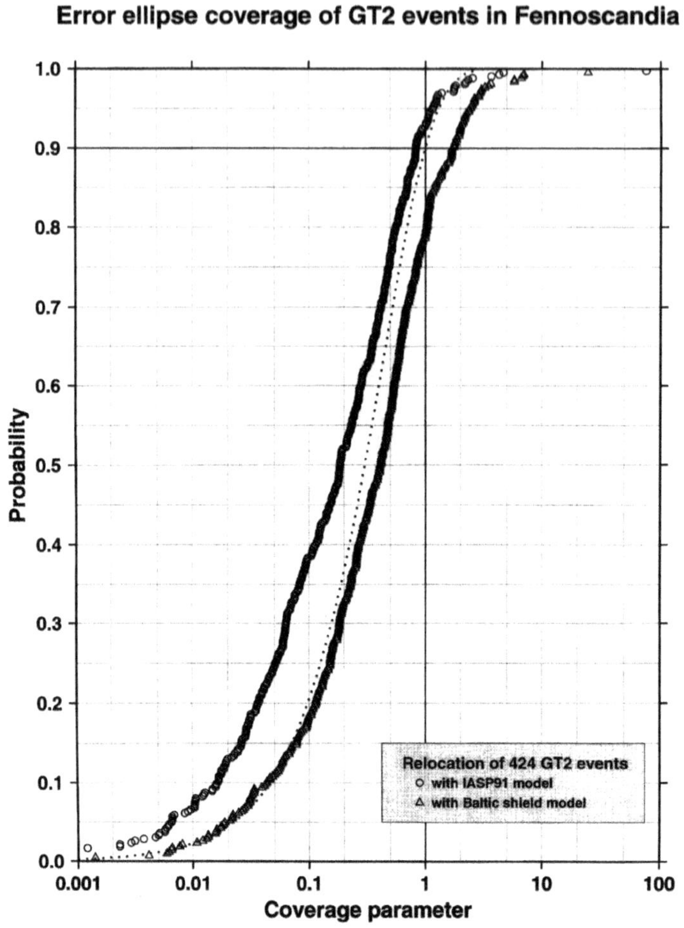

Figure 6
Cumulative probability distribution of the coverage parameter with (triangles) and without (circles) applying the Baltic Shield model. The theoretical 90 percentile χ^2 distribution with 2 degrees of freedom is plotted with dotted lines. A coverage parameter less than 1 means that the ground-truth location is inside the error ellipse, while a coverage parameter larger than 1 indicates that the true location is not covered by the error ellipse.

only slightly underestimated the error ellipses, the actual coverage of the North Atlantic ridge events is only about 70%. With the application of SSSCs and their better error estimates, the actual coverage exceeds 90% and at the same time the average size of the error ellipse decreases from 82,300 km^2 to 73,700 km^2. The different metrics obtained from events within and outside Fennoscandia illustrate well the difficulties of regional location calibration.

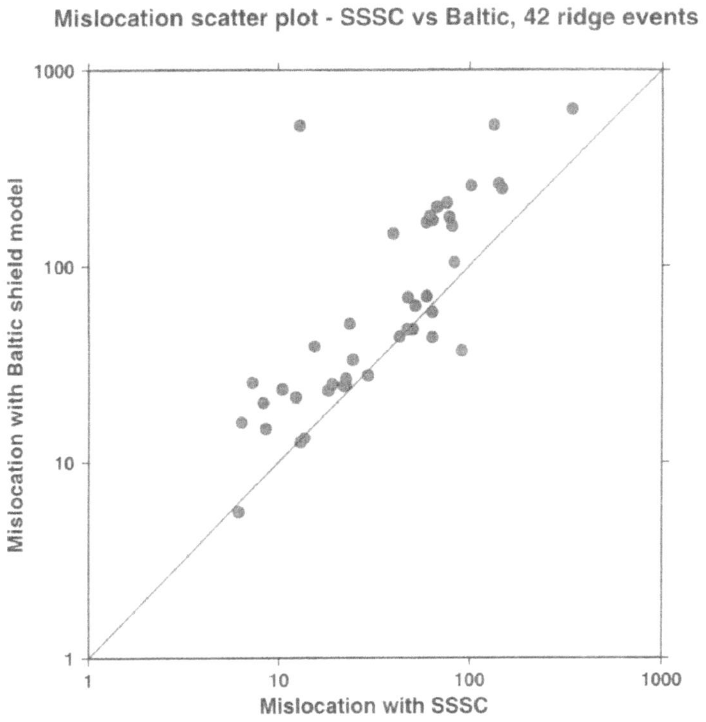

Figure 7
Mislocation scatter plot of North Atlantic ridge events located with Baltic Shield model travel times and with applying SSSCs.

6. Conclusions

Collecting ground-truth information is an essential part of location calibration and requires broad international cooperation. The PIDC maintains several databases that provide sources for developing and testing various calibration techniques. These data are all publicly available at the PIDC web site, www.pidc.org.

In the course of the location calibration effort, several new techniques have been developed and successfully applied, such as the novel representation of uncertainties and the representation of empirical teleseismic slowness-azimuth station corrections. Existing methods, such as the application of 1-D regional travel-time curves have also been tested and applied. Recently many researchers focused on the development of path-dependent regional corrections. One approach has been fully tested and implemented in routine operations (YANG *et al.*, 2001). In the future further methods for providing path-dependent corrections will be investigated. Studies for improving depth determination and depth error estimates, extending slowness-azimuth corrections to regional distances, and investigating the phase-dependence of measurement errors are envisioned.

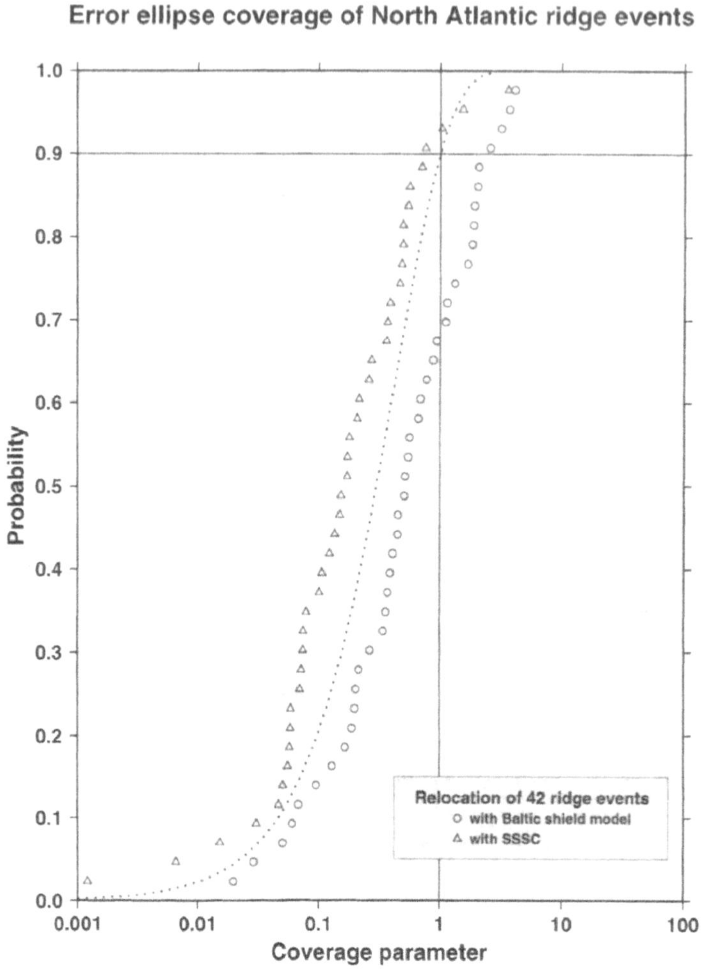

Figure 8
Cumulative probability distribution of the coverage parameter with SSSCs (triangles) and with Baltic Shield model travel times (circles). The theoretical 90 percentile χ^2 distribution with 2 degrees of freedom is plotted with dotted lines.

The location calibration metrics developed at the PIDC have proven to be useful for assessing the performance of various calibration techniques (e.g., YANG *et al.*, 2001). The metrics are stringent enough to ensure that only techniques that irrefutably improve both locations and error estimates are accepted for implementation in the operational system of the PIDC. To date the scope of location calibration has encompassed only seismic and hydroacoustic data. It is important to extend the calibration effort to incorporate the collection of ground truth information for the infrasonic technology so that those networks can also be calibrated.

The PIDC primarily focuses on developing and exploring methods that can be proven to be useful for location calibration. The systematic calibration of the IMS network using these techniques will be a task of the Comprehensive Nuclear-Test-Ban Treaty Organization (CTBTO), International Data Centre (IDC). As part of the transfer of technology and knowledge from the PIDC to the CTBTO IDC, all databases described in this paper have been delivered to the IDC.

Acknowledgments

The authors are grateful to Hans Israelsson, Walt Nagy, Henry Swanger, Keith McLaughlin, Jin Wang, Ronald Cook, Anna Gault, Daud Shah and the representatives of National Data Centers for their invaluable contribution to the location calibration effort. The figures in this paper were generated using GMT (WESSEL and SMITH, 1998).

REFERENCES

BACHE, T. C., and BRATT, S. R. (1988), *Locating Events with a Sparse Network of Regional Arrays*, Bull. Seismol. Soc. Am. *78*, 780–798.

BONDÁR, I., and NORTH, R. G. (1999), *Development of Calibration Techniques for Potential Use by the CTBT International Monitoring System*, Phys. Earth Planet. Int. *113*, 11–24.

BONDÁR, I., NORTH, R. G., and BEALL, G. (1999), *Teleseismic Slowness-azimuth Station Corrections for the International Monitoring System Seismic Network*, Bull. Seismol. Soc. Am. *89*, 4, 989–1003.

BONDÁR, I., YANG, X., MCLAUGHLIN, K., NORTH, R. G., RYABOY, V., and NAGY, W. (1998), *Source Specific Station Corrections for Regional Phases at IMS Stations in North America and Fennoscandia*, EOS, AGU *79*, 45, F555.

DEWEY, J. (1991), *Joint epicenter determination for earthquakes occurring over several decades: A case history from Northern Algeria*. In *Seismicity and Seismic Risk of the Ibero-Maghrebian Region* (eds. J. Mezcua and A. Udias), Instituto Geográfico Nacional Monografia 8, pp. 51–63.

DOUGLAS, A. (1967), *Joint Epicentre Determination*, Nature *215*, 47–48.

ENGDAHL, E. R. (1998), *Development of an Archive of Seismic Ground Truth Events Globally in Support of Monitoring Under the CTBT*, Proceedings of the 20th Annual Seismic Research Symposium on Monitoring a Comprehensive Test Ban Treaty (CTBT), Santa Fe, NM, Sept. 1998, pp. 11–18.

ISRAELSSON, H., SWANGER, H., and BEALL, G. (1997), *Independent Modeling of Time Measurement and Model Errors*, Center for Monitoring Research, CCB-PRO-97/24.

JORDAN, T. H., and SVERDRUP, K. A. (1981), *Teleseismic Location Techniques and Their Application to Earthquake Clusters in the South-central Pacific*, Bull. Seismol. Soc. Am. *71*, 1105–1130.

JURKEVICS, A. (1988), *Polarization Analysis of Three-component Data*, Bull. Seismol. Soc. Am. *78*, pp. 1725–1743.

KENNETT, B. L. N. (1991, ed.), *IASPEI 1991 Seismological Tables*, Res. School of Earth Sci. Austr. Natl. Univ., Canberra, Australia.

MYKKELTVEIT, S., and BUNGUM, H. (1984), *Processing of Regional Seismic Events Using Data from Small Aperture Arrays*, Bull. Seismol. Soc. Am. *74*, 2313–2334.

NORTH, R. G., BONDÁR, I., and RYABOY, V. (1997), *New Regional Travel-time Tables for the Baltic Shield Inferred from DSS Observations for Calibration of the International Monitoring System*, Proceedings of the 19th Annual Research Symposium on Monitoring a Comprehensive Test Ban Treaty, 115–124.

SWEENEY, J. (1998), *Criteria for Selecting Accurate Event Locations from NEIC and ISC Bulletins*, Lawrence-Livermore National Laboratory Report, UCRL-JC-130655.

YANG, X., and MCLAUGHLIN, K. (1999), *PIDC Calibration Event Bulletin (CEB) Database*, Center for Monitoring Research, Technical Report, CMR-99/17, 1999.

YANG, X., and ROMNEY, C. (1999), *PIDC Ground Truth Event (GT) Database*, Center for Monitoring Research, Technical Report, CMR-99/15, 1999.

YANG, X., NORTH, R. G., and ROMNEY, C. (1999), *PIDC Nuclear Explosion Database (Revision 2)*, Center for Monitoring Research, Technical Report, CMR-99/16, 1999.

YANG, X., MCLAUGHLIN, K., and NORTH, R. G. (2001), *Source Specific Station Corrections for Regional Phases at Fennoscandian Stations*, this issue.

WESSEL, P., and SMITH, W.H.F. (1998), *New, Improved Version of the Generic Mapping Tools Released*, EOS Trans. AGU *79*, 579.

(Received April 30, 1999, revised November 17, 1999, accepted February 9, 2000)

Pure appl. geophys. 158 (2001) 35–57
0033-4553/01/020035-23 $ 1.50 + 0.20/0

Pure and Applied Geophysics

Source Specific Station Corrections for Regional Phases at Fennoscandian Stations

XIAOPING YANG,[1] ISTVÁN BONDÁR, KEITH MCLAUGHLIN,[1] and ROBERT NORTH[1]

Abstract—The IASPEI91 global travel-time curves are used as the default for event location at the Prototype International Data Center (PIDC). In order to improve event location, a 1-D Baltic travel-time model was implemented at the PIDC in 1997 for locating events using regional phases from Fennoscandian stations. Where a single model is insufficient for characterizing the regional geology, path-dependent corrections, or Source Specific Station Corrections (SSSCs), are more appropriate for event locations. We have developed SSSCs for regional phases at the Fennoscandian stations by interpolating travel times through different 1-D models. SSSCs for stations NRIS and SPITS are also derived, given the fact that paths from both stations to high latitude events are within the Fennoscandia regionalization as Baltic.

Validation testing of the SSSCs demonstrates that using SSSCs in event location is superior to not using SSSCs, and, in most cases, to using the 1-D model directly when locating events. For a ground-truth data set which includes events in the Baltic Shield with location accuracy better than 2 km, the average improvement in location due to SSSCs is 9 km, and the median coverage ellipse is reduced by 2710 km^2 (from 3830 to 1120 km^2). These results are similar to those obtained using the 1-D Baltic model. For a CEB (Calibration Event Bulletin) data set which includes events along the North Atlantic oceanic ridge and in central/southern Europe, using SSSCs the ridge events move closer to the ridge axis, and the European events move closer to CEB locations than 1-D Baltic locations. For a constrained JHD (Joint Hypocenter Determination) data set of events in the Novaya Zemlya region, when using SSSCs or the 1-D Baltic model, relative to the JHD locations mislocations are less or similar to those without SSSCs. All coverage ellipses are smaller but still contain the JHD solutions.

Our SSSCs are strongly dependent on the 1-D regional models and regionalization. Future development in 1-D velocity models and travel-time curves should improve such SSSCs, event locations, and uncertainties. It is hoped that the implementation and demonstration of SSSCs in the PIDC software will encourage these further developments. These SSSCs were implemented at the PIDC for Reviewed Event Bulletin (REB) location in April 1999.

Key words: SSSCs, Fennoscandia, travel time, event location, IMS, PIDC.

1. *Introduction*

The software at the Prototype International Data Center (PIDC, in Arlington, VA) allows for a hierarchy of calibrations relevant to location improvement. Since the beginning of the GSETT3 (Group of Scientific Experts' Third Technical Test)

[1] Science Applications International Corporations, Center for Monitoring Research, 1300 N. 17th Street, Arlington, VA 22209, USA. E-mail: xiaoping@cmr.gov

experiment on January 1 1995, the IASPEI91 model (KENNETT, 1991) has been used as the reference travel-time set. Ellipticity and elevation corrections are made for each arrival, followed by an optional set of Bulk Time Station Corrections (BTSC) for each station and travel-time phase. So far BTSCs have not been implemented at the PIDC. Since slownesses and azimuths are very critical when locating events with only a few stations, there is a provision for tabulated Slowness and Azimuth Station Corrections (SASC) for each station and array (see BONDÁR *et al.*, 1999). Separate regional travel-time curves (distance less than 20 degrees) may be designated for *Pn*, *Pg*, *Sn*, and *Lg* for each IMS station. In the hierarchy of location calibration the PIDC software may also use tabulated path corrections, or Source Specific Station Corrections (SSSC), to apply corrections relative to IASPEI91 as a function of source location for any station and phase.

The IASPEI91 global travel-time tables deviate considerably from true travel times in shields and platforms at regional distances. In order to improve event locations for CTBT (Comprehensive Nuclear-Test-Ban Treaty) monitoring using the International Monitoring System (IMS) seismic network, either regional velocity models or travel-time corrections may be applied when locating events. A layered (1-D) Baltic model for the Fennoscandian region was developed and implemented at the PIDC in September 1997 (BONDÁR and NORTH, 1999). As a result, locations in that region improved 1.5–3.5 km on average. However, regional paths to Fennoscandian stations extend well into the North Atlantic Ocean, the Eurasian platforms, and tectonic regions such as southern Europe. Consequently as a side effect of the 1-D Baltic model, mislocations of events outside of the shield have resulted (e.g., North Atlantic Mid-Ocean Ridge and central/southern Europe). Event locations in the Barents Sea, Kara Sea, and Novaya Zemlya region are also biased because the 1-D Baltic model was not used at stations NRIS and SPITS although both paths are fast relative to IASPEI91. Different from the 1-D model approach, SSSCs can be applied for event locations more

Table 1

Fennoscandian and other IMS stations with SSSCs developed

Station	Location	Latitude	Longitude	Type	Network
ARCES	ARCESS Array, Norway	69.5349	25.5058	array	PRIMARY
FINES	FINESS Array, Finland	61.4436	26.0771	array	PRIMARY
HFS	Hagfors Array, Sweden	60.1344	13.6968	array	AUXILIARY
KAF	Kangasniemi Station, Finland	62.1112	26.3095	3C	GSETT-3
NOA	NORSAR Array, Norway	61.0397	11.2148	array	PRIMARY
NORES	NORESS Array, Norway	60.7353	11.5414	array	GSETT-3
NRIS	Norilsk, Russia	69.0061	87.9964	3C	PRIMARY
SPITS	Spitsbergen Array, Norway	78.1777	16.3700	array	AUXILIARY
VAF	Ylistaro Station, Finland	63.0422	22.6715	3C	GSETT-3

Network: PRIMARY: IMS primary station; AUXILIARY: IMS auxiliary station; GSETT-3: Group of Scientific Experts' Third Technical Test (not IMS station).

effectively. This paper reports on development and testing of regional SSSCs for Fennoscandian stations. Regional SSSCs for North American IMS stations are under development and testing (BONDÁR *et al.*, 1998; YANG *et al.*, 1999).

The SSSCs are corrections, relative to base travel-time tables, for a given phase at a given station. These corrections are specified on a rectangular latitude/longitude grid in the PIDC location software (NAGY, 1996). Both a travel time-correction and a modeling error are given at each grid point. We have adopted an approach developed by Bondár (BONDÁR *et al.*, 1998). This approach is based on regionalization and regional travel times for different tectonic regions surrounding each station. The predicted travel times are calculated as the weighted sum of total travel times in each region along the source-station path.

In Fennoscandia, Barents Sea, Kara Sea, and Novaya Zemlya region's, there are abundant ground-truth/calibration events (BONDÁR *et al.*, this issue) in the Baltic Shield area, along the oceanic ridge, and around Novaya Zemlya. SSSCs can be tested there by comparing various locations with the ground-truth information. Fennoscandia is also the only region where the 1-D travel times (Baltic model) were implemented in location processing at the PIDC. SSSCs for IMS station NRIS and SPITS (Table 1) are also developed and evaluated here because paths to high latitude events are fast with respect to IASPEI91.

We demonstrate in this paper that event locations in the region can be improved using regional travel-time corrections. Event mislocations in the North Atlantic and central/southern Europe are reduced by replacing the 1-D Baltic model in Fennoscandia with more general SSSCs. Event locations at high latitudes are improved by introducing SSSCs at stations NRIS and SPITS.

2. *Regional SSSCs and Modeling Errors*

We generate regional phase SSSC grid files using Bondár's method (BONDÁR *et al.*, 1998). Grid spacing is chosen as one degree for both latitude and longitude. The SSSC grids may be specified over a region which contains several tectonic regions covered by convex polygons. Each polygon is specified by clockwise latitude-longitude pairs, and the polygons are numbered in the input file. A polygon is characterized by a single travel-time table that includes distance-dependent modeling errors. Even though the SSSC domain does not necessarily contain the station, there must be a polygon that contains the station and at the same time touches or is inside the SSSC rectangular domain. For areas not covered by a polygon or where nodes fall outside of the bounding edges of the domain, zero correction and zero modeling error are assigned to the SSSC grid.

The estimated total travel time is constructed as a linear combination of the total travel time provided by the travel-time tables in the polygons transected by the event-station path. Modeling errors are constructed in a similar manner. SSSCs are defined

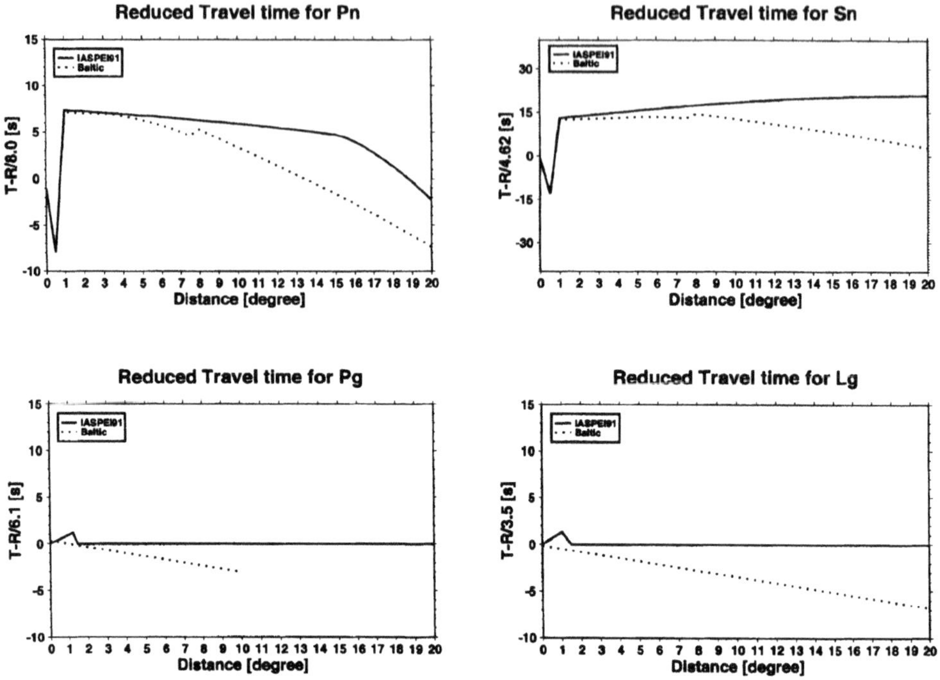
Reduced Travel time for Pn
IASPEI91
Baltic
T-R/8.0 [s]
Distance [degree]
Reduced Travel time for Sn
IASPEI91
Baltic
T-R/4.62 [s]
Distance [degree]
Reduced Travel time for Pg
IASPEI91
Baltic
T-R/6.1 [s]
Distance [degree]
Reduced Travel time for Lg
IASPEI91
Baltic
T-R/3.5 [s]
Distance [degree]

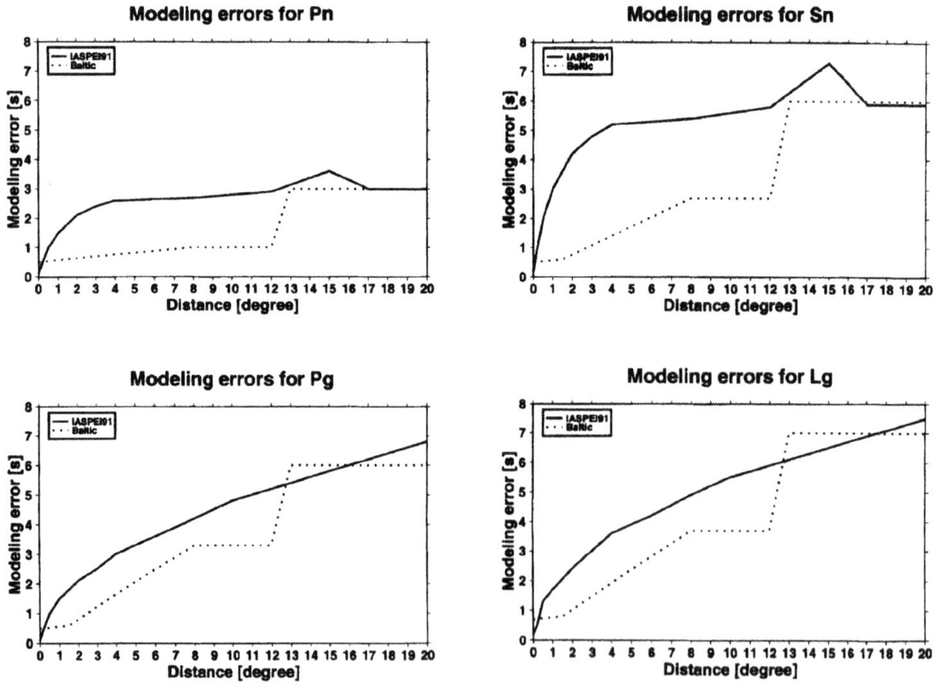
Modeling errors for Pn
IASPEI91
Baltic
Modeling error [s]
Distance [degree]
Modeling errors for Sn
IASPEI91
Baltic
Modeling error [s]
Distance [degree]
Modeling errors for Pg
IASPEI91
Baltic
Modeling error [s]
Distance [degree]
Modeling errors for Lg
IASPEI91
Baltic
Modeling error [s]
Distance [degree]

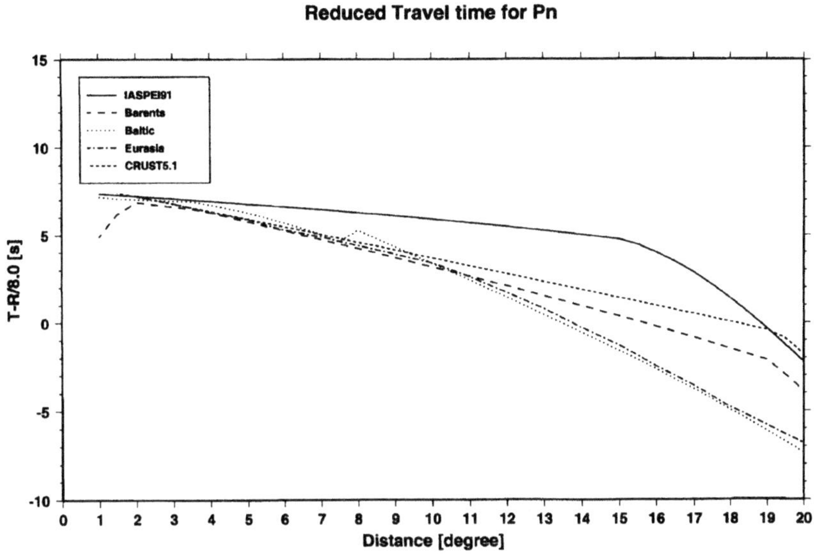

Figure 2

Comparisons of reduced travel times for *Pn*: IASPEI91 (KENNET, 1991), Barents (KREMENNETSKAYA and ASMING, 1999), Baltic (BONDÁR and NORTH, 1999), Eurasian (RF/US WORKING GROUP, 1998), and revised CRUST5.1 (MOONEY *et al.*, 1998; STEVENS and MCLAUGHLIN, 1997). The regional travel times are in general agreement within 12 degrees, and they are all faster than IASPEI91.

as the difference between the predicted total travel time and the one provided by the reference model (IASPEI91), i.e. $T_{\mathrm{SSSC}} = T_{\mathrm{MODEL}} - T_{\mathrm{IASPEI91}}$.

The estimated total travel time is

$$T_{\mathrm{MODEL}} = \sum_i \frac{x_i \cdot T_i(\Delta)}{\Delta} \quad ,$$

where $\Delta = \sum_i x_i$, and the corresponding estimated modeling error is

$$\sigma^2 = \sum_i \frac{x_i \cdot \sigma_i^2(\Delta)}{\Delta}$$

where x_i is the horizontal distance along the ray path in the *i*-th tectonic region, $T_i(\Delta)$ is the travel time if only the *i*-th 1-D velocity model is taken into account, and $\sigma_i(\Delta)$ is the corresponding distance-dependent modeling error in the *i*-th region. Such definitions ensure the smooth transition of travel time and modeling error curves

◀

Figure 1

Travel times and modeling errors for *Pn*, *Sn*, *Pg*, and *Lg* phases from the IASPEI91 and Baltic Shield models.

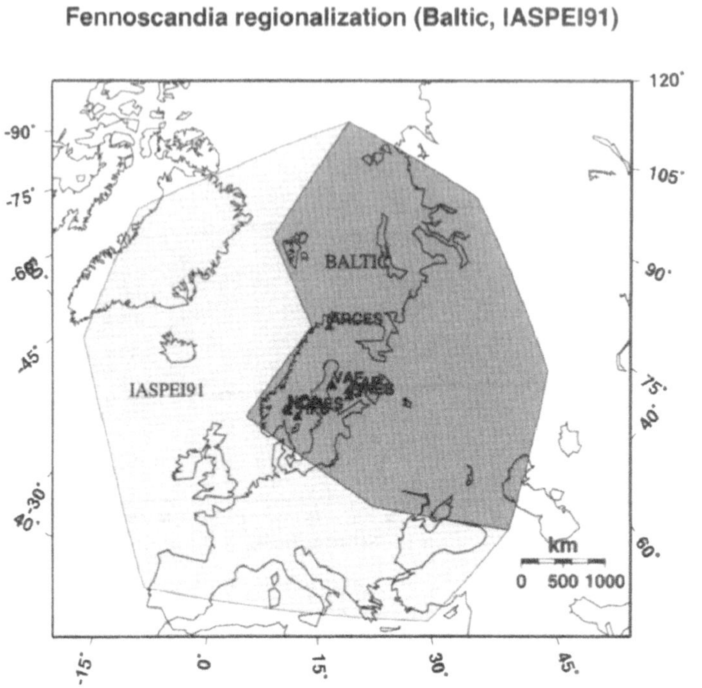
Fennoscandia regionalization (Baltic, IASPEI91)
BALTIC
ARCES
IASPEI91
km
0 500 1000

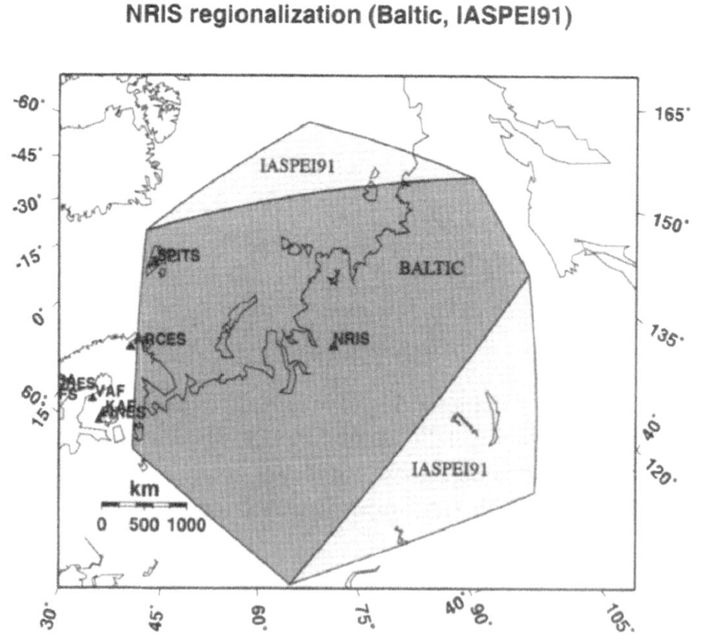
NRIS regionalization (Baltic, IASPEI91)
IASPEI91
SPITS
BALTIC
ARCES
NRIS
km
0 500 1000
IASPEI91

Figure 3

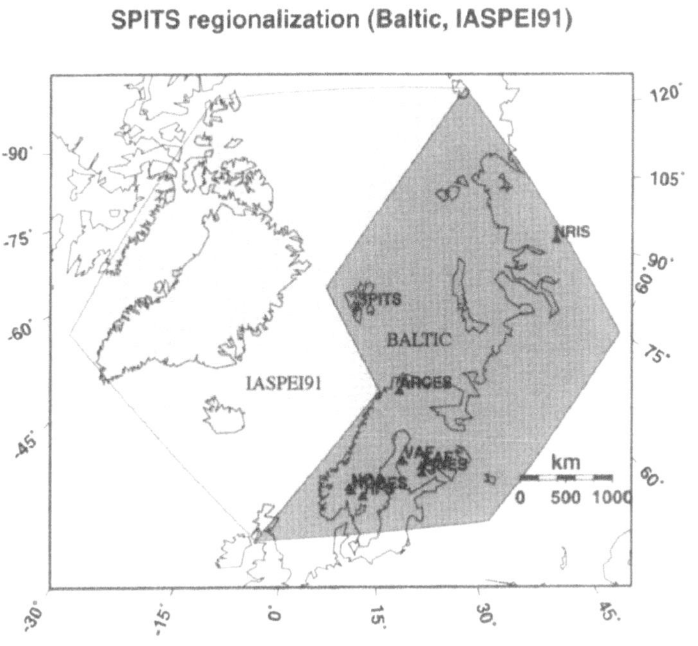

Figure 3
Regionalizations for Fennoscandian stations, NRIS, and SPITS. The 1-D Baltic model is used in Fennoscandia, eastern Europe, Barents Sea, Kara Sea, and Novaya Zemlya region. The IASPEI91 model is used in other areas. The travel times and modeling errors for both models are shown in Figure 1.

across tectonic boundaries. Therefore the definitions given above provide reasonable approximation to a series of head waves.

3. Pn, Sn, Pg, and Lg Travel Times for 1-D Models

The 1-D Baltic model (BONDÁR and NORTH, 1999) is employed to derive SSSCs relative to the IASPEI91 model for the stations given in Table 1. Figure 1 shows the reduced travel times and modeling errors for *Pn*, *Sn*, *Pg*, and *Lg* phases from the Baltic and IASPEI91 models. The large modeling errors for the Baltic shield at distances further than 12 degrees reflect the uncertainty due to the lack of data when deriving the model and extrapolation to distances beyond 12 degrees.

Comparisons were made between the Baltic *Pn* travel times and those from other studies (Fig. 2). The Baltic, revised CRUST5.1 (MOONEY *et al.*, 1998; STEVENS and MCLAUGHLIN, 1997), Barents (KREMENETSKAYA and ASMING, 1999), and Eurasian (RF/US WORKING GROUP, 1998) *Pn* travel times are in general agreement within 0–12 degree distance. The Baltic and Eurasian travel times are slightly faster at distances beyond 12 degrees than the other two. In comparison with the global

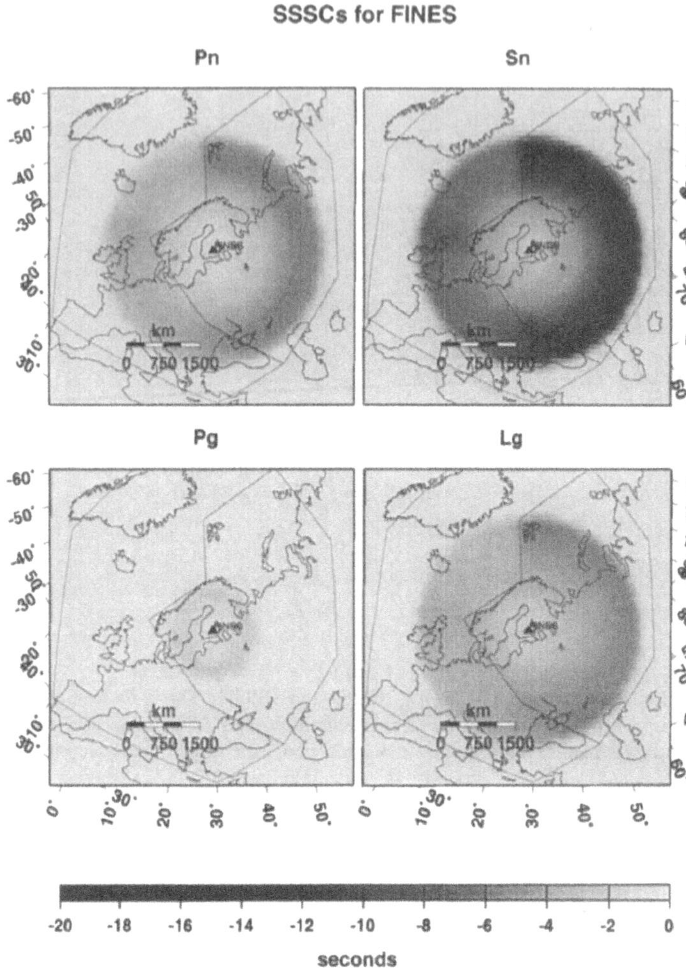

Figure 4
Pn, *Sn*, *Pg*, and *Lg* SSSCs for IMS station FINES.

average (IASPEI91), the regional travel times inferred from explosion seismology observations are about 5 seconds faster beyond 12 degrees. We note that the Baltic and Barents *Pn* travel times are in excellent agreement within 12 degrees however the former is about 3 seconds faster beyond 12 degrees, which are within modeling errors assumed for the Baltic model (Fig. 1).

4. SSSCs for Fennoscandian and Other High Latitude IMS Stations

We derived SSSCs and modeling errors for the Fennoscandian and other IMS stations listed in Table 1. Figure 3 shows the Fennoscandia regionalization based on

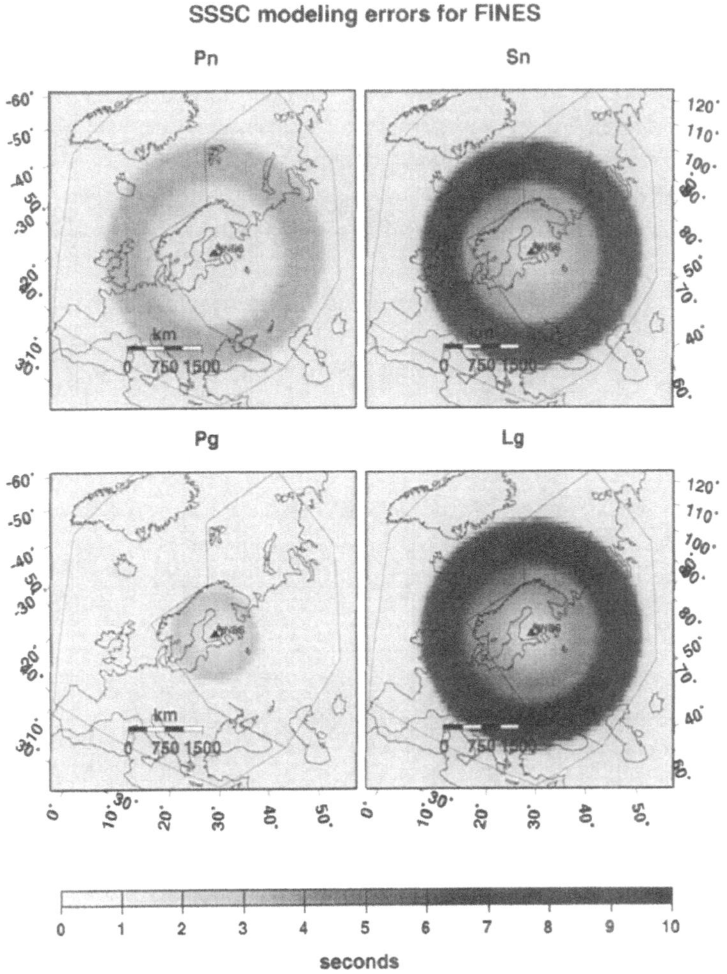

Figure 5
Pn, *Sn*, *Pg*, and *Lg* SSSC modeling errors for IMS station FINES.

several studies referenced in Figure 2. The 1-D Baltic model is applied to the Fennoscandian shield, east European platform, and the Barents region. This same regionalization is applied to all SSSCs grids for Fennoscandian stations. The regionalizations for stations NRIS and SPITS are simplified extensions to cover the regional distances for these two stations. For station SPITS the IASPEI91 region is extended further north, and for station NRIS the Baltic region is extended further west.

Figures 4 and 5 show examples of the SSSCs and modeling errors for Fennoscandian station FINES. The negative travel-time corrections in the ocean occur because a significant portion of the path is in the shield where the travel time is fast.

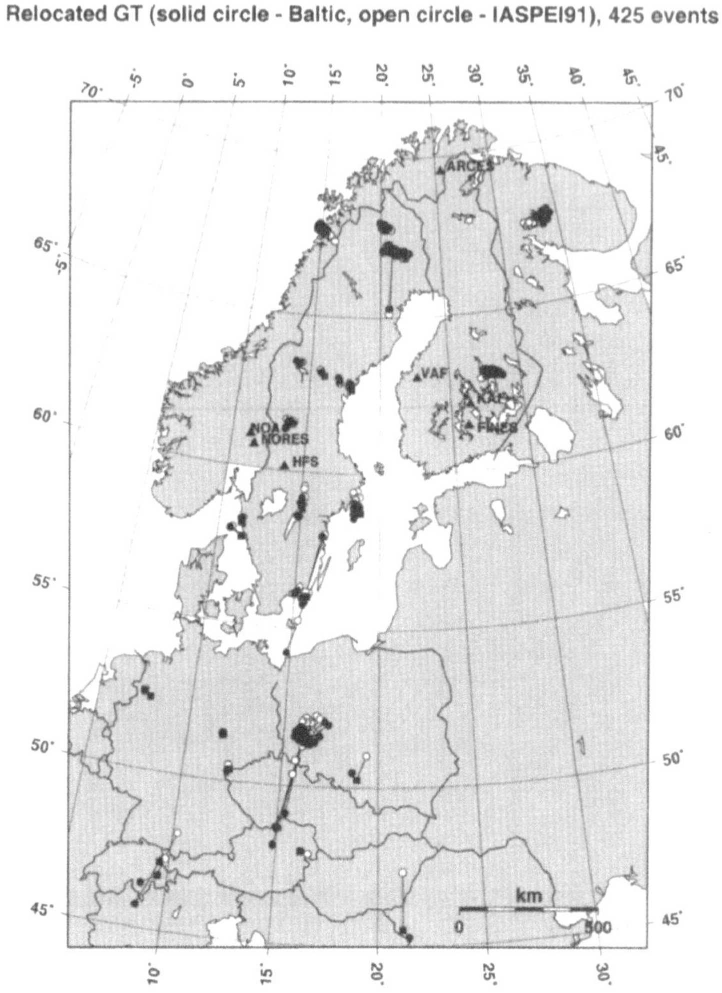

Figure 6
Location using the Baltic 1-D model (solid circle) and the IASPEI91 model (open circle) for the GT events (square). Line segments link different locations for the same event. While event locations are significantly improved within the shield by using the 1-D Baltic model, mislocations are increased in central Europe because the regional paths extend beyond the shield.

5. Event Locations Using SSSCs

For SSSC validation testing we evaluate the improvement on hypocenter locations and error ellipses by applying SSSCs to three data sets of surface to shallow events. Such events are directly relevant to the CTBT monitoring. While origin time is an important parameter for an event, available GT information on origin time is too scarce to warrant a valid evaluation at this time.

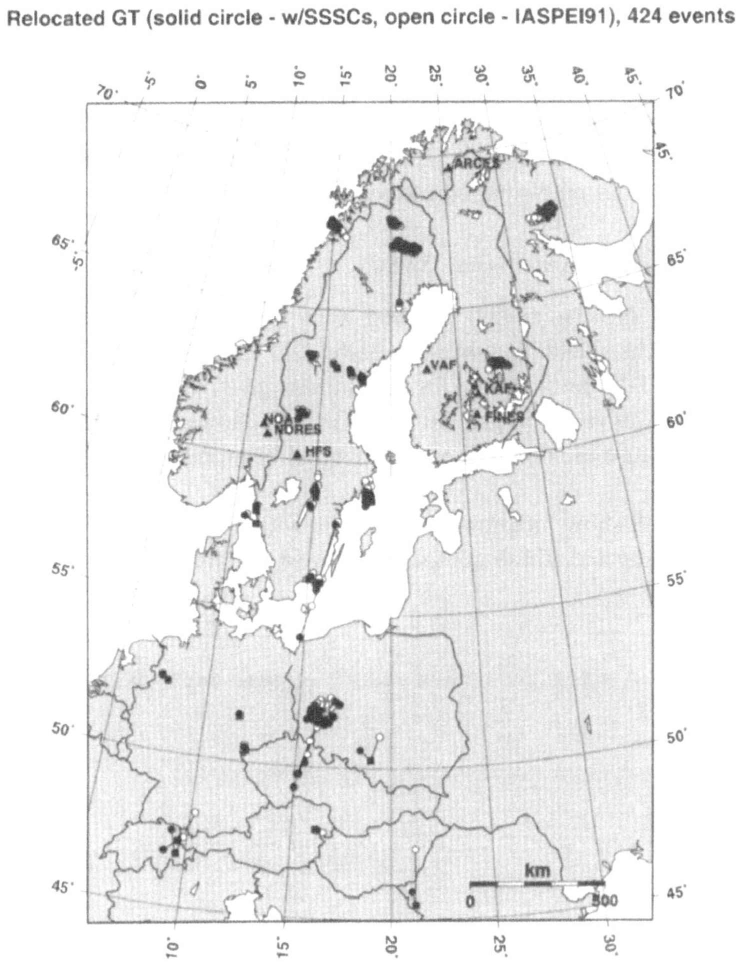

Figure 7
Location comparisons with SSSCs (solid circle) and without SSSCs (i.e., IASPEI91; open circle) for the ground-truth events (square). Line segments link different locations for the same event. Event locations within the shield are similarly improved using SSSCs and Baltic model, relative to the IASPEI91 model. Outside of the shield, event locations using SSSCs are closer to the GT locations than using the 1-D Baltic model.

We employed a ground-truth (GT) data set that includes events (mostly mining events in the shield) with location accuracy better than 2 km (BONDÁR and NORTH, 1999). This data set is independent from the one used in deriving the 1-D Baltic model, which was based on long-range seismic profile observations. We selected another data set from the PIDC Calibration Event Bulletin (CEB; BONDÁR *et al.*, this issue) to include additional events outside the shield area, particularly along the North Atlantic Mid-Ocean Ridge. A third data set was selected to include

Joint Hypocenter Determination (JHD; ISRAELSSON *et al.*, 1997) events in the Barents Sea, Kara Sea, and Novaya Zemlya region. These data sets are drawn from the GT and CEB databases designated for the calibration effort at the PIDC. The CEB database includes well located events and the GT database includes events with known location accuracy. The JHD locations were constrained using ground-truth information and empirical corrections (DOUGLAS, 1967; DEWEY, 1972).

5.1. Relocations of GT Events Using SSSCs

BONDÁR and NORTH (1999) showed that event locations are significantly improved within the shield by using the Baltic 1-D model instead of the IASPEI91 model. Using the 1-D Baltic model for the stations given in Table 1, 62% of the 425 events have locations closer to ground truth (Fig. 6). However, mislocations resulted in some areas because the regional paths extended beyond the shield, e.g., in central/southern Europe.

Figures 7 and 8 show location comparisons with SSSCs and without them (i.e., IASPEI91). As expected, these comparisons yield results similar to those given in

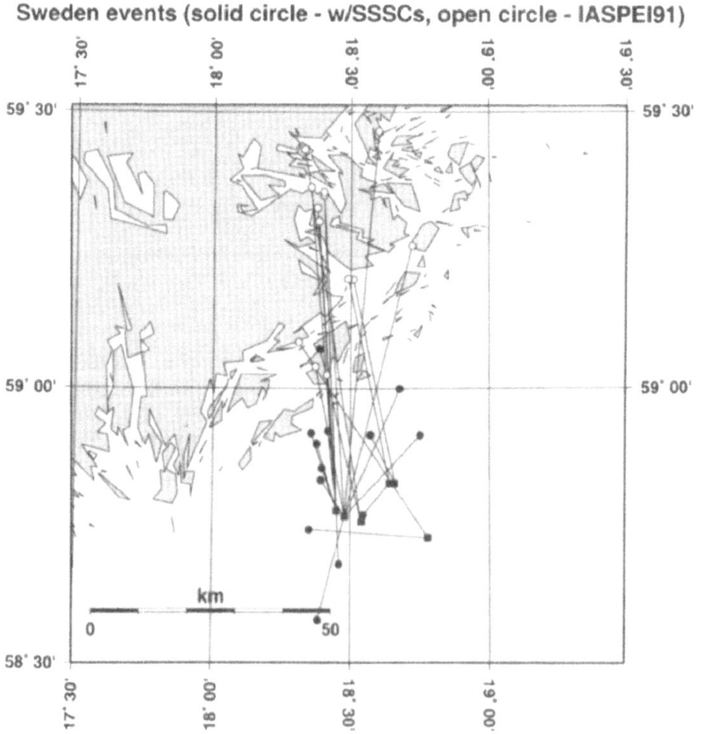

Figure 8

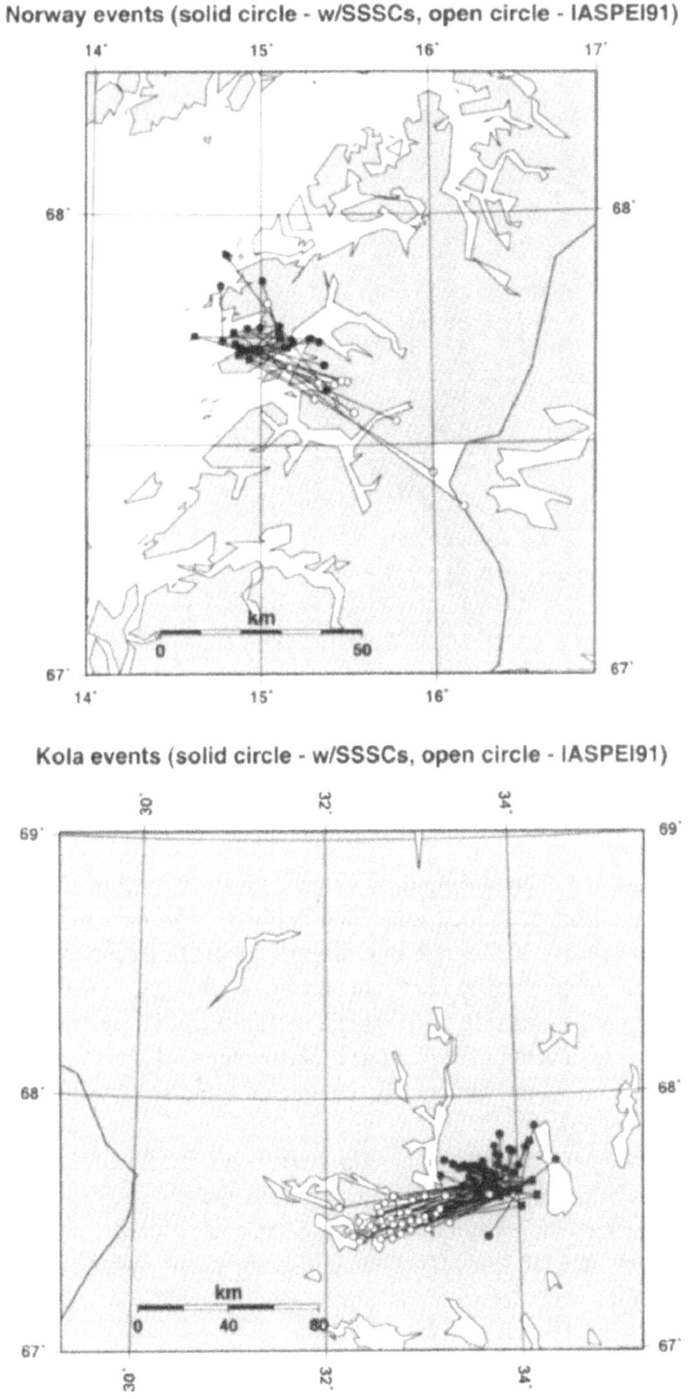
Norway events (solid circle - w/SSSCs, open circle - IASPEI91)
14°
15°
16°
17°
68°
67°
km
0
50
Kola events (solid circle - w/SSSCs, open circle - IASPEI91)
30°
32°
34°
69°
68°
67°
km
0
40
80

Figure 8

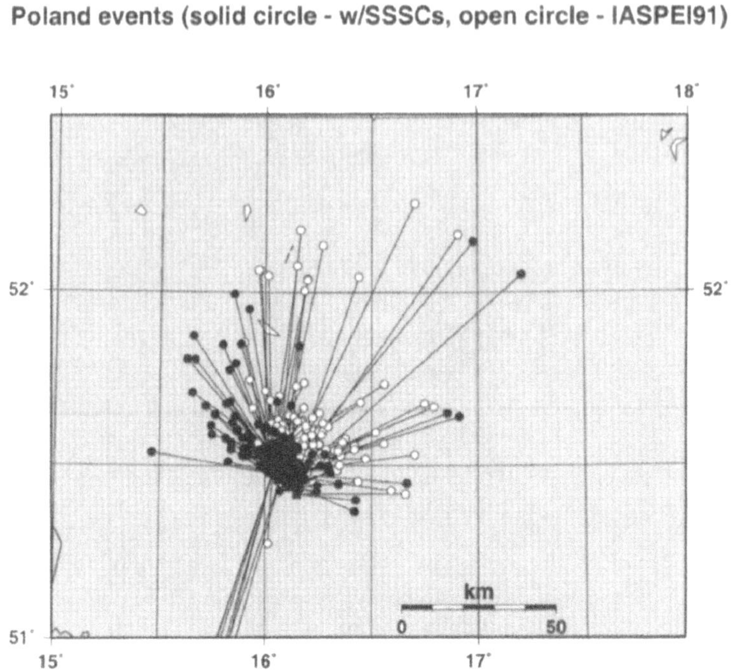

Figure 8
Enlarged map from Figure 7 for Sweden, Norway, Kola, and Poland. Biases are significantly reduced when using SSSCs.

Bondár and North (1999). Compared with the locations without SSSCs, nearly 2/3 of the events have improved locations. The majority of events moved closer to the ground truth by at least 20% while only about 1/4 of the events moved away from the ground-truth locations by 20% or more. Using the SSSCs, however, the improvement for events outside of the shield is significantly improved (e.g., compare Fig. 6 and Fig. 7 for events in Poland and Switzerland). In particular, as shown in Section 5.2, location improvement for events on the North Atlantic Mid-Ocean Ridge is clearly evident.

To measure the performance of modeling errors we examined the area of the 90% coverage ellipses and whether the ellipses contain the ground truth locations. The area of the coverage ellipses obtained using SSSCs is reduced for all events with a median reduction of 2710 km^2. The percentage of events whose coverage ellipses contain the ground-truth locations is only reduced by 5% although is still better than 90%.

The following metrics resulted from applying SSSCs to the ground-truth data set:

- 63% of events moved closer to the GT locations with a median improvement of 9 km;

- 38% of events moved away from the GT with a median deterioration of 5 km;
- 52% of events moved closer to the GT by 20% or more with a median improvement of 10 km;
- 27% of events moved away from the GT by 20% or more with a median deterioration of 7 km;
- Coverage ellipse areas decreased for all events, with a median decrease of 2710 km^2 (from 3830 km^2) to 1120 km^2);
- The 90% ellipse area coverage decreased by only 5% (from 98% to 93%).

As expected, these statistics are similar to those obtained using the 1-D Baltic model alone (Fig. 6) since the GT data set used contained mostly events within the shield.

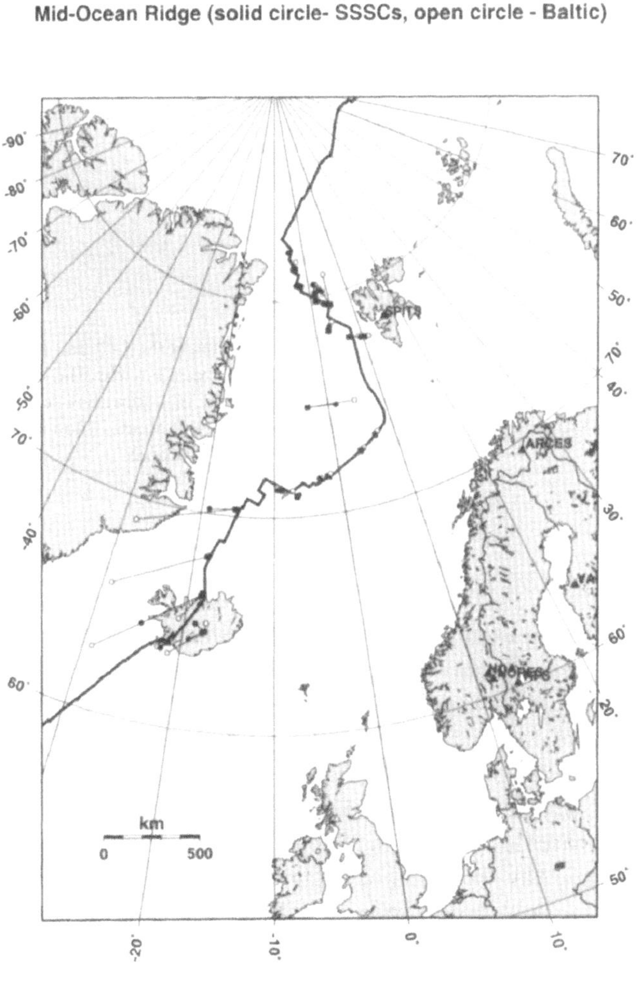

Figure 9

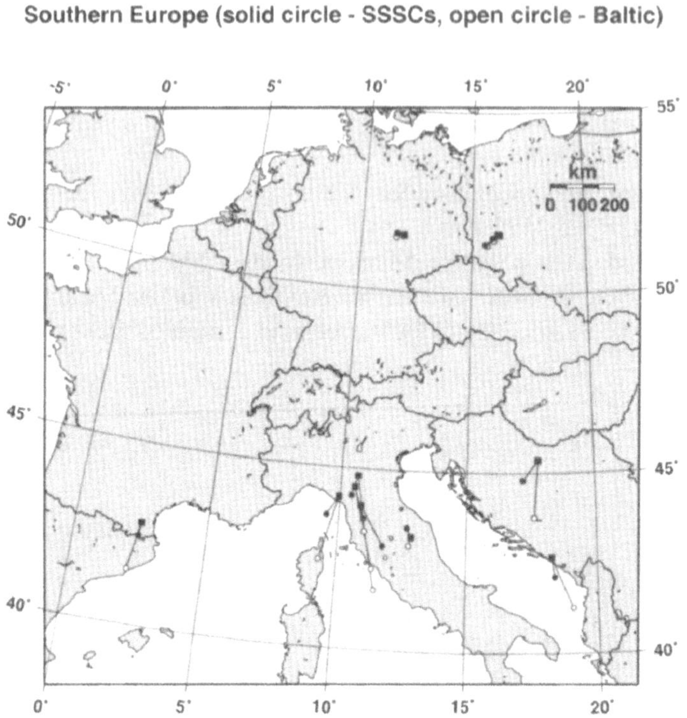

Figure 9
Location comparisons between using SSSCs (solid circle) and using the 1-D Baltic model (open circle) for the CEB events (square). Line siegments link different locations for the same event. Only the Fennoscandian stations, NRIS, and SPITS (Table 1) are used in relocations. Using SSSCs, the North Atlantic Mid-Ocean Ridge events move closer to the ridge axis, and events in southern and central Europe are closer to CEB than Baltic locations.

5.2. Relocations of CEB Events Using SSSCs

In order to test for events outside the Baltic Shield, particularly along the North Atlantic Mid-Ocean Ridge, we relocated CEB events in this region which include stations in Table 1 at regional distances. We used only this set of stations in relocations in order to screen out the effect of teleseismic stations since some IMS teleseismic stations play dominant roles in fixing the locations. 42 CEB events were relocated using SSSCs and using the 1-D Baltic model, respectively.

Figure 9 compares locations using SSSCs and the 1-D Baltic model. Compared to the CEB locations, the locations of events along the North Atlantic Mid-Ocean Ridge and events in southern Europe using SSSCs are significantly better than those obtained using the 1-D Baltic model. When using SSSCs, the North Atlantic Mid-

Ocean Ridge events move closer to the ridge axis, and events in southern and central Europe move closer to CEB locations.

5.3. Relocations of JHD Events Using SSSCs

Station NRIS and SPITS play important roles in event locations at high northern latitudes. Figure 10 presents location results for the Kara Sea event on August 16, 1997 (Table 2). Since all paths from these stations are within the Baltic model region (Fig. 3), the event locations are similar when using SSSCs and using the 1-D Baltic model. The JHD solution is about 15 km from the REB location which was obtained using the IASPEI91 model and unrealistic error estimates. In Figure 10a the SSSC or 1-D Baltic model is applied to *Pn* phases at station FINES, SPITS, and NRIS. Using

(a) Event 97/08/16 (solid circle - w/SSSCs, open circle - w/oSSSCs)

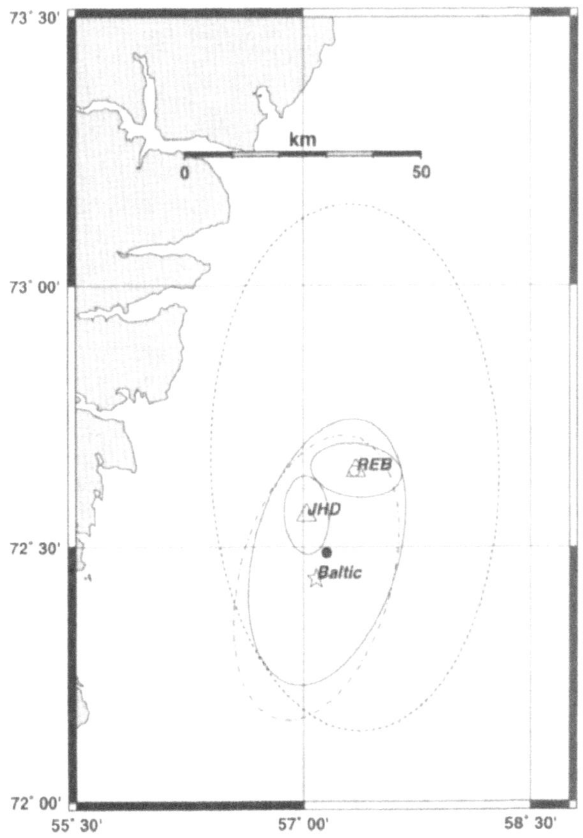

Figure 10a

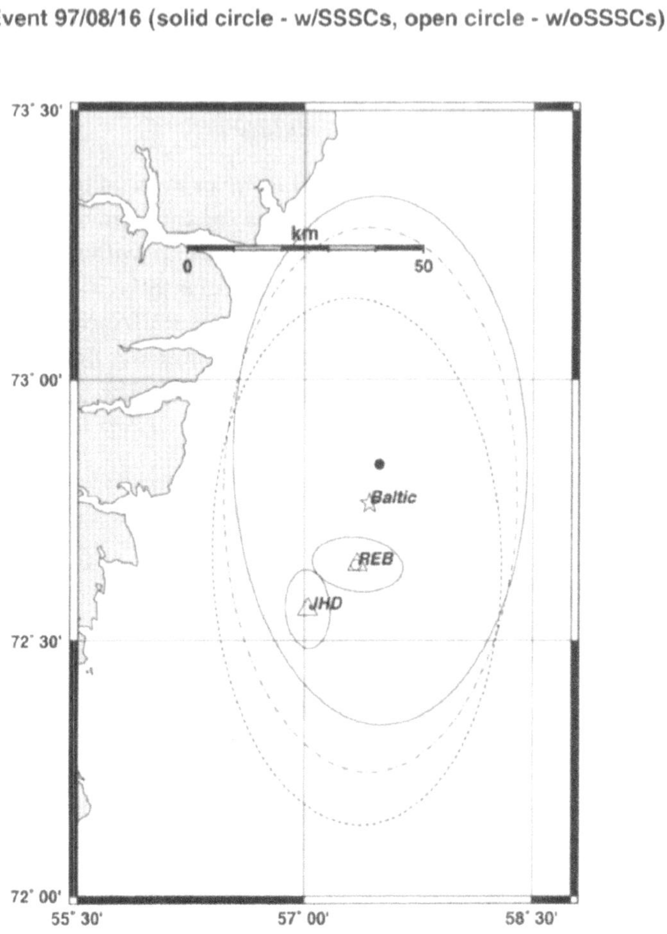

Figure 10
Location comparisons using all stations for the Kara Sea event on August 16, 1997 when the SSSC or 1-D Baltic model is applied only to Pn phases at Fennoscandian station FINES (b) vs. at SPITS and NRIS as well (a). The error ellipses obtained using the 1-D Baltic model, with and without SSSCs are shown in long dashed, solid, and dashed lines, respectively. The error ellipse (solid line) of the REB result (triangle) is unrealistically small. The shield paths to NRIS and SPITS play important roles for event locations in this region. Using SSSCs (solid circle) or the 1-D Baltic model (star) for all stations in Table 1, event locations move closer to the JHD (triangle) solution and the reduced error ellipses contain the JHD solution.

the SSSC or the 1-D Baltic model, event locations move closer to the JHD solution and the reduced coverage ellipses contain the JHD solution. In Figure 10b the SSSC or 1-D Baltic model are not used for NRIS and SPITS. Large bias with respect to the JHD location resulted when using slow travel times (IASPEI91) for these paths.

We further relocated four other events in the region (Table 2) which have JHD locations (ISRAELSSON *et al.*, 1997) and arrival data from IMS stations in Table 1.

Table 2

JHD events in Novaya Zemlya region

Date	Time	Latitude	Longitude	M_b	Depth	Source
10/24/90	14:58:00	73.317	54.803	5.7	0.0	ISRAELSSON *et al.* (1997)
12/31/92	09:29:23	73.600	55.200	2.7	0.0	ISRAELSSON *et al.* (1997)
06/13/95	19:22:38	75.200	56.700	3.4	0.0	ISRAELSSON *et al.* (1997)
01/13/96	17:17:23	75.200	56.700	2.4	0.0	ISRAELSSON *et al.* (1997)
08/16/97	02:11:03	72.562	57.023	3.9	0.0	ISRAELSSON *et al.* (1997)

These JHD locations include constraining ground truth and empirical corrections derived from a suite of earthquakes and explosions in the Novaya Zemlya region. They are our best available locations for these events at this time. Figure 11 indicates the location comparisons between different solutions, using all available stations. The locations are similar when using SSSCs and the 1-D Baltic model. In comparison with the results without SSSCs, using SSSCs or the 1-D Baltic model all coverage ellipses are smaller but contain the JHD solutions. The locations are also quite similar with and without SSSCs because of the fairly good azimuthal coverage.

For the event on 90/10/24 (southern one in Fig. 10a), the location using SSSCs or the 1-D Baltic model is closer to the JHD solution than that without SSSCs. The SSSC or 1-D Baltic model is applied to Pn and Sn phases at station ARCES, FINES, and NRIS. Compared to the locations not using SSSCs or 1-D Baltic model for NRIS, the location does not appear to be affected substantially by the lack of SSSCs at NRIS because it is a well-recorded large event (mb 5.7). The shield paths to NRIS and SPITS have significant effect on locations for small events in this region. For the event on 92/12/31 (northern one in Fig. 10a), the location using SSSCs or the 1-D Baltic model is also closer to the JHD solution than that without SSSCs. The SSSC or 1-D Baltic model is applied to *Pn* and *Sn* phases at station ARCES and SPITS. For the event on 95/06/13 (Fig. 10b) and 96/ 01/13 (Fig. 10c), the locations using SSSCs and the 1-D Baltic model are similar. Relative to the JHD solution, the mislocation using SSSCs or the 1-D Baltic model is similar to that without SSSCs.

6. *Conclusions*

SSSCs for Fennoscandian and other high latitude IMS stations in Table 1 have been developed and tested. Validation testing reflects significant improvement in event locations when applying SSSCs. Within the Fennoscandian shield, similar to using the 1-D Baltic model, locations are improved for the majority of events and the coverage ellipse area is reduced for all events with a median reduction of 2710 km^2 (from 3830 to 1120 km^2). Mislocations due to side effects of the 1-D Baltic model used in Fennoscandia are significantly reduced in central and southern Europe and

(a) Events 90/10/24, 92/12/31 (solid circle- w/SSSCs; open circle- w/oSSSCs)

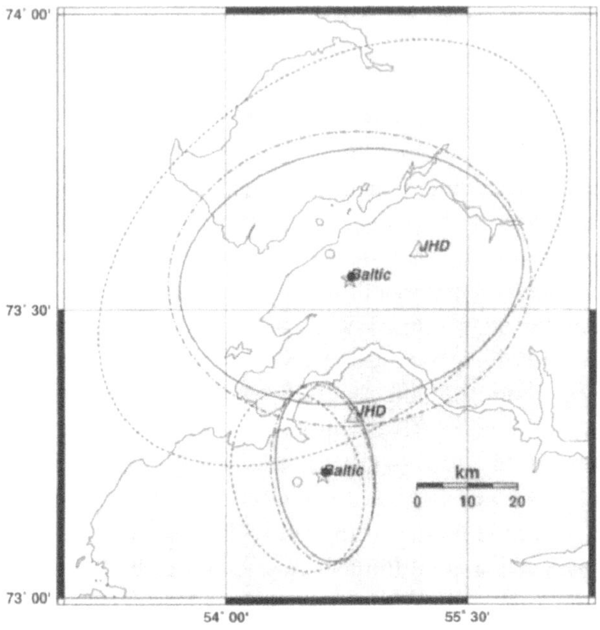

(b) Event 95/06/13 (solid circle- w/SSSCs; open circle- w/oSSSCs)

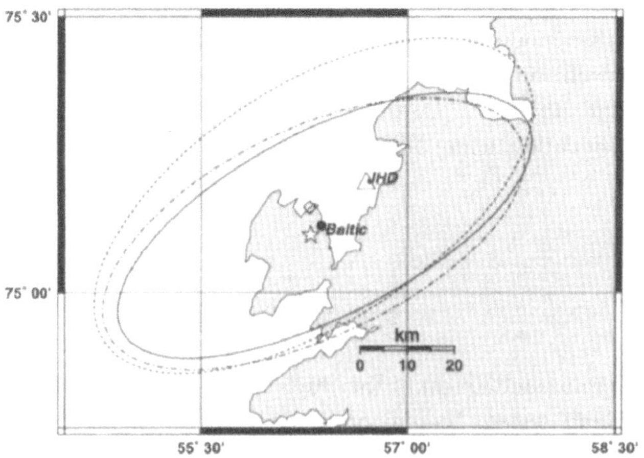

Figure 11a, b

(c) Event 96/01/13 (solid circle- w/SSSCs; open circle- w/oSSSCs)

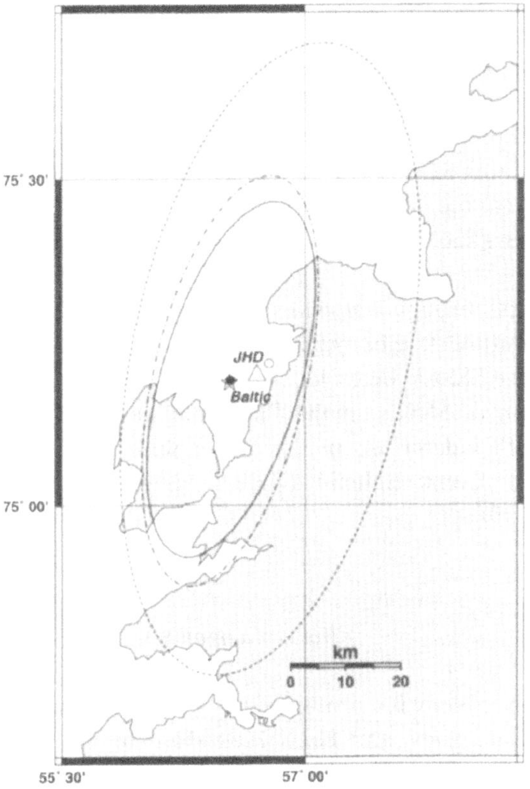

Figure 11

Location comparisons for the events in Novaya Zemlya region (Table 2). The error ellipses obtained using the 1-D Baltic model, with and without SSSCs, are shown in long dashed, solid, and dashed lines, respectively. (a) 90/10/24 and 92/12/31. Relative to JHD solutions (triangle), locations are improved using SSSCs (solid circle) or the 1-D Baltic model (star) in comparison with the locations without SSSCs (open circle). The SSSC or 1-D Baltic model is applied to *Pn* and *Sn* at ARCES, FINES, and NRIS for the event on 90/10/24 and at ARCES and SPITS for the event on 92/12/31. (b) 95/06/13. Relative to JHD the mislocation using SSSCs/1-D Baltic model is similar to that without SSSCs. The SSSC/1-D Baltic model is applied to *Pn*/*Sn* phases at station FINES (no *Sn*) and SPITS. (c) 96/01/13. Relative to JHD solution the mislocation using SSSCs/1-D Baltic model is similar to that without SSSCs. The SSSC/1D Baltic model is applied to *Pn* and *Sn* phases at ARCES (no *Sn*), NRIS, and SPITS.

along the North Atlantic Mid-Ocean Ridge. Locations around Novaya Zemlya region are improved in many cases by introducing faster travel times than IASPEI91 for stations NRIS and SPITS.

Station NRIS and SPITS play important roles in event locations at high northern latitudes. Both paths to such events are within the Fennoscandia regionalization as Baltic. Using SSSCs or the 1-D Baltic model for these two stations, coverage ellipses are reduced but contain the JHD solutions for all events. Relative to the JHD locations, mislocations are less or similar to those without SSSCs.

The SSSCs employ the 1-D Baltic model (implemented in PIDC Operations in 1997) for the shield and the IASPEI91 model for the tectonic and oceanic regions. The SSSC approach provides a more robust, accurate, flexible, effective, and elegant way to implement the 1-D models so that side effect mislocations outside the region are significantly reduced. Test results show that using SSSCs in event location is superior to not using SSSCs, and, in most cases, to using the 1-D model directly when locating events.

Our SSSCs are strongly dependent on the input 1-D regional models and regionalization. Future development in 1-D velocity models and travel-time curves should improve the SSSCs, event locations, and uncertainties. The implementation and demonstration of SSSCs in the PIDC software will encourage these further developments. SSSCs described in this paper were implemented at the PIDC for routine Reviewed Event Bulletin (REB) analysis in April 1999 (YANG and McLAUGHLIN, 1999).

Acknowledgments

We thank Walter Nagy for providing the travel-time calculation tool and helpful discussions on SSSC grids, and Hans Israelsson for the JHD data and insightful discussions. The maps in this paper were generated using GMT (Wessel, P. and W. H. F. Smith, New, improved version of the Generic Mapping Tools released, EOS Trans. AGU, 79, 579, 1998).

References

BONDÁR, I., and NORTH, R. G. (1999), *Development of Calibration Techniques for Potential Use by the CTBT International Monitoring System*, Phys. Earth Planet. Int. *113*, 11–24.

BONDÁR, I., NORTH, R. G., and BEALL, G. (1999), *Teleseismic Slowness-azimuth Station Corrections for the International Monitoring System Seismic Network*, Bull. Seismol. Soc. Am. *89*, 989–1003.

BONDÁR, I., YANG, X., NORTH, R. G., and ROMNEY, C. (2001), *Location Calibration Data for CTBT Monitoring at the Prototype International Data Center*, this issue.

BONDÁR, I., YANG, X., MCLAUGHLIN, K., NORTH, R. G., RYABOY, V., and NAGY, W. (1998), *Source Specific Station Corrections for Regional Phases at IMS Stations in North America and Fennoscandia*, EOS Trans. AGU, 79(45), Fall Meet. Suppl., F839.

DEWEY, J. (1972), *Seismicity and Tectonics of Western Venezuela*, Bull. Seismol. Soc. Am. *62*, 1711–1751.

DOUGLAS, A. (1967), *Joint Epicentre Determination*, Nature *215*, 47–48.

ISRAELSSON, H., Fisk, M., YANG, X., and NORTH, R. G. (1997), *The August 16, 1997 Event in the Kara Sea*, Center for Monitoring Research, Technical Report CMR-97/38.

KENNETT B. L. N. (1991, ed.), *IASPEI 1991 Seismological Tables*, Research School of Earth Sciences. Australian National University, Canberra, Australia.

KREMENETSKAYA, E., and ASMING, V. (1999), *Location Calibration of the Barents Region*, Proceedings of the Calibration Workshop, January 12–14, Oslo, Norway.

MOONEY, W., LASKE, G., and MASTERS, T. (1998), *CRUST 5.1: A Global Crustal Model at 5X5*, J. Geophys. Res. *103*, 727–747.

NAGY, W. (1996), *New Region-dependent Travel Time Handling Facilities at the IDC; Functionality, Testing and Implementation Details*, Center for Monitoring Research, CCB-PRO-96/33.

RUSSIAN FEDERATION/UNITED STATES CALIBRATION WORKING GROUP (1998), *Evaluation of the IMS Location Accuracy of Large Chemical and Nuclear Explosions in Northern Eurasian and North America Using Regional and Global Pn Travel-time Tables*, Proceedings of the 20th Annual Seismic Research Symposium on Monitoring CTBT, 104–113.

STEVENS, J., and MCLAUGHLIN, K. (1997), *Improved Methods for Regionalized Surface Wave Analysis*, *Maxwell Technologies*, MFD-TR-97-15887.

YANG, X., MCLAUGHLIN, K., and NORTH, R. G. (1999), *Source Specific Station Corrections for Regional Phases at IMS Stations*, Proceedings of the 21th Annual Seismic Research Symposium on Monitoring CTBT, 333–343.

YANG, X., and MCLAUGHLIN, K. (1999), *SSSCs for Regional Phases at Fennoscandian and Other Stations*, Center for Monitoring Research, CCB-PRO-99/03.

(Received April 30, 1999, revised October 29, 1999, accepted February 9, 2000)

Pure appl. geophys. 158 (2001) 59–77
0033–4553/01/020059–19 $ 1.50 + 0.20/0

Pure and Applied Geophysics

Analysis of the IMS Location Accuracy in Northern Eurasia and North America Using Regional and Global *Pn* Travel-time Tables

RUSSIAN FEDERATION/UNITED STATES CALIBRATION WORKING GROUP[1]

Abstract—Joint Research Program of Seismic Calibration of the International Monitoring System (IMS) in Northern Eurasia and North America has been signed by the Nuclear Treaty Programs Office (NTPO), Department of Defense USA, and the Special Monitoring Service (SMS) of the Ministry of Defense, Russian Federation (RF). Under the Program historical data from nuclear and large chemical explosions of known location and shot time, together with appropriate geological and geophysical data, has been used to derive regional *Pn/P* travel-time tables for seismic event location within the lower 48 States of the USA and the European part of the RF. These travel-time tables are up to 5 seconds faster in shields than the IASPEI91 tables, and up to 5 seconds slower in the Western USA. Relocation experiments using the regional *Pn* travel-time curves and surrogate networks for the IMS network generally improved locations for regional seismic events. The distance between true and estimated location (mislocation) was decreased from an average of 18.8 km for the IASPEI91 tables to 10.1 km for the regional *Pn* travel-time tables. However, the regional travel-time table approach has limitations caused by travel-time variations inside major tectonic provinces and paths crossing several tectonic provinces with substantially different crustal and upper mantle velocity structure.

The RF members of the Calibration Working Group (WG): Colonel Vyacheslav Gordon (chairman); Dr. Prof. Marat Mamsurov, and Dr. Nikolai Vasiliev. The US members of the WG: Dr. Anton Dainty (chairman), Dr. Douglas Baumgardt, Mr. John Murphy, Dr. Robert North, and Dr. Vladislav Ryaboy.

Key words: The International Monitoring System (IMS), travel-time tables, on-site inspection.

Introduction

The International Monitoring System (IMS) is being developed for global monitoring under the Comprehensive Nuclear-Test-Ban Treaty (CTBT). To fulfill this task seismic events should be located precisely. A location accuracy of 1000 sq.km in continental areas has been proposed for the IMS based on the area allowed for On-Site Inspections. The globally averaged IASPEI91 Seismological Tables (KENNETT, 1991) are now being used by the prototype International Data Center (PIDC) to locate seismic events. However, the required accuracy cannot be achieved for events recorded only at regional distances with globally averaged travel times because of large variations of travel times of regional phases due to lateral

[1] Dr. Vladislav Ryaboy, corresponding author, Center for Monitoring Research, 1300 N, 17th Street, Suite 1450, Arlington, VA 22209, USA. E-mail: ryaboy@cmr.gov

velocity variations in the crust and upper mantle (e.g., JEFFREYS, 1976). To achieve the CTBT goal of attaining 1000 sq.km location accuracy the IMS must be calibrated, i.e., travel times of seismic waves used in the location must take into account variations in the structure of the earth.

The most precise data which can be used to calibrate the IMS are large chemical and nuclear explosions with known locations and origin times (so-called "GT0" events). Deep Seismic Sounding (DSS) data, long-range seismic profiles and well-located earthquakes can also make important contributions to IMS calibration. Relevant seismic and geological data should also be collected and analyzed. In the Joint Research Program a preliminary tectonic regionalization and preliminary analysis of travel times of *Pn* seismic phases recorded from selected nuclear and chemical explosions detonated in Northern Eurasia and North America (SPRINGER and KINNAMAN, 1971; BOCHAROV *et al.*, 1989; MIKHAILOV, 1994, 1996 and other data) have been completed and are described here. Specifically, we discuss for the lower 48 States of the USA and the European part of the RF, the following: 1) regionalization based on tectonic and seismic data; 2) development of regional *Pn* travel-time tables; 3) IMS location accuracy evaluation using regional *Pn* travel-time tables.

Regionalization Based on Tectonic and Seismic Data

Tectonic maps of North America (KING, 1977) and Northern Eurasia (SPIZHARSKIY, 1966; PAVLENKOVA, 1996; ZONENSHAIN *et al.*, 1990) were generalized and digitized to allow more reliable interpolation and comparison of seismic data (travel times of seismic waves, crustal and upper mantle velocity models, location of seismic stations and seismic events) for different geological provinces. Detailed seismic studies of the earth's crust and upper mantle have been carried out in different regions of Northern Eurasia and North America and published in numerous reports, papers and monographs. This information was collected and analyzed. Seismic maps, compiled for the US and RF territory and digitized at Cornell University (BARAZANGI and STEER, 1998) were also used for IMS calibration.

The generalized tectonic map of North America on which are plotted the locations of eight large explosions used for IMS calibration and relocation experiments (solid stars), locations of five CEB events used for relocation experiments (open stars), and locations of IMS stations (triangles) are shown in Figure 1. The main tectonic provinces of North America are:

- the tectonically active Western areas (transition zone from Pacific Ocean to North American continent) (Fig. 1, I and IA);
- the Paleozoic platform (Fig. 1, II) and Paleozoic folded systems (Appalachian, Wichita, and Ouachita Systems) (Fig. 1, III);
- the North American Pre-Cambrian Platform (Fig. 1, IV) including the Canadian Shield (North American Craton) (Fig. 1, V).

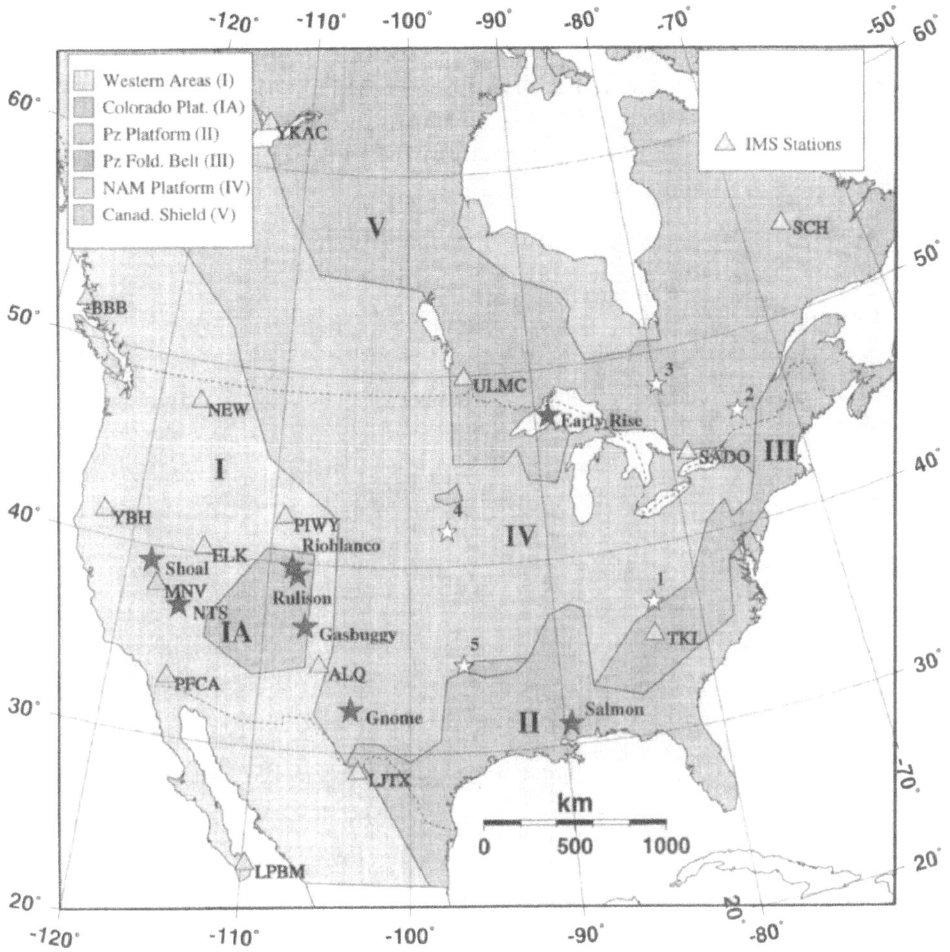

Figure 1
Generalized tectonic map of North America, locations of historical chemical and nuclear explosions (solid stars) used for development of regional *Pn* travel-time tables and relocation experiments, CEB events (open stars) used for relocation experiments, and the IMS stations (triangles).

Refraction profiles show that these tectonic regions have substantially different crustal and upper mantle velocity structures (PAKISER and MOONEY, 1989; ROMANOWICZ, 1979; VAN DER LEE and NOLET, 1997 and many other published works). The greatest variation of crust and upper mantle velocity structures is between western areas of North America and the Pre-Cambrian platform. When crossing the boundary from west to east between these tectonic provinces, values of mean velocity in the crust, Moho depth, and *Pn* velocity increase from approximately 6.2–6.3 km/s, 30–35 km, and 7.7–7.9 km/s to 6.5–6.6 km/s, 40–50 km, and 8.1–8.2 km/s, respectively.

Figure 2 shows the generalized tectonic map of the western part of Northern Eurasia. The locations of nuclear explosions (solid stars) and stations (triangles) used for the IMS calibration, and ray-path projections on the earth's surface are indicated on Figure 2. The main tectonic provinces shown on Figure 2 are:

- the East-European Pre-Cambrian platform (EEPLAT) (Fig. 2, 1) including the Baltic (Fig. 2, 2) and Ukrainian shields;

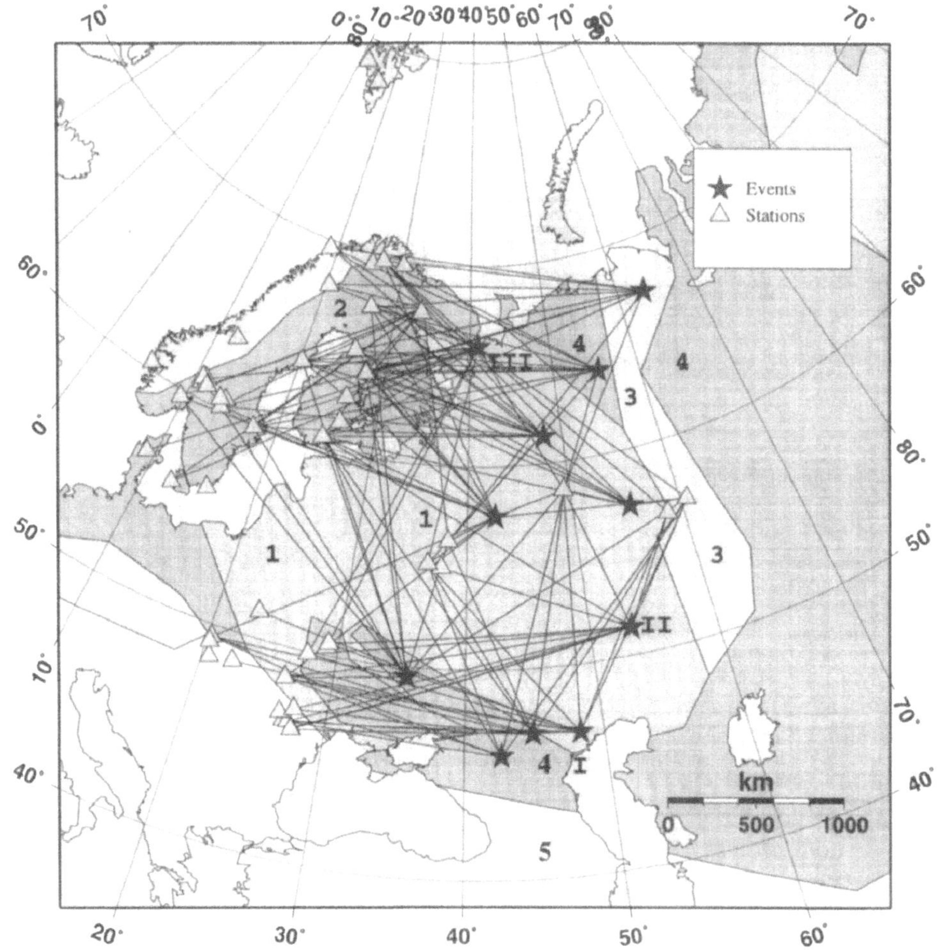

Figure 2
Generalized tectonic map of the European part of RF (1 – East European Pre-Cambrian platform, 2 – Baltic shield, 3 – Urals, 4 – Paleozoic platforms, 5 – Tectonic active regions), location of PNEs (solid stars) and seismic stations (triangles) used for development of the regional *Pn* travel-time table and relocation experiments. Four PNEs used for relocation experiments: I – Astrakhan region, II – Uralsk region (Uralsk-1 and Uralsk-2), and III – Arkhangelsk region.

- the Paleozoic folded system of the Urals (Fig. 2, 3); and the West-Siberian (east of the Urals), Timan-Pechora (northwest of the Urals), and Scythian (south of the EEPLAT) Paleozoic platforms (Fig. 2, 4);
- the Caucasus Alpine folded system is located to the south of the Scythian platform.

Detailed geophysical maps of the RF were compiled by the GEON Center (RF Ministry of Natural Resources, Moscow). These maps were digitized (BARAZANGI and STEER, 1998) for our use. A brief summary and analysis of the RF seismic maps with reference to the generalized tectonic map of Northern Eurasia (Fig. 2) is given below.

The crustal thickness (Moho depth) within Northern Eurasia varies from 55–60 km beneath the Ural mountains to 32–35 km beneath the North-Caspian sedimentary basin. The depth to Moho is greater beneath the ancient Pre-Cambrian platforms of Northern Eurasia than for the young Paleozoic platforms. Crossing the boundaries between the Pre-Cambrian platforms (East-European and Siberian) and the young Paleozoic platforms (Timan-Pechora and West-Siberian platforms) is typically accompanied by a decrease of crustal thickness from approximately 45–50 km to 35–40 km, respectively. In general the crustal thickness decreases towards the Arctic Ocean.

Pn velocity within the RF varies from approximately 7.8–8.0 km/s to 8.4 km/s and greater. The largest values of *Pn* velocity were measured beneath the East-European and Siberian Pre-Cambrian platforms (8.1–8.4 km/s and greater) (PAVLENKOVA *et al.*, 1996). The smallest values of *Pn* velocity (less than 8.0 km/s) were measured beneath the Baikal rift zone, the Trans-Baikal region, tectonically active regions of Southern Siberia (Altay-Sayany folded zone) and the central part of the Paleozoic West-Siberian platform. This last tectonic unit is considered by many geologists to be an ancient rift zone (ZONENSHAIN *et al.*, 1990). The *Pn* velocity values beneath the Ural mountains, North-Caspian sedimentary basin, and the central part of the Caspian Sea are 7.9–8.0 km/s, 8.2–8.3 km/s, and 8.0 km/s, respectively. *P*-wave velocity values at the basement surface vary from approximately 5.4–5.6 km/s to 6.4–6.5 km/s and greater. The largest values of basement velocity were measured beneath the East-European and Siberian Pre-Cambrian platforms (6.2–6.5 km/s and greater). The smallest values of basement velocity (less than 6.0 km/s) were measured beneath the Paleozoic West-Siberian platform.

The primary and auxiliary IMS seismic stations are located in a variety of tectonic provinces of Northern Eurasia and North America. Ray paths of seismic waves from regional events to the IMS network will cross different tectonic provinces having substantially different crustal and uppermost mantle velocity structures. The work presented here seeks to provide a sound basis for regionalization of *Pn* travel times to improve locations of regional seismic events in these regions.

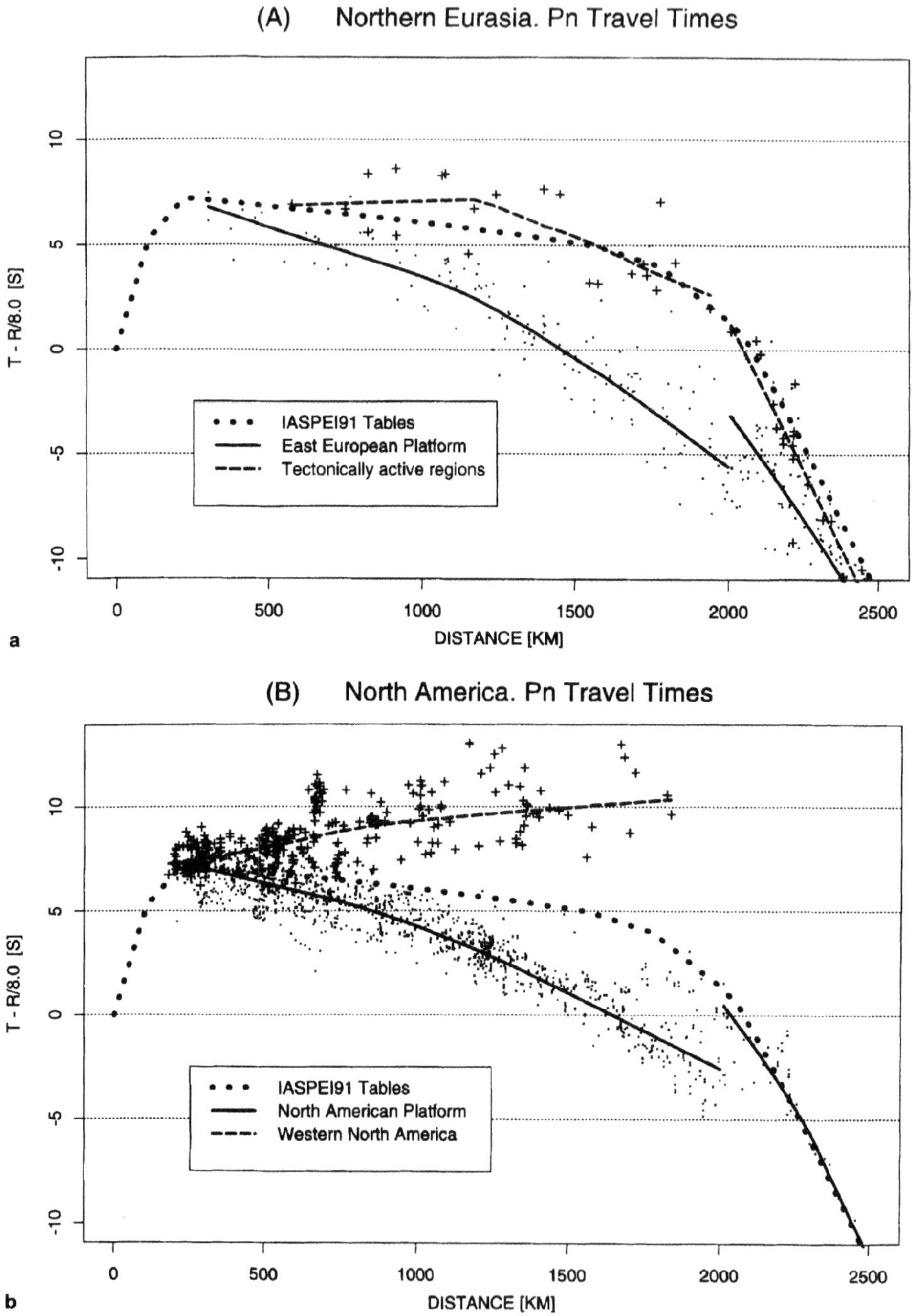

Figure 3

Regional *Pn* travel-time curves for major tectonic provinces of Northern Eurasia (A) and North America (B).

Regional Pn Travel Times for Northern Eurasia and North America

Pn travel times were taken from various archives, reports, published papers and from the US Geological Survey and International Seismological Centre bulletins. Maps of the variation of travel-time residuals from the IASPEI91 tables, and subsequently travel-time curves, were constructed for explosions detonated in North America and the European part of the RF. The locations of these explosions are shown on Figures 1 and 2. The methods of mapping and contouring of seismic wave travel-time residuals are based on the Generic Mapping Tools (GMT) software package (WESSEL and SMITH, 1995). Contoured travel-time residuals were used to study travel-time variations vs. distance and azimuth within different geological provinces of North America and Northern Eurasia. Examples of maps of *Pn* residuals for regions of Europe and North America are shown in RF/US WORKING GROUP (1997); FIRBAS (1999); and RYABOY *et al.* (1999). Identification and removal of outliers is an important task for travel-time analysis; the abrupt and unusual variations of the observed travel-times were used as a criterion to identify outliers.

Preliminary analysis of observed *Pn* travel-time data for North America and Northern Eurasia reveals that large regional travel-time variations, of the order of 5–10 sec, are measured at distances of 1000–2000 km. Based on the tectonic regionalization and the residual maps, a preliminary division into major provinces of *Pn* travel-time variations was made. Figure 3 shows *Pn* travel times measured for ray paths within these major tectonic provinces in Northern Eurasia (Fig. 3A) and North America (Fig. 3B). *Pn* travel times for the ancient Pre-Cambrian provinces (North American and East European platforms, including the Canadian, Baltic, and Ukrainian shields) are fast, and *Pn* travel times are up to 5 sec earlier than predicted by the IASPEI91 tables for the distance range 1000–2000 km. Observed travel times of P-wave first arrivals are fit by the IASPEI91 tables in the North American platform for the distance range 2000–2500 km. The tectonically active regions of the western US are characterized by substantially slower *Pn* travel times, up to 5 sec late in comparison with the IASPEI91 tables. *Pn* travel-time residuals relative to the IASPEI91 tables for the southern regions of the European part of the RF, including the Alpine tectonic zone, are approximately $\pm(1.0\text{–}2.0)$ sec. The IASPEI91 tables were used as regional travel-time tables for relocation experiments in the southern regions of the RF.

Regional *Pn* travel-time curves for the East European and North American Pre-Cambrian platforms and for western regions of North America were constructed by applying a smoothing procedure to the observed data. The procedure is based on a polynomial approximation, and was performed using the S-PLUS software package (BECKER *et al.*, 1988). Figure 3 shows the smoothed regional *Pn* travel-time curves developed for the major tectonic provinces of Northern Eurasia and North America (BONDÁR and RYABOY, 1997; NORTH *et al.*, 1997): the East-European platform, the North-American platform, and western areas of North America (Table 1).

The *Pn* travel-time curves within the Pre-Cambrian platforms of Northern Eurasia, including the Baltic Shield (VINNIK and RYABOY, 1981; BONDÁR and RYABOY, 1997; NORTH *et al.*, 1997), and North America are very similar. Figure 4 shows regional *Pn* travel-time curves for the East European and North American Pre-Cambrian platforms and for western tectonically active regions of North America. The IASPEI91 and Baltic Shield reference curves are presented for comparison. The *Pn* travel-time curve for the North American Pre-Cambrian platform is characterized by slightly slower travel times at distances greater than 1500 km. This may be due to specific features of the upper mantle velocity structure or to misinterpretation within this distance range of *P* as *Pn*, and should be investigated through waveform analysis.

Pn travel times measured from explosions with known location and origin time and from well located earthquakes for other Pre-Cambrian provinces such as the Australian shield (BOLT *et al.*, 1958; DENHAM *et al.*, 1972; FINLAYSON *et al.*, 1974; BOWMAN and KENNETT, 1993) and the Siberian platform (EGORKIN *et al.*, 1987; PAVLENKOVA *et al.*, 1996) are in good agreement with the *Pn* curves for Pre-Cambrian platforms in Figure 4. This suggests that the travel times of *Pn* waves recorded within these regions might be transportable to other Pre-Cambrian tectonic provinces where we do not have reliable *Pn* travel-time measurements. More detailed consideration of this topic is beyond the scope of this work.

Table 1

Regional Pn travel-time tables (depth = 0.0)

Distance (deg.)	East European Platform (s)	North American Platform (s)	Western areas of North America (s)
0	0	0	0
1	21.3*	21.3*	21.3*
2	35.0	35.0	35.2
3	48.5	48.6	49.3
4	61.9	62.1	63.5
5	75.4	75.6	77.7
6	88.7	89.2	92.0
7	102.2	102.6	106.2
8	115.6	116.0	120.4
9	129.0	129.3	134.4
10	142.4	142.7	148.5
11	155.5	155.8	162.5
12	168.5	169.1	176.6
13	181.5	182.1	190.6
14	194.3	195.2	204.6
15	207.2	208.5	218.7
16	219.9	221.3	232.7
17	232.7	234.3	
18	245.4	247.1	

* Extrapolated data

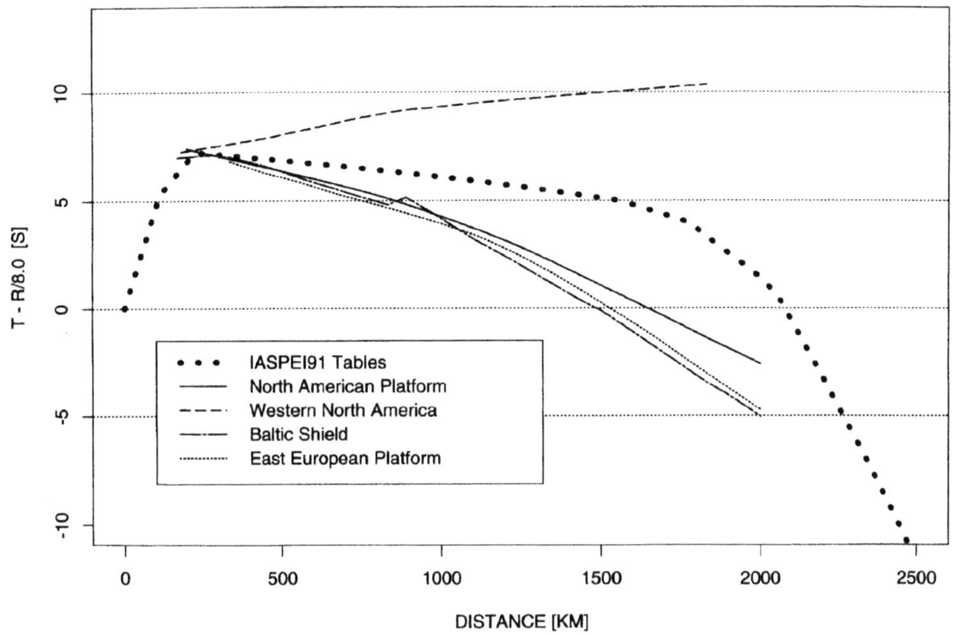

Figure 4
Regional *Pn* travel-time curves for ancient Pre-Cambrian platforms and tectonic provinces.

The *Pn* travel-time observations scatter from the smoothed curves by up to three or four seconds when the travel times are plotted without regard to the azimuth from the source. Measurement errors, deduced from the measured signal-to-noise ratio (SNR) and estimated by ISRAELSSON (1997) to be 0.2–0.5 sec, are considerably smaller than this. Thus the scatter is likely caused by factors such as lateral structural variations of the earth's crust and upper mantle. We assume the scatter is due to the model of a single travel-time curve being too simple, and call it "modeling error." The modeling error can be represented as a seismic phase-dependent function of distance and depth. Network configuration, measurement error and modeling error determine the ellipse error parameters in the current CMR bulletin.

Figure 5 and Table 2 show the *Pn* modeling errors for different regional travel-time curves calculated using *Pn* travel-time deviations from the smoothed regional *Pn* travel-time curves. These modeling errors of the regional *Pn* travel-time curves increase from approximately 1 sec (distance range 200 km to 1400 km) up to 2 sec at distances of 1800–2000 km. The increase in modeling errors at distances greater than 1300–1500 km could be due to the misinterpretation of *P* as *Pn* waves. Estimates of the modeling uncertainties for the globally-averaged IASPEI91 model (ISRAELSSON, 1997 and other CMR reports) were derived from a data set of over 3000 large events from the Group of Scientific Experts third technical test (GSETT-3) (Fig. 5). The modeling errors apply to the currently used IASPEI91 travel-time tables.

IMS Location Accuracy Evaluation Using Regional Pn Travel-time Tables

Simulation studies of the error ellipse area have shown that, although increasing the number and azimuthal coverage of the recording network stations will reduce location uncertainties, the goal of achieving 1000 sq.km error ellipses

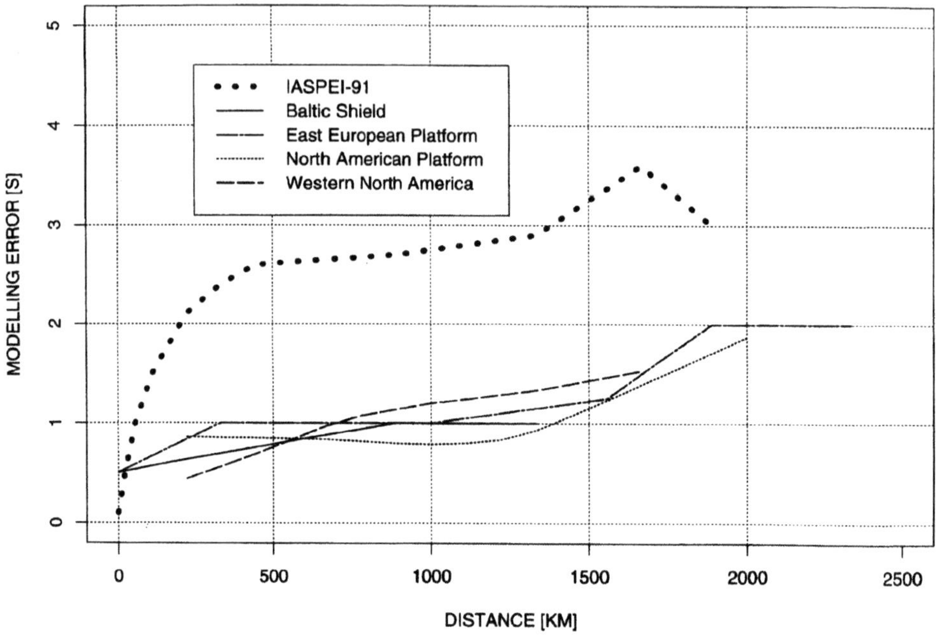

Figure 5
Variations of *Pn* modeling errors with epicentral distance for large tectonic provinces.

Table 2

Distance-dependent modeling errors

Distance (deg.)	East European Platform (s)	North American Platform (s)	Western areas of North America (s)
0.0	0.0	0.0	0.0
0.2	0.1	0.1	0.1
0.5	0.2	0.2	0.2
1.0	0.3	0.3	0.3
2.0	0.7	0.9	0.4
3.0	1.0	0.9	0.6
4.0	1.0	0.9	0.7
8.0	1.0	0.9	1.1
12.0	1.2	0.9	1.3
15.0	1.5	1.3	1.5
17.0	1.8	1.6	1.6

cannot be achieved with global-averaged travel-time curves, like IASPEI91, which have large modeling errors (RF/US CALIBRATION WORKING GROUP, 1998). To reliably reduce an error ellipse area to 1000 sq.km, regional travel-time curves that can predict *Pn* travel times with errors not more than 1.0–1.5 sec must be used.

The regional *Pn* travel-time curves (Table 1 and Fig. 3) and modeling errors (Table 2 and Fig. 5) derived in this work were used to relocate nuclear and large chemical explosions and events within North America from the PIDC Calibration Event bulletin (CEB). The CEB includes reliably located mine explosions and shallow earthquakes. The LocSAT code (BRATT and BACHE, 1988) was used for seismic events location. Different regional travel-time tables can be used simultaneously in this code, allowing large-scale lateral velocity heterogeneities of the crust and upper mantle to be taken into consideration. Locations calculated for the relocated events were compared to the true locations.

The following procedures were applied for selection of seismic stations and travel-time tables in the relocation experiments:

- Seismic station designation — the "surrogate" seismic network used for the relocation experiments consisted of IMS stations or substitutes located at distances less than 300–500 km from the corresponding IMS stations. All stations were less than 2000 km from the epicenter.
- *Pn* travel-time tables — a *Pn* travel-time table was defined for each "event-station" pair according to the tectonic province type crossed by not less than 70% of the line connecting each station to the epicenter. Observations were not used for event relocations if at least 70% of the line connecting a station and epicenter did not lie within one of the tectonic province types. IASPEI91 travel-time tables and modeling errors were used as regional tables for event locations within southern regions of the European part of Northern Eurasia.
- PIDC locations — for comparison, all epicenters were also located using the regular PIDC method: IASPEI91 travel-time tables and modeling errors estimated for the IASPEI91 model (Figs. 3 and 5). For all relocation experiments depth was constrained and set to 0.0 km.

As examples, we will describe in detail relocation experiments for two selected nuclear explosions. One peaceful nuclear explosion (PNE) was detonated in Northern Eurasia (PNE, Astrakhan region) and the other in North America (Gnome nuclear explosion).

PNE Detonated 24 September 1983 (origin time 05:00:00.0) within North-Caspian Sedimentary Basin, Astrakhan Region (RF/US WORKING GROUP, 1997)

Figure 6 shows the generalized tectonic map of the European part of Northern Eurasia, IMS stations (light triangles) and a network used for the event relocation (solid triangles). The event was located by eight stations (SUKH, DST, MLR, MKR, OBN, KIRV, ARU, and BRVK) using the IASPEI91 travel-time tables and by six

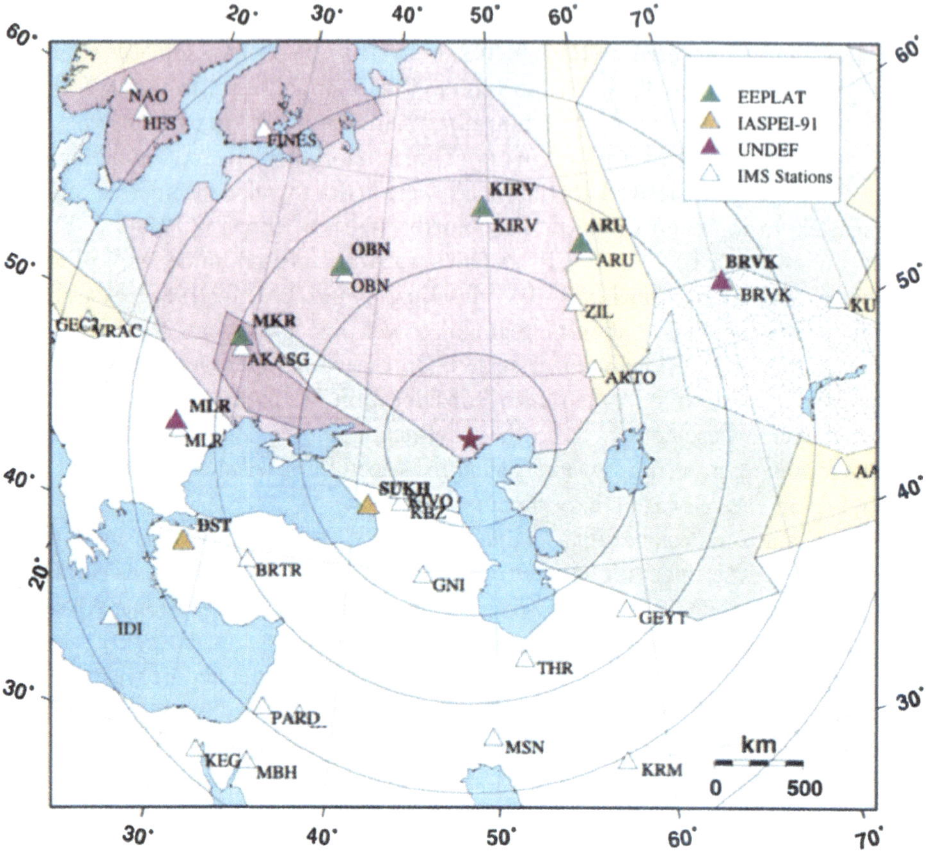

Figure 6
Map of the IMS network simulated for relocation of the PNE detonated on 24 September 1983 in the Astrakhan region, North Caspian depression, (red star — PNE; green and brown triangles — stations for which a *Pn* travel-time table for the East European platform and the IASPEI91 table were used for relocation experiments, respectively; magenta triangles — stations which were not used for relocation experiments with regional travel-time tables; grey triangles — IMS stations). In the background the simplified tectonic map is shown (see Fig. 2).

stations using regional travel-time tables: the *Pn* travel-time curve for the East-European platform was used for four stations (MKR, OBN, KIRV, and ARU) and the *Pn* IASPEI91 travel times were used instead of the regional *Pn* travel-time curve for two stations (SUKH and DST). Two stations (MLR and BRVK) were not used for the event location using regional *Pn* travel-time tables because the procedures were not met (see the procedures above). Despite the small number of defining stations, the solution using the regional *Pn* travel-time table for the East European platform (EEPLAT) and the IASPEI91 table for tectonically active southern regions is considerably closer to the true epicenter than the IASPEI91 solution, and the error ellipse includes the true epicenter (Fig. 7).

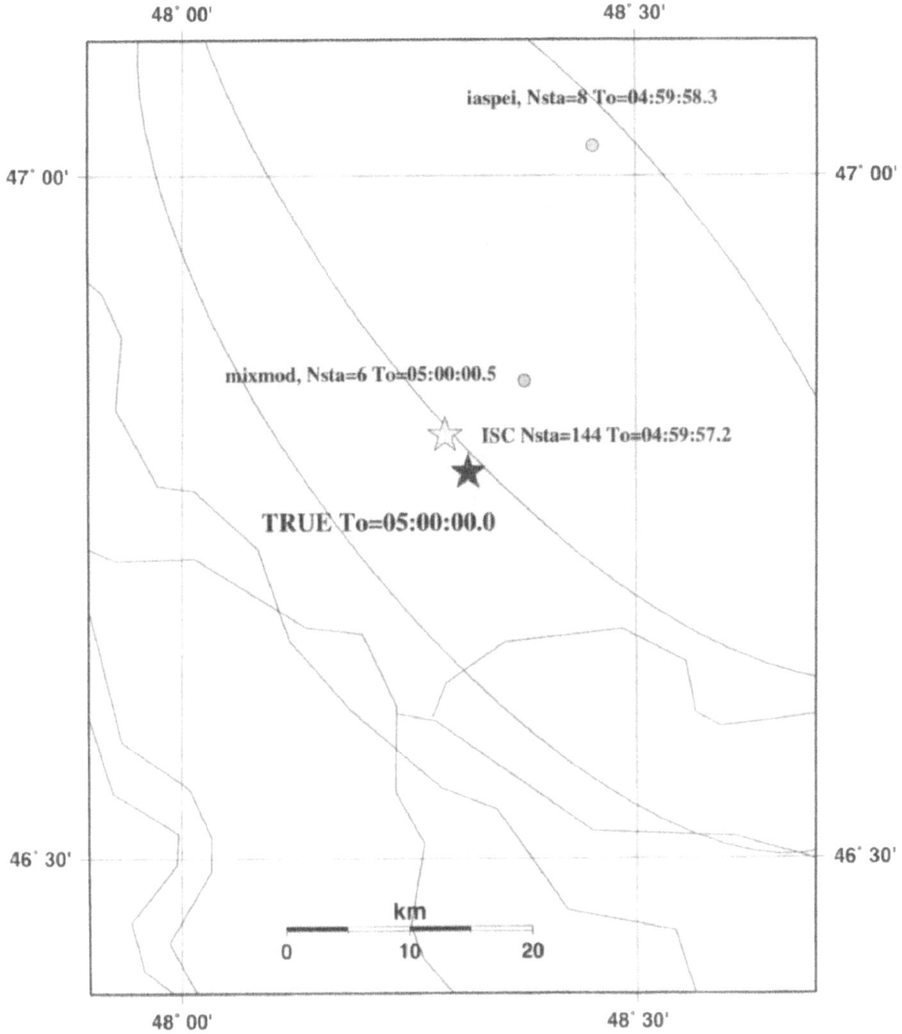

Figure 7
Location estimations of a PNE detonated on 24 September 1983 in the Astrakhan region (North Caspian depression). The error ellipse area calculated for the regional case is still large due to the inclusion of two arrival-time measurements (SUKH and DST) with large IASPEI91 modeling errors.

GNOME Nuclear Explosion Detonated 10 December 1961 (origin time 19:00:00.0) within the Southwestern Marginal Part of the North American Platform (ROMNEY et al., 1962)

The location was estimated using eight seismic stations. Figure 8 shows the generalized tectonic map of North America, IMS stations (light triangles), and the IMS network simulated for the Gnome explosion relocation (solid triangles). The regional *Pn* travel-time curve developed for the North American platform (NAM-

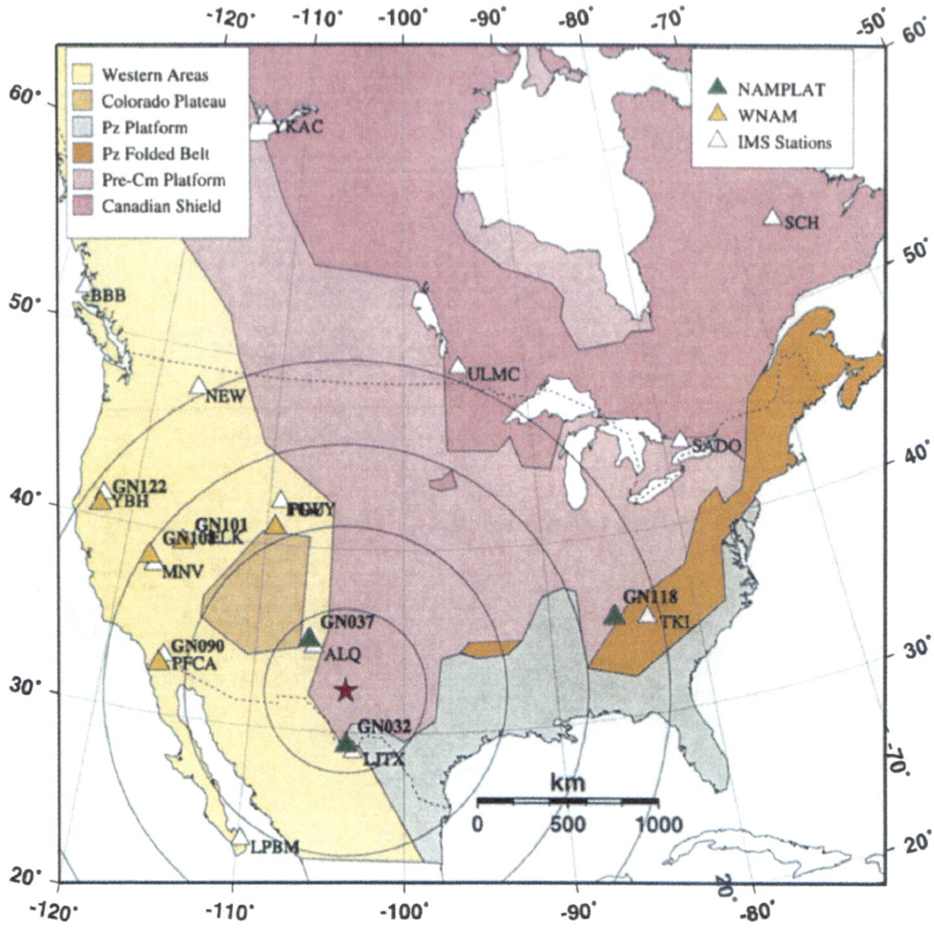

Figure 8
Map of the IMS network simulated for relocation of the Gnome nuclear explosion detonated on 10 December 1961 in southwestern areas of the North American platform (red star — nuclear explosion; green and brown triangles — stations for which regional *Pn* travel-time tables for the North American platform and for Western areas of US were used for relocation experiments, respectively; grey triangles — IMS stations). In the background the simplified tectonic map is shown.

PLAT) was used for ray paths to three stations (GN032, GN037, and GN118), and the regional *Pn* curve developed for western North America (WNAM) was used for ray paths to five stations (GN090, GN101, GN108, GN122, and FGU stations). Figure 9 shows the results of the relocation experiment for Gnome. The distance between the true epicenter and the estimated location decreased from 28.3 km to 10.6 km using the two regional *Pn* travel-time tables ("mixmod" solution) in comparison with IASPEI91 tables ("iaspei" solution). The calculated error ellipse for the location based on the regional *Pn* travel-time tables is substantially smaller than for the "iaspei" solution (Fig. 9).

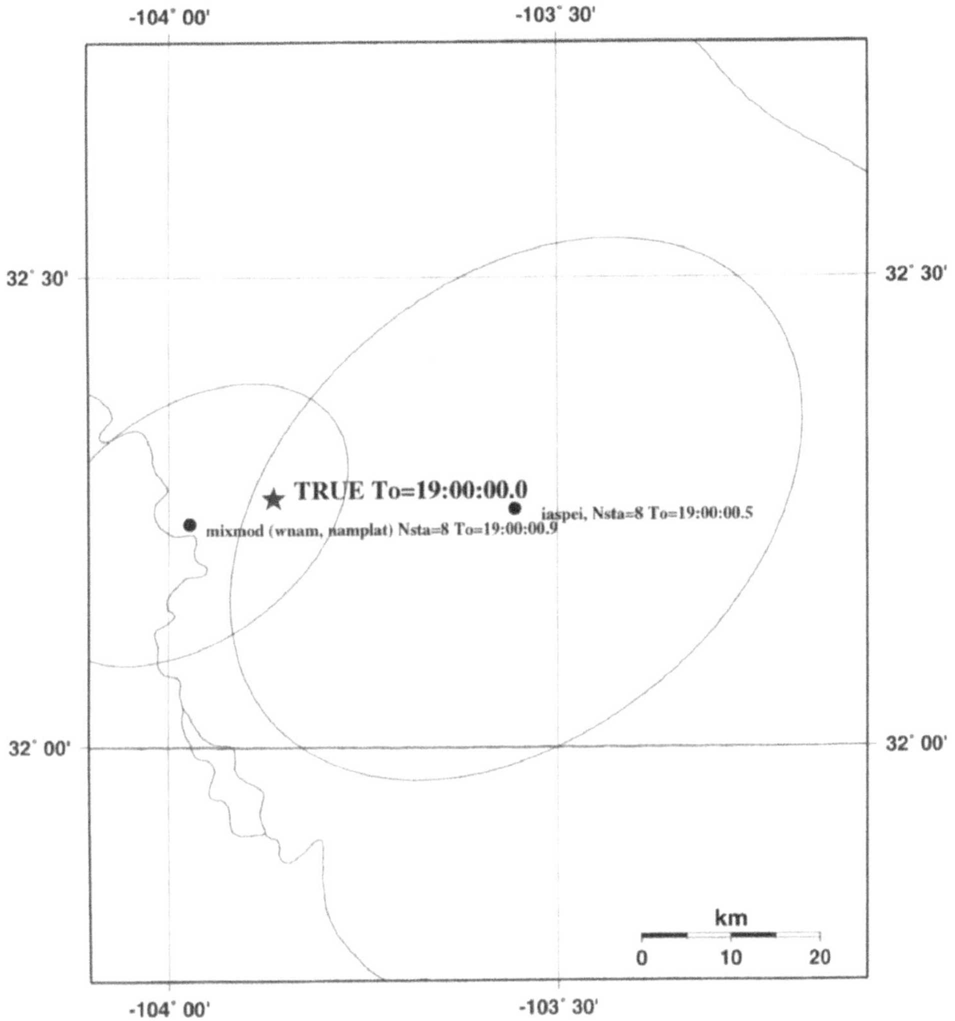

Figure 9
Location estimations of the Gnome nuclear explosion detonated on 10 December 1961 in southwestern areas of the North American platform.

Seismic events used for relocation experiments are shown on Figures 1 and 2, and the results of the relocation experiments with 14 events in Northern Eurasia and North America are shown in Table 3 and Figure 10. In Figure 10, the mislocations of the IASPEI91 tables are plotted against the mislocation for the regional *Pn* travel-time curves. The diagonal line across the plot is equal performance, with points above the line indicating the regional curves performed better whereas points below the line indicate that better results were obtained using the IASPEI91 model.

Table 3

Regional event mislocation

Event Number	Events	Latitude (deg.)	Longitude (deg.)	Mislocation using IASPEI91 tables (km)	Mislocation using regional travel-time curves (km)
North America					
1	Early Rise	47.5505	−88.9390	8.6	5.3
2	Gnome	32.2636	−103.8658	28.3	10.6
3	Salmon	31.1421	−89.5699	30.9	23.2
4	Shoal	39.2003	−118.3803	10.3	9.6
5	Gasbuggy	36.6778	−107.2083	14.4	13.4
IDC CEB					
6	CEB 1	36.9963	−83.110	27.1	8.2
7	CEB 2	45.9562	−74.2821	10.6	7.0
8	CEB 3	48.2452	−80.0234	14.0	15.9
9	CEB 4	41.8121	−97.200	10.8	3.1
10	CEB 5	34.6742	−96.3752	8.3	15.0
Northern Eurasia					
11	Uralsk-1	51.3583	53.3194	45.7	8.2
12	Uralsk-2	51.3625	53.3061	11.6	1.4
13	Astrakhan	46.7831	48.3152	28.6	8.8
14	Arkhangelsk	65.9939	41.0381	14.3	11.9
Events 1–14/Mean				18.8	10.1

The relocated events can be split into three groups. The first group consists of five events (Astrakhan, Uralsk-1, Gnome, Salmon, and CEB no. 1) with clear location improvement. Mislocations were decreased for this group of events, from approximately 30–45 km (using IASPEI91 tables for relocations) to 10–25 km (using regional travel-time tables for relocations). The second group of events, which includes Arkhangelsk, Uralsk-2, Shoal, Early Rise, CEB no. 2 and CEB no. 4, had less pronounced mislocation improvement, from approximately 10–15 km (using IASPEI91 tables) to 2–12 km (using regional travel-time tables). This group of events was mainly located by networks with good azimuthal coverage, and locations using regional travel-time curves would not be expected to improve substantially on locations using global travel-time curve, such as IASPEI91. The third group of events (CEB no. 3 and CEB no. 5) did not show location improvement. Possible causes of this lack of improvement include poor signal-to-noise ratio, and influence of outliers, or perhaps factors related to significant lateral heterogeneity not accounted for in our regional location method.

Results

A number of nuclear and large chemical explosions with known location and origin time was selected within central and southern regions of North America and Northern Eurasia, and *Pn* travel times recorded from these explosions were

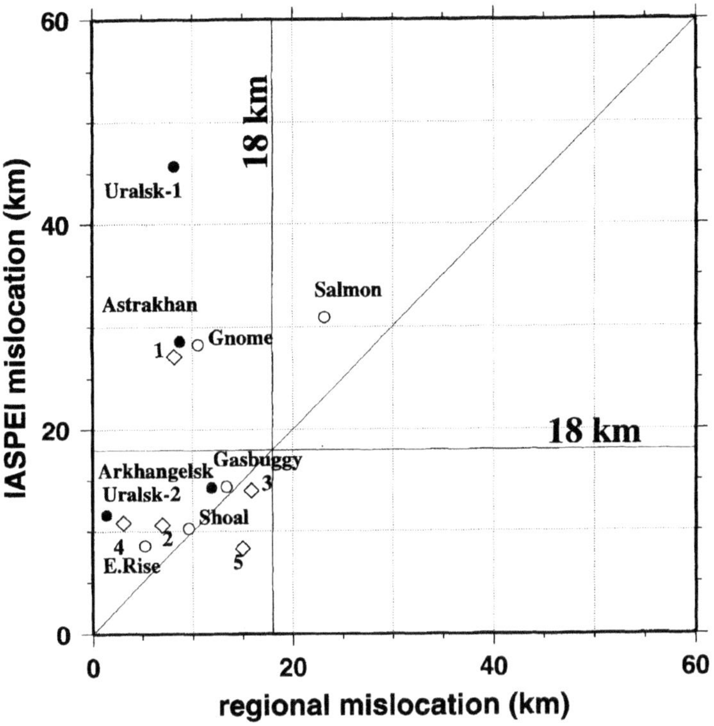

Figure 10
Location evaluation of regional seismic events which occurred within the European part of Northern Eurasia and North America (open diamonds are the IDC CEB events, and open circles and solid circles are nuclear and chemical explosions detonated within North America and the European part of the RF, respectively).

analyzed to derive regional travel-time tables and modeling errors. These tables and modeling errors were derived for the East European Platform, the North American Platform, and the Western United States. The regional 1-D *Pn* travel-time tables and modeling errors were used to relocate 14 events (nuclear and large chemical explosions, and the PIDC CEB events) in the European part of the RF and within the USA.

The application of calibrated regional *Pn* travel-time tables improves location estimation and decreases error ellipse size in most cases. The mean distance between true and calculated epicenters decreased for all events analyzed from 18.8 km for the IASPEI91 model to 10.1 km for the regional travel-time models. An important result from this work is that to meet a location accuracy of 1000 sq.km (mislocation approximately 18 km) when only a few regional stations with less than ideal azimuthal coverage record, true travel-time errors of around $\pm(1.0–1.5)$ sec are needed. However, the regional travel-time table approach has limitations caused by travel-time variations inside major tectonic provinces and paths crossing several tectonic provinces with substantially different crustal and upper mantle velocity structure. Seismic event locations based on 3-D velocity models probably have the potential to overcome these limitations.

References

BARAZANGI, M., and STEER, D. *Knowledge-based visualization of multidisciplinary databases for IDC interactive analysis: Region specific and global,* Cornell University, Institute for the Study of the Continents. R&D Status Report, February 1998.

BECKER, R. A., CHAMBERS, J. M., and WILKS, A. R., *The New S Language. A Programming Environment for Data Analysis and Graphics* (Wadsworth and Brooks, 1988) p. 702 pp.

BOCHAROV, V. S., ZELENTSOV, S. A., and MIKHAILOV, V. N. (1989), *Characteristics of 96 Underground Nuclear Explosions at the Semipalatinsk Experimental Test Site,* Atomnaya Energiya *67*(3), 210–214.

BONDÁR, I., and RYABOY, V. (1997), *Regional Travel-time Tables for the Baltic Shield Region,* Center for Monitoring Research, Technical Report CMR-97/24, 34 pp.

BOLT, B. A., DOYLE, H. A., and SUTTON, D. J. (1958), *Seismic Observations from the 1956 Atomic Explosions in Australia,* Geophys. J. R. astr. Soc. *1*, 135–145.

BOWMAN, J. R., and KENNETT, L. N. (1993), *The Velocity Structure of the Australian Shield from Seismic Travel Times,* Bull. Seismol. Soc. Am. *83*(1), 25–37.

BRATT, S. R., and BACHE, T. C. (1988), *Locating Events with a Sparse Network of Regional Arrays,* Bull. Seismol. Soc. Am. *78*, 780–798.

DENHAM, D., SIMPSON, D. W., GREGSON, P. J., and SUTTON, D. J. (1972), *Travel Times and Amplitudes from Explosions in Northern Australia,* Geophys. J. R. astr. Soc. *28*, 225–235.

EGORKIN, A. V., ZUGANOV, S. K., PAVLENKOVA, N. A., and CHERNYSHEV, N. M. (1987), *Results of Lithospheric Studies from Long-range Profiles in Siberia,* Tectonophysics *140*, 29 47.

FINLAYSON, D. M., CULL, J. P., and DRUMMOND, B. J. (1974), *Upper Mantle Structure from the Trans-Australia Seismic Survey (TASS) and Other Seismic Refraction Data,* J. Geol. Soc. Australia *24*, 447–458.

FIRBAS, P. (1999), *Calibration Based on Ground Truth Data and 3-D Modeling: Framework and Feasibility Study for Western Europe,* Pure appl. geophys., submitted.

ISRAELSSON, H., *Estimation of measurement errors of arrival times.* In *Evaluation and Tuning Studies at the Prototype IDC July–December 1996* (Center for Monitoring Research) Technical Report CMR-97/01, 1997.

JEFFREYS, Sir H., *The Earth,* 6th ed. (Cambridge Univ. Press, 1976) 346 pp.

KENNETT, B. L. N. (ed.), *IASPEI 1991 Seismological Tables,* Research School of Earth Sciences (Australian National University, 1991) 167 pp.

KING, P. B., *The Evolution of North America* (Princeton Univ. Press, 1977) 197 pp.

MIKHAILOV, V. N. (ed.), *Underground Nuclear Explosions for Peaceful Uses* (VNIPIPpromtech-nologii, 1994).

MIKHAILOV, V. N. (ed.), *USSR Nuclear Weapons Tests and Peaceful Nuclear Explosions, 1949 through 1990* (RFNC-VNIEF, 1996).

NORTH, R. G., BONDÁR, I., and RYABOY, V., *New regional travel-time tables for the Baltic shield inferred from DSS observations for calibration of the International Monitoring System (IMS).* In Proceedings of the 19th Annual Seismic Research Symposium on Monitoring a Comprehensive Test Ban Treaty, 115–124, 1997.

PAKISER, L. C., and MOONEY, W. D. (ed.), *Geophysical Framework of the Continental United States* (Geological Society of America, 1989) Memoir *172*, 826 pp.

PAVLENKOVA, N. I., PAVLENKOVA, G. A., and SOLODILOV, L. N. (1996), *High Velocities in the Uppermost Mantle of the Siberian Craton,* Tectonophysics *262*, 261–278.

PAVLENKOVA, N. I., *Crust and upper mantle structure in Northern Eurasia from seismic data. Advances in Geophysics,* v. 37 (Academic Press, 1996) 1–121 pp.

ROMANOVICZ, B. A. (1979), *Seismic Structure of the Upper Mantle beneath the United States by Three-dimensional Inversion of Body-wave Arrival Times*, Geophys. J.R. astr. Soc. *57*(2), 479–506.

RYABOY, V. Z., BAUMGARDT, D. R., FIRBAS, P., and DAINTY, A. M. (1999), *Application of 3-D Crustal and Upper Mantle Velocity Model of North America for Location of Regional Seismic Events*, Pure and appl. geophy. submitted.

RF/US CALIBRATION WORKING GROUP, *Joint RF/US Research Program of Seismic Calibration of the International Monitoring System in Northern Eurasia and North America*, In Proceedings of the 19th Annual Seismic Research Symposium on Monitoring a Comprehensive Test Ban Treaty, 1997, 152–161 pp.

RF/US CALIBRATION WORKING GROUP, *Evaluation of the IMS location accuracy of large chemical and nuclear explosions in Northern Eurasia and North America using regional and global Pn travel-time tables.* In *On Monitoring CTBT*, Proceedings of the 20-th Research Symposium, 1998, 92–102 pp.

ROMNEY, C. F., BROOKS, B. G., MANSFIELD, R. H., CARDER, D. S., JORDAN, J. N., and GORDON, D. W. (1962), *Travel Times and Amplitudes of Principal Body Phases Recorded from Gnome*, BSSA *52*(5), 1057–1074.

SPIZHARSKIY, T. N. (ed.), *Tectonic map of the USSR* (Ministry of Geology of the USSR Press, Moscow, 1966).

SPRINGER, D. L., and KINNAMAN, R. L. (1971), *Seismic Source Summary for U.S. Underground Nuclear Explosions, 1961–1970*, BSSA *61*(4), 1073–1098.

VAN DER LEE, S., and NOLET, G. (1997), *Upper Mantle S Velocity Structure of North America*, J. Geophys. Res. *102*(B10), 22,815–22,838.

VINNIK, L. P., and RYABOY, V. Z. (1981), *Deep Structure of the East European Platform According to Seismic Data*, Phys. Earth and Planet. Inter. *25*, 27–37.

WESSEL, P., and SMITH, W. H. F., *The Generic Mapping Tools (GMT)*, Version 3, Technical Reference and Cookbook. Geodynamics Branch, Geosciences Laboratory NOAA, NOES12, 1995, 80 pp.

ZONENSHAIN, L. P., KUZMIN, M. I., and NATAPOV, L. M., *Geology of the USSR: A plate-tectonic synthesis*. In *Geodynamics* series, (ed., B. M. Page) vol. 21 (AGU, Washington, D.C. 1990) 242 pp.

(Received April 20, 1999, revised December 13, 1999, accepted February 9, 2000)

Pure appl. geophys. 158 (2001) 79–103
0033–4553/01/002079–25 $ 1.50 + 0.20/0

Pure and Applied Geophysics

Application of 3-D Crustal and Upper Mantle Velocity Model of North America for Location of Regional Seismic Events

VLADISLAV RYABOY,[1] DOUGLAS R. BAUMGARDT,[2] PETR FIRBAS[3] and ANTON M. DAINTY[4]

Abstract—Seismic event locations based on regional 1-D velocity-depth sections can have bias errors caused by travel-time variations within different tectonic provinces and due to ray-paths crossing boundaries between tectonic provinces with different crustal and upper mantle velocity structures. Seismic event locations based on 3-D velocity models have the potential to overcome these limitations. This paper summarizes preliminary results for calibration of IMS for North America using 3-D velocity model. A 3-D modeling software was used to compute Source-Station Specific Corrections (SSSCs(3-D)) for *Pn* travel times utilizing 3-D crustal and upper mantle velocity model for the region. This research was performed within the framework of the United States/Russian Federation Joint Program of Seismic Calibration of the International Monitoring System (IMS) in Northern Eurasia and North America.

An initial 3-D velocity model for North America was derived by combining and interpolating 1-D velocity-depth sections for different tectonic units. In areas where no information on 1-D velocity-depth sections was available, tectonic regionalization was used to extrapolate or interpolate. A Moho depth map was integrated. This approach combines the information obtained from refraction profiles with information derived from local and regional network data. The initial 3-D velocity model was tested against maps of *Pn* travel-time residuals for eight calibration explosions; corrections to the 3-D model were made to fit the observed residuals. Our goal was to find a 3-D crustal and upper mantle velocity model capable predicting *Pn* travel times with an accuracy of 1.0–1.5 seconds (r.m.s.).

The 3-D velocity model for North America that gave the best fit to the observed travel times, was used to produce maps of SSSCs(3-D) for seismic stations. The computed SSSCs(3-D) vary approximately from +5 seconds to −5 seconds for the western USA and the Pre-Cambrian platform, respectively. These SSSCs(3-D) along with estimated modeling and measurement errors were used to relocate, using regional data, an independent set of large chemical explosions (with known locations and origin times) detonated within various tectonic provinces of North America. Utilization of the 3-D velocity model through application of the computed SSSCs(3-D) resulted in a substantial improvement in seismic event location accuracy and in a significant decrease of error ellipse area for all events analyzed in comparison both with locations based on the IASPEI91 travel times and locations based on 1-D regional velocity models.

Key words: Upper mantle, 3-D velocity model, IMS calibration, North America.

[1] Center for Monitoring Research, 1300N. 17th Street, Suite 1450, Arlington, VA 22209, USA. E-mail: ryaboy@cmr.gov
[2] ENSCO, 5400 Port Royal Road, Springfield, VA 22151, USA.
[3] CTBTO, International Data Centre, P. O. Box 1250, A-1400 Vienna, Austria.
[4] Threat Reduction Agency, 45045 Aviation Drive, Dulles, VA 20166, USA.

Introduction

The International Monitoring System (IMS) is being developed for global monitoring under the Comprehensive Nuclear-Test-Ban Treaty (CTBT). To help fulfill this task, seismic events should be located with high accuracy. The globally averaged IASPEI91 seismological travel-time tables (KENNETT, 1991) have been used by the prototype International Data Center (PIDC) to locate seismic events.

A location accuracy of 1000 km^2 in continental areas has often been proposed for the IMS based on the limitation of On-Site Inspections to this area under the Treaty. However, the Third Technical Test (GSETT-3) of the Group of Scientific Experts (GSE) of the UN Conference on Disarmament in Geneva has demonstrated that this accuracy cannot be achieved with globally averaged travel times because of large variations of travel times of regional phases. These travel time variations are caused by lateral velocity variations in the crust and upper mantle. To achieve the CTBT goal of 1000 km^2 location accuracy the IMS has to be calibrated, i.e., travel times of seismic waves used in the location must take into account the 3-D structure of the earth (FIRBAS *et al.*, 1997).

The most precise data that can be used to calibrate the IMS, are large chemical and nuclear explosions with known locations and origin times. Deep Seismic Sounding (DSS) data, long-range seismic profile observations, well located mine explosions and earthquakes can also make important contributions to IMS calibration.

Regional seismic event location accuracy is critically dependent on the joint contribution of several factors, including seismic network configuration, errors in arrival time measurements, phase (mis)identification, and accuracy of reference travel-time curves based on crustal and upper mantle velocity structures. Seismic event locations based on regional 1-D velocity models can have bias errors caused by travel-time variations within different major tectonic provinces and by ray-paths crossing boundaries between tectonic provinces with different crustal and upper mantle velocity structures. Seismic event locations based on 3-D velocity models have the potential to overcome these limitations.

The main task of the calibration effort reported here is to improve the IMS location accuracy of the regional seismic events. This paper builds upon the methodology of a feasibility study for 3-D modeling presented in FIRBAS (1997a,b, 1999); FIRBAS *et al.* (1998). The effort is focused on calibration of IMS stations in North America using 3-D crustal and upper mantle velocity models developed for this region. The main objectives of this study are the following:

- Description of *Pn* travel-times data collection and analysis for different tectonic provinces;
- Investigation of the feasibility of 3-D modeling for location of regional seismic events;
- Demonstration of improvements achieved in location accuracy when 3-D models are used.

The 3-D approach for the IMS calibration includes the following main steps:

- Regionalization based on tectonic data, travel times, and crustal and upper mantle velocity structures;
- Collection and analysis of *Pn* travel times;
- Collection and analysis of 1-D crustal and upper mantle velocity-depth sections;
- Development of a 3-D crustal and upper mantle velocity model;
- Computation of Source-Station Specific Corrections (SSSCs(3-D)) for seismic stations using the derived 3-D crustal and upper mantle velocity model;
- Regional seismic events relocation to demonstrate the improvement in location accuracy.

These steps are described in detail further in this paper.

Development of the 3-D Velocity Model

Regionalization Based on Tectonic and Seismic Data

The tectonic map of North America (KING, 1977) was taken as a starting point for delineating various tectonic provinces. This map was simplified and digitized to allow more reliable interpolation and comparison of seismic data for different tectonic provinces (Fig. 1) and to help create a simple starting 3-D velocity model of crust and upper mantle for North America.

Ray paths of seismic waves traveling from regional seismic events to the IMS stations quite frequently cross the boundaries of substantially different tectonic provinces. Detailed seismic studies of the earth's crust and upper mantle have been carried out in most regions of North America and published in numerous reports, papers and monographs (e.g., PAKISER and MOONEY, 1989). This information was collected and carefully analyzed. Maps of crustal thickness and *Pn* velocity, compiled for the USA territory (BRAILE *et al.*, 1989) were digitized (BARAZANGI and STEER, 1998) and used for this study. These data provided a good basis for regionalization based on tectonic and seismic data.

Seismic Travel-time Data Collection and Analysis

Several large explosions with known locations were selected to serve as reference events for location calibration. In the following we will refer to all these events collectively as "calibration events." Travel-time data for these calibration events were utilized for IMS calibration predominantly concentrating on the continental area between latitudes 55 N and 30 N degrees. The results of this study may provide a good starting model for areas north of latitude 55 N degrees, an area to which we plan to extend our calibration study. Figure 1 shows locations of the calibration explosions listed in Table 1. In addition, locations of IMS primary and auxiliary seismic stations based on the CTBT Protocol are shown.

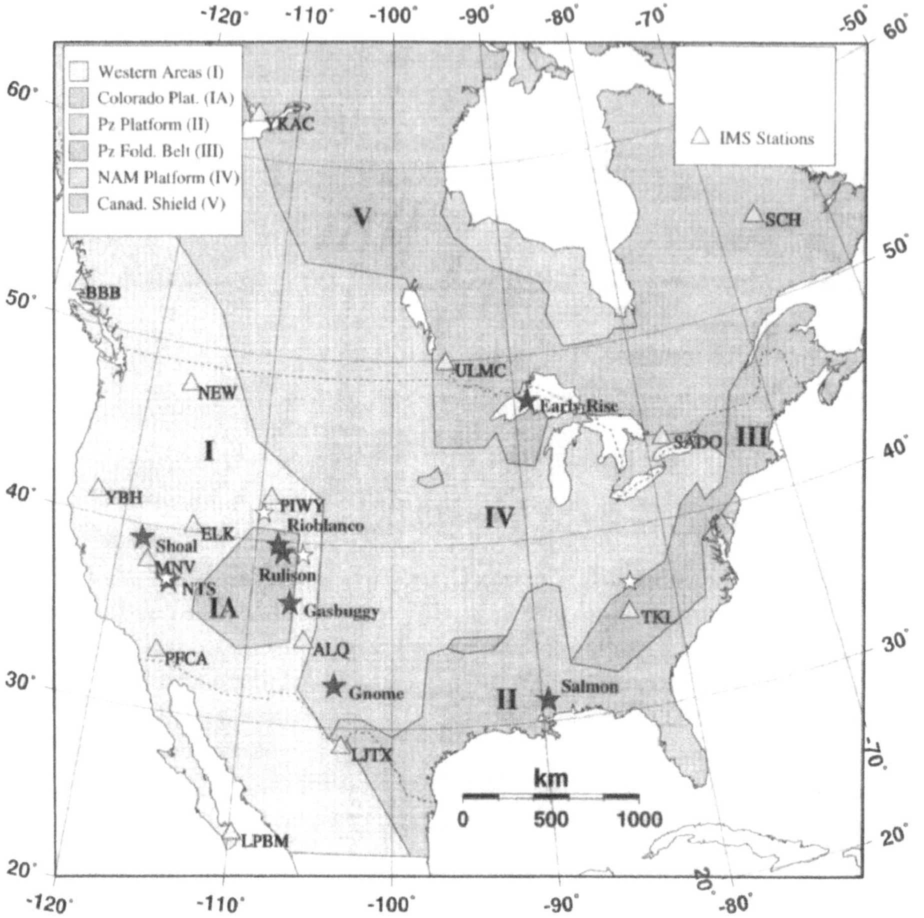

Figure 1

Location of large explosions used for the IMS calibration and relocation experiments. Triangles are the IMS stations, solid stars are calibration explosions used for the IMS calibration and relocation experiments, and open stars are explosions used only for relocation experiments. In the background the simplified tectonic map is shown.

The data were obtained for nuclear explosions mainly from reports on long-range seismic measurements (LRSM) (LRSM Shoal, 1963; LRSM Salmon, 1964; LRSM Gasbuggy, 1967; LRSM Rio Blanco, 1974) and partially from literature (LEHMAN, 1964; ROMNEY *et al.*, 1962; JORDAN *et al.*, 1966; SPRINGER and KINNAMAN, 1971; RF/US CALIBRATION WORKING GROUP, 1998). All of these events were very shallow, and thus zero depth was used for all of them. *Pn*/*P* travel-time data were also obtained from the mentioned LRSM reports and published data and partially were calculated using time arrivals from the International Seismological Center (ISC) Bulletin. In accordance with the LRSM reports magnitude values (M_b) of nuclear explosions listed in Table 1 vary approximately from 4.5 to 5.5.

Table 1

Hypocenters and Origin Times (where applicable) of the Used Calibration Events

Calibration Event	Origin Latitude [Degrees]	Origin Longitude [Degrees]	Origin Time	M_b or yield
Early Rise (summary of 30 events)	47.5505	-88.9390	30 origin times used	5 t
Salmon	31.1421	−89.5699	22 Oct. 1964; 16:00:00.0	4.6
Gnome	32.2636	−103.8658	10 Dec. 1961; 19:00:00.0	3.1 kt
Gasbuggy	36.6778	−107.2083	10 Dec. 1967; 19:30:00.1	4.5
Rio Blanco	39.7930	−108.3666	17 May 1973; 16:00:00.1	4.9
Rulison	39.4058	−107.9481	10 Sept. 1969; 21:00:00.1	4.6
Shoal	39.2003	−118.3803	26 Oct. 1963; 17:00:00.1	4.9
NTS (summary of 40 events)	37.10	−116.04	40 origin times used	4.5–5.5

The Early Rise experiment provided very important data for North American platform region calibration (IYER *et al.*, 1969; BORCHERDT *et al.*, 1969). Six historical nuclear explosions (Salmon, Gnome, Gasbuggy, Rio Blanco, Rulison, and Shoal) with known source locations and origin times were used in this study to cover different provinces in the western, central, and southern parts of the USA. To get the best information for the calibration study using Nevada Test Site (NTS), travel-time data for 40 nuclear explosions with known locations and origin times, all within a small region of the NTS (of a radius of 1.5 km), were summarized and collectively used.

Pn travel times recorded from the calibration events were collected and carefully evaluated. Data from all available seismic stations were used in this study to derive maps of *Pn* travel-time deviations from the IASPEI91 travel times. These maps were plotted by GMT software (WESSEL and SMITH, 1995) and used to study *Pn* travel-time variations as a function of distance and azimuth within different geological provinces of North America. Identification and removal of outliers was an important task. Abrupt and unreasonable variations of travel times from reversed and overlapping observations were used as criteria to identify outliers. Waveform analysis (not performed here) would also be extremely helpful to solve this problem.

Figures 2 and 3 show examples of maps of observed *Pn* travel-time deviations (in seconds) from the IASPEI91 travel times (interpolated/extrapolated to cover the range up to 17 degrees). Seismic stations used are shown by crosses. The maps are given for the Lake Superior Early Rise experiment and for the Gnome nuclear explosion, respectively. The shot points are shown as a star. Maps were derived for

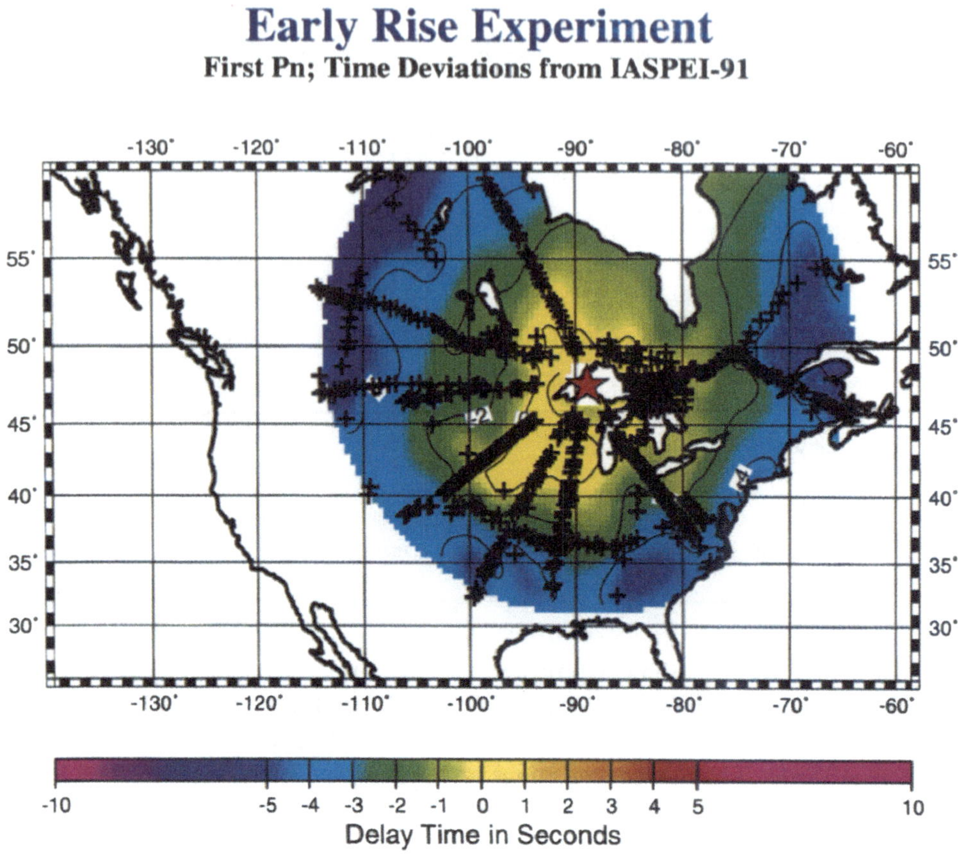

Figure 2
Early Rise experiment, a set of chemical explosions (red star is a shot point). Map of *Pn* travel-time deviations from the IASPEI91 travel-time tables (isolines labelled by sec) with stations shown by crosses.

all the eight calibration events and used for iterative development of 3-D velocity model for North America.

Figure 2 shows pronounced negative *Pn* travel-time deviations from the IASPEI91 global travel times (up to −5 sec and faster) that were observed at regional distances within the North American platform and Canadian shield from the Early Rise explosions.

The nuclear explosion Gnome is located near the tectonic boundary between the North American platform and the western regions of North America. Figure 3 shows very pronounced variations of the *Pn* travel-time deviations ranging from approximately −5 sec for the North American Pre-Cambrian platform to +5 sec for active tectonic zones of the western regions of North America.

It is worth noting that results of interpolation/extrapolation of travel-time anomalies are dependent on the spatial density of seismic stations. The travel-time

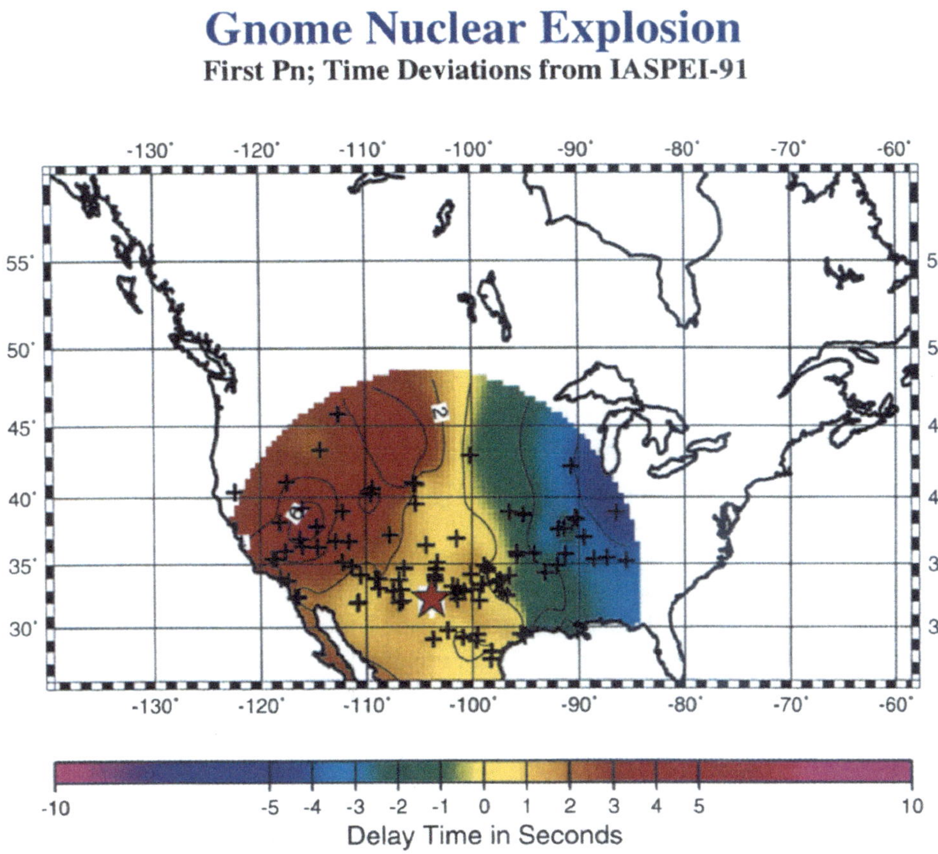

Figure 3
Gnome nuclear explosion (red star is a shot point). Map of *Pn* travel-time deviations from the IASPEI91 travel-time tables (isolines labelled by sec) with stations shown by crosses.

anomalies are determined with high accuracy in areas where the station density is relatively high. In the areas with lower density of stations or in the areas without stations, the interpolated/extrapolated travel times are obviously of lower reliability and this has to be taken into account when making judgements on the calculated travel-times fit quality. In many cases these problems can be resolved due to multiple coverage of the same area by crossing rays for several calibration events.

Analysis of explosion seismology observations shows that the *Pn* travel-time variations are caused by lateral inhomogeneities of the earth's crust and upper mantle velocity structure, both on large and small scales. In the development of our 3-D model of North America we started fitting the large anomalies first. Where necessary, we continued to fit smaller scale anomalies. Both types of these anomalies had to be taken into account to achieve accurate location of regional seismic events as required by the IMS calibration task.

3-D Modeling Software Used

The computer program, called "sssc" (FIRBAS, 1997b, 1999), was used to compute Source-Station-Specific Corrections (SSSCs) for *Pn* travel times by utilizing 3-D velocity earth models. The utility of this code has been tested in a feasibility study for IMS calibration in Europe (FIRBAS, 1997a). The study concluded that success of 3-D modeling for regional distances depends on several factors. First, the 3-D model representation has to be flexible enough to incorporate many different kinds of model formats. Second, the algorithm used for calculation of travel times of regional phases has to be reasonably accurate and robust. Third, the approach should also allow a consistent calculation of SSSCs for teleseismic phases so that both regional and teleseismic phases can be consistently used for location of larger events.

The parameterization of the 3-D model employs an approach that does allow combination of the best velocity information available from multiple sources. Block parameterization is combined with parameterization based on interpolation and extrapolation and permits substantial features of the earth's heterogeneities to be represented, including a laterally heterogeneous crust, an arbitrary curved Moho discontinuity, and a heterogeneous upper mantle. The base model covers the entire earth and thus SSSCs for both regional and teleseismic phases can be computed.

All elements of the model (crust, Moho, uppermost mantle, etc.) are represented independently. An advantage of this approach is that the number of parameters varied in an iterative search to improve the base model can be small and these parameters have a clear physical interpretation. For example, the global Moho map can be easily extended by a detailed Moho map for particular areas, in our case for North America. In fact, such a "high quality information" as Moho depth may not be necessary to be varied during the 3-D models iterative search process.

The computation of travel times in 3-D heterogeneous structures with curved interfaces is generally quite a complex and time-consuming task. Both the shooting and bending methods have their pitfalls, which makes their direct application for our purpose not quite practical. Also, the linearization approach, which can be very successfully applied in global tomography, is not directly applicable to our problem because it does not guarantee the accuracy needed for regional phases, whose ray paths are passing close to the curved Moho surface. A further complication is that the IASPEI91 travel-time curve phases have travel-time values assigned for specific distance ranges in degrees. When using a perturbation approach such as SSSC, standard location procedures expect travel-time values for all distances within the IASPEI91 phase ranges.

The 3-D modeling program (FIRBAS, 1997b, 1999) tries to strike a balance between the accuracy needed for the IMS calibration, speed, and robustness. In our case, since we want to represent travel times with an accuracy of 1.0–1.5 seconds r.m.s. at all ranges, the algorithm for travel-times calculation must provide travel times of regional phases with a precision a tenth of a second.

The "sssc" algorithm combines the ray-shooting (interval ray-tracing) and linearization approach. In the global background model, each branch of any travel-time curve is represented by a unique code. The ray reaching the station is found by paraxial approximation, bending (morphing) the particular ray by the Fermat principle, and in the last stage, applying linearization to get the best estimate of the travel time. The IASPEI91 travel times are also calculated by the program "sssc" and are accurate to at least several hundredths of a second. When the IASPEI91 travel time is subtracted from the full travel time in the 3-D velocity model, the deviation or SSSC(3-D) is obtained.

The program is able to employ both globally averaged velocity models for the entire mantle and results of global tomography. Currently it uses the Harvard model SP12WM13 (SU and DZIEWONSKI, 1993). This comprehensive approach permits computation of not only travel times for regional and crustal phases, but also travel times for teleseismic phases.

Deriving 3-D Velocity Models for Location Calibration

This section summarizes the results achieved in developing a preliminary 3-D velocity model for North America. The goal of our effort was to define a 3-D crustal and upper mantle velocity model capable of predicting *Pn* travel times with accuracy of approximately 1.0–1.5 second (r.m.s.). The simplest 3-D velocity models were sought from a family of 3-D models that simultaneously fit the observed *Pn* travel times for all eight calibration events.

The first step was to examine the observed travel-time deviation and tectonic maps. Inspection of the maps of the observed *Pn* travel-time deviations from IASPEI91 travel times revealed two groups of *Pn* travel-time anomalies. The first group is characterized by a linear size of the order of 1000–3000 km (e.g., western areas of North America and the Pre-Cambrian platform) and was strongly correlated with the tectonic regionalization. The second group of the anomalies is characterized by a smaller linear size of approximately 100 km to 500 km. We concentrated on the large-scale anomalies first. During the modeling process, however, it was found that at least some of the smaller-scale anomalies must be represented in the 3-D model to achieve the required *Pn* travel-time accuracy.

As the first step, we created an initial 3-D model for North America consisting of a global model of the earth's crust (MOONEY *et al.*, 1998) and regionalized upper mantle (RUM) global velocity model of (GUDMUNDSSON and SAMBRIDGE, 1998). These global models have 5 × 5 degrees and 2 × 2 degrees resolution, respectively. Moho depth was taken from the map of BRAILE *et al.* (1989) for North America with a resolution 1 × 1 degrees. The Moho depth map was married to the crustal models by varying the depth of the lowermost crustal layer, and to the RUM by varying the top of the uppermost layer. Testing this initial 3-D velocity model revealed that the crustal model and the Moho depth surface variation had only minor effects on travel times, while the velocities in the upper mantle for North American platform based on the RUM model were generally too low. To better match the observations we fixed the 3-D crustal velocity model and the Moho depth and changed the 3-D upper mantle velocity model in a series of forward iterations.

We started the iterative process by combining and interpolating regional 1-D velocity-depth sections for the upper mantle. In the first iterations only three regions of 1-D velocity-depth sections were used. The initial 1-D velocity sections used are described below. Later this number was increased to nine 1-D velocity-depth sections. In areas where no information on 1-D velocity-depth sections was available, tectonic regionalization was used to extrapolate and/or interpolate seismic data. Forward calculations of *Pn* residuals from the 3-D velocity models were tested against the maps of observed travel-time deviations described above, and corrections were made where necessary. This was an iterative procedure, which was, however, quite straightforward, as the number of varied parameters could be kept relatively low. First, the 1-D velocity-depth sections were kept fixed and just the boundaries between tectonic units in the upper mantle were moved as needed. When the movement of the boundaries in the 2 × 2 degrees resolution model did not bring any further substantial improvement, new 1-D velocity-depth sections were added and/or old 1-D velocity-depth sections were modified. To develop the best 3-D model, we used the suite of eight calibration events listed above. Ray paths of these explosions cross all major tectonic provinces of the USA territory and neighboring areas of Canada and this system of crossing rays and multiple coverage does help to create a robust 3-D model.

The starting 1-D velocity-depth section for the North American platform (temporarily including the Canadian shield) was derived using regional *Pn* travel-time curve (RF/US CALIBRATION WORKING GROUP, 1998, 1999) and 1-D velocity models adapted from the literature (MASSE, 1973; LEWIS and MEYER, 1968; HALES, 1991; GREEN and HALES, 1968; MEREU and HUNTER, 1969). The starting 1-D velocity-depth section for the western regions (the Rocky Mountains and Basin and Range province) was a modified 1-D velocity model adapted from studies of regional travel times for this area and derived from the literature (RF/US CALIBRATION WORKING GROUP, 1998, 1999; ARCHAMBEAU *et al.*, 1969; BURDICK and HELMBERGER, 1978; MASSE *et al.*, 1972). This velocity-depth section included a lid in the uppermost mantle with velocity 7.9 km/s and a well-developed low velocity layer. The smallest values of the *P*-wave velocity in the uppermost mantle within the USA territory were derived beneath Colorado plateau (LEE and GRAND, 1996) and subduction zone of the Juan de Fuca plate (Pacific border). The other seven 1-D velocity-depth sections were derived step-by-step during the modeling process, using results of tomographic studies and models derived from the literature (ROMANOVICZ, 1979; VAN DER LEE and NOLET, 1997; WALCK, 1984; RUPPERT *et al.*, 1998).

Comparisons of observed and calculated maps of the *Pn* travel-time residuals for all the calibration explosions were performed by computing maps of differences between the maps of observed and computed *Pn* travel times. The density of recording stations was taken into account when making judgments and proposing step-by-step 3-D model modifications.

A total of forty-five iterative 3-D velocity models were developed and evaluated. The current model (revision 4.5) fits the observed travel times with an average precision of 1.0–1.5 seconds r.m.s residuals. Figure 4 shows 1-D velocity models for *P* waves derived by this process in comparison with the IASPEI91 model. Figure 5 shows upper mantle regionalization and numbers on the maps on the Figure 5 refer to numbers on the velocity models on Figure 4. Maps of observed travel-time residuals (Figs. 2 and 3) and maps of calculated *Pn* travel-time residuals (Figs. 6 and 7) agree rather well. Both maps of observed and calculated *Pn* travel-time anomalies show that *Pn* waves beneath North American platform are faster up to 10 sec in comparison with western areas of North America. Figures 8 and 9 show maps of the difference between observed and calculated *Pn* travel times for the Early Rise and Gnome explosions. The differences are mainly within ± 1.0 sec, increasing in some relatively small areas up to ± 2.0 sec.

The mean and r.m.s. values of differences between observed and computed *Pn* travel-times were calculated for all calibration events (Table 2). The areas covered by observations were used to derive mean and r.m.s. values. The r.m.s values for the eight explosions vary from approximately 0.7 sec to 1.4 sec. The model residuals fit the observed residuals within 1.0–1.5 sec r.m.s. The average r.m.s. value for all calibration explosions is approximately 1.0 sec. The value of 1.0 sec was therefore used in relocation experiments (described below) as the *Pn* travel-time modeling error for estimating the size of the error ellipses.

The 3-D velocity model developed for the USA (revision 4.5) can be used to produce maps of SSSCs for the all IMS stations and other stations needed for the relocation experiments. Figure 10 shows, as an example, the SSSC map calculated for the TXAR IMS station. This station is located near the boundary between tectonically active western regions of North America and the Pre-Cambrian platform. The computed SSSCs vary from $+5$ seconds to -5 seconds for the western USA and the North American Pre-Cambrian platform, respectively.

Relocation Experiments

The purpose of the relocation experiments was to evaluate influence of SSSCs on the accuracy of location of seismic events recorded at regional distances and thus to test the usefulness of the derived preliminary 3-D model. The calculated SSSCs along with the estimated modeling error (1.0 sec) and the estimated measurement error (0.5 sec) were first used to locate all the eight calibration events listed in Table 1 (and used to derive the 3-D model). This step can be also viewed as a test of the internal consistency of the 3-D model process development.

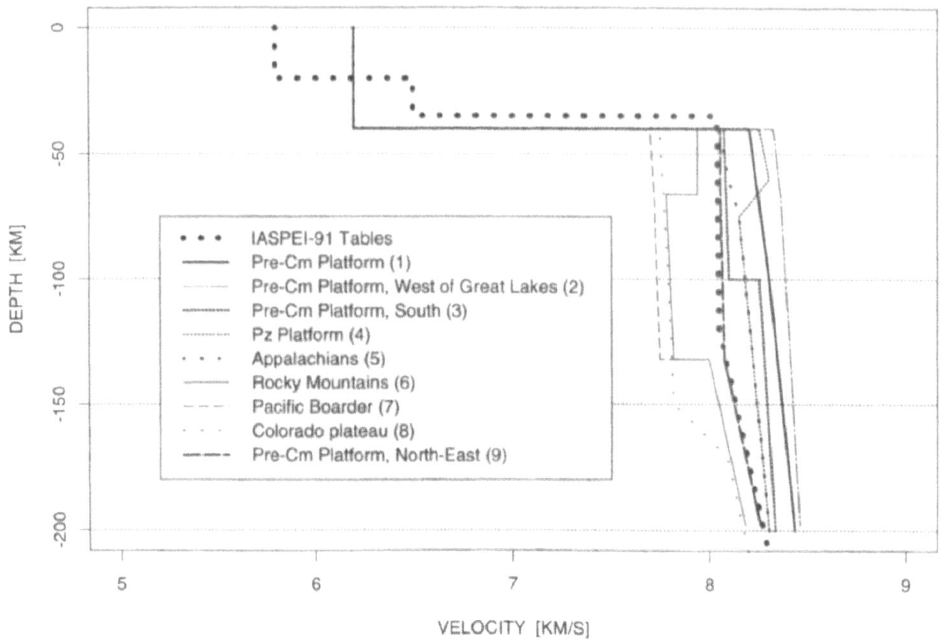

Figure 4
Regional upper mantle 1-D *P*-velocity models of the USA territory and neighboring areas of Canada.

As an independent test, four large chemical explosions with known locations and origin times, not used in the model development, were relocated using derived SSSCs. The coordinates of the hypocenters and origin times of the explosions are given in Table 3. These events shown in Figure 1 as small open stars:

- The Rockville Dam chemical explosion ($M_b = 4.5$), 315 tons, detonated on 3 April 1966, approximately 130 km to the east of the Rulison nuclear explosion (LRSM Rockville Dam, 1966).
- The Southern Nevada chemical explosion, 1.3 kt, detonated by US DOE on 22 September 1993 in the southern part of Nevada Test Site, approximately 20 km South of the NTS calibration explosions used in this study (ISC Bulletin).
- The Green River large mine explosion ($M_b = 5.3$) occurred on 3 February 1995. It was approximately 200 km northwest of Rio Blanco nuclear explosion. Up to one meter of surface subsidence occurred in a 1 by 2 km area above the mine. This mine explosion was recorded at local, regional, and teleseismic distances by a dense network consisting of 344 stations. The Green River mine explosion meets requirements of the ground-truth data quality with known location within 1–2 kms (ISC Bulletin).
- The "041996" explosion ($M_b = 3.9$) was detonated on 19 April 1996 in eastern USA. This event was kindly provided us from the calibration database which is being created by Dr. E.R. Engdahl.

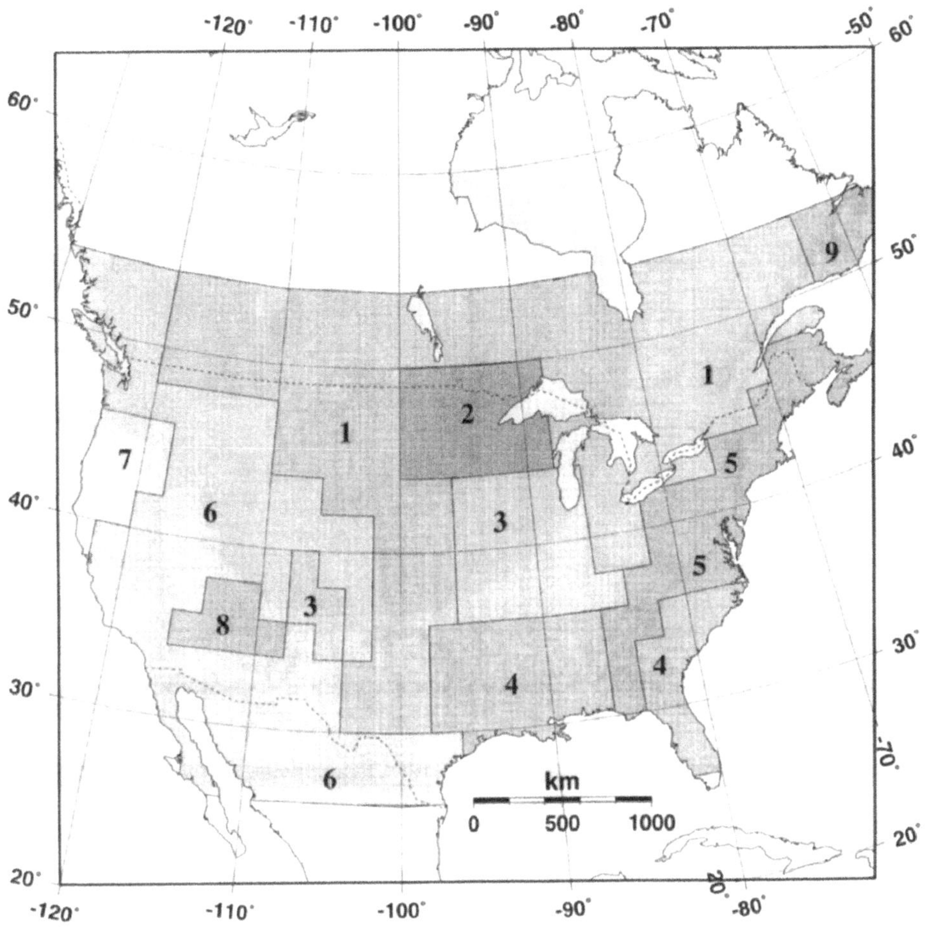

Figure 5

Upper mantle *P*-velocity regionalization of the USA territory and neighboring areas of Canada (3-D model, revision 4.5). Numbers on the map refers to numbers on the velocity models on Figure 4.

The LocSAT program (BRATT and BACHE, 1988) was used for seismic event relocation experiments. Only *Pn* travel times were used. The events were relocated using "surrogate networks" to the IMS network and number of stations used was intentionally limited only to 5–10 stations for each event to mimic the sparse IMS network. Surrogate stations were used to substitute for IMS stations for which data were not available. SSSCs calculated for each station using 3-D velocity model were applied.

As example, we will describe the relocation experiment for the Rockville Dam chemical explosion. The location was estimated using six stations. Figure 11 shows the generalized tectonic map of North America, the explosion location (star), IMS stations (light triangles) and stations of the IMS network simulated for the relocation

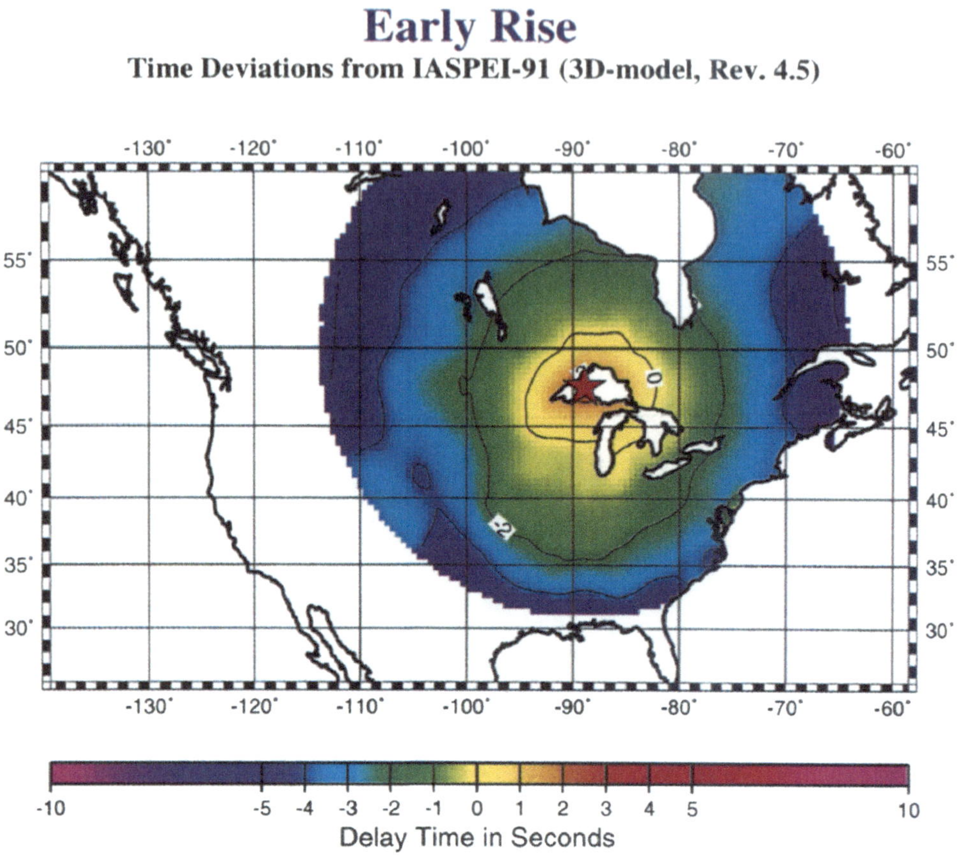

Figure 6
Early Rise, a set of chemical explosions (red star is a shot point). Calculated *Pn* travel-time deviations (in sec) for 3-D model, revision 4.5, deviations from the IASPEI91 travel-time tables.

experiment (solid triangles). Two stations (RK-ON and CPSO) are located within the North American Platform and four stations (ALQ, GCA, MN-NV, and EUR) are located within western USA. Figure 12 shows the results of the relocation experiment for the Rockville Dam chemical explosion. The SSSCs(3-D) were calculated for each of the six mentioned stations and used for the relocation experiment. The distance between the ground-truth epicenter and the estimated location (mislocation) decreased from 20.7 km, in the case using the IASPEI91 travel times, to 9.0 km, using SSSCs(3-D). The calculated error ellipse for the location based on SSSCs(3-D) is substantially smaller than for the IASPEI91 solution and includes the ground-truth epicenter. Mislocation based on the International Seismological Center (ISC) Bulletin solution using 25 stations is 7.1 km.

The main results of the relocation experiments are summarized in Tables 4 and 5 and Figures 13 and 14. Table 4 shows values of mislocation of regional seismic events

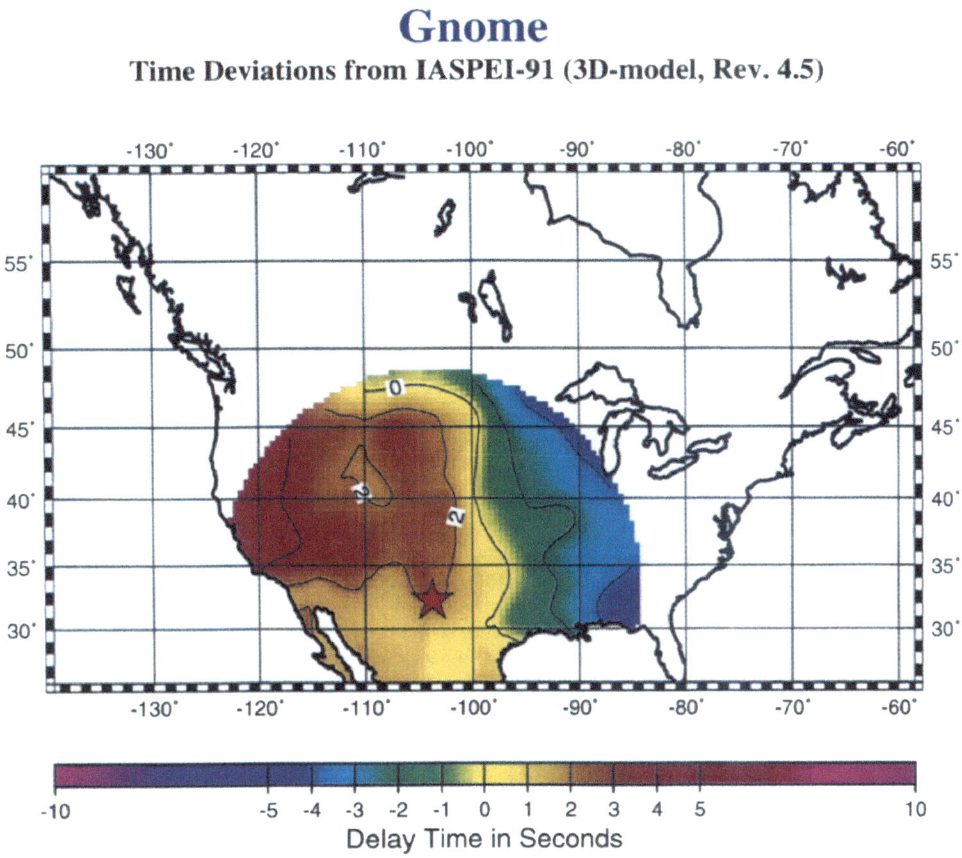

Figure 7
Gnome nuclear explosion (red star is a shot point). Calculated *Pn* travel-time deviations (in sec) for 3-D model, revision 4.5, deviations from the IASPEI91 travel-time tables.

for three different methods of location using 1-D globally averaged IASPEI91 model, two regional travel-time curves (Dainty, 1999; RF/US Calibration Working Group, 1999), and the SSSCs calculated for the 3-D velocity model.

The main results were as follows:

- The mean mislocations of the first five large historical explosions (Table 4), using the SSSCs(3-D) method, improved to 4.5 km, compared to mean mislocations of 18.5 km for the IASPEI91 travel-time table location and 12.4 km for the regional travel-time method of the RF/US Calibration Working Group (1999).
- Mean mislocation for eight historical nuclear and chemical explosions used for development of the 3-D crustal and upper mantle velocity model decreased from 18.0 km to 4.8 km, using IASPEI91 travel times and SSSCs(3-D), respectively.

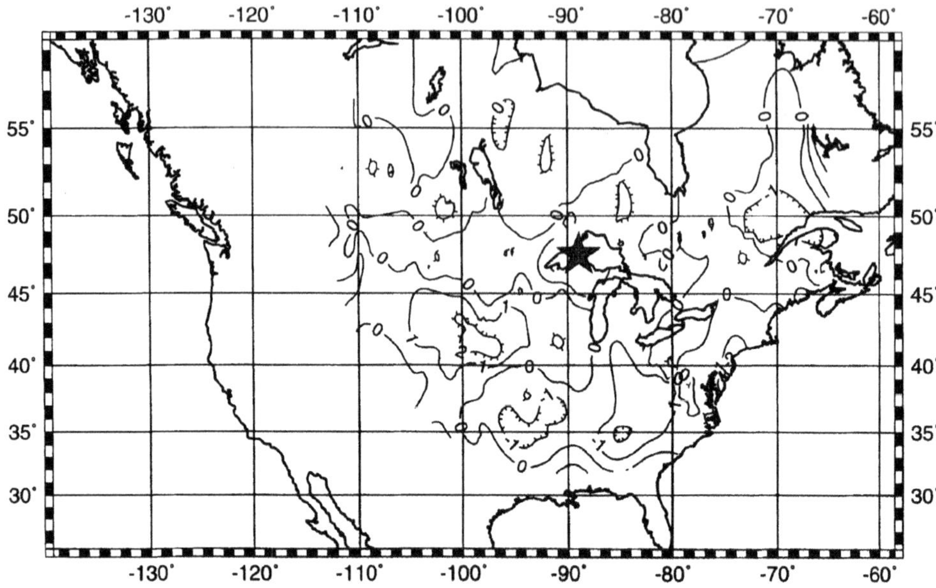

Figure 8
Early Rise experiment (solid star is a shot point). Difference between observed and calculated *Pn* travel times (in sec) for 3-D model, revision 4.5.

- Mean mislocation for four chemical explosions used only for relocation experiments decreased from 22 km to 7.3 km, using IASPEI91 travel times and SSSCs(3-D), respectively.

The improvement of location accuracy obtained for the events not included in the development of the 3-D model is consistent with the improvement in location accuracy for the calibration events.

Figure 13 compares the mislocations obtained when applying the IASPEI91 travel times with mislocations obtained when SSSCs inferred from 3-D model are applied. The diagonal line across the plot is line of equal performance. All points above the line indicate that location using SSSCs(3-D) is better in comparison with the IASPEI91 location. All data points are above this line, which means that the improvement in location accuracy was reached for 100% of events.

Figure 14 shows a boxplot to display the distribution of mislocation values obtained for regional events for three methods of location, i.e., IASPEI91 travel time (12 events), two regional *Pn* travel-time curves (5 events), SSSCs inferred from 3-D model (12 events). The horizontal line in the interior of the boxes is located at the median of the mislocation values and estimates the center of the data distribution. The height of the box indicates the width of the distribution of the data. The horizontal bars show the min and max values for the distribution. Figure 14 shows that locations using SSSCs(3-D) are characterized by the smallest average misloca-

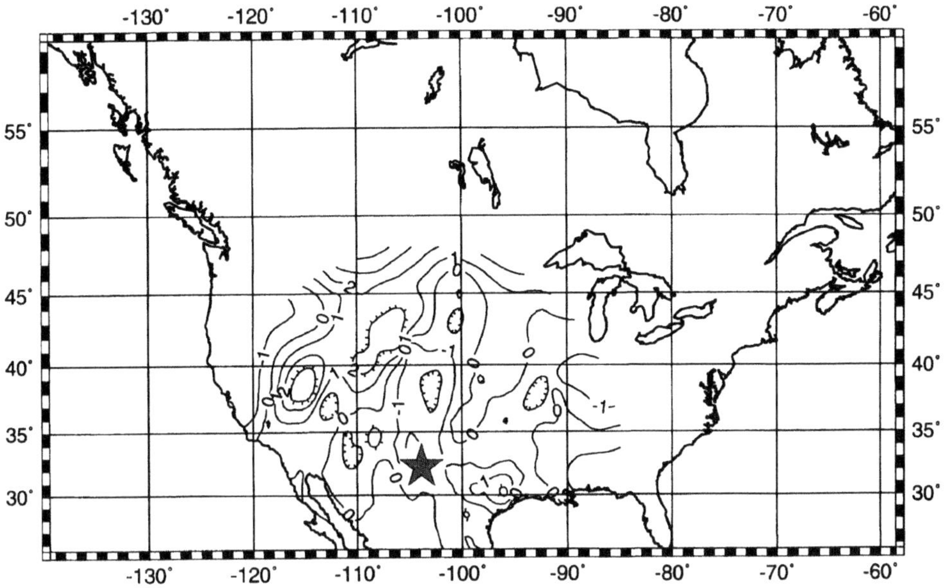

Figure 9
Gnome nuclear explosion (solid star is a shot point). Difference between observed and calculated *Pn* travel times (in sec) for 3-D model, revision 4.5.

tion and by the smallest width of the mislocation distribution, i.e., they are most accurate and most robust of the three approaches.

Table 5 presents error ellipse areas calculated for all 12 calibration and test events located in different areas of the USA (Fig. 1). Areas of the error ellipses calculated for the events located using the IASPEI91 travel times and using prototype IDC modeling errors vary from 1,886.7 km^2 (the NTS nuclear explosions) to 13,759.5 km^2 (the Salmon nuclear explosion), the mean value is 4,268.7 km^2. However, the error ellipse areas calculated for locations based on SSSCs inferred from 3-D model (modeling error assumed 1.0 sec) decreased for all events, and the error ellipse areas

Table 2

Difference between observed Pn travel times and calculated for 3-D velocity model (Rev. 4.5)

Explosion	Mean Misfit [seconds]	RMS Misfit [seconds]
Early Rise	0.00	0.69
Salmon	0.18	0.91
Gnome	0.05	1.00
Gasbuggy	−0.57	1.01
Rio Blanco	−0.19	1.27
Rulison	−0.81	1.19
Shoal	1.18	1.42
NTS	0.28	1.03

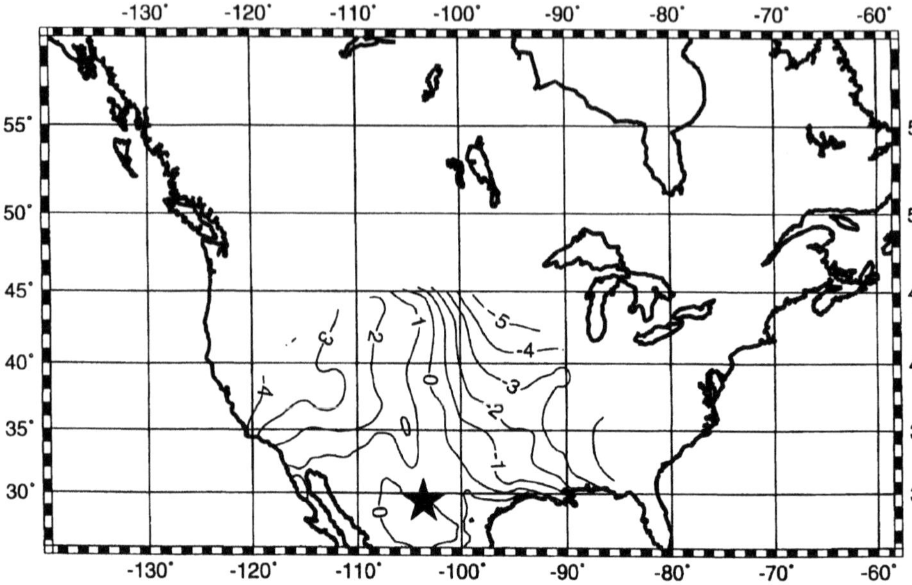

Figure 10
SSSCs calculated using 3-D model, revision 4.5, for the TXAR station at 29.33°N 103.67°W. *Pn* travel-time deviations (in sec) from the IASPEI91 travel-time tables. Solid star is the TXAR station location.

vary from 278.1 km^2 (the Rulison nuclear explosion) to 2,082.3 km^2 (the Salmon nuclear explosion), and with mean value of 641.3 km^2. Analysis of Table 5 showed that the larger error ellipses for Salmon and 04/19/96 explosions are mainly caused by larger azimuthal gap (220 to 240 degrees) compared to azimuthal gap of 100–150 degrees only for other explosions listed in Table 5.

It is worth stressing that for 100% of the calibration events relocated using SSSCs(3-D) the ground-truth locations (GT) are inside the error ellipses calculated at

Table 3
Hypocentres and Origin Times of the Events used for the relocation experiments only

Event	Origin Latitude [Degrees]	Origin Longitude [Degrees]	Origin Time	M_b or yield
Rockville Dam	39.363	−106.462	3 Apr. 1966; 16:21:33.6	4.5
Southern Nevada	37.202	−116.210	22 Sep. 1993; 7:01:00.08	1.3 kt
Green River	41.52	−109.64	3 Feb. 1995; 15:26:10.7	5.3
"041996"	36.932	−83.025	19 Apr. 1996; 8:50:16.31	3.9

90% confidence level. This confirms that the 3-D model used is able to predict the SSSCs with high accuracy resulting in 1.0 sec r.m.s.

Table 4

Regional events mislocation (nuclear and large chemical explosions)

Event Number	Explosion	IASPEI91 Mislocation [km]	Two Regional Travel-time Curves [km] (% of IASPEI91 Mislocation)	SSSCs Derived from 3-D Model [km] (% of IASPEI91 Mislocation)
1	Early Rise	8.6	5.3 (62%)	5.6 (65%)
2	Salmon	30.9	23.2 (75%)	4.6 (15%)
3	Gnome	28.3	10.6 (37%)	4.5 (16%)
4	Gasbuggy	14.4	13.4 (93%)	5.1 (35%)
5	Shoal	10.3	9.6 (93%)	2.5 (24%)
	Events 1–5/Mean	18.5	12.4 (67%)	4.5 (24%)
6	Rio Blanco	21.5		9.0 (42%)
7	Rulison	19.7		4.6 (23%)
8	NTS	10.6		2.8 (26%)
	Events 1–8/Mean	18.0		4.8 (27%)
9	Rockville Dam	20.7		9.0 (43%)
10	South Nevada	14.4		6.9 (48%)
11	Green River Mine	20.9		6.1 (29%)
12	Explosion 4/19/96	31.9		7.1 (22%)
	Events 9–12/Mean	22.0		7.3 (33%)
	Events 1–12/Mean	*19.4*		*5.7 (29%)*

Table 5

Error ellipse areas for calibration explosions within North America

Explosion	Error ellipse area using IASPEI91 [km^2]	Error ellipse area using SSSCs [km^2]	GT location within an area of SSSCs error ellipse
Early Rise	3,387.1	463.2	Yes
Salmon	13,759.5	2,082.3	Yes
Gnome	3,281.0	478.7	Yes
Gasbuggy	2,205.7	414.1	Yes
Shoal	2,675.4	463.0	Yes
Rio Blanco	1,911.7	294.3	Yes
Rulison	1,921.1	278.1	Yes
NTS	1,886.7	319.3	Yes
Rockville Dam	3,386.0	471.0	Yes
South Nevada	2,464.1	428.6	Yes
Green River Mine	2,529.3	337.4	Yes
Explosion 04/19/96	11,816.5	1,665.9	Yes
Average	*4,268.7*	641.3	100% in ellipse

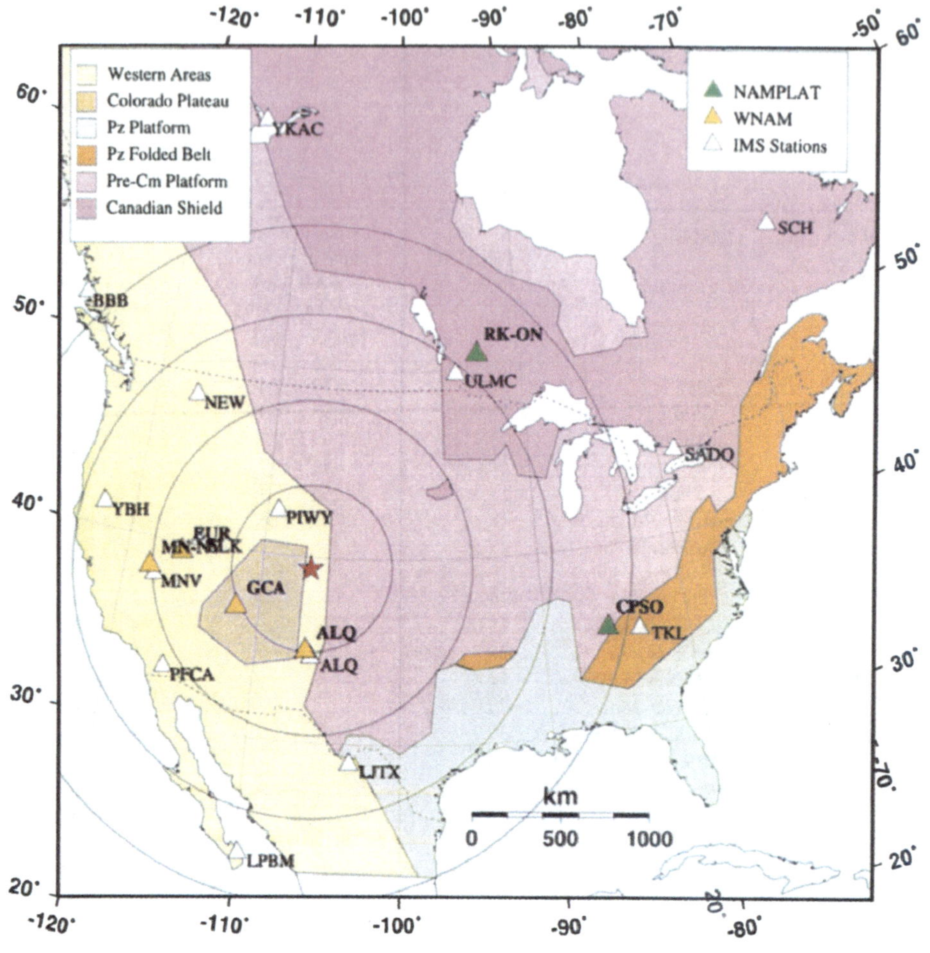

Figure 11
Map of the IMS network simulated for relocation of the Rockville Dam chemical explosion detonated of 3 April 1966 in the Colorado region (red star is a true location, light triangles are IMS stations, and solid triangles are six stations of the IMS network simulated for the relocation experiment).

Conclusions

The application of the SSSCs(3-D) with estimated modeling error of 1.0 second calculated for the preliminary 3-D velocity model derived in this study led to a clear improvement in seismic event location accuracy and substantial decrease in error ellipse area compared to locations based on the IASPEI91 travel times and locations based on two regional travel-time curves (1-D velocity models).

The average distance between true and calculated epicenters decreased from 19.4 km (IASPEI91 travel times) to 5.7 km (SSSCs(3-D)) for all 12 events analyzed.

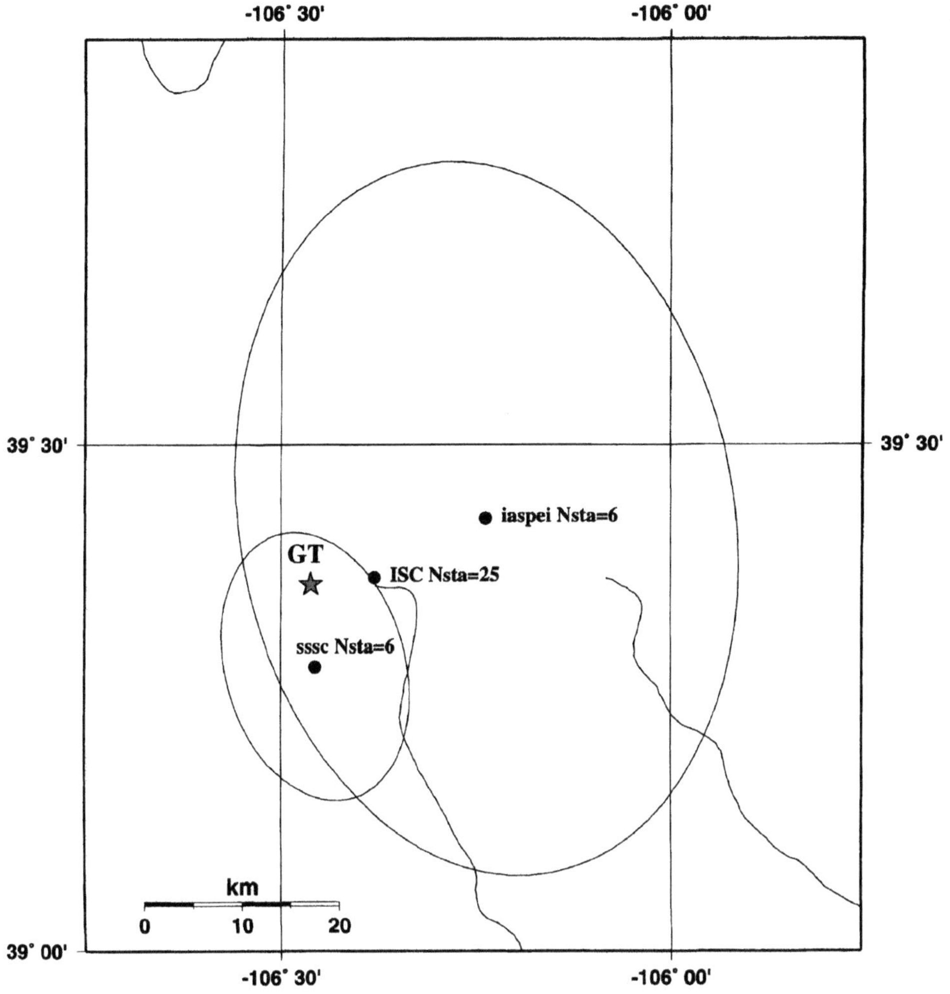

Figure 12
Location estimations for simulated IMS network of the Rockville Dam chemical explosion detonated of 3 April 1966 in the Colorado region, using SSSCs(3-D).

The results were strengthened by the observation that all computed event locations were inside the computed 90% error ellipses. The improvement in accuracy of epicenter locations and decrease of error ellipse areas while keeping true locations inside the computed error ellipses show the promise of the application of the 3-D crustal and upper mantle velocity model approach for regional seismic event locations.

The 3-D model derived in this study is still considered preliminary and there is still room for improvement. The model will be extended to cover the areas north

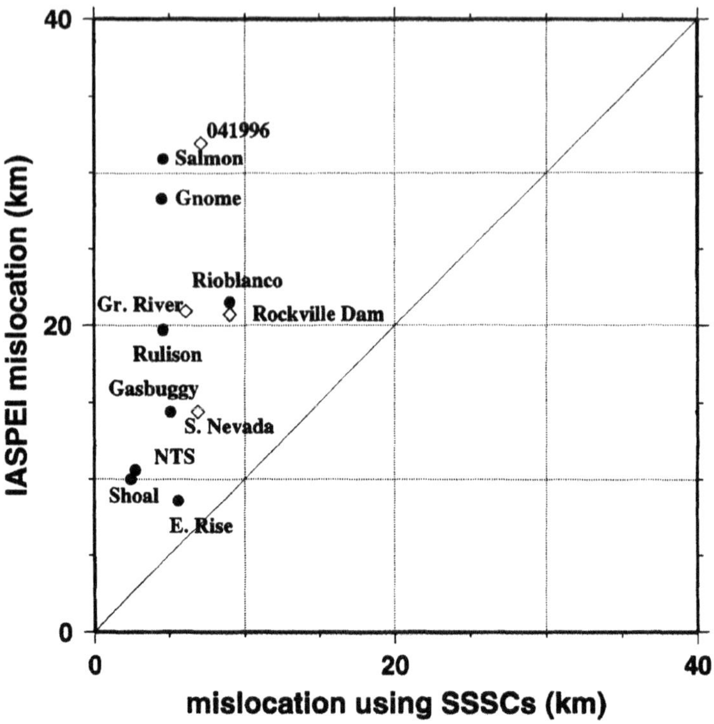

Figure 13
Mislocation evaluation. Circles are explosions used for the 3-D velocity model calibration and relocation experiments, diamonds are explosions used for the relocation experiments only.

of latitude 55°N. Also, more extensive tests are planned to further validate the improvement in location accuracy when applying this approach.

Other studies have investigated SSSCs for calibrating locations based on 1-D velocity models (BONDÁR, 1998; BONDÁR *et al.*, 1998; YANG *et al.*, 1998). The SSSCs in these studies were derived empirically from regional 1-D velocity models developed for two or more tectonic regions. For each station in the network, 2-D SSSC time corrections are generated by linear interpolation between the times obtained from the 1-D regional models, and distance-dependent modeling errors are also derived by interpolating modeling errors for the 1-D models. Our approach provides an alternative, more geophysically-based, method of interpolation for SSSCs derived from empirical data, i.e., using a 3-D earth model to derive SSSCs for any station near the reference events. In our planned future research, we will compare these two methods applied to the same data set to determine if the 3-D modeling method provides more accurate location calibrations than interpolating between 1-D models.

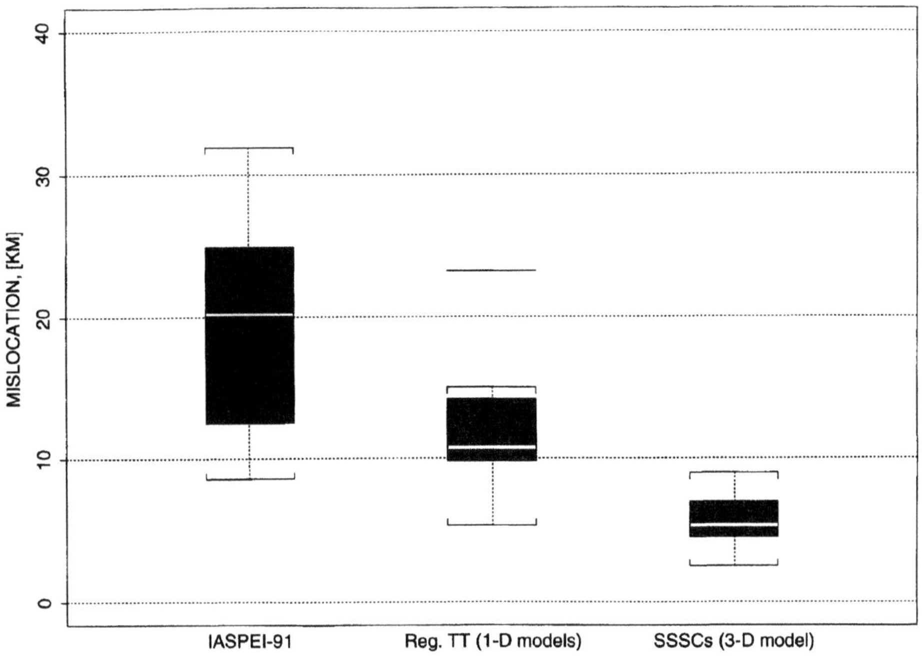

Figure 14
Statistical evaluation of the mislocations of regional seismic events in North America using IASPEI91 travel-time tables, two regional travel-time curves (1-D models), and SSSCs calculated for 3-D model, revision 4.5. The horizontal lines in the interior of the boxes are located at the median of the data. The height of the boxes is equal the difference between the third quartile of the data and the first quartile. The vertical dotted lines extend to the extreme values of the data. Horizontal line at the top of the graph for 1-D model represents an outlier.

Acknowledgments

The authors are grateful to Dr. Carl Romney for his constant helpfulness during this work and for his valuable criticism. Chris Shneider helped with software and calculations used in this study. This research was sponsored by the Defense Threat Reduction Agency, USA, under Contract No. DSWA01-98-C-0032.

References

Archambeau, C. B., Flinn, E. A., and Lamb, D. G. (1969), *Fine Structure of the Upper Mantle*, J. Geophys. Res. *74* (*25*), 5825–5865.

Barazangi, M., and Steer, D., *Knowledge-based visualization of multidisciplinary databases for IDC interactive analysis: Region specific and global* (Cornell University, Institute for the Study of the Continents, R&D Status Report, 1998).

Bondár, I. (1998), *Location Calibration Efforts at the Prototype International Data Center*, EOS, *79* (17), 210.

BONDÁR, I., YANG, X., MCLAUGHLIN, K., NORTH, R., RYABOY, V., and NAGY, W. (1998), *Source Specific Station Corrections for Regional Phases at IMS Stations in North America and Fennoscandia*, EOS *79* (45), F555.

BORCHERDT, C. A., MEYER, R. P., and VAN SCHAACK, J. R. (1969), *Project Early Rise: Shot Times and Locations*, J. Geophys. Res. *74* (17), 4442–4452.

BRAILE, L. W., HINZE, W. J., VON FRESE, R. R. B., and KELLER, G. R., *Seismic properties of the crust and uppermost mantle of the conterminous United States and adjacent Canada*. In *Geophysical Framework of the Continental United States* (eds. Pakiser, L.C., and Mooney, W. D.) (The Geological Society of America 1989), pp. 655–680.

BRATT, S. R., and BACHE, T. C. (1988), *Locating Events with a Sparse Network of Regional Arrays*, Bull. Seismol. Soc. Am. *78*, 780–798.

BURDICK, L. J., and HELMBERGER, D. V. (1978), *The Upper Mantle P Velocity Structure of the Western United States*, J. Geophys. Res. *83* (B4), 1699–1711.

DAINTY, A. M., *US(DOD) – Russia(MOD) Joint Calibration Project: US Summary*. In *Workshop on IMS Location Calibration, Technical Documentation, Oslo, Norway, 12–14 January 1999* (NORSAR,1999), 3, pp. 1–8.

FIRBAS, P., PESHKOV, A. B., and RYABOY, V., *From IASPEI91 global model to a 3-D model for CTBT monitoring*. In *Upper Mantle Heterogeneities from Active and Passive Seismology* (ed. Fuchs, K.) (Kluwer, 1997) pp. 199–214.

FIRBAS, P., *IMS Calibration for Europe* (Technical Report CMR-97/02, Center for Monitoring Research, Arlington, 1997a).

FIRBAS, P., *Program SSSC for computation of station-source specific time corrections for 3-D velocity models* (Technical Report CMR-97/08, Center for Monitoring Research, Arlington, 1997b).

FIRBAS, P., FUCHS, K., and MOONEY, W. (1998), *Calibration of Seismograph Network May Meet Test Ban Treaty's Monitoring Needs*, EOS *79* (35), 413.

FIRBAS, P. (1999) *Calibration Based on Ground Truth Data and 3-D Modeling: Framework and Feasibility Study for Western Europe*, Pure appl. geophys, submitted.

GREEN, R. W. E., and HALES, A. L. (1968), *The Travel Times of P Waves to 30° in the Central United States and Upper Mantle Structure*, Bull. Seismol. Soc. Am. *58* (1), 267–289.

GUDMUNDSSON, O., and SAMBRIDGE, M. (1998), *A Regionalized Upper Mantle (RUM) Seismic Model*, J. Geophys. Res. *103* (B4), 7121–7136.

HALES, A. L. (1991), *Upper Mantle Models and the Thickness of the Continental Lithosphere*, Geophys. J. Int. *105*, 355–363.

IYER, H. M., PAKISER, L. C., STUART, D. J., and WARREN, D. H. (1969), *Project Early Rise: Seismic Probing of the Upper Mantle*, J. Geophys. Res. *74* (17), 4409–4441.

JORDAN, J. N., MICKEY, W. V., HELTERBRAN, W., and CLARK D. M. (1966), *Travel Times and Amplitudes from the Salmon Explosion*, Bull. Seismol. Soc. Am. *71* (14), 3469–3482.

KENNETT, B. L. N. (editor), *IASPEI 1991 Seismological Tables* (Australian National University, 1991).

KING, P. B. *The Evolution of North America* (Princeton Univ. Press, 1977).

LEE, D-K, and GRAND, S. P. (1996), *Upper Mantle Shear Structure Beneath the Colorado Rocky Mountains*, J. Geophys. Res. *101* (B10), 22,233–22,244.

LEHMANN, I. (1964), *On the Travel Times of P as Determined from Nuclear Explosions*, Bull. Seismol. Soc. Am. *54* (1), 123–139.

LEWIS, B. T. R., and MEYER, R. P. (1968), *A Seismic Investigation of the Upper Mantle to the West of Lake Superior*, Bull. Seismol. Soc. Am. *58* (2), 565–596.

LRSM SHOAL Report, *VELA-UNIFORM Project*, 26 October 1963.

LRSM SALMON Report, *VELA-UNIFORM Project*, 22 October 1964.

LRSM ROCKVILLE DAM Report, *VELA-UNIFORM Project*, 19 August 1966.

LRSM GASBUGGY Report, *VELA-UNIFORM Project*, 10 December 1967.

LRSM RIO BLANCO Report, *VELA-UNIFORM Project*, 19 June 1974.

MASSE, R. P., LANDISMAN, M., and JENKINS, J. B. (1972), *An Investigation of the Upper Mantle Compressional Velocity Distribution Beneath the Basin and Range Province*, Geophys. J. R. astr. Soc. *30*, 19–36.

MASSE, R. P. (1973), *Compressional Velocity Distribution Beneath Central and Eastern North America*, Bull. Seismol, Soc. Am. *63* (3), 911–935.
MOONEY, W. D., LASKE, G., and MASTERS, T. G. (1998), *Crust 5.1: A Global Crustal Model at* $5^{o} \times 5^{o}$, J. Geophys. Res. *103* (B1), 727–747.
MEREU, R. F., and HUNTER, J. A. (1969), *Crustal and Upper Mantle Structure Under the Canadian Shield from Project Early Rise Data*, Bull. Seismol. Soc. Am. *59*, (1), 147–165.
PAKISER, L. C., and MOONEY, W. D. (ed.), *Geophysical Framework of the Continental United States* (the Geological Society of America, 1989).
RF/US CALIBRATION WORKING GROUP, *Evaluation of the IMS location accuracy of large chemical and nuclear explosions in Northern Eurasia and North America using regional and global Pn travel-time tables*. In *Monitoring CTBT, Proceedings of the 20-th Research Symposium* (Washington, 1998) pp. 92–102.
RF/US CALIBRATION WORKING GROUP (1999), *Analysis of the IMS Location Accuracy in Northern Eurasia and North America Using Regional and Global Pn Travel-time Tables*, Pure appl. geophys. submitted.
ROMANOWICZ, B. A. (1979), *Seismic Structure of the Upper Mantle Beneath the United States by Three-dimensional Inversion of Body Wave Arrival Times*, Geophys. J. R. astr. Soc. *57*, 479–506.
ROMNEY, C., BROOKS, B. G., MANSFIELD, R. H., CARDER, D. S., JORDAN, J. N., and GORDON, D. W. (1962), *Travel Times and Amplitudes of Principal Body Phases Recorded from Gnome*, Bull. Seismol. Soc. Am. *52* (5), 1057–1074.
RUPPERT, S., FLIEDNER, M. M., and ZANDT, G. (1998), *Thin Crust and Active Upper Mantle Beneath the Southern Sierra Nevada in the Western United States*, Tectonophysics *286*, 237–252.
SPRINGER, D. L., and KINNAMAN, R. L. (1971), *Seismic Source Summary for U.S. Underground Nuclear Explosions, 1961–1970*, Bull. Seismol. Soc. Am. *61* (4), 1073–1098.
SU, W., and DZIEWONSKI, A. M. (1993), *Joint 3-D Inversion for P- and S-velocities in the Mantle*, EOS, Trans. Am. Geophys. Soc. *74*, 557.
VAN DER LEE, and NOLET, G. (1997), *Upper Mantle S Velocity Structure of North America*, J. Geophys. Res. *102* (B10), 22,815–22,838.
WALCK, M. C. (1984), *The P-wave Upper Mantle Structure Beneath an Active Spreading Centre: the Gulf of California*, Geophys. J. R. astr. Soc. *76*, 697–723.
WESSEL, P., and SMITH, W. H. F., *The generic mapping tools (GMT). Version 3. Technical reference and cookbook* (NOAA, Geodynamics branch, Geosciences laboratory NOAA, N/OES12, 1995).
YANG, X., MCLAUGHLIN, K., and NORTH, R., *SSSCs for regional phases at IMS stations in North America and Fennoscandia* (Technical report CMR-97/08, Center for Monitoring Research, 1998).

(Received April 28, 1999, revised November 18, 1999, accepted February 9, 2000)

Pure appl. geophys. 158 (2001) 105–116
0033–4553/01/020105–12 $ 1.50 + 0.20/0

Pure and Applied Geophysics

Relocation of Wyoming Mine Production Blasts Using Calibration Explosions

CAROL A. FINN,[1] GORDON D. KRAFT,[2] MATTHEW S. SIBOL,[2] RONALD L. JONES,[2] and MARK E. PULASKI[2]

Abstract — An important requirement for a comprehensive seismic monitoring system is the capability to accurately locate small seismic events worldwide. Accurate event location can improve the probability of determining whether or not a small event, recorded predominantly by local and regional stations, is a nuclear explosion. For those portions of the earth where crustal velocities are not well established, reference event calibration techniques offer a method of increased locational accuracy and reduced locational bias.

In this study, data from a set of mining events with good ground-truth data in the Powder River Basin region of eastern Wyoming are used to investigate the potential of event calibration techniques in the area. Results of this study are compared with locations published in the prototype International Data Center's Reviewed Event Bulletin (REB). A Joint Hypocenter Determination (JHD) method was applied to a set of 23 events. Four of those events with superior ground-truth control (mining company report or Global Positioning System data) were used as JHD reference events. Nineteen (83%) of the solutions converged and the resulting set of station-phase travel-time corrections from the JHD results was then tested. When those travel-time corrections were applied individually to the four events with good ground-truth control, the average locational error reduced the original REB location error from 16.1 km to 5.7 km (65% improvement). The JHD locations indicated reduced locational bias and all of the individual error ellipses enclosed the actual known event locations.

Given a set of well-recorded calibration events, it appears that the JHD methodology is a viable technique for improving locational accuracy of future small events where the location depends on arrival times from predominantly local and/or regional stations. In this specific case, the International Association of Seismology and the Physics of the Earth's Interior (IASPEI) travel-time tables, coupled with JHD-derived travel-time corrections, may obviate the need for an accurately known regional velocity structure in the Powder River Basin region.

Key words: Joint hypocenter locations, mining seismicity, calibration, Wyoming.

Introduction

With a comprehensive monitoring system capable of detecting small nuclear test explosions worldwide, it is becoming increasingly more important that suspicious

[1] Air Force Technical Applications Center, United States National Data Center, now with U.S. Geological Survey Mid-Continent Mapping Center, 1400 Independence Road, Rolla MO 65401, USA
[2] ENSCO, Inc., United States National Data Center, 5400 Port Royal Road, Springfield VA 22151, USA

seismic events be located accurately. Accurate event location can enhance the probability of determining that the event is or is not a nuclear explosion. Small events are recorded predominately by local and regional stations where travel paths are concentrated in the inhomogeneous crust and upper mantle. Regional departures from a worldwide average velocity structure model can result in location bias. Since it is not always practical to determine individual local/regional velocity structures, a more plausible method for reducing location bias is to determine region-specific station-phase travel-time corrections using reference events. This study explores the possibility of using such a method in eastern Wyoming. Mining-related activity in western Wyoming and eastern Kentucky make excellent additional candidates for extending this technique to other parts of the United States.

Studies of seismic signals from mining sources in the United States have concentrated on the Powder River Basin (PRB) in northeastern Wyoming, where extensive coal mining operations allow the close monitoring of mining explosions (PEARSON *et al.*, 1995). In 1995, 21 mining blasts in the PRB and two refraction shots in central Wyoming were recorded by the Group of Scientific Experts Technical Test 3 (GSETT-3) network of stations and reported on in the prototype International Data Center (pIDC) Reviewed Event Bulletin (REB). Those events provide a useful calibration data set when combined with information provided by mine company personnel and on-site observations of the events (Fig. 1 and Table 1).

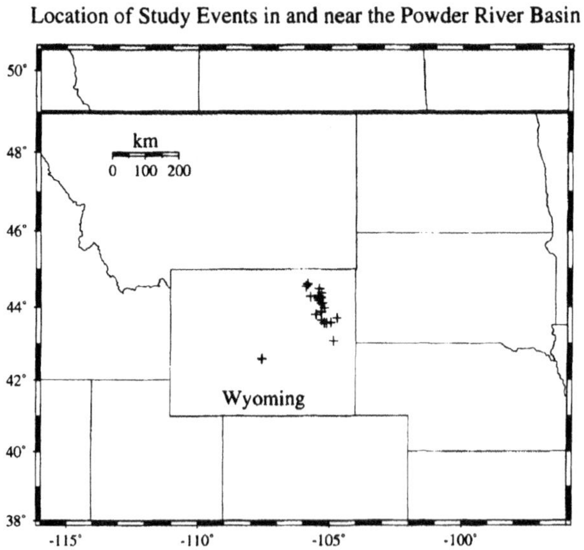

Figure 1
Events (+) used in this study include mine explosions in the Powder River Basin of northeastern Wyoming and two refraction shots in central Wyoming.

Table 1

Central and Northeastern Wyoming (Powder River Basin) mine production blasts and refraction explosions (depths fixed at 0.0 km)

Date	Time	REB Location				JHD Location			
		Lat.	Lon.	E^1_{max}	E^1_{min}	Lat.	Lon.	E^1_{max}	E^1_{min}
950718	194742.4[2]	43.580	−104.980	12.1	10.2	43.644	−105.217	11.4	2.8
950126	234342.8[2]	43.640	−105.320	24.9	10.9	43.699	−105.365	12.2	7.9
950317	203727.3[2]	43.560	−105.130	25.7	9.4	43.671	−105.276	14.1	7.8
950817	112958.3[2]	42.570	−107.580	32.4	18.5	42.722	−107.591	38.0	12.1
950101	203752.5	44.230	−105.440	17.5	9.6	44.254	−105.518	16.6	5.4
950211	201835.4	43.560	−105.230	26.3	10.0				
950211	221238.4	44.270	−105.740	34.8	10.9	44.342	−105.873	14.2	2.8
950211	222355.6	43.790	−105.540	26.1	11.9	43.749	−105.600	24.5	7.9
950212	223145.0	43.070	−104.870	40.2	17.2				
950216	194328.3	44.480	−105.410	13.0	8.1	44.543	−105.560	13.3	8.4
950224	173810.4	44.260	−105.330	18.6	8.7	44.316	−105.501	8.9	4.6
950302	224209.9	44.610	−105.840	14.4	12.1	44.704	−105.927	13.2	7.5
950320	213231.4	43.970	−105.210	11.5	8.9	44.047	−105.361	10.0	5.4
950413	200531.8	43.860	−105.330	10.7	9.4	44.025	−105.486	10.5	5.7
950424	211905.7	44.300	−105.440	12.5	8.0	44.314	−105.583	12.3	5.5
950506	170440.5	44.170	−105.420	11.6	7.6	44.408	−105.689	13.8	2.4
950520	205052.6	44.100	−105.280	22.3	15.2	44.157	−105.394	10.8	6.6
950619	213825.9	44.090	−105.350	40.1	10.0	44.223	−105.416	5.9	1.9
950629	235142.2	44.380	−105.320	34.8	12.2	44.433	−105.544	30.3	5.5
950707	164237.3	43.700	−104.740	30.8	10.8	43.236	−104.539	25.2	6.9
950707	175416.7	44.540	−105.910	20.9	11.6				
950725	173821.3	44.600	−105.840	24.2	10.1	44.589	−105.936	19.5	5.5
950809	112958.7	42.590	−107.600	27.1	15.2				
		Average[3]		22.0	10.7			16.0	5.9
		Std. Dev.[3]		9.4	2.7			8.1	2.5

[1] 90% confidence error ellipse semi-major (E_{max}) and semi-minor (E_{min}) axes.
[2] Calibration events.
[3] Does not include events which have no JHD solution.

Relocations with Calibration Events

The Joint Hypocenter Determination (JHD) method of DEWEY (1995) was used for 23 PRB events reported in the REB. For this data set, four calibration events were held fixed to their known locations as reported by the mine company or determined from Global Positioning System observations. The depths of all events were held fixed at zero, since they are all assumed to be mining events located at or very near the surface. In the absence of a well-determined velocity structure for this area, International Association of Seismology and the Physics of the Earth's Interior (IASPEI) travel-time curves (KENNETT, 1991) were used for *P*, *Pn*, *S*, and *Sn* wave arrivals. Constant velocities of 6.1 km/sec and 3.5 km/sec were used for *Pg* and *Lg* phases, respectively. Four events did not converge, leaving a total of

19 events with final JHD solutions. Using the JHD station-phase travel-time corrections developed from that data set, the four ground-truth events were relocated individually and compared to the REB locations. For the small events in this study, it is not surprising that convergence was not always achieved. The JHD method attempts to solve for a travel-time correction for each phase (*P*, *Pn*, *Pg*, *Sn*, and *Lg*) at each station, in addition to the standard hypocenter parameters. With the sparse data set available for study, the observed station set for each event was different. Very few events were observed with a common station set. Furthermore, for each station, the phase set differed from event to event, e.g., a station may record a *Pg* for one event and a *Pn* and *Sn* for another event. As a result, the solution of N unknowns with M observations led to numerical instabilities when M was not much larger than N.

Because of the sparse data set, as many restrictions as possible were needed to constrain the solutions. This included constraining the depths to 0.0 km (a plausible constraint in this case). Removing this constraint caused a larger number of hypocenter solutions not to converge, which in turn reduced the number of station-phase travel-time corrections that could be determined and reduced the overall significance of any findings. Thus it was impossible to assess how the relocation scheme would deal with a deep earthquake.

Location error analysis was performed for the large data set of 19 events (4 calibration events and 15 additional events throughout the PRB), resulting in error ellipses calculated at the 90% confidence level. Figure 2 shows the REB and JHD locations and their associated error ellipses. Table 1 contains summary information

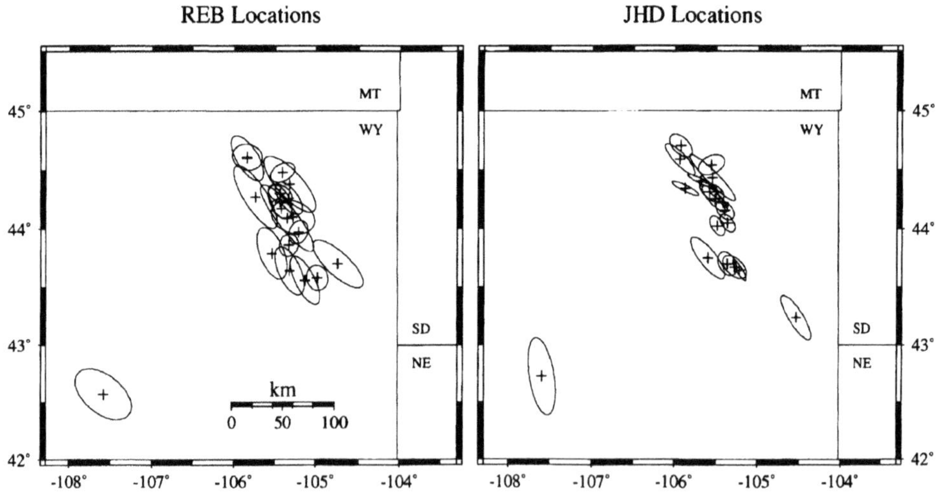

Figure 2
A comparison of REB (left) and JHD (right) computed locations (+) and error ellipse estimates (90% confidence level). On average, the REB error ellipses are larger than the JHD error ellipse estimates.

for each event, including a comparison between the REB and JHD error ellipse semi-major and semi-minor axes. The first four events listed in that table are the calibration events. On average, the JHD error ellipses are smaller than the REB error ellipses by approximately 5.4 km (computed by averaging major and minor axes). This reduction in the error ellipse estimates can be seen in Figure 2.

Figure 3 shows known active coal mines in northeastern Wyoming where the bulk of the coal-mining activity exists, along with the same REB and JHD locations and error ellipses shown in Figure 2. It appears that in addition to the reduction in error ellipse size, the JHD locations, as a group, are more tightly clustered around the active mines than are the REB locations. It should be noted that error ellipses are derived from *a priori* assumptions about uncertainties. The reduction in size of error ellipses is not always meaningful, however it is assumed that the tighter cluster of events around known mine locations is indirect evidence that the JHD methodology is an improved method for relocation. It should also be pointed out that only known (to the authors) mining locations are shown in Figure 3. Sites other than those marked by a "+" symbol could also be active mining areas.

The one exception is an event to the southeast of the main cluster. The JHD solution locates this event even further to the southeast, away from known mining areas. Examination of the locational data reveals that this anomalous behavior is due to an extremely large P-wave travel-time correction (> 7 seconds), determined by the

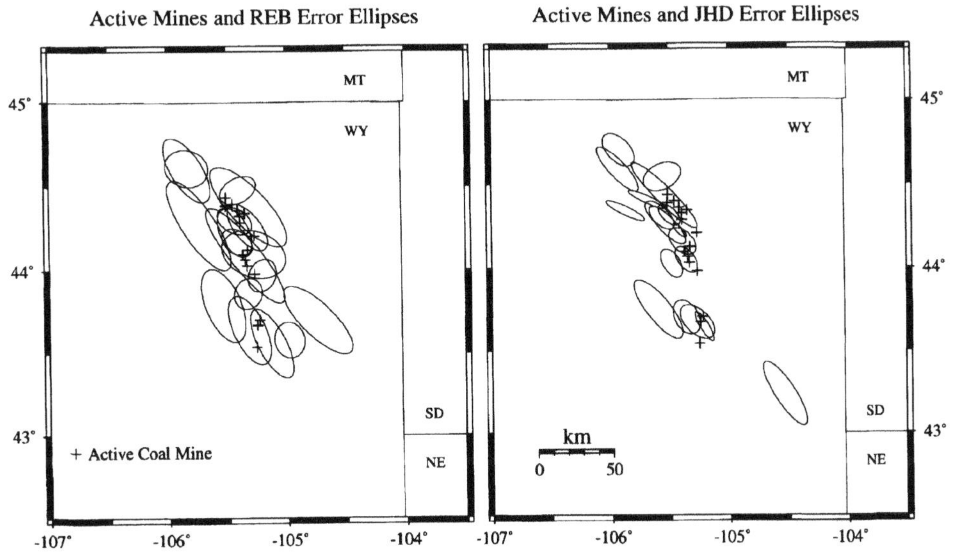

Figure 3
Active coal mine locations (+) in northeastern Wyoming with REB (left) and JHD (right) error ellipse estimates. In general, the JHD locations tend to cluster more tightly around known mining locations and their error ellipse estimates tend to be smaller.

JHD for station PFO (located in southern California). PFO does not contribute to any other event in our data set therefore comparison with other PFO data is not possible. Waveform data for PFO were not available for review of the detection onset pick. However, it is suspected that the arrival pick is good since the time residual for the REB location is less than one second. This particular event is observed at stations PDAR, ULM, PFO, and FCC, none of which are located in the northwest or southeast epicentral quadrants which would have better constrained the solution. The largest time residual (relative to the REB location) for this event is observed for Sn at PDAR. This +1.16 sec travel-time residual, in combination with the poor azimuthal coverage, may have caused the JHD solution to converge at a more southeasterly location.

The remaining 17 events appear to be better located. However, no additional information is available for the majority of these events which would allow us to tie

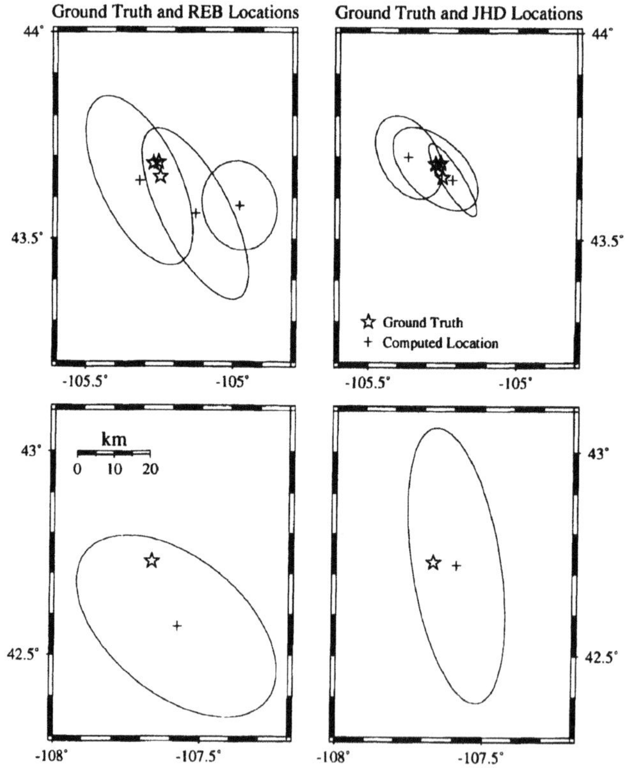

Figure 4
REB (left) and JHD-tt corrected (right) computed locations (+) and ground-truth locations (☆) for the four calibration events. The JHD locations tend to cluster more tightly around the ground-truth locations and the error ellipse estimates for the JHD-tt corrected solutions are smaller and enclose the confirmed locations for all four events.

an event to a specific coal mine. Enough information is, however, available for the four calibration events to tie these events to specific coal mine explosions.

Figure 4 shows the four calibration events for which ground-truth locations were available. The top two figures display the REB locations and the JHD station-phase travel-time corrected (JHD-tt corrected) locations for three events in the PRB. The two lower figures are for a refraction shot farther south and west of the bulk of the events. In all cases, the station-phase travel-time corrections derived during the JHD location process have brought the individual locations closer to the known locations and eliminated some of the systematic bias in the error ellipses. Tables 2 and 3 indicate the JHD-tt corrected and REB locations, respectively, along with the known locations for the four calibration events. The JHD-tt corrected locations average 5.7 km away from the ground-truth locations, and the error ellipses overlap those locations in all four cases. Table 3 presents similar comparisons for REB locations, which average 16.1 km away from ground truth and the error ellipses overlap the true locations in three out of four of the cases. This represents an average locational error reduction of over 10 km (65%).

Table 2

Comparison of Ground-truth and JHD Travel-time Corrected Locations (depths fixed at 0.0 km)

Date	Time	Ground Truth		JHD Location		Azimuth[1] (deg)	Distance (km)
		Lat.	Lon.	Lat.	Lon.		
950126	234342.8	43.683	−105.274	43.699	−105.365	283.7	7.6
950317	203727.3	43.650	−105.250	43.671	−105.276	318.1	3.1
950718	194742.4	43.685	−105.256	43.644	−105.217	145.4	5.5
950817	112958.3	42.729	−107.668	42.722	−107.591	97.0	6.4
						Average	5.7

[1] Ground-truth location to JHD location.

Table 3

Comparison of Ground-truth Locations and REB Locations (depths fixed at 0.0 km)

Date	Time	Ground Truth		REB Location		Azimuth[1] (deg)	Distance (km)
		Lat.	Lon.	Lat.	Lon.		
950126	234342.8	43.683	−105.274	43.640	−105.320	217.8	6.0
950317	203727.3	43.650	−105.250	43.560	−105.130	135.9	13.9
950718	194742.4	43.685	−105.256	43.580	−104.980	117.5	25.2
950817	112958.3	42.729	−107.668	42.570	−107.580	157.7	19.1
						Average	16.1

[1] Ground-truth location to REB location.

Another point that should be noted regarding these event locations is that, in most cases, the event solutions could have benefited from the addition of a station to the southeast or to the northwest of the PRB. It is in those directions that the error ellipses are the largest and where the largest station azimuthal gaps occur. Figure 5 shows the locations of the stations that were most frequently associated to the events in the PRB. Also presented in the figure are three additional stations (TXAR, MIAR, and WHY) which were used very infrequently. It is believed that the addition of arrival times from those stations could improve the overall locational accuracy of future events in the PRB.

The only pIDC auxiliary station to contribute a defining phase to events in the data set is station FCC. A defining *P* phase was recorded at FCC for 10 of the events. To examine the importance of that station in the location of those events, FCC arrival-time data were removed and locations were recomputed using the JHD station-phase corrections. The elimination of FCC data caused one event not to converge due to insufficient data. However, the results of this exercise demonstrated that for the other nine events, the locations shifted an average of 1.2 km. The largest shift was 3.6 km and all other locations moved less than 3 km. Four event locations moved about 0.1 km.

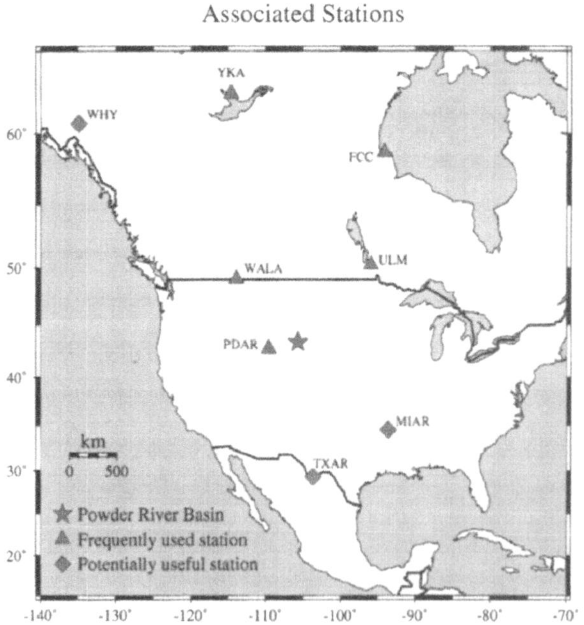

Figure 5
Stations which frequently observe PRB mining blasts are shown along with stations which would be potentially useful for improving event location estimates in the region.

Powder River Basin Events not in Reviewed Event Bulletins

Prior to performing the present study, it was noticed that many of the large mining explosions recorded by the mining companies were not recorded by the GSETT-3 monitoring network. In particular, there were 10 explosions between January and September of 1995 which were sufficiently large in magnitude that one would have expected them to be recorded in the REBs. Other events of similar size had been detected by the GSETT-3 network. To determine the reason those events were missed, available station and communication status files from the pIDC and the United States National Data Center were examined. The pIDC databases were searched to see if either associated or unassociated detections were observed at candidate stations within predicted *P*-wave arrival time windows. Phase arrival time windows were determined for auxiliary stations which should have had a high probability of detecting these events. Queries were then submitted to the Canadian National Data Center and the United States Geological Survey requesting arrival information for these auxiliary stations within the predicted windows. It was found that for about one third of the missed events, data outages and communication problems simultaneously occurring at key stations accounted for events not being detected. For another third of the events, information about station outages and/or the time of the event was not known with sufficient certainty to state whether or not critical stations were operational at the time of the event. For the remaining third of the events, it was not clear why these events were not detected since critical stations appeared to be operational at the time. Perhaps the detection or non-detection was related, in part, to the shot geometry or delay pattern employed by the mine operator for those blasts.

Other Large Mining Events in the United States

In early 1995, two seismic events occurred in the United States which were recorded by the GSETT-3 network. These events are of particular interest because they occurred in mining districts and, in fact, both events have been determined to have been caused by roof collapses in the area being mined (PECHMANN *et al.*, 1995; FINN and WOODS, 1996). The first event occurred on February 3, 1995 in the Trona mining district of southwestern Wyoming. The m_b 5.2 event was reported in the United States Geological Survey Supplementary Event Lists (SEL) as a suspected rockburst, and was later confirmed by mine operators to be caused by a collapse. The event was reported in the pIDC REB with 26 defining phases from 22 stations at distances ranging from 1 to 138 degrees from the mine.

A pair of events occurred on March 11, 1995 at a subsurface coal mine on the Kentucky/Virginia border. According to mine operators, a large roof collapse (m_b 3.6, Mn 3.8) was followed by a smaller collapse (Mn 3.3) approximately one hour later. Only the first event was reported in the REB, with 19 defining phases from 12 stations at distances ranging from 8 to 145 degrees from the mine. Figure 6 shows the known mine location, the REB location, and error ellipse for the larger first event and the SEL locations, and error ellipses for both events.

Through contacts with the mining companies, the United States Bureau of Mines, and the Mine Safety and Health Administration, United States researchers were able to obtain specific information about the conditions surrounding these events, including accurate "ground-truth" locations. The ground truth consists of latitude, longitude, and depth, but not origin time, since these events occurred unexpectedly. Since both events were well-recorded by the GSETT-3 network of stations, they make good candidates for calibration events for each region, respectively.

Summary and Conclusions

A study was undertaken to examine epicenter location accuracy using events with known locations. The study area chosen was the Powder River Basin region of

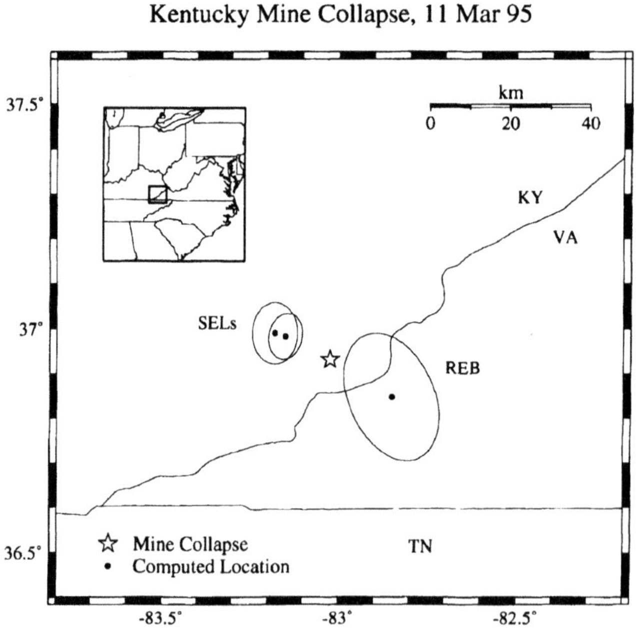

Figure 6
Ground-truth location of two events (☆) from a mine on the Kentucky/Virginia border. The single REB solution for the larger first event and the SEL solutions for both events are indicated by REB and SELs, respectively.

Wyoming, since this area is the site of mining operations which provide an ample data set of mining-related events. Using a Joint Hypocenter Determination (JHD) method, station-phase specific travel-time corrections were computed using data from 23 events. We were able to obtain location convergence on a set of nineteen events (including the four calibration events). The JHD determined, arrival-time corrections were then applied to the set of ground-truth events to determine if actual locational accuracy was improved.

A comparison of known locations, REB locations, and the locations determined using JHD station-phase travel-time corrections showed that the latter are on average 10 km (65%) closer to the known locations than are the corresponding REB locations. The JHD travel-time corrected locations also appear to have reduced the location bias observed in the REB locations. The error ellipses of the JHD solutions enclose all the ground-truth locations while the REB error ellipses enclosed three of the four true locations. In addition to the general reduction in error ellipse size, the 19 JHD travel-time corrected locations are more tightly clustered around known active mines than are the REB locations.

In the PRB region the requirement of an accurately-known regional velocity structure does not appear to be necessary. Instead, IASPEI curves and JHD determined station-phase travel-time corrections appear to be adequate in this area. Thus, in areas where calibration events are available, it appears that the JHD methodology is a viable technique for improving locations of small events in which detection is predominately from regional stations.

Acknowledgments

The plots in this paper were generated using the Generic Mapping Tools (GMT) software package developed by WESSEL and SMITH (1991). This research was supported in part by Department of the Air Force contract Number F08650-95-D-0033.

REFERENCES

DEWEY, JAMES W. (1992), *Program JHD89*, Program Documentation Files, April 23, 1992.

FINN, C. A., and WOODS M. T. (1996), *Mine Collapse Identification Using P-Lg Ratios and Rg Phase Modeling*, Seism. Res. Lett. *67*(2), p. 38.

KENNETT, B. L. N. (1991), *IASPEI 1991, Seismological Tables*, Research School of Earth Sciences, Australian National University, 167 pp.

PEARSON, D. C., STUMP, B. W., BAKER, D. F., and EDWARDS, C. L. (1995). The LANL, LLNL, AFTAC Black Thunder Mine regional mining blast experiment. In *Proceedings of the 17th Annual Seismic Research Symposium on Monitoring a Comprehensive Test Ban Treaty*, 12–15 September 1995 (J. F. Lewkowicz, J. M. McPhetres, and D. T. Reiter, eds.), 14 August 1995, pp. 562–571.

PECHMANN, J. C., WALTER, W. R., NAVA, S. J., and ARABASZ, W. J. (1995), *The February 3, 1995 M_L 5.1 Seismic Event in the Trona Mining District of Southwestern Wyoming*, Seism. Res. Lett. *66*(3), pp. 25–34, May–June 1995.

WESSEL, P., and SMITH, W. H. F. (1991). *Free Software Helps Map and Display Data*, EOS Trans. Am. Geophys. Union *72*, pp. 441, 445–446.

(Received July 20, 1999, revised February 4, 2000, accepted February 9, 2000)

Pure appl. geophys. 158 (2001) 117–128
0033-4553/01/020117-12 $ 1.50 + 0.20/0

Pure and Applied Geophysics

Seismic Location Calibration of the European Arctic

ELENA KREMENETSKAYA,[1] VLADIMIR ASMING,[1] and FRODE RINGDAL[2]

Abstract—A crustal velocity model has been developed for Fennoscandia, the Baltic shield and adjacent areas. This model represents a simplified average of various models developed for parts of this region. We show that *P*-wave travel times calculated with this model provide an excellent fit to observations at the Fennoscandian, KRSC and IRIS station networks for a set of seismic events with known or very well-constrained locations. The station-event paths cover large parts of Western Russia and the Barents Sea, thus indicating that this model, which we denote the Barents model, is appropriate for this entire region. We show by examples that significant improvements in event location precision can be achieved compared to using the IASPEI model. We finally use the Barents model to calculate locations of recent small seismic events in the Novaya Zemlya region of interest in a CTBT monitoring context.

Key words: Location, crustal models, travel times, calibration.

Introduction

Kola Regional Seismological Centre (KRSC) of the Russian Academy of Sciences have for many years cooperated with NORSAR in the continuous monitoring of seismic events in Northwest Russia and adjacent sea areas. This work has been based on a network of sensitive regional arrays which was installed in northern Europe during the last decade, in preparation for the global seismic monitoring network under a comprehensive nuclear-test-ban treaty (CTBT).

KRSC began its seismic network processing in 1982. Initially, this was done primarily by processing data from the KRSC network of seismological stations, however in recent years the analysis has been supplemented with data from IRIS stations (KBS, LVZ, KEV, ARU, ALE, NRI, etc.) and the Scandinavian seismic arrays (ARCESS, SPITS, FINESS, HFS, NORESS) for analyzing of the most interesting events.

As part of a project aimed at improving seismic monitoring capabilities under a CTBT, Kola Regional Seismological Centre (KRSC) and NORSAR are conducting a comprehensive study of seismicity, seismic wave propagation and seismic event

[1] Kola Regional Seismological Centre, Apatity, Russia.
[2] NORSAR, Kjeller, Norway.

location in the European Arctic. For Fennoscandia, excellent velocity models have previously been developed, and one such model is currently used at the prototype IDC (BONDÁR and RYABOY, 1997). The velocity model in use at KRSC for the past several years (the Barents model) is very similar to the model used at the prototype IDC, and is given in Table 1. In this paper, we have studied the improvements that can be achieved when applying the Barents model to seismic events in NW Russia and the Barents Sea region, when compared to the IASPEI91 model (KENNETT, 1991). While the IASPEI91 is an excellent average model for the entire globe, it is well known that regional velocity models can provide improvements in many cases, and we will show that this also applies for event location in the European Arctic.

Data Base

As a data base for this study we have selected seven well-recorded events in the region, as listed in Table 2. For three events we have been able to obtain accurate ground-truth information (i.e., location in the category GT0 according to the definition by BONDÁR and NORTH, 1999), as specified in the table. This includes the calibration explosion in Khibiny on 29 September 1996 (RINGDAL *et al.*, 1996) and

Table 1

Barents regional velocity model

Depth (km)	V_P (km/s)	V_S (km/s)	Comment
0–16	6.20	3.58	
16–40	6.70	3.87	
40–55	8.10	4.60	
55–210	8.23	4.68	
>210			Same as IASPEI91

Table 2

Calibration events used in this study

Region	Date	Origin time	Latitude (N)	Longitude (E)
Arkhangelsk**	18.07.1985	21.15:00.3	65.9939	41.0381
Solikamsk*	5.01.1995	12.46:02.1	59.59	56.80
NW from Spitsbergen	26.04.1995	8.55:59.9	85.128	8.58
Zapolyarny	7.06.1995	11.09:42.7	69.43	30.835
Barents Sea	11.06.1995	19.27:14.0	75.745	34.727
Novaya Zemlya	13.06.1995	19.22:39.0	75.175	56.627
Khibiny**	29.09.1996	6.05:46.19	67.675	33.728

*Accurate location

**Accurate location and origin time

the nuclear explosion near Arkhangelsk on 18 July 1985, for which both the exact location and origin time is known. The Solikamsk event on 5 January 1995 was caused by a collapse associated with a mine, and we have provided an exact location. For four events in Table 2 we have recomputed the location using available stations in the GSETT-3, the Kola and the IRIS networks.

Station Network

The KRSC regional seismic network in the Kola Peninsula comprises a total of 7 seismic stations, as described by KREMENETSKAYA and ASMING (1999). In addition, several stations in Fennoscandia, some IRIS stations, as well as stations contributing to the prototype IDC have been available. We have only used data from stations within an epicentral distance of approximately 30 degrees for each event, and concentrated on station-epicenter combinations that cross parts of the Barents region. The station network used in this study together with the station-event paths is shown in Figure 1.

Barents Velocity Model

The Barents velocity model in Table 1 is a crustal model that has been in use at KRSC for many years. It represents a simplified average over various models developed for parts of Fennoscandia and NW Russia. Much of these developments is

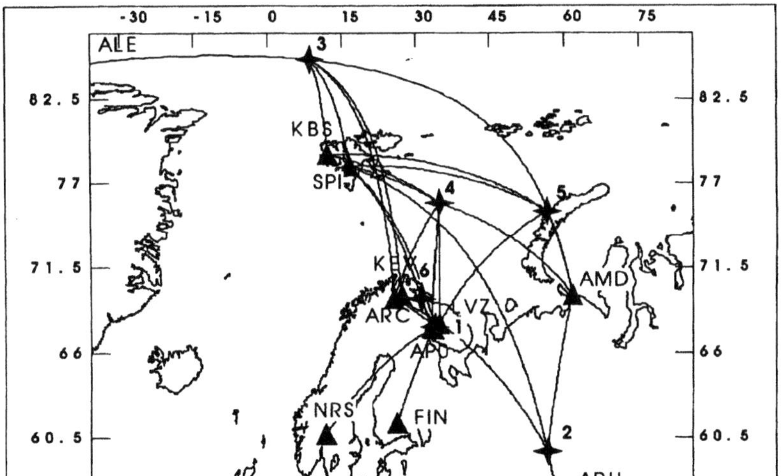

Figure 1
Map showing the calibration events and the station-event paths forming the data base for this study.

based upon profiles from explosion seismology in this area. We also note that the Barents model is similar (although not identical) to that described by MYKKELTVEIT and RINGDAL (1981), and it is also quite close to the model currently used at the prototype IDC.

In the data analysis of this paper, it is an important point to make that the model is derived independently of the data. Therefore, the fit of the data to the model will be a true validation of how the model works in practice for this region.

Data Analysis

Using the Barents model, we have located all the reference events apart from those with ground-truth information, and calculated the estimated *P*- and *S*-phase arrival times using the Barents model. Note that the *P* phase represents the first arriving *P*-type phase, which is Pn in most cases, except for the shortest distances. The results are shown in Figure 2, both for the ground truth events (triangles) and the calculated solutions (circles). The Barents model as well as the IASPEI91 model are shown for comparison. It is evident that the Barents model provides the best fit for the ground-truth data, and that this model in fact represents the data very accurately. By

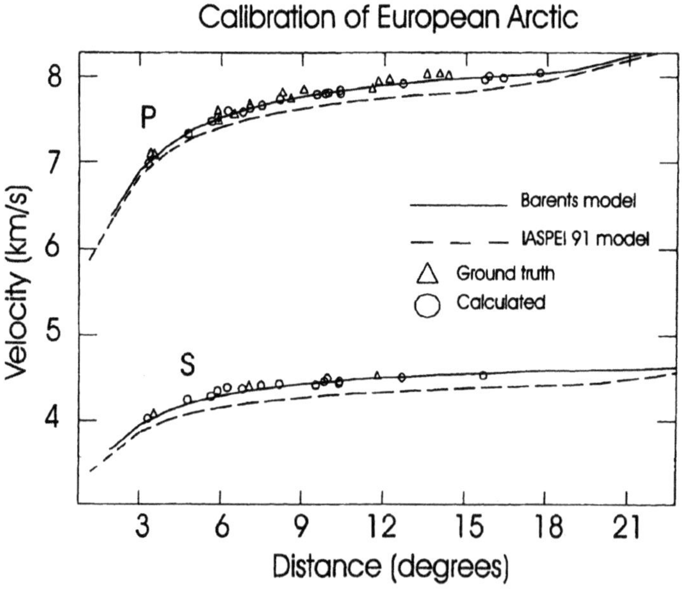

Figure 2
Observed *P*- and *S*-wave velocity as a function of epicentral distance for the events in the data base. Ground-truth observations are shown as triangles, whereas the circles represent observations using calculated epicenters. The Barents model (solid line) and the IASPEI91 model (stippled line) are shown for comparison. See text for details.

definition, the calculated data points will fit the Barents model better than the IASPEI model, but more importantly, it seems to be a good consistency with the Barents model over the entire set of observations, both for *P* and *S* phases. Our results therefore indicate the validity of the Barents model for the entire region under consideration.

Location Experiments

We have used the two models to calculate epicenters for those events with known ground-truth information, and compared the results. Two examples are shown in Figures 3 and 4. For the 18 July 1985 Arkhangelsk explosion (Fig. 3), the location errors were 17.2 km (IASPEI91 model) and only 4.4 km (Barents model). For the 5 January 1995 event (Fig. 4), the errors were 19.5 km (IASPEI91 model) and 8.5 km (Barents model). It is also noteworthy (although not shown in the figure) that the USGS/ISC global network location errors for both events were about 10 km, thus the regional model shows location improvement in spite of using only a fraction of the available stations.

To validate the model further we have relocated several previous seismic events for which we do not have ground truth, but which are located very accurately by the joint hypocentral determination method (see Table 3). As can be seen from this table, and further illustrated in Figure 5, the locations by the regional network are within 5–10 km of the locations obtained by joint hypocentral determination (JHD) using worldwide data.

The model therefore seems to be quite adequate for event location in the Barents region. In addition, the documented consistency with precise global network locations is especially important since we are able to use the network to locate regional events far smaller than those which can be detected teleseismically. For example, the KRSC network was the only network with sufficient data to reliably locate the smallest recorded nuclear explosion on the Novaya Zemlya test site (m_b = 3.8) on August 26, 1984 (MIKHAILOV *et al.*, 1996). The result is shown in Figure 6. Our estimated epicentral coordinates of this explosion are 73.326 N, 54.763 E, thus placing the event within the group of explosions shown in Figure 5. While we have no other network solution with which to compare our result, we believe this explosion to be rather accurately located.

We have finally used the Barents model to calculate locations of recent small events near Novaya Zemlya, of interest in seismic monitoring (see Table 4). These events, which were not included in the paper by RINGDAL (1997), are shown in Figures 7–9 and comprise a small seismic event near Novaya Zemlya on 23 February, 1995 and the two seismic events in the Kara Sea on 16 August, 1997. While we do not have ground-truth reference for any of these events, we observe that our location of the first event on 16 August, 1997 is close to the location published by several other authors (e.g., RINGDAL *et al.*, 1997). The second event on that day is less accurately

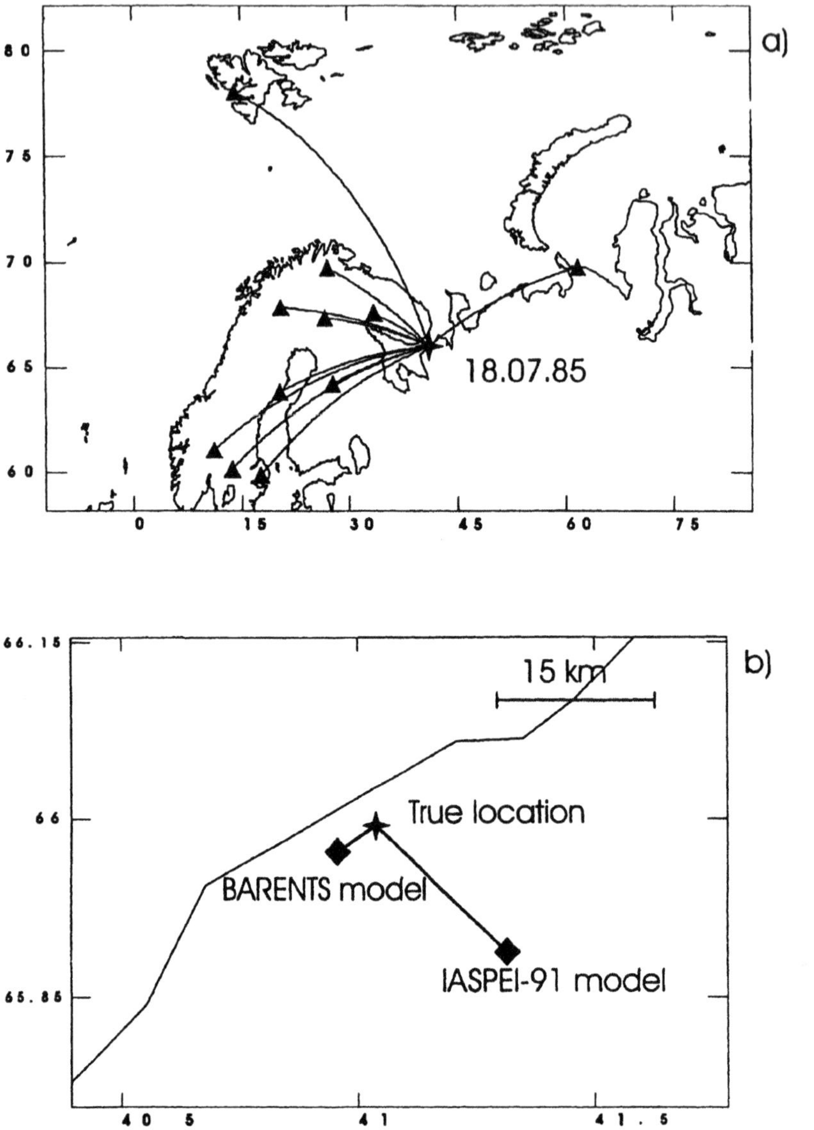

Figure 3
Location comparison using the Barents and IASPEI91 models for the 18.07.85 nuclear explosion near Arkhangelsk. Part a) shows the station-epicenter paths and part b) shows the location estimates relative to the true epicenter.

located, since it was much smaller. Our estimated location, based on three stations, is about 30 km east of the first event. However, from a detailed analysis of the signals recorded at Spitsbergen and Amderma, we believe that the two events that day were approximately colocated. The difference in our location estimates of these two events

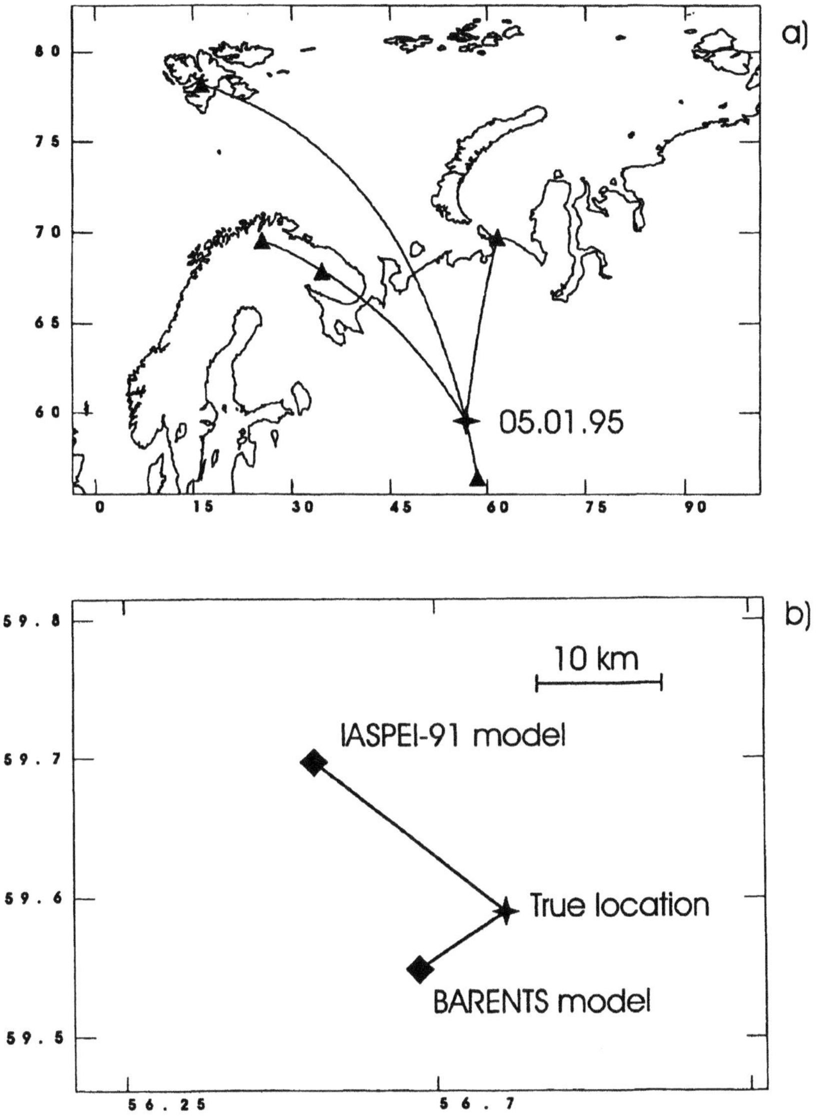

Figure 4
Location comparison using the Barents and IASPEI91 models for the 05.01.95 mine collapse near Solikamsk. Part a) shows the station-epicenter paths and part b) shows the location estimates relative to the true epicenter.

is indicative of the problems arising when the signal-to-noise ratio of one or more key stations is low. The event on 23 February, 1995 has to our knowledge not been reported earlier, and it is difficult to ascertain the accuracy of our location estimate. We note that in all of these location experiments, we have restricted the solution to zero depth. It is interesting to notice that both the 23 February, 1995 event and the

Table 3

Location comparison – regional versus global network

Date	KRSC location	JHD location	Comment
18.08.83	73.289 N, 54.893 E	73.358 N, 54.943 E	
01.08.86	72.945 N, 56.549 E	73.031 N, 56.726 E	Marshall *et al.* (1989)
02.08.87	73.298 N, 54.398 E	73.324 N, 54.597 E	
07.05.88	73.275 N, 54.436 E	73.314 N, 54.557 E	
24.10.90	73.304 N, 54.634 E	73.317 N, 54.803 E	

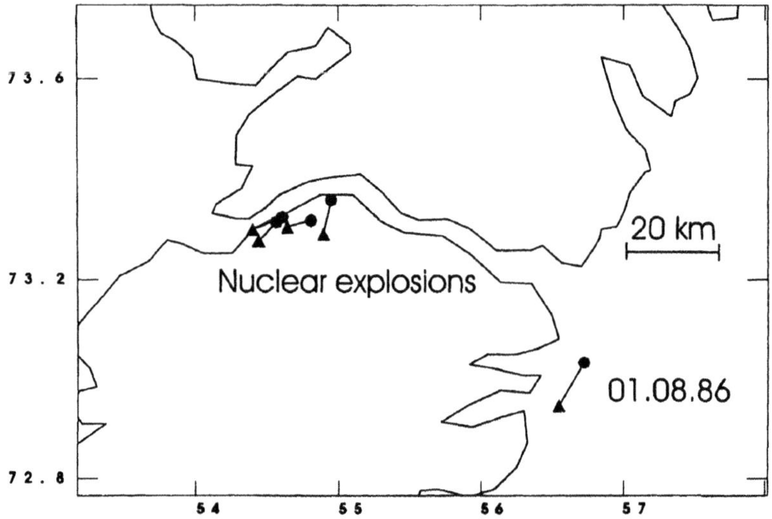

Figure 5
Location comparison of JHD epicenter estimates using a global network (filled circles) and regional location estimates using the Barents model with a regional network (filled triangles). The figure corresponds to the data in Table 3.

first event on 16 August, 1997 had estimated origin times very close to the minute (using the Barents model and a restricted (zero) depth). The estimate of the origin time is of course a tradeoff between location, depth and travel-time model uncertainties, and we do not wish to speculate on this issue.

Effects of Reading Errors

In practical IDC operation, the location accuracy will be determined not only from the quality of the velocity model, but also from the quality and accuracy of the phase readings used in the location algorithm. We have carried out a preliminary study, using a set of 52 Khibiny explosions (Kola Peninsula) detected and located by

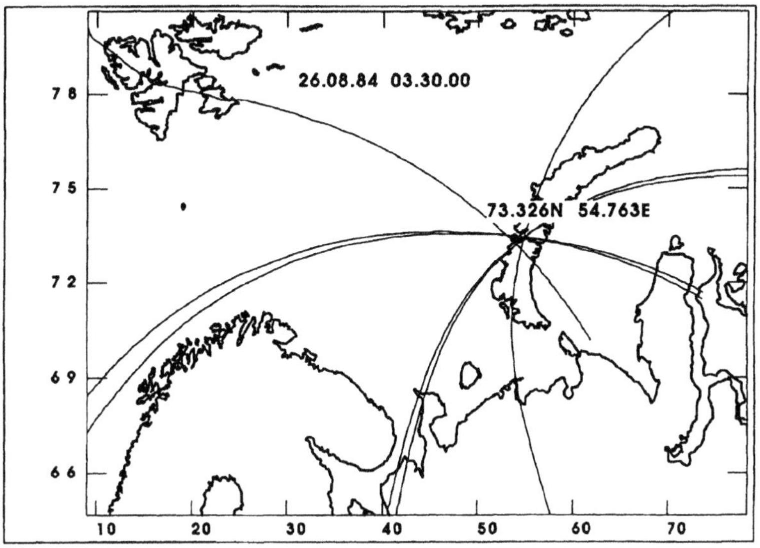

Figure 6
Location of a small nuclear explosion on Novaya Zemlya, using the Barents model with a regional network.

Table 4

Four low magnitude seismic events located by the regional network

Date/time	Location	m_b	Comment
26.08.84/03.30.00	73.326 N, 54.763 E	3.8	Nuclear explosion. Located in this paper
23.02.95/21.50.00	71.856 N, 55.685 E	2.5	Located in this paper
16.08.97/02.11.00	72.510 N, 57.550 E	3.5	Location estimate close to epicenter reported in REB
16.08.97/06.19.10	72.681 N, 58.453 E	2.6	Uncertain estimate. Probably very close to preceding event

at least 4 stations (with *P* detections) in the GSETT-3 network. All 52 events have known ground-truth locations. We used phase readings exactly as provided in the prototype IDC Reviewed Event Bulletin (REB). Most of these readings have been based on automatic timing, but were sometimes adjusted by the analyst producing the bulletin. In some cases, the analysts have added phase readings not detected automatically.

For each event, we compared locations using the IASPEI91 model with locations based on the same observations, although with the Barents model. To obtain a simple measure of the results, we calculated the percentage of these 52 events that were located within 18 km of the true epicenter. It should be noted that a circular

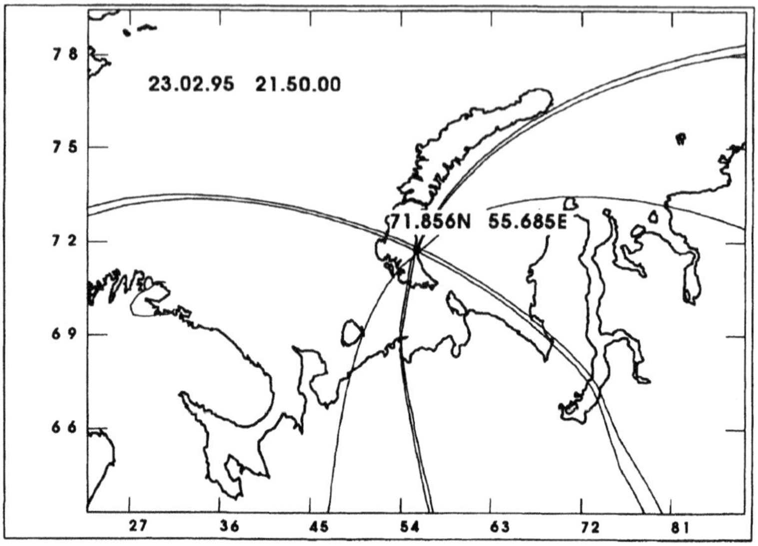

Figure 7
Location of a small seismic event on 23 February 1995, using the Barents model with a regional network.

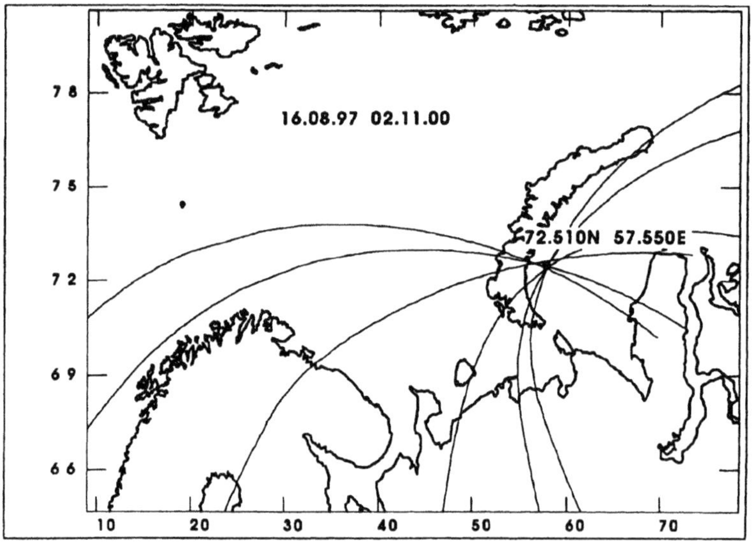

Figure 8
Location of the first seismic event on 16 August 1997, using the Barents model with a regional network.

area of 18 km represents an area of approximately 1000 sq.km, which is a generally accepted target for location precision in the GSETT-3 network.

It turned out that this percentage was 21% when using the IASPEI91 model compared to 37% when using the Barents model for the same data. However, we

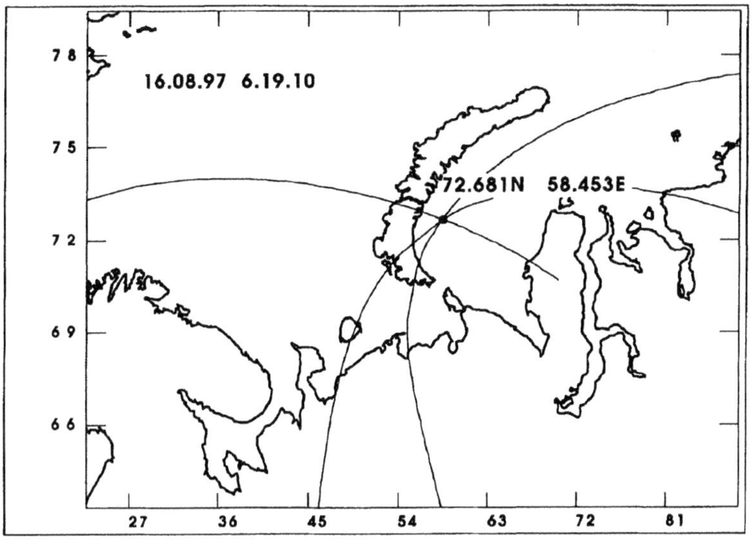

Figure 9
Location of the second seismic event on 16 August 1997, using the Barents model with a regional network.

observed that the S residuals were rather large with the Barents model, and therefore attempted to locate the events using the P-phase data only (with the Barents model). This resulted in a percentage as high as 62%, which is a significant improvement over both of the other approaches. It appears from this result that the S-phase readings used in the GSETT-3 bulletins might be less accurate than desirable. The reasons for this are unknown, but will be further investigated.

Conclusions

We conclude that the Barents model is appropriate not only for Fennoscandia, but for the general Barents region from Spitsbergen to Novaya Zemlya, and also for northwestern Russia. Use of this model would be expected to improve seismic event location accuracy considerably compared to the use of IASPEI91, especially when both P and S phases are used in the location procedure.

We have also observed in this paper that in the absence of a well-calibrated velocity model, it might seem preferable to make epicenter estimates based on P phases only, since these location estimates are less sensitive to model errors than locations based on a combination of P and S phases. However, it must be noted that the S phases, even in the absence of a good velocity model, do place important constraints on the distance to the epicenters. The use of S therefore in many cases reduces the likelihood of gross error, which might occur if there are only few P readings with poor azimuthal distribution. We plan to conduct more detailed studies of this problem.

Acknowledgements

This research was supported by the Nuclear Treaty Program Office of the U.S. Department of Defense and was monitored by AFTAC, Patrick AFB, Florida, under Contract No. F08650-96-C-0001.

References

BONDÁR, I., and NORTH, R. G. (1999), *Development of Calibration Techniques for the Comprehensive Nuclear-Test-Ban Treaty (CTBT) International Monitoring System*, Phys. Earth Plan. Int. *113*, 11–24.

BONDÁR, I., and RYABOY, V. (1997), *Regional Travel-time Tables for the Baltic Shield Region*, Technical Report CMR-97/24, Center for Monitoring Research, Arlington, Virginia, USA.

KENNETT, B. L. N. (1991) (ed.), *IASPEI 1991 Seismological Tables*, Res. School of Earth Sci. Austr. Natl. Univ., Canberra, Australia.

KREMENETSKAYA, E. O., and ASMING, V. E. (1999), *Seismology on Kola: Monitoring Earthquakes and Explosions in the Barents Region*, IRIS Newsletter *XVI*(2), 11–13.

MARSHALL, P. D., STEWART, R. C., and LILWALL, R. C. (1989), *The Seismic Disturbance on 1986 August 1 near Novaya Zemlya: A Source of Concern?* Geophys. J. *98*, 565–573.

MYKKELTVEIT, S., and RINGDAL, F. (1981), *Phase identification and event location at regional distance using small-aperture array data.* In *Identification of Seismic Sources – Earthquake or Underground Explosion*, (E.S. Husebye and S. Mykkeltveit, eds.) NATO Advanced Study Institute Series 467–481.

MIKHAILOV, V. N. *et al.* (1996), *USSR Nuclear Weapons Tests and Peaceful Nuclear Explosions, 1949 through 1990*, RFNC – VNIIEF, Sarov, 1996, 63 pp.

RINGDAL, F. (1997), *Study of Low-magnitude Seismic Events near the Novaya Zemlya Test Site*, Bull. Seismol. Soc. Am. *87(6)*, 1563–1575.

RINGDAL, F., KVÆRNA, T., KREMENETSKAYA, E., and ASMING, V. (1997), *The seismic event near Novaya Zemlya on 16 August, 1997.* Semiannual Technical Summary, 1 April–30 September 1997, NORSAR Sci. Rep. 1–97/98, Kjeller, Norway.

RINGDAL, F., KREMENETSKAYA, E., ASMING, V., KUZMIN, I., EVTUHIN, S., and KOVALENKO, V. (1996), *Study of the Calibration Explosion on 29 September 1996 in the Khibiny Massif, Kola Peninsula*, Semiannual Technical Summary 1, April–30 September 1996, NORSAR Sci. Rep. 1–96/97, Kjeller, Norway.

(Received June 30, 1999, revised October 15, 1999, accepted February 9, 2000)

Pure appl. geophys. 158 (2001) 129–141
0033–4553/01/020129–13 $ 1.50 + 0.20/0

Pure and Applied Geophysics

Earthquake Location Accuracies in Norway Based on a Comparison between Local and Regional Networks

E. C. Hicks,[1,2] H. Bungum[1,2] and F. Ringdal[1]

Abstract—Detailed studies of the low to intermediate seismicity in two coastal regions of Norway have been used in a comparison between earthquake locations from local high-precision networks on the one side and locations using a sparse regional array network on the other side. To this end, a reference set of 32 low-magnitude earthquakes have been located using two local temporary networks in northern and western Norway, with estimated epicenter accuracies better than 5 and 10 km, respectively. Comparisons are made between the local network solutions and the NORSAR Generalized Beamforming (GBF) system, which provides automatic phase association and location estimates using the Fennoscandian regional array network. The median automatic GBF location error is of the order of 20–30 km when four or more arrays detect the event, increasing to about 80–100 km when only two arrays are available, and the automatic GBF bulletin is essentially complete down to magnitude $M_L = 2.0$. Most of the mislocation vectors of the NORSAR GBF solutions are oriented perpendicular to the Norwegian coast, and with a tendency to pull the location in a southeasternly direction. The GBF performance is clearly better, both in terms of accuracy and completeness, than the performance of the automatic bulletin of the Prototype International Data Center (PIDC) which uses data from essentially the same network. The analyst reviewed NORSAR and PIDC bulletins show, not unexpectedly, an improvement in location accuracy compared to the automatic solutions and appear to be of similar quality for the few common events, with an average mislocation of about 20 km. The NORSAR reviewed bulletin is more complete at low magnitudes compared to PIDC, and there appears to be a potential for significant improvements in the PIDC processing of small seismic events in this region.

Key words: Earthquake location, local networks, regional networks, seismic arrays.

1. Introduction

A considerable effort is currently taking place to develop and apply location calibration information for seismic events recorded by the International Monitoring System (IMS) for the Comprehensive Nuclear-Test-Ban Treaty (CTBT) (e.g., Bondár and North, 1999). One important source of such calibration information is sets of explosions or earthquakes with very accurate locations (so-called Ground Truth information). Earthquakes monitored by local microseismic networks with

[1] NORSAR, POB 51, N-2027 Kjeller, Norway.
[2] Department for Geology, University of Oslo, Oslo, Norway.

high location precision are in many cases appropriate as calibration events, and in this paper we present a number of earthquake hypocenters calculated from local networks in northern and western Norway (located within the two boxes in Figure 1,

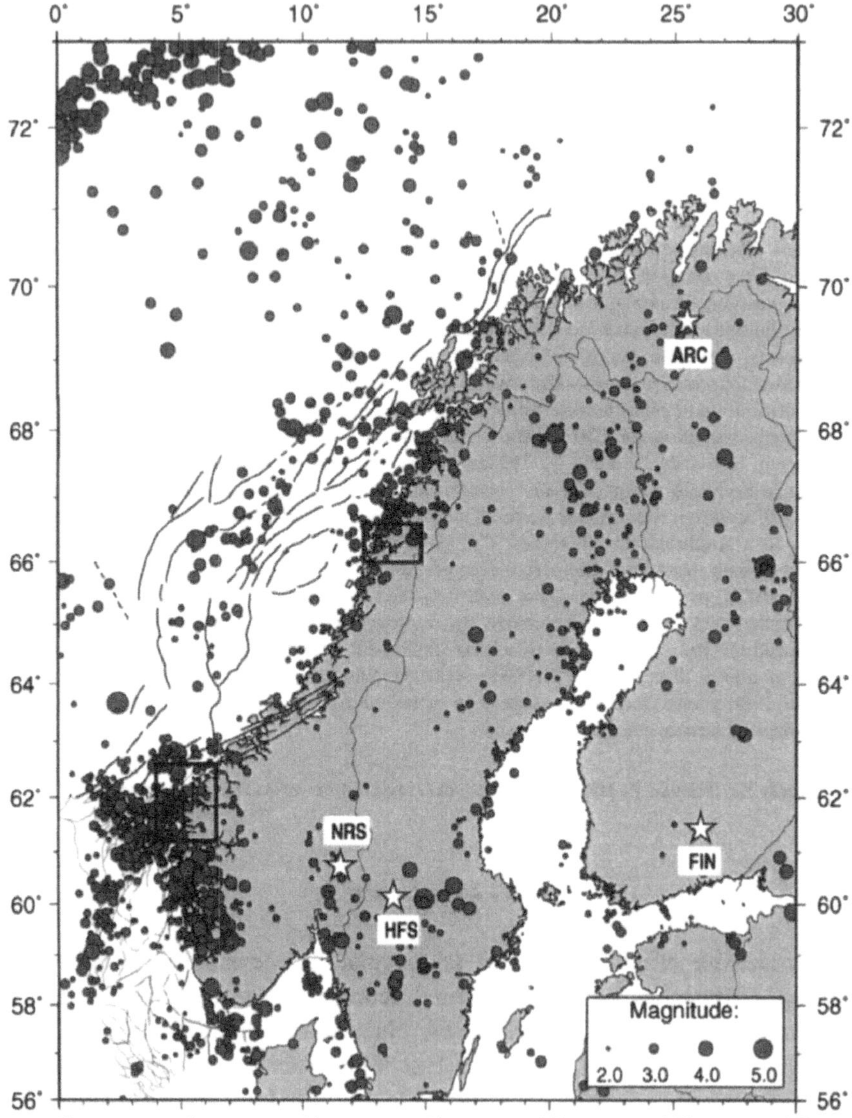

Figure 1
Seismic activity in Norway 1980–1999, $M_{\omega} > 2.0$. The locations of the two study areas in this paper, Rana and Bremanger, are shown by the two rectangles in northern and western Norway, respectively. Structural information is from Blystad *et al.* (1995). Seismic arrays are shown by white stars (NRS — NORES, ARC — ARCES, HFS — Hagfors, FIN — FINESS).

where also the regional seismicity is shown). We use these results to evaluate the accuracy of automatic and interactive location estimates using a sparse regional array network, essentially comprising the IMS seismic stations in Fennoscandia. While most of our emphasis is on evaluating the NORSAR automatic processing system, we also compare the results with those of the automatic PIDC process, as well as with analyst-reviewed results at NORSAR and the PIDC.

2. NORSAR and PIDC Detection and Location Processing

The NORSAR automatic system makes use of the Generalized Beamforming (GBF) method, which was developed by RINGDAL and KVÆRNA (1989), and which has been applied routinely at NORSAR since 1990. The GBF algorithm is based upon processing data from a sparse network of regional arrays, and associates detected phases by forming a regional grid system and "steering" the network towards each individual grid point. Each detected phase at one of the arrays is treated as a 0–1 valued function, where the value 1 is assigned if the detection corresponds in azimuth and slowness to the grid point. By simply adding these functions, suitably delayed in time, one obtains an efficient phase association of seismic events as well as a preliminary location. By using a denser grid ("beampacking") around this initial epicenter, the location estimates are subsequently refined.

The automatic association and event definition procedure at the PIDC makes use of the same basic principles as the NORSAR GBF, but introduces a set of weighting criteria (different from the 0/1 weightings used at NORSAR) in the beamforming procedure. A seismic event is defined when the weighted sum of detected phases exceeds a predefined threshold. The threshold setting represents a tradeoff between the desire for completeness (no missed events) and for avoiding spurious events (false associations). In contrast to the NORSAR GBF, the threshold at the PIDC is set to a relatively high value in order to minimize the number of spurious (false) events. On the other hand, this results in several real seismic events being missed by the PIDC automatic procedure, as will be further shown by examples in this paper.

3. The Local Networks and their Capabilities

The two local networks studied in this paper are located in the Ranafjord area (northern Norway) and the Bremanger area (western Norway). Figure 1 shows the two study areas along with a seismicity map of Fennoscandia. The technical installations for both networks are similar, relying on radio links to transmit data to a central station, where the data are digitized, and a real-time STA/LTA analysis is performed by a local, PC-based acquisition system (HICKS *et al.*, 2000a). Triggers are stored locally and downloaded to NORSAR on a daily basis. Sampling rates are 40 Hz for both networks. Whenever convenient, the readings from the two networks have been supplemented by readings from nearby permanent seismic stations part of the National Norwegian Seismic Network. However, these additional stations did not provide any significant improvement of the locations, but rather acted as confirmation.

The Ranafjord network (Fig. 2) comprised initially six seismic stations installed in June 1997 as part of a research project (NEONOR, Neotectonics in Norway), with the main purpose of monitoring possibly seismic activity along potentially active local faults (HICKS *et al.*, 2000a). The network was reduced to four stations in September 1998, but without significant loss in location precision. The Bremanger network (Fig. 3) has been in operation since October 1998, but has not been operating continuously since it was installed, and less data are therefore available here than one otherwise should have expected in view of the proximity to seismically active areas of the North Sea (BUNGUM *et al.*, 1991).

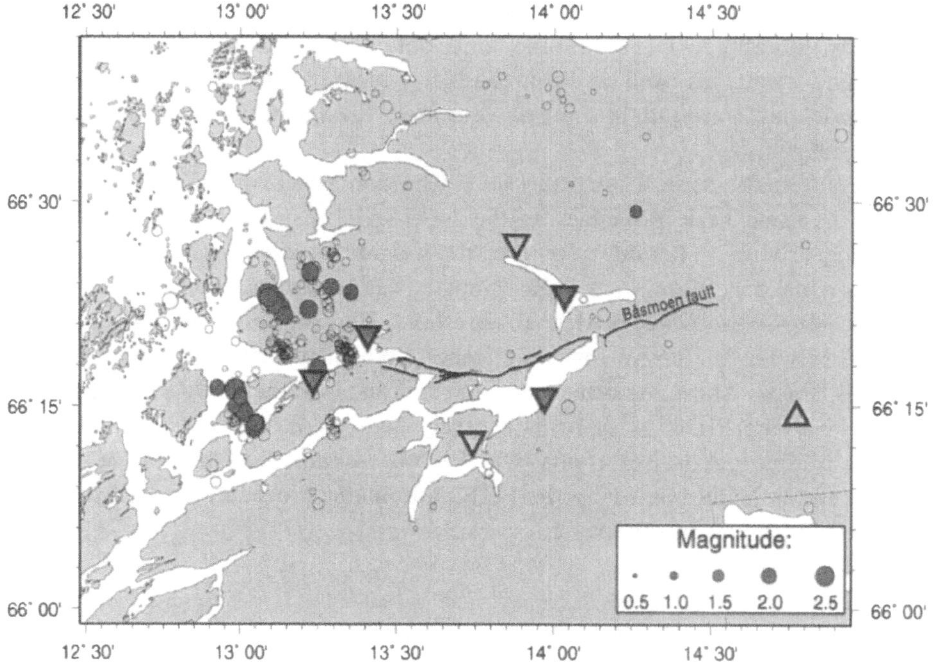

Figure 2
Local seismic activity located by the Rana network 1997–1999. The filled circles represent the events also located by the NORSAR GBF system. The stations are indicated by inverted triangles, the four stations remaining after September 1998 are filled. The three-component MOR8 station that is part of the national Norwegian seismic network, operated by the University of Bergen, is shown by a triangle to the east in the figure. The solid black line is the postglacial Båsmoen fault, considered to be potentially seismically active.

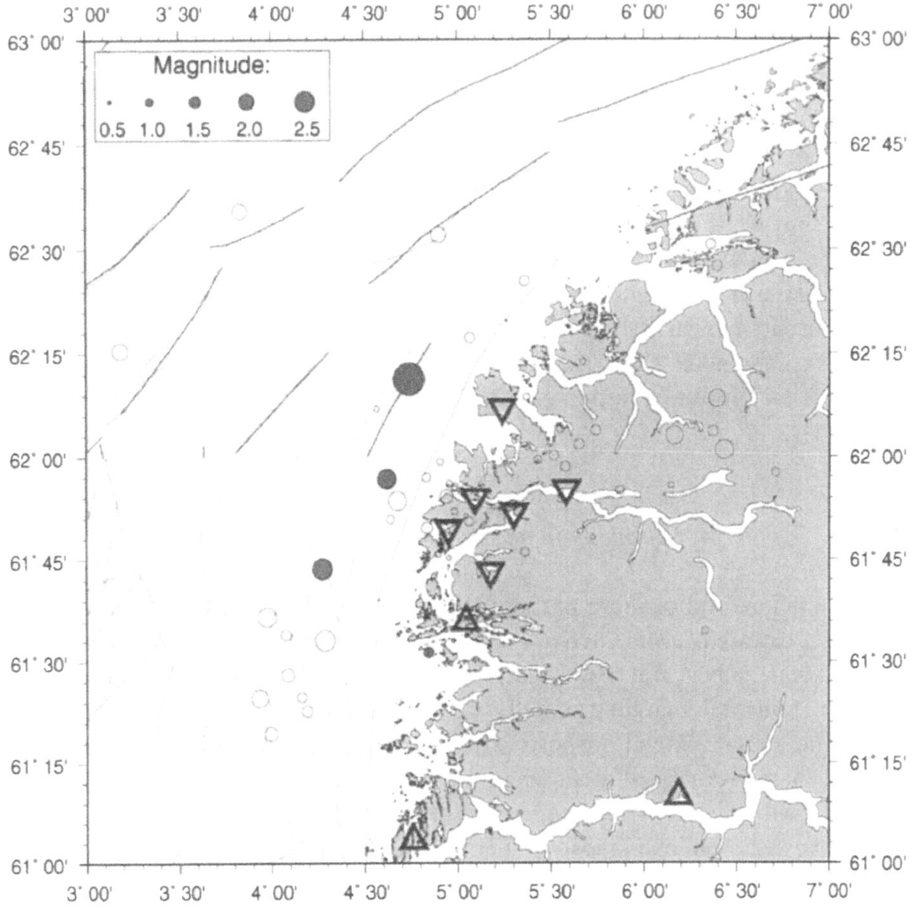

Figure 3

Local seismic activity located by the Bremanger network, 1998–1999. The filled circles represent the earthquakes also located by the NORSAR GBF system. The stations in the local network are indicated by inverted triangles. The NNSN stations in the area are shown as triangles.

The location algorithms used for both networks is a version of the Hypocenter program (LIENERT *et al.*, 1986), which uses scaled, adaptively damped least-squares to determine hypocenter location. Due to the close distances from hypocenter to station (5–40 km for the Rana network), the phase arrivals for events of this magnitude can pe picked with a high accuracy. This is especially the case for the 'larger' events ($M_L > 1.5$) such as those used in this study. The short distances also mean that first arrivals are direct waves, so the only velocities used in the location are the upper 15 km of the crust. Consequently, RMS travel-time residuals are less than 0.1 s for all earthquakes in the Rana network, and generally less than 0.5 s for the earthquakes in western Norway, using arrival times from the local and permanent networks. The velocity models used are fairly accurate for the relatively consistent geophysical properties in the areas surrounding the networks.

Based on a detailed analysis of the location error ellipsoids, we estimate that the accuracy of the epicenter locations in Rana is better than 5 km, and most likely within 2–3 km for the events near the network. The events located in the Bremanger area are somewhat farther from the network itself, and of higher magnitude. In this case there is, however, a more significant contribution from the permanent network, which is better in western than in northern Norway. A conservative estimate of the location accuracy near Bremanger would be between 5 and 10 km. This means that the local earthquakes from Rana and Bremanger will qualify as GT5 and GT10 events (GT = Ground Truth), i.e., with location accuracies better than 5 and 10 km, respectively.

4. Seismicity in and around the Local Networks

The offshore and onshore parts of northern Norway have long been considered an area of elevated seismic activity with regard to the rest of the Baltic Shield and margin areas, albeit not particularly high as compared to some other passive (rifted) continental margins globally (BUNGUM *et al.*, 1991; BYRKJELAND *et al.*, 2000). The largest known onshore earthquake in Fennoscandia in historical times occurred in the Ranafjord area, where the Rana network is located, on August 31, 1819, and with an estimated magnitude of M_S 5.8–6.2 (MUIR WOOD, 1989). This earthquake was felt over most of Fennoscandia, as far away as Stockholm and Oslo.

The northern parts of the North Sea, adjacent to the Bremanger network, are among the most seismically active areas in northern Europe, in a structurally very complex region. The largest recent earthquakes in this area occurred on August 8, 1988, and on January 23, 1989, with respective magnitudes of M_L 5.1 and 4.9 (HANSEN *et al.*, 1989). The 1988 earthquake occurred around 200 km northwest of the current network, while the 1989 earthquake was around 50 km due west of the current network.

Of the 420 events located by the Rana network, 340 are located in the immediate vicinity of the network, and 40 of these are confirmed explosions or probable explosions, leaving around 300 as probable earthquakes. Magnitudes range from M_L 0.1 to 2.8, with most events in the M_L 1.0 to 1.5 range. The hypocenter depths are shallow, mainly from 4 to 12 km, thereby indicating that this is essentially a swarm activity of the type seen also elsewhere in this region (BUNGUM *et al.*, 1979; ATAKAN *et al.*, 1994).

Figure 2 shows the 1997–1999 microseismic activity in the Rana region plotted according to magnitude (explosions removed). Five main groups of events are visible in the western part of the network. These groups occur as swarms, having well defined activity periods and hypocenter depths. The largest events within the network occurred within the two westernmost groups, which are also located in the vicinity of many of the reported phenomena concerning the 1819 earthquake. The easternmost group has hypocenter depths predominantly around 4–6 km, while the other three mainland groups mainly have depths in the 10–12 km range. The depth estimate for the large westernmost group is slightly more uncertain since these events lie further outside the network, but since several of the earthquakes were noticed as loud bangs/cracking noises the depths are most likely less than five km for this group also.

Focal mechanism solutions determined using data from the Rana network show an σ_{Hmax} orientation parallel to the coast, which is a 90° rotation with respect to the regional, ridge push dominated, stress field (HICKS *et al.*, 2000b). This implicates a strong local stress influence on the seismic activity in the area. The northern North Sea is a geologically very complex area, with an intricate system of rifted basins and highs. However, focal mechanisms in this area do largely comply with the expected direction of the ridge push force, although some mechanisms southwest of the network (LINDHOLM *et al.*, 2000) do have a similar inversion of the σ_{Hmax} direction as seen in the Rana area.

The Bremanger area has of yet not shown any clear patterns of seismicity, the activity appears to be fairly well distributed, as shown in Figure 3. The largest earthquake (M_L 3.9) located occurred in an area where there has been no earlier known activity. The other two earthquakes with magnitudes larger than 2.0 occurred within the areas known to have the highest seismic activity from earlier instrumental data.

5. Location Results Using the Regional Array Network

A total of 32 of the local earthquakes detected by the local networks were also detected and located by NORSARs automatic GBF system. Of these, 21 were reviewed by the NORSAR analysts. Eight events were detected by the PIDC automatic bulletin, six of which were reviewed. All of these events are listed in Table 1, together with locations and location differences. For the Rana region the details of the magnitude information are given in Figure 4, where it is seen that the PIDC system has a detection threshold near M_L 2.5 (but with two missing M_L 2.7 events), while the GBF system seems to have a detection threshold of about M_L 2.0 (but with one missing M_L 2.1 event).

The location differences are plotted vs. number of stations used in the solution in Figure 5, with a second-order regression line for the GBF solutions, and the median location differences are also shown in the same figure. It can be inferred from these results that both the GBF and PIDC location accuracies are quite sensitive to

Table 1

Locations and location differences for the 32 events in this study. The reference locations are from the solutions determined by the local networks, where solutions north of 65°N are from Rana (six stations, four since September, 1998), the remaining from Bremanger (also six stations). Solutions using a sparse regional network (NORSAR, automatic (GBF) and reviewed; PIDC, automatic and reviewed) are given by location difference (Δ) in km, and azimuth, both compared to the local solution. The number of stations used in each case is also included. The events are sorted by number of stations in the GBF solution.

Date & time	Local solutions				NORSAR GBF			NORSAR reviewed			PIDC automatic (SEL1)			PIDC reviewed (REB)		
	Mag.	Depth	Lat.	Lon.	Δ (km)	azi	nsta	Δ (km)	azi	nsta	Δ (km)	azi	nsta	Δ (km)	azi	nsta
99.05.29–00:31:44	3.9	8.7	62.189	4.741	22.6	234.2	7	20.5	65.2	8	52.3	355.8	3	24.5	210.0	8
98.06.18–22:54:00	2.7	11.4	66.376	13.111	22.1	97.3	5				39.3	111.3	1	59.0	78.5	3
98.12.16–20:57:46	2.7	8.7	66.270	12.983	36.8	131.6	5	30.5	102.4	5						
98.10.26–13:17:22	2.5	4.0	66.227	13.050	31.4	128.7	5	28.5	116.5	5	98.8	170.9	2	9.2	28.1	4
99.02.25–14:11:43	2.3	4.9	66.295	13.247	17.0	68.7	5	39.9	112.1	5	31.4	74.9	2			
98.02.09–12:59:05	2.8	10.7	66.385	13.088	56.6	93.4	4	22.3	114.7	6	55.5	94.4	2	24.6	100.2	4
98.03.09–14:19:57	2.8	6.6	65.854	13.529	37.1	107.8	4	5.6	93.4	6	54.7	68.3	3	2.1	190.5	4
97.11.25–22:24:17	2.7	11.0	66.500	12.403	88.5	100.0	4	15.4	118.4	5						
97.11.21–18:00:09	2.3	6.3	66.413	13.222	31.2	328.1	4	2.7	163.5	5						
98.10.13–22:21:59	2.3	3.0	66.241	13.015	14.2	328.7	4	34.7	101.3	4	36.8	98.6	2	38.2	112.0	4
99.01.07–14:04:13	2.3	0.0	66.856	13.894	21.4	60.5	4	45.5	108.0	4	11.9	83.4	2			
99.04.13–21:31:40	2.3	0.1	66.368	13.218	35.5	332.7	4	9.4	72.6	4						
98.10.29–21:07:29	2.3	3.0	66.222	13.045	21.1	204.8	4	36.2	102.6	4						
98.01.08–08:04:46	2.2	12.8	66.368	13.134	33.9	338.2	4	11.7	105.9	4						
98.01.11–20:01:18	2.2	12.3	66.373	13.110	33.0	339.6	4	16.0	105.6	4						
98.08.11–18:52:27	2.1	12.3	66.360	13.144	13.7	265.5	4									
98.12.24–07:50:10	2.1	16.6	66.395	13.288	14.8	109.6	4	18.6	79.1	4						
98.10.26–22:56:19	2.0	3.0	66.224	13.041	44.5	288.8	4	28.7	115.5	5						
99.06.26–21:30:46	2.5	23.7	61.728	4.274	23.4	47.2	3	10.2	167.3	2						
98.02.04–14:31:40	2.3	10.6	66.382	13.091	56.5	93.0	3	18.3	128.4	3						
98.12.04–12:39:29	2.3	22.2	66.163	12.275	103.0	115.6	3									
98.10.29–05:59:53	2.0	3.0	66.228	13.048	16.3	326.9	3	36.2	103.7	4						
98.12.05–22:14:53	1.9	13.0	66.751	13.812	26.3	115.1	3	23.4	102.2	4						
99.06.15–01:12:57	2.4	13.7	61.947	4.621	24.7	249.9	2	7.3	236.5	2						
98.02.28–16:53:26	2.0	11.6	66.701	13.316	179.3	90.0	2									
98.12.16–19:57:46	2.0	0.0	66.272	12.924	67.7	110.8	2									
99.04.09–08:04:28	2.0	8.0	66.389	13.351	159.0	115.2	2									
98.10.09–05:30:14	1.8	3.0	66.249	12.981	45.5	118.8	2									
98.12.17–18:33:24	1.8	0.0	66.262	13.000	181.7	118.3	2									
98.10.23–02:50:21	1.8	3.0	66.259	13.000	64.0	110.8	2									
98.12.21–17:09:47	1.6	33.0	66.488	14.259	161.3	119.3	2									
99.05.15–05:50:29	1.3	8.2	61.523	4.844	31.5	102.6	1									

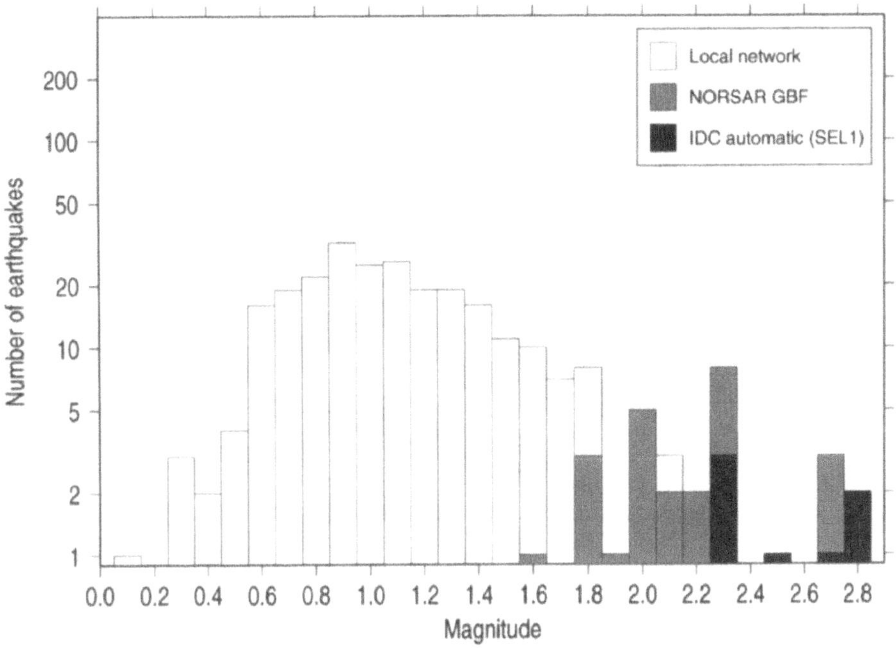

Figure 4
Histogram of frequency-magnitude distribution for the Rana local seismic network (based on about two years of operation) together with events reported by the automatic NORSAR GBF (Generalized Beamforming) system (shaded) and the automatic PIDC (Prototype International Data Center) system (black).

number of stations used (and thereby event magnitude). In contrast, the NORSAR analyst-reviewed solutions retain good accuracy (median error about 20 km) even for the smaller events, although events detected on only 1 or 2 stations are usually not reviewed. It is, however, because of the large scatter, difficult to use Figure 5 in comparing the performance of the two systems in more detail, except that the analyst review causes a clear improvement in the location accuracies for both systems. Table 2 shows average location 'errors' for the five events for which all four types of solutions are available (all from Rana, see Table 1), and it appears that the GBF has better automatic solutions than the PIDC (32 versus 60 km) while the reviewed solutions are quite similar (22 versus 20 km). However, the low number of events in the PIDC bulletins combined with the large scatter makes it difficult to conclude very clearly here.

It can be seen from Table 1 (see also Fig. 5) that the automatic GBF system provides epicenters with median location error about 20–30 km for events that are detectable on four or more stations. For the smaller events, detectable at only 1 or 2 stations, the GBF location accuracy deteriorates, with a median error of

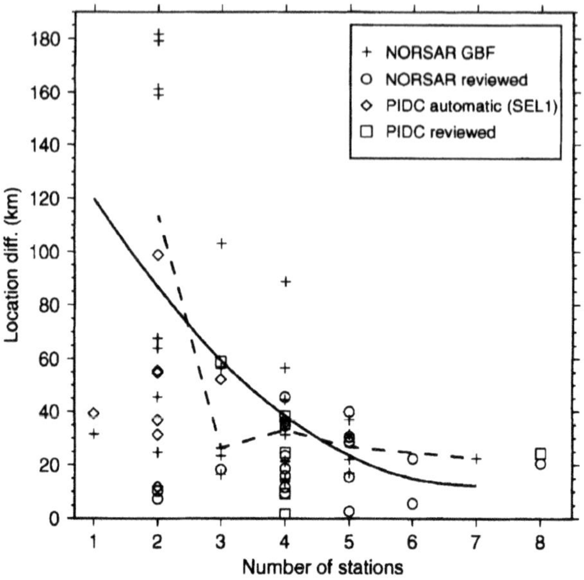

Figure 5
Scatter plot of the location differences between the local network locations and the NORSAR GBF (automatic) solutions (circles), the PIDC automatic (diamonds) and the PIDC reviewed (squares) solutions, respectively, plotted vs. number of stations used in the solution. A second-order regression analysis for the GBF solutions is shown by the solid line, while the dashed line represents the median differences for the GBF solutions.

Table 2
Average location difference, with standard deviations, for five events from the Rana region covered by all solutions (see Table 1), between the local network solutions (location error less than 5 km) and the four regional network solutions, NORSAR and PIDC, automatic and reviewed.

Regional network location	Average location difference (km)
NORSAR GBF automatic	32 ± 16
PIDC automatic	60 ± 23
NORSAR reviewed	22 ± 11
PIDC reviewed (REB)	20 ± 14

about 80–100 km and with a large scatter. The azimuthal distribution of the location differences for the GBF solutions are shown in Figure 6. The mislocation vectors are generally oriented NW-SE (perpendicular to the coast), and it is also apparent that the GBF system tends to bias the solutions towards the southeast, in particular for events with the largest uncertainties (fewer detecting stations).

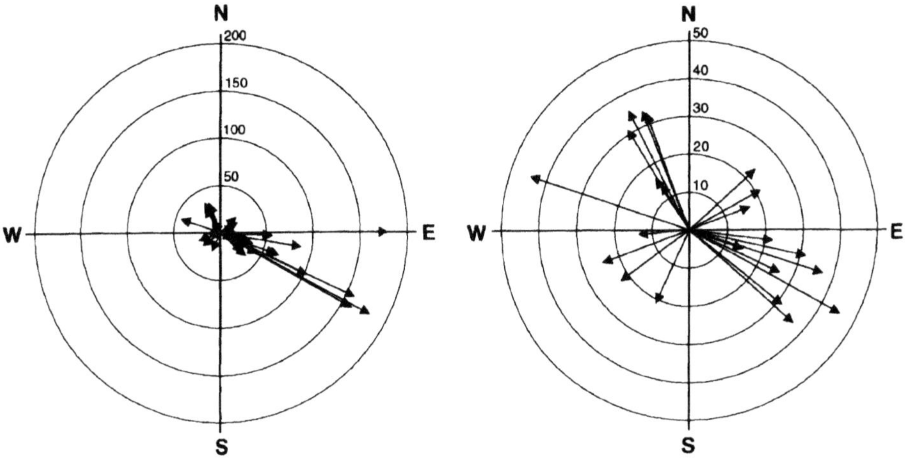

Figure 6
Location differences (in km) and azimuthal directions of the NORSAR GBF solutions with regard to the local solutions. The plot to the left contains all solutions, extending up to 200 km, while the plot to the right contains only events with location differences less than 50 km.

6. Concluding Remarks

The following main conclusions can be drawn from this study of detection and location of small events in northern and western Norway:

- Using essentially the same network of seismic stations (arrays), the automatic NORSAR GBF is significantly better than the automatic PIDC system both in terms of location accuracy (~30 and ~60 km for common events) and detectability (M_L 2.0 and 2.5).
- The quality of the automatic GBF locations deteriorates quite rapidly when fewer stations are used in the solution, whereas the accuracy of the NORSAR analyst-reviewed solutions remains high.
- The analyst-reviewed NORSAR and PIDC bulletins have similar location accuracies (~20 km) for the few common events, but the NORSAR bulletin is more complete at low magnitudes.

In this paper we have sometimes used the term 'location error' or 'location accuracy' when comparing results from the local and regional networks. We note that the local network solutions, which have been used as reference, may themselves be mislocated by up to 5 or 10 km. The real performance of the regional network locations should therefore be slightly better than evaluated here. We should also note that the grid spacing for the GBF system is 33.3 km (0.3°), so the solutions within 20 km are therefore in general located to the closest grid point. Potentials for further improvements here are apparent. With respect to the PIDC solutions we finally note that, for the region considered in this paper, the only significant difference between the arrays and stations used by the two systems is that the PIDC system uses the large-aperture NOA array instead of the regional small-aperture NORES array. While this could explain some of the difference between the two systems, it seems that there should still be potentials for the PIDC to detect events at a lower magnitude level than what is done today, and possibly also with better precisions in its automatic solutions which are quite important in an operational situation within a CTBT context.

In closing we note that events located by local networks as analyzed here provide an interesting potential for extending the database of Ground Truth events.

Acknowledgments

The NEONOR project, providing the local microearthquake networks used in this paper, has been supported by the Research Council of Norway, Amoco, Norsk Hydro, Phillips Petroleum, Statkraft, the Norwegian Petroleum Directorate, the Norwegian Geological Survey and NORSAR.

References

ATAKAN, K., LINDHOLM, C. D., and HAVSKOV, J. (1994), *Earthquake Swarm in Steigen, Northern Norway: An Unusual Example of Intraplate Seismicity*, Terra Nova *6*, 180–194.

BLYSTAD, P., BREKKE, H., FÆRSETH, R. B., LARSEN, B. T., SKOGSEID, J., and TØRUDBAKKEN, B. (1995), *Structural Elements of the Norwegian Continental Shelf, Part II: The Norwegian Sea Region*, NPD Bulletin, 8.

BONDÁR, I., and NORTH, R. G. (1999), *Development of Calibration Techniques for the Comprehensive Nuclear-Test-Ban Treaty (CTBT) International Monitoring System*, Phys. Earth. Plan. Int. *113*, 11–24.

BUNGUM, H., ALSAKER, A., KVAMME, L. B., and HANSEN, R. A. (1991), *Seismicity and Seismotectonics of Norway and Surrounding Continental Shelf Areas*, J. Geophys. Res. *96*, 2249–2265.

BUNGUM, H., HOKLAND, B. K., HUSEBYE, E. S., and RINGDAL, F. (1979), *An Exceptional Intraplate Earthquake Sequence in Meløy, Northern Norway*, Nature *280*, 32–35.

BYRKJELAND, U., BUNGUM, H., and ELDHOLM, O. (2000), *Seismotectonics of the Norwegian Margin*, J. Geophys. Res. *105*, 6221–6236.

HANSEN, R. A., BUNGUM, H., and ALSAKER, A. (1989), *Three Recent Larger Earthquakes Offshore Norway*, Terra Nova *1*, 284–295.

HICKS, E., BUNGUM, H., and LINDHOLM, C. D. (2000a), *Seismic Activity, Inferred from Crustal Stresses and Seismotectonics in the Rana Region, Northern Norway*, Quaternary Sci. Rev., *19*, 1423–1436.

HICKS, E., BUNGUM, H., and LINDHOLM, C. D. (2000b), *Crustal Stresses in Norway and Surrounding Areas as Derived from Earthquake Focal Mechanism Solutions and in-situ Stress Measurements*, Nor. Geol. Tidsskr., in press.

LIENERT, B. R. E., BERG, E., and FRAZER, L. N. (1986), *Hypocentre: An Earthquake Location Method Using Centered, Scaled and Adaptively Damped Least Squares*, Bull. Seismol. Soc. Am. *76*, 771–783.

LINDHOLM, C. D., BUNGUM, H., HICKS, E., and VILLAGRAN, M. (2000), *Crustal Stress and Tectonics in Norwegian Regions Determined from Earthquake Focal Mechanisms*, Geol. Soc. Lond., Spec. Publ. *167*, 429–439.

MUIR WOOD, R. (1989), *The Scandinavian Earthquakes of 22 December 1759 and 31 August 1819*, Disasters *12*, 223–236.

RINGDAL, F., and KVÆRNA, T. (1989), *A Multichannel Processing Approach to Real Time Network Detection, Phase Association and Threshold Monitoring*, Bull. Seismol. Soc. Am. *79*, 1927–1940.

(Received June 20, 1999, revised December 10, 1999, accepted February 9, 2000)

Pure appl. geophys. 158 (2001) 143–171
0033-4553/01/020143-29 $ 1.50 + 0.20/0

Pure and Applied Geophysics

A Study of Small Magnitude Seismic Events During 1961–1989 on and near the Semipalatinsk Test Site, Kazakhstan

VITALY I. KHALTURIN,[1] TATYANA G. RAUTIAN[1] and PAUL G. RICHARDS[2]

Abstract—Official Russian sources in 1996 and 1997 have stated that 340 underground nuclear tests (UNTs) were conducted during 1961–1989 at the Semipalatinsk Test Site (STS) in Eastern Kazakhstan. Only 271 of these nuclear tests appear to have been described with well-determined origin time, coordinates and magnitudes in the openly available technical literature. Thus, good open documentation has been lacking for 69 UNTs at STS.

The main goal of our study was to provide detections, estimates of origin time and location, and magnitudes, for as many of these previously undocumented events as possible. We used data from temporary and permanent seismographic stations in the former USSR at distances from 500 km to about 1500 km from STS. As a result, we have been able to assign magnitude for eight previously located UNTs whose magnitude was not previously known. For 31 UNTs, we have estimated origin time and assigned magnitude — and for 19 of these 31 we have obtained locations based on seismic signals. Of the remaining 30 poorly documented UNTs, 15 had announced yields that were less than one ton, and 13 occurred simultaneously with another test which *was* detected. There are only two UNTs, for which the announced yield exceeds one ton and we have been unable to find seismic signals.

Most of the newly detected and located events were sub-kiloton. Their magnitudes range from 2.7 up to 5.1 (a multi-kiloton event on 1965 Feb. 4 that was often obscured at teleseismic stations by signals from an earthquake swarm in the Aleutians).

For 17 small UNTs at STS, we compare the locations (with their uncertainties) that we had earlier determined in 1994 from analysis of regional seismic waves, with ground-truth information obtained in 1998. The average error of the seismically-determined locations is only about 5 km. The ground-truth location is almost always within the predicted small uncertainty of the seismically-determined location.

Seismically-determined yield estimates are in good agreement with the announced total annual yield of nuclear tests, for each year from 1964 to 1989 at Semipalatinsk.

We also report the origin time, location, and seismic magnitude of 29 chemical explosions and a few earthquakes on or near STS during the years 1961–1989.

Our new documentation of STS explosions is important for evaluating the detection, location, and identification capabilities of teleseismic and regional arrays and stations; and how these capabilities have changed with time.

Key words: Nuclear explosions, ground-truth data, chemical explosions, seismic discrimination, Semipalatinsk Test Site.

[1] Institute of Physics of the Earth, Russian Academy of Sciences, Moscow.

[2] Lamont-Doherty Earth Observatory, Palisades, NY 10964, and Department of Earth and Environmental Sciences, Columbia University, U.S.A. E-mail: richards@LDEO.columbia.edu

Introduction

It has been reported in recent official Russian publications (MIKHAILOV *et al.*, 1996; USSR NUCLEAR TESTS, 1997) that a total of 340 underground nuclear tests (UNTs) were carried out on the Semipalatinsk Test Site (STS) from 1961 to 1989. Only 279 of them had been included in previously published lists of Soviet underground nuclear explosions that included purportedly accurate origin times and locations (specifically, the lists contained in BOCHAROV *et al.*, 1989; RINGDAL *et al.*, 1992; and LILWALL and FARTHING, 1990). For eight of these 279 explosions, the magnitudes have not been available. So accurate epicenter parameters of 61 UNEs and the magnitudes of 69 UNEs appear not to have been given previously for this test site.

The main goal of this paper is to estimate the origin time and location, and to assign the magnitude, for as many of the 69 hitherto undocumented UNTs at STS as possible. Our analysis is based principally upon seismic observations using regional stations located in Kazakhstan and elsewhere in Central Asia and southwest Siberia. We also evaluate the accuracy of locations for small UNTs at STS, as determined from regional seismic signals.

Besides underground nuclear explosions, our paper includes information about chemical explosions and earthquakes which have been detected on and near STS. Their parameters also were obtained from data of regional stations, and in some cases teleseismically. It is of interest, that some of these earthquakes and chemical explosions were included in some lists of Soviet underground nuclear explosions published in the West in the mid-1980s before Russian announcements about Soviet nuclear explosions were made beginning in 1989.

It is important to develop thorough documentation of all nuclear explosions, and especially for small explosions, as an aid in evaluating the detection and identification capability of monitoring stations. Of course, explosion monitoring in the present and the future will typically be done using stations that differ from those we have used to document small explosions at STS. Nevertheless our database of small explosions (chemical and nuclear), and nearby earthquakes, can provide guidance in estimating the capability of current networks, which can be expected to be better than the capability that was available for much of the period of active nuclear testing.

We shall use the distinction between a nuclear test and a nuclear explosion that was adopted in the revised protocol of 1990 for the Threshold Test Ban Treaty. Thus, a single underground nuclear test (UNT) can consist of a number of different underground nuclear explosions (UNEs) provided these are carried out within a time interval not exceeding 0.1 s and within an area delineated by a circle whose diameter is less than 2 km. Explosions with a time interval longer than 0.1 s, or a distance greater than 2 km, are counted as separate tests. We note that this distinction between UNTs and UNEs has been followed in official

Russian documentation of the Soviet test program at STS, but with one exception, namely the nuclear test which was conducted at Degelen in a tunnel on January 30, 1974. (It consisted of three separate explosions. Western lists typically have given two tests separated by 4.4 seconds, but the official Russian publication states, in translation, that "This test can be classified as one test with several subexplosions, however the time difference between subexplosions was more than 0.1 sec" and we have followed the Russian listing in counting this as one UNT.)

In the following sections, first we summarize the information from Russia officially available on STS UNTs. Second we describe the seismic data and methods of analysis that we have used in the study of small events. Third we describe our regional seismic detections of small events on and near STS, and their locations and magnitudes. Fourth we discuss the accuracy of their seismically-determined locations by making comparisons with ground-truth information given by LEITH (1998). Fifth we describe the agreement between seismically-estimated yields, and announced yield information; and we comment on detection thresholds for a small number of different networks.

1. Summary of Available Official Information about UNTs from STS

1.1. General Information about all UNTs from STS

The boundaries of the Semipalatinsk Test Site (STS) were defined and communicated to the US by the USSR in 1990 upon entry into force of the Threshold Test Ban Treaty. The information the USSR gave the US (now filed in the library of the US Department of State) consists of a set of 152 marker locations (latitude, longitude) on the perimeter of this test site, together with a map showing the boundary line from one marker to the next. As indicated in Figure 1, STS is about 110 km from east to west, and 170 km from north to south.

The 340 UNTs at STS listed by MIKHAILOV *et al.* (1996) and USSR NUCLEAR TESTS (1997) were each associated with one of three subareas of the test site. Thus, 209 UNTs were in the Degelen subarea, 105 at Balapan (sometimes referred to as Shagan), and 26 at Murzhik (sometimes referred to as Konystan). These explosions covered a wide range in yield, from less than 1 ton up to 165 kilotons (kt). Among 96 UNTs with magnitude m_b less than 5.0, 84 were at Degelen, only 7 at Balapan, and 5 at Murzhik, so about 88% of the smaller yield events were in tunnels at Degelen. All Balapan and Murzhik tests were in shafts.

The origin time and coordinates (latitude, longitude and depth) of STS UNTs have so far been announced by Soviet/Russian sources only for a group of 96 events that were conducted during a period from October 1961 to December 1972 (BOCHAROV *et al.*, 1989; see also VERGINO, 1989). Among these 96 UNTs, 6 were small and were not mentioned by LILWALL and FARTHING (1990) or by RINGDAL *et al.* (1992). With the official Russian announcements of 1996–1997 it became clear

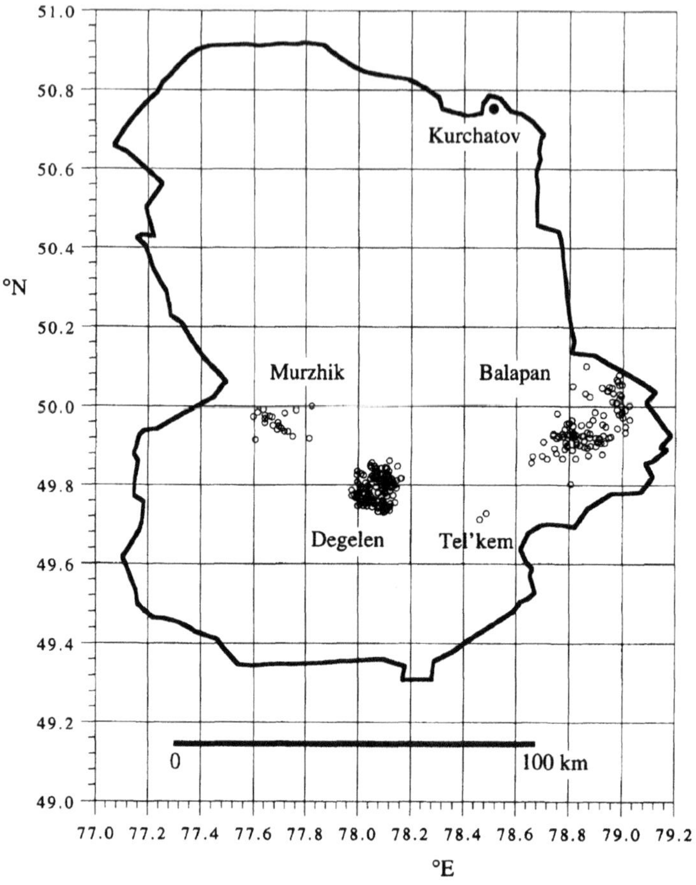

Figure 1
The boundary of the Semipalatinsk Test Site (STS) is shown, together with the location of 279 previously well-documented underground nuclear tests (UNTs) at this test site, in the three subareas of Degelen, Murzhik, and Balapan. The two UNTs shown near (49.7°N, 78.4°E) were tests of the capability to build canals, and are counted with the Murzhik events. The town of Kurchatov, headquarters for many activities on this test site, lies on the northeast.

that 20 small magnitude tests at STS during this 1961–1972 time period had not been included in the list of BOCHAROV *et al.* (1989).

In 1992, the Russian Federation declassified information about the dates on which Soviet UNTs had occurred, the number of tests and number of nuclear explosions carried out within one nuclear test, the yield range, the subarea, and the purpose of these UNEs. However, the origin time, coordinates, and yield of most Soviet UNEs are still unavailable; and their seismic magnitudes, as determined from Soviet seismographic networks have not been announced, either by the network operated by scientists, known as ESSN, or by the military network, known as SSK.

A preliminary list of Soviet nuclear explosions was published by GORIN *et al.* (1994) in a scientific journal. Two official documents, MIKHAILOV *et al.* (1996) and USSR NUCLEAR TESTS (1997), contain information about the date, environment, purpose and yield range and identification numbers (from #1 to #715) for all 715 nuclear tests carried out by the Soviets, including for 340 UNTs conducted at the STS. In these latter two publications, yields were given for nuclear explosions conducted off the two weapons test sites at Novaya Zemlya and Semipalatinsk. Yields were also given for 27 UNTs at STS. For each of the remaining 313 STS UNTs, one of three yield ranges is given for the test: either, less than 1 ton (of TNT equivalent); or, from 1 ton to 20 kt; or, from 20 kt to 150 kt. Yields at STS are announced as greater than 150 kt, for only two tests: November 2, 1972, with yield 165 kt and m_b 6.16; and July 23, 1973, with yield range 150–1,500 kt and m_b 6.17. These magnitudes were assigned by RINGDAL *et al.* (1992) using the procedure of LILWALL *et al.* (1988) developed by the British Atomic Weapons Establishment, and we refer later to such magnitudes as m_b (AWE). Both these tests took place prior to the date (March 31, 1976) given in the Threshold Test Ban Treaty, negotiated in 1974, for imposition of a 150 kt threshold.

Apart from the explosion locations given by BOCHAROV *et al.* (1989), Soviet and Russian publications have not listed UNT coordinates. But within the framework of Kazakhstan – US cooperation, coordinates of tunnel portals at Degelen have become available (LEITH, 1998). Separately, the coordinates of Balapan shafts have been obtained through field work conducted by the National Nuclear Centre of the Republic of Kazakhstan, and these locations are also now available (NNCRK, 1999). We use ground-truth information in Tables below whenever these locations are available for specific explosions. When ground truth is absent (for example for chemical explosions) we give coordinates determined by seismological methods – which are shown to be quite accurate in a later section of this paper.

1.2. Small UNTs at STS Previously Undocumented by Western Seismologists in the Open Literature

In this section we identify 61 UNTs at STS, out of the 340 now officially announced, for which accurate location information has not been given in openly available publications so far as we are aware; and we identify 69 for which accurate magnitude information has not been given. We assign each of these 69 UNTs to a category that indicates why their documentation has been poor (for example, low yield, or occurrence at the same time as another UNT). The following sections then report our own efforts to acquire and generate additional information, including locations and magnitudes, for as many of these 69 UNTs as possible.

Thus, the International Seismological Centre (ISC) has reported the seismically-determined location and magnitudes of 271 UNTs at STS. Many researchers have carried out additional levels of analysis based upon ISC data for subsets of the STS events listed by the ISC. One of the largest such efforts, by the British Atomic Weapons Establishment (AWE), has applied the Joint Epicenter Determination method described by DOUGLAS (1967) to ISC data, using several UNTs at STS as master events for which ground truth information was given by BOCHAROV *et al.* (1989) (and in English translation by VERGINO, 1989). The AWE location estimates are given by LILWALL and FARTHING (1990). AWE has also obtained maximum likelihood m_b's for 239 STS UNTs and has made them widely available on an informal basis. AWE m_b's for 100 UNTs and one chemical explosion in the Balapan subarea were published by RINGDAL *et al.* (1992). An additional 8 UNTs, not mentioned by LILWALL and FARTHING (1990) or RINGDAL *et al.* (1992), are given with locations but not magnitudes by BOCHAROV *et al.* (1989) and VERGINO (1989). On the basis of comparisons with SPOT locations (e.g., THURBER *et al.*, 1993) and recently available ground-truth information, we believe the AWE locations for 271 UNTs at STS, based on re-analysis of ISC data, are accurate to within a few km. In Figure 1, we show a map of the Semipalatinsk Test Site boundaries (as reported by the Soviet Union at the time of TTBT entry-into-force in 1990), together with the locations of 279 UNTs with coordinates given by the publications cited in this paragraph.

We can give three reasons why specific UNTs at STS were not included in the ISC and subsequent AWE listings. First, some UNTs have now been announced as having had yield less than 1 ton; such tests would generally be too small for either regional or teleseismic detection. Second, some UNTs were carried out at essentially the same time as another UNT and only one test was reported. Third, some UNTs have now been announced as having had yield greater than 1 ton, but they may still have been too weak for teleseismic detection with high confidence, given the networks in operation at the time. For these events, we can inquire as to the possibility of regional detection as discussed in the following sections. [Note that the papers by LILWALL and FARTHING (1990), and RINGDAL *et al.* (1992) were not intended as reports on teleseismic detection capability, and they characterized UNTs only for which the teleseismic data were of high quality. We comment below on papers by SYKES and RUGGI (1986, 1989) and RINGDAL (1990) which reported the occurrence of several UNTs at STS additional to those listed by the ISC. Such detections were useful and important but they were not associated with accurate location estimates so we continue to include them in this paper with what we call "previously undocumented" events.]

Using information from the official Russian publications, we can tentatively see for each of these three reasons how many UNTs were not included in lists of events accurately located by seismic methods:

1.2.1. Weak UNTs with yield Y announced as less than 1 ton

This category consists of the 15 UNTs listed in Table 1. They would not be detected by standard instruments at distances more than 100–150 km. One of these

Table 1

List of weak UNTs at STS with $Y < 1$ ton, which could not be detected even at typical regional distances.

N	Date	N	Date
1	1968 May 23	9	1979 Apr 10
2	1970 Feb 18	10	1979 Jun 12
3	1972 Apr 20	11	1980 Mar 14
4	1973 Sep 20	12	1981 Mar 25
5	1974 Feb 28	13	1981 Jun 04
6	1978 May 24	14	1981 Oct 16
7	1978 Jun 02	15	1983 Mar 11
8	1979 Mar 23		

small UNTs, with yield *Y* less than 1 ton, was carried out at Balapan (1973, Sep. 20); the other 14 were carried out in the Degelen subarea.

1.2.2. Pairs of UNTs exploded simultaneously

This category is concerned with pairs of tests carried out within a short time interval, or even simultaneously, but with a spatial and/or temporal interval that requires them to be listed as different tests. In order to discuss specific UNTs with the same date, we use the number for each test that appears on official Russian lists.

Thus, we note that in the official lists there are 19 pairs exploded on the same day. Two of them were on the same day but are known to be separated by a long time interval: #414 and #415 (December 16, 1974, Degelen) with more than three hours time interval; and #440 and #441 (April 21, 1976, at Degelen and Balapan and hence with a significant spatial separation) with a four-minutes interval.

For the remaining 17 pairs of UNTs, only for four pairs were both tests detected and reported in the standard western publications. These tests were conducted on Dec. 10, 1972 (#376 and #377, time interval 10 sec), on Oct. 29, 1977 (#473 and #474, time interval 4.9 sec), on Aug. 29, 1978 (#493 and #494, time interval 8.8 sec) and Nov. 29, 1978 (#506 and #507, time interval 4.8 sec). In all four cases, the two tests were carried out in different subareas (Balapan or Degelen) at separations of 50 km and more, which significantly facilitated their identification as test pairs.

The last 13 pairs of UNTs are shown in Table 2. Among them are 12 pairs for which just one UNT (for each pair) was reported by ISC; and one pair (Dec. 5, 1980, #561 and #562) for which neither test was reported. In the last column of Table 2 are the numbers of the 14 UNTs unreported by the ISC. Only for one pair of UNTs (1980 Dec. 5) were both tests unreported by the ISC, so one event from this pair potentially can be detected. The remaining 13 UNTs could not be detected by standard methods.

1.2.3. Small UNTs with yield more than 1 ton, not reported by the ISC

These events, listed in Table 3, are most interesting for us because potentially they can be detected at regional distances. They are the main object of our investigation.

Table 2

13 pairs of UNTs at STS which were exploded simultaneously or with very short time interval.

N	Date	Subarea	Detected test #	m_b	Undetected test #
1	1970 Jun 28	Both Degelen	321	5.7	322
2	1970 Sep 06	Both Degelen	325	5.4	326
3	1971 Mar 22	Both Degelen	333	5.7	334
4	1971 Dec 30	Both Degelen	354	5.7	353
5	1972 Jun 07	Both Degelen	360	5.4	361
6	1975 Feb 20	Both Degelen	417	5.7	418
7	1976 Dec 07	Both Balapan	454	5.9	455
8	1977 Mar 29	Degelen	457	5.4	–
		Murzhik	–	–	458
9	1977 Dec 26	Both Degelen	479	4.9	480
10	1979 Jul 18	Murzhik	524	5.2	–
		Delegen	–	–	525
11	1980 Dec 05	Both Degelen	–	–	561 & 562
			Both tests were undetected		
12	1983 Nov 29	Both Degelen	629	5.4	630
13	1987 Apr 03	Balapan	671	6.1	–
		Degelen	–	–	672

Table 3

List of 33 separate UNTs (Y > 1 ton) which were not reported in standard western publications, but which potentially can be detected at regional distances.

N	Date	Subarea	N	Date	Subarea
1	1964 Jun 06	Deg	18	1974 Jul 29	Bal
2	1964 Aug 18	Deg	19	1974 Nov 28	Mur
3	1964 Sep 30	Deg	20	1975 Jul 15	Deg
4	1965 Feb 04	Deg	21	1975 Oct 05	Deg
5	1965 Mar 27	Deg	22	1976 Mar 17	Deg
6	1966 Oct 29	Deg	23	1976 Apr 10	Deg
7	1966 Nov 19	Deg	24	1976 Aug 04	Mur
8	1967 Sep 02	Deg	25	1977 Nov 12	Bal
9	1968 Oct 29	Deg	26	1977 Nov 27	Deg
10	1969 Apr 04	Deg	27	1980 Jun 25	Deg
11	1969 Apr 13	Deg	28	1980 Oct 23	Deg
12	1969 Oct 30	Deg	29	1980 Dec 05	Deg
13	1969 Nov 27	Deg	30	1983 Nov 02	Deg
14	1971 Jan 29	Deg	31	1985 Jul 11	Deg
15	1971 Apr 09	Deg	32	1985 Jul 19	Deg
16	1973 Nov 04	Bal	33	1988 Dec 28	Deg
17	1973 Dec 31	Deg			

We note that a total of 61 previously undocumented UNTs are given in Tables 1, 2, and 3, namely, 15 UNTs with yield less than 1 ton; 13 UNTs that occurred at essentially the same time as another UNT, including one of the pair #561 and #562; and 33 UNTs which potentially can be detected at regional distances, including the other of the pair #561 and #562. In section 3 below, we report our locations and magnitudes, based on regional detections, for most of these 33 UNTs.

2. Regional Data and Methods of Analysis for Small-magnitude Seismic Events from STS

Our work on this subject was carried out in two stages. The first, in late 1993, resulted in a technical report (KHALTURIN *et al.*, 1994), produced prior to the publication of the first Russian preliminary list of Soviet UNTs (GORIN *et al.*, 1994). Information about the dates on which UNTs occurred, and ground-truth locations, were not then available for us. We used regional data, and tried to detect and locate all seismic events at STS which could be UNTs, chemical explosions or earthquakes. The second stage was carried out in 1997–1999, in light of official information on UNT dates and acquisition of ground-truth locations. Naturally, the magnitude threshold of detected signals in the first stage was significantly higher than in the second stage.

2.1. System of Observations. Stations and Instrumentation

Our results are based mainly on seismic data acquired by the Complex Seismological Expedition (CSE) of the Institute of the Physics of the Earth, Russian Academy of Sciences. Also we used bulletins of other regional stations of Central Asia including the Altai region. In total we used the records or bulletins of more than 50 seismographic stations. Most of them were operated on a temporary basis, and typically we used data from about 3 to 6 stations for each event to estimate its location. Most useful for detecting and locating small magnitude UNTs, were seismograms of narrow-band short-period instruments installed in several stations by CSE in North Kazakhstan at distances of 500 km to 1200 km from STS. Long-term CSE observations in this region show that high-frequency regional phases propagate very efficiently. Also very helpful for detection and discrimination were records of "multichannel frequency selected stations," known from their Russian acronym as "ChISS," installed at the base stations Zerenda, Talgar, Novosibirsk (all in the distance range 740–780 km from STS) and Garm (1350 km from STS). These stations included the ChISS set of 8 or 12 channels from 0.5 to 45 Hz or from 0.025 to 45 Hz. Base stations also operated instruments as characterized in Table 4 and Figure 2.

Two different types of high-gain short-period instruments – RVZT and CSE (see Table 4 and Fig. 2) – were developed for detection of low-magnitude UNEs. Several

Table 4

Main parameters of instrumentation installed at base stations and temporary stations of CSE.

Type	Abr.	T1–T2, sec	Magnification
Broadband, long period	SKD	0.2–18	1,000 to 1,500
Broadband, short period	SKM	0.07–1.4	30,000 to 60,000
Narrow-band	RVZT	0.2–1.2	100,000 to 300,000
Narrow-band, high gain	CSE	0.7–1.1	500,000 to 1,100,000

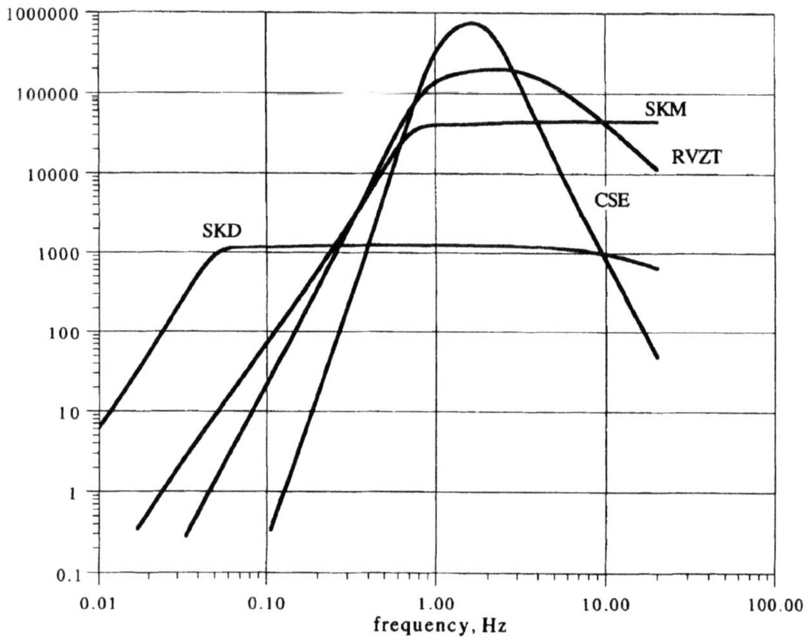

Figure 2
Responses of standard seismographic instruments used by the Complex Seismological Expedition for regional recordings of UNTs at STS and other regional seismic events.

places were found (the best ones were on the Kokchetav massif in northern Kazakhstan) where magnification could be set as high as one million, with noise amplitudes only about 1 mm on the paper record, even though the attenuation of regional waves was very low. Figure 3 gives an example of such a high gain record, for an event not previously documented by standard western publications. It is the UNT of January 29, 1971. This signal was recorded by a CSE instrument located near the site of the present-day Zerenda broadband station (ZRN) on the Kokchetav massif, at a distance of 720 km with S/N ratio about 150–200.

ChISS instruments were installed in Talgar and Garm in September 1961, prior to the first UNT conducted at STS. Later, ChISS instruments were installed by the

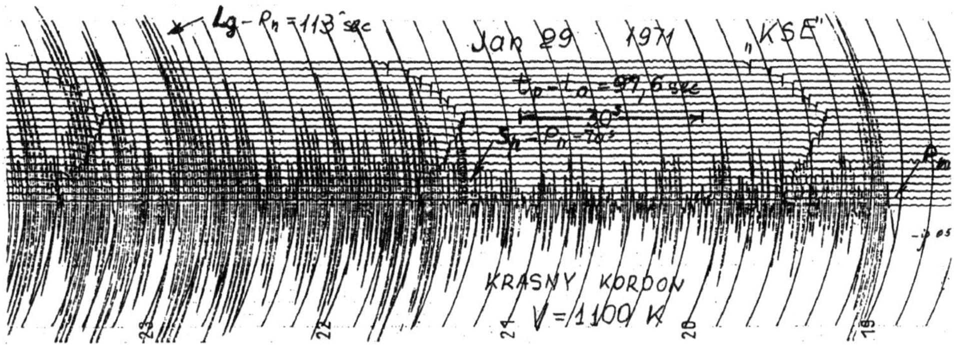

Figure 3

Example of a low-noise short-period record with a gain of 1.1 million at Krasny Kordon, of a previously undocumented UNT at Degelen on Jan. 19, 1971; m_b 4.47 at a distance of 720 km. Time runs from right to left in these and other analog records from the Soviet Union.

Complex Seismological Expedition (CSE) in Zerenda and Novosibirsk. They operated until the end of 1990. Figure 4 gives a ChISS record of the first Soviet underground nuclear explosion, at STS on October 11, 1961. These historic data are of remarkably high quality. ChISS records such as these are very effective even today for quantifying frequency-dependent features of regional seismic wave propagation. After 1969, ChISS instruments with ink-pen recording were installed by CSE for military seismic surveys in several places including Semipalatinsk, Makanchi (East Kazakhstan), Mongolia, Mayli-Say (East Fergana), and Malin (Northern Ukraine). The main goal was detection and prompt discrimination of the signals from underground and atmospheric nuclear explosions from non-Soviet test sites (Chinese nuclear testing in the atmosphere continued up to 1980). Discrimination was based on measurement of *P*-wave amplitudes on high frequency channels, and *S*, *Lg*, and *Rg* on low frequency channels. Figure 5 shows regional phases on the ChISS record at Talgar of a Degelen UNT, $m_b = 5.08$. On the different narrowband channels the calibration signal can be clearly seen, with frequency changing slowly from 40–45 Hz down to 0.3–0.5 Hz and thus appearing on different channels at different times.

2.2. Regional Phases Observed from UNEs at STS

The following regional phases are observed from the Semipalatinsk underground nuclear tests at distances up to 1200 km:

- *Pg* (6.15 km/s), *Sg* (3.55 km/s) and *Rg* (2.6–2.8 km/s) observed up to 230–250 km;
- *Pn*, appearing beyond 230–250 km as the first arrival with velocity 8.1 km/s. Its velocity stays constant up to distances of 800–900 km and then starts to increase slightly. Following *Pn* out to 800–900 km, *Pg* is observed with velocity 6.2 km/s.
- *Sn* is also observed (4.7 km/s) beyond about 240 km, but it is weak and can be clearly detected only on 60–70% of the records.

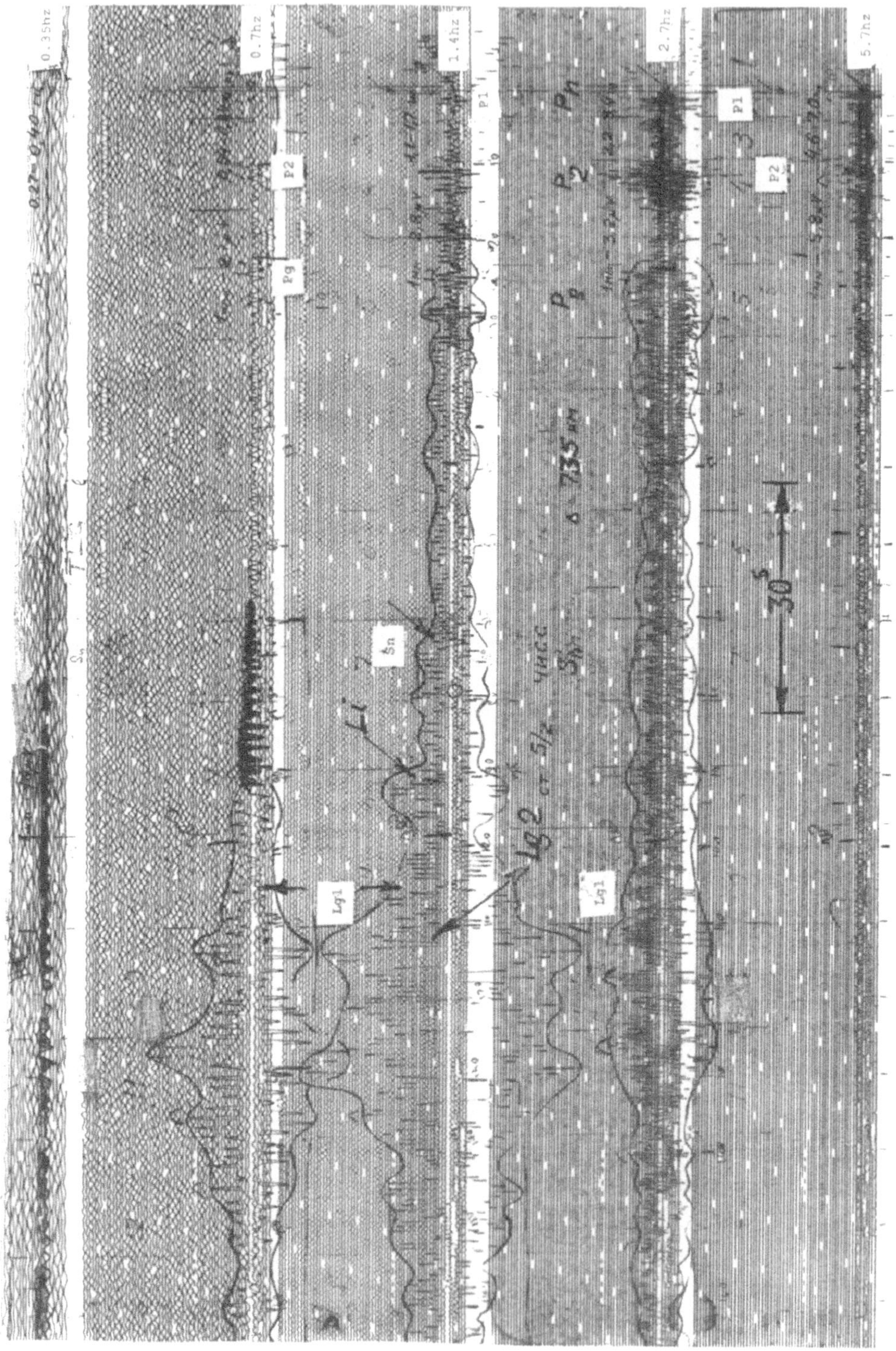
0.35hz
0.7hz
1.4hz
2.7hz
5.7hz
P1
P2
Pg
Sn
Lg1
Lg2
30s

• A very intensive *Lg* group is observed beyond about 240 km with an impulsive onset. Usually the amplitude of *Lg* is 4 to 8 times bigger than amplitudes of *Pn* or *Pg*. Beyond 400–500 km the *Lg* wave train consists of two groups, denoted as Lg_1 and Lg_2, with velocities 3.55 and 3.40 km/s.
• An intensive *Rg* wave train is observed for UNEs with magnitudes greater than 5 on the long-period records. The train is short and consists of 1.5 to 2 cycles without clear dispersion. The velocity of the apparent first arrival is 3.0 km/s for periods 7–10 s.

Figure 5, showing five ChISS channels recorded at distance 730 km for a Degelen UNT of magnitude 5.08, indicates that the phases *Pg*, *Pn* (marked as P_1), *Sn*, Lg_1 and Lg_2 can be clearly seen, often with impulsive arrivals.

KHALTURIN *et al.* (1994) describe how these regional travel-time distance relations were used to estimate the location and location accuracy of small seismic events at STS.

2.3. Magnitude Determination

Two methods were used to assign magnitudes (m_b) from regional data. The first method was based on the *K* scale (energy class), which is still used in the former Soviet Union by analysts at all local networks, for characterizing the size of seismic events using data acquired at distances up to 3000 km. The energy class *K* is calculated from the sum of max *P* amplitude, and max *S* or *Lg* amplitude, on records of the short-period instrument SKM. For UNEs and chemical explosions at STS, KHALTURIN *et al.* (1998) report the following relationship between *K* values, and m_b values as given by the British Atomic Weapons Establishment and by NORSAR:

$$m_b(K) = 0.46\ K - 0.64.$$

We used this relationship in the present study, to assign m_b values when the *K* value was available.

◄

Figure 4
The ChISS record at Talgar of the first underground nuclear explosion at the Semipalatinsk Test Site, on October 11, 1961. Distance is 735 km. The phase marked here as P_1 is the *Pn* wave, and is strong at high frequencies. It even appears clearly on the 11 Hz channel (not shown here). The phase marked as P_2, between P_1 and *Pg*, is strong at Talgar for Degelen explosions. P_2 is three times smaller at Talgar for Balapan explosions. *Lg* waves are strong, particularly on the 1.4 Hz and 0.7 Hz channels, with amplitude about 6 to 10 times that of the *P* wave – a typical ratio for UNE records at many stations around STS at distances 500–1300 km. *Lg* is weak on channels (not shown here) of frequency lower than 0.7 Hz, in contrast to the strong low-frequency *Lg* of typical earthquakes. The magnitude value 4.78 is supported by several stations. Using the m_b – yield relation of RINGDAL *et al.* (1992), this magnitude corresponds to a yield of 2.7 kilotons – somewhat higher than the announced yield of 1 kiloton for this event.

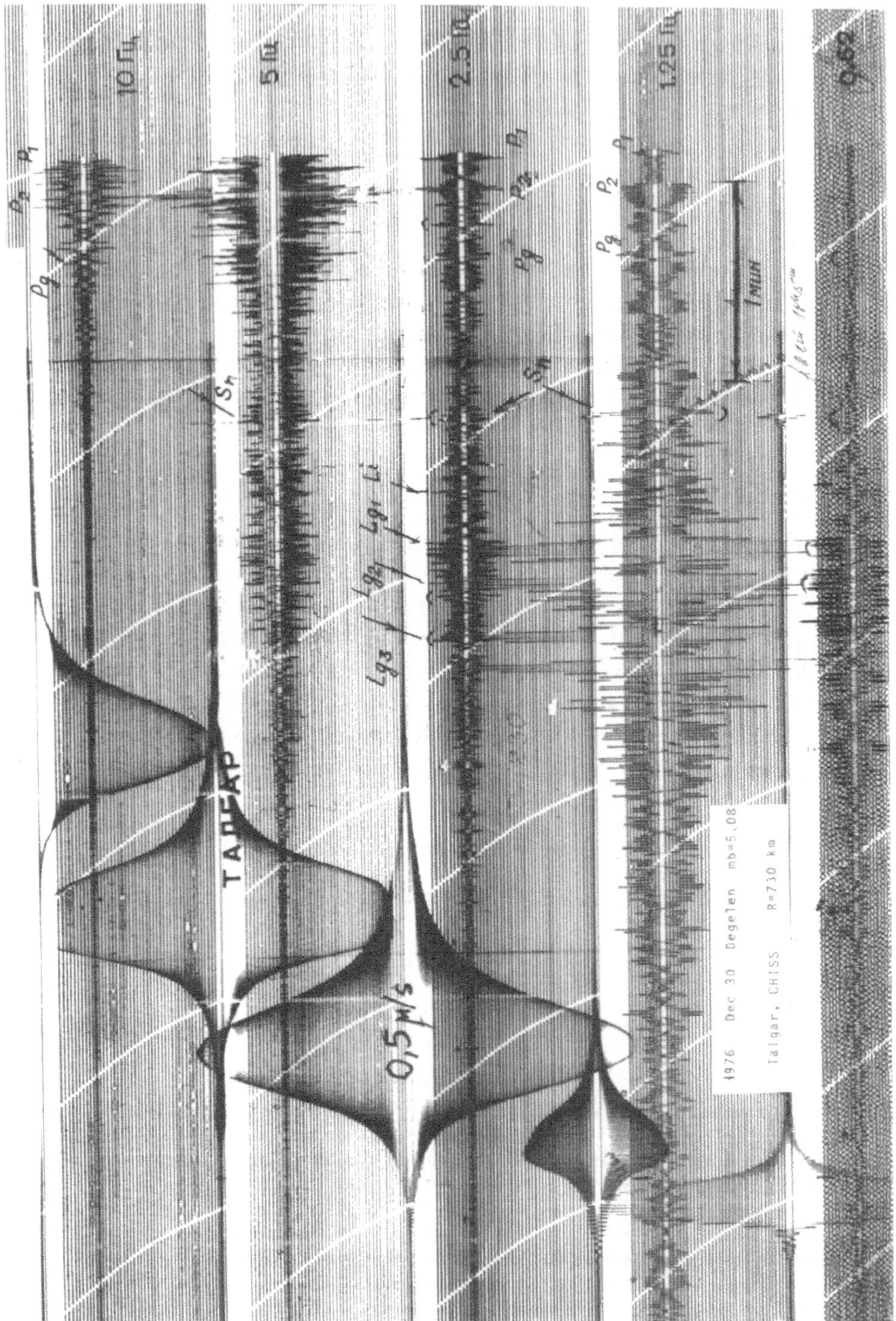

10 Гц
5 Гц
2.5 Гц
1.25 Гц
ТАЛГАР
0,5 м/s
1 мин
1976 Dec 30 Degelen mb=5.08
Talgar, CHISS R=730 km

The second method to assign m_b was based on calibration of a measurement of the maximum amplitude of *Lg* waves. This scale was typically applied to the narrow-band records. The resulting magnitude is denoted as $m_b(Lg)$.

3. Detection of Small Events from STS from Regional Recordings in Kazakhstan and Central Asia

3.1. Monitoring of Seismic Signals from STS: Detection and Identification

The seismographic network operated by CSE was used to acquire observations in the Kazakhstan region throughout the period of UNT activity at STS – from 1961 to 1989. During the long-term monitoring effort, besides the well-known intermediate and large magnitude UNTs from STS, several tens of small magnitude events were detected that were not mentioned by LILWALL and FARTHING (1990) or RINGDAL *et al.* (1992). These events can be UNTs, or they could be chemical explosions used for military experiments and for construction. Few of these signals can be from earthquakes, which are very rare in the Semipalatinsk region since it is located on the far western flank of the Altai seismic zone.

Our first stage of study (KHALTURIN *et al.*, 1994) examined data for 57 of these events that were on or near STS; estimated their coordinates, origin time, and magnitude; and made a preliminary identification as to the nature of each event (nuclear or chemical explosion, or earthquake). We now know that these 57 events consisted of 19 UNTs, 27 chemical explosions, 8 small magnitude UNTs known from BOCHAROV *et al.* (1989), and three earthquakes. Our first stage identified all of the UNTs and earthquakes correctly, but wrongly listed two of the chemical explosions as UNTs, and two other chemical explosions as "either UNE or chemical explosion."

Our second stage of study, in this paper, done following the release of UNT date information, has examined data for 71 events on or near STS, and has resulted in estimates of the origin time and magnitude of an additional 12 small UNTs which were missed in the first stage. So, from the 33 previously undocumented UNTs of Table 3, our first stage of study uncovered 19 UNTs and the present paper documents another 12. The present paper also includes two more chemical explosions than our first study: that of September 15, 1984 (previously listed by

◄

Figure 5

The ChISS record at Talgar, of the Degelen UNE of December 30, 1976, with m_b 5.08. Note the strong *Pn* arrival (marked as P_l) on 10 Hz and 5 Hz channels, whereas *Lg* is strongest on the 1.25 Hz channel. A calibration signal is given for all these narrowband channels, corresponding to a sinusoidal ground velocity (with slowly varying period) whose amplitude has the constant value of 0.5 micron/s.

some papers as a UNT), and the chemical explosion carried out in 1987 as part of a Joint US-Soviet Experiment of the USSR Academy of Sciences and the US Natural Resources Defense Council (see GIVEN *et al.*, 1990).

3.1.1. Earthquakes near the Semipalatinsk Test Site

Among the 71 regionally-recorded events were three earthquakes, given in Table 5, which occurred near the border of STS or in the surrounding area. This Table also lists three more recent earthquakes in the region.

In Table 5, event #1 was described by KHALTURIN *et al.* (1994). Event #2 was widely detected, and it is still sometimes listed as a UNT though the seismological basis for identifying it as an earthquake was given several years ago (POOLEY *et al.*, 1983). Event #3 was reported by the ISC (47.9 N, 83.5 E, m_b 4.5) and listed by SYKES and RUGGI (1989) as a Soviet UNE at (50 N, 79 E) with m_b 3.6. Events ##4–6 occurred near STS after the period of our study. Event #5 was reported in the US Geological Survey's Preliminary Determination of Epicenters. Events #4 and 6 have been reported by the Altay-Sayan Seismological Expedition, based in Novosibirsk.

3.1.2. Chemical explosions at the Semipalatinsk Test Site

The remaining 60 regionally-recorded events have all the characteristics of explosions. Comparing their origin times with the list of UNT dates contained in Russian official publications issued in 1996–1997 we now conclude that 29 were chemical explosions at STS, and 31 were indeed UNTs at STS. Parameters of the 29 chemical explosions are given in Table 6. More than half of these chemical explosions were assumed in some publications during the late 1980s to be UNTs from STS. It is of course appropriate to list all candidate UNEs in projects that set out to evaluate monitoring capability.

In Table 6, origin time, coordinates and *K* values were obtained from CSE observations; the magnitude value *m*(NOR) refers to magnitudes obtained from F. Ringdal (personal communication, 1994); magnitudes in parentheses ( ) were

Table 5

Earthquakes located near the Semipalatinsk Test Site during 1961–1989 (##1–3) and examples of more recent earthquakes (##4–6).

#	Date	Time	Lat.	Long.	K	m_b(K)
1.	1966 Dec 26	17:39:38.5	49.52	78.71	10.7	4.28
2.	1976 Mar 20	04:03:36	50.00	77.25	12.9	5.34
3.	1981 Mar 31	07:51:30	47.80	81.00	12.2	4.96
4.	1995 Mar 6	05:27:13	48.4	78.1	9.2	3.6
5.	1996 Mar 26	13:58:12	50.085	77.097	11.3	4.2
6.	1996 Sep 25	19:24:01	47.7	80.2	10.9	4.0

Table 6

Twenty-nine chemical explosions at the Semipalatinsk Test Site during 1961–1989 which were detected and located from CSE observations.

Date	Time	Lat.	Long.	K	m(NOR)	m_b(K)	Notes
1961 Jun 05	03:50:00	49.77	77.98	11.0	–	4.42	A
1973 Mar 23	06:30:00	49.94	79.06	9.53	3.7	3.74	
1974 Sep 27	07:34:00	49.98	79.00	10.47	4.0	4.16	
1978 Jul 31	08:00:00	50.42	77.87	10.2	3.9	4.05	B
1979 May 24	04:07:00	49.94	78.79	10.33	3.9	4.05	SR-1
1979 Sep 14	07:33:00	49.95	78.84	10.75	4.4	4.30	SR-2
1979 Sep 15	04:07:00	49.94	78.82	8.85	3.8	3.44	
1980 Jul 13	08:10:00	49.91	78.84	10.33	(5.0)	4.10	SR-3
1980 Sep 20	10:40:01	49.96	78.88	9.83	3.8	3.88	SR-4
1980 Sep 30	05:57:12	49.95	78.40	–	3.6	–	SR-5
1980 Sep 30	05:57:17	49.95	78.40	11.03	4.4	4.42	SR-6
1980 Nov 06	17:42:58	50.14	78.76	9.17	3.9	3.56	
1981 May 28	04:08:28	50.00	78.00	7.70	–	2.90	
1981 Jun 05	03:22:18	49.84	78.72	10.30	4.0	4.10	SR-7
1981 Jul 05	03:59:14	49.87	78.99	10.47	(4.6)	4.17	SR-8
1981 Sep 30	12:55:10	49.94	78.90	10.70	4.3	4.28	SR-9
1981 Nov 19	05:57:14	50.11	78.95	9.60	4.0	3.78	
1982 Jun 11	10:59:07	49.90	77.90	10.65	4.1	4.26	SR-10
1982 Jul 12	10:29:18	49.90	77.90	10.67	3.9	4.27	SR-11
1982 Sep 04	05:47:17	50.06	78.56	9.47	3.6	3.72	SR-12
1982 Sep 15	4:33:19	49.85	78.85	10.86	4.2	4.36	SR-13
1983 July 28	03:41:28	50.07	78.60	10.74	4.3	4.34	SR-14
1984 Jun 23	02:57:16	49.92	78.93	11.06	4.4	4.44	SR-15
1984 Sep 15	06:15:10	49.99	78.88	11.17	–	4.48	C and SR-16
1985 Jun 27	11:57:04	49.78	77.97	8.5	–	3.27	D
1987 Jun 29	04:55:08	49.78	77.97	8.5	–	3.27	D
1987 Sep 02	09:27:05	50.00	70.34	–	2.7	–	E
1987 Sep 16	07:30:01	49.86	78.73	10.64	4.3	4.26	SR-17
1988 Sep 26	07:45:04	50.08	78.80	10.41	4.3	4.15	F

Notes:

A — Fully contained explosion with a yield of 600 tons of TNT, carried out in a Degelen Mountain tunnel prior to the start of nuclear testing (SULTANOV *et al.*, 1995). Goals included calibration of a seismic network and estimation of the expected seismic signal strength at different distances.

B — Large experimental chemical explosion on the surface with a yield of 5000 tons.

C — This chemical explosion has been wrongly listed as an underground nuclear test, in some cases with $m_b = 5.04$; for example see LILWALL and FARTHING (1990), and RINGDAL *et al.* (1992). For this chemical explosion, $m_b(ISC) = 4.7$, $m_b(HFS) = 5.2$.

D — These two experimental chemical explosions, each of 500 tons of TNT, were conducted by the Institute of Dynamics of the Geospheres (IDG) at the same place on the surface of Degelen Mountain near the mouth of tunnel #160(49.7841°N, 77.96722°E). See ADUSHKIN *et al.* (1997).

E — Chemical explosion (yield 20 tons, depth 25 m) carried out in the Degelen subarea during the Joint US-Soviet Experiment of the USSR Academy of Sciences and the US Natural Resources Defense Council (see GIVEN *et al.*, 1990).

F — HANSEN *et al.* (1990) assumed this chemical explosion was an underground nuclear test in their study of the stability of RMS *Lg*.

SR — SYKES and RUGGI (1986, 1989) list all these 15 chemical explosions as UNTs from STS with the following m_b values: 1 – 4.9; 2 – 5.2; 3 – 5.0; 4 – 4.9; 5 – 4.6; 6 – 5.2; 7 – 4.7; 8 – 4.6; 9 – 4.6; 10 – 4.6; 11 – 4.6;

Table 6 (continued)

12 – 4.1; 13 – 5.0; 14 – 5.0; 15 – 4.4; 16 – 5.2; 17 – 5.0. Some if not all of these magnitudes are from the Hagfors Observatory (HFS). On average HFS magnitudes are larger than NORSAR magnitudes by about 0.6–0.8 magnitude units. A very weak regional signal originating on October 20, 1989 at 13:22:45 from about (50°N, 78°E) may be a chemical explosion or a collapse of the cavity from a previous UNE.

assigned by Hagfors (HFS) and are known typically to be significantly higher than m(NOR) values; and $m_b(K)$ is the body wave magnitude calculated from energy class K using the relationship $m_b(K) = 0.46\ K–0.64$.

More recently than the period of our study, chemical explosions ranging in size from a few kilograms up to 100 tons have been carried out in Kazakhstan in a cooperative program between the National Nuclear Centre of the Republic of Kazakhstan and the US Defense Threat Reduction Agency. Times and locations of such explosions larger than one ton are listed in Table 7. Both calibration explosions listed with $Y = 100$ tons were carried out at Degelen in the tunnels. Their m_b values (3.7–3.8) correspond to the upper limit of $m_b(Y)$ relationship described in KHALTURIN *et al.* (1998). All other explosions were made in Balapan in shafts.

3.2. Detection and Analysis of small UNTs from STS

Here we give the origin times, locations and magnitudes of small UNTs whose detection and or location was not previously well documented by western seismologists in the open literature. We also give magnitude values based on regional detections, for 8 small events whose hypocenter coordinates have been available from BOCHAROV *et al.* (1989).

Table 7

Calibration chemical explosions at the Semipalatinsk Test Site since 1997.

Date	Time	Lat.	Long.	Depth, m	Y, tons
1997 Jul 13	08:11.08.799	49.8786	78.7601	630	5.00
1997 Aug 03	08:07:20.04	49.9781	78.8200	50	25
1997 Aug 31	07:08:39.26	49.8837	78.8148	360	25
1997 Sep 28	07:30:15.13	49.8802	78.7587	550	25
1998 Jul 13	10.44.56.363	49.8801	78.6921	20	2.03
1998 Jul 14	05.11.35.570	49.9315	78.7871	20	2.03
1998 Jul 14	08.19.39.296	49.9129	78.7488	20	2.03
1998 Aug 14	04.26.52.813	50.0357	79.0114	13	2.03
1998 Aug 14	05.39.24.970	50.0576	78.9387	2.5	2.03
1998 Aug 15	02.40.59.116	49.8724	78.6478	14	2.03
1998 Aug 15	05.05.11.156	49.8786	78.7601	630	2.03
1998 Aug 22	05:00:18.90	49.7667	77.9908		100 ($m_b = 3.8$)
1998 Sep 17	07:19:40.44	49.9810	78.7559	30	25
1999 Sep 25	05:00:05.7	49.7841	77.8240		100 ($m_b = 3.7$)

3.2.1. Main parameters of previously undocumented small UNTs

For the 33 small UNTs listed in Table 3, we have found regional seismic detections at temporary and permanent stations of CSE for all but two events. Parameters for the 33 events are listed in Table 8. Origin times were estimated for all 31 of them. For the 19 largest of these small events, we obtained location estimates based on seismic signals, and K values and hence $m_b(K)$. For the 12 smallest events we give estimated $m_b(Lg)$ values – which range from 2.2 to 3.7.

During the course of our work, ground-truth information on locations became available from LEITH (1998) and NNCRK (1999) for 30 of the UNTs, and in section 4 we discuss the accuracy of our seismically-estimated locations. Thus, origin time, K and $m_b(Lg)$ of all UNTs, and coordinates of the 1976 Aug. 4 UNE, are our estimates based on regional observations of CSE. Coordinates of all other UNTs are ground-truth values.

One relatively large UNT, with $m_b(K)$ 5.1 (Feb. 4, 1965), was not reported by standard western publications as it was obscured teleseismically by a swarm of Aleutian earthquakes. We can be sure this was a coincidence rather than an effort to obscure the event, because the origin time (06:00:00) was typical for UNTs of the mid-1960s. But even if we exclude this large event, the m_b value (calculated from K) for missed events ranges up to 4.55, and during 1964–1989, about 10 Soviet UNTs at STS, with magnitude 4.0 or more, had teleseismic signals that were too weak or too noisy to lead to publication of good location estimates. Some of these events were detected teleseismically at particular arrays (RINGDAL, 1990).

3.2.2. Magnitude estimation of small known UNTs

Among the analyzed signals were 8 small UNEs known from BOCHAROV *et al.* (1989) but listed there without magnitudes. Four of these events had been reported using teleseismic signals by SYKES and RUGGI (1986, 1989), who also listed a magnitude for three of the events.

For these 8 events the energy class K is known from regional records at several stations, allowing us to give values of $m_b(K)$ which we list in Table 9. The other m_b values are from SYKES and RUGGI (1989).

4. Comparison of Ground-truth and Seismologically-determined Locations of Small Magnitude UNTs from STS

Our earlier study (KHALTURIN *et al.*, 1994) determined coordinates of 18 small-magnitude UNTs (one additional UNT was detected only by one station) on the basis of arrival times of regional waves, and estimated the location uncertainty, which was typically an area of the order of 100 km^2. One of those UNTs was at

Table 8

Parameters of small announced UNTs studied in this paper.

Date	Time	Subarea	Lat.	Long.	K	$m_b(K)$	$m_b(Lg)$ or m(NOR)	Note
1964 Jun 06	00:00:00	Deg	49.7747	77.9881	11.0	4.42	–	
1964 Aug 18	06:00:00	Deg	49.8206	78.0819	8.5	3.27	–	
1964 Sep 30	Not detected							
1965 Feb 04	06:00:00	Deg	49.7731	77.9914	12.5	5.10	–	A
1965 Mar 27	06:30:00	Deg	49.7747	77.9881	8.4	3.22	–	
1966 Oct 29	03:58:00	Deg	49.7847	77.9994	9.0	3.50	–	
1966 Nov 19	03:58:00	Deg	49.8297	78.0575	8.7	3.36	–	
1967 Sep 02	04:04:00	Deg	49.7419	78.0256	10.3	4.10	–	
1968 Oct 29	03:54:00	Deg	49.8333	78.0928	10.8	4.33	–	
1969 Apr 04	04:57:00	Deg	49.7533	78.0536	9.2	3.60	–	
1969 Apr 13	04:04:00	Deg	49.7356	78.1047	11.3	4.55	–	
1969 Oct 30	Not detected							
1969 Nov 27	05:02:00	Deg	49.8367	78.0597	10.3	4.10	–	
1971 Jan 29	05:03:00	Deg	49.8053	78.1686	11.1	4.47	–	
1971 Apr 09	02:33:00	Deg	49.8322	78.0386	9.6	3.78	–	B
1973 Nov 04	03:57:00	Bal	50.0716	78.0362	–	–	2.6-Lg	
1973 Dec 31	04:03:00	Deg	49.7394	78.0863	10.6	4.24	4.0-Nor	
1974 Jul 29	03:28:00	Bal	49.9375	78.9358	–	–	3.3-Lg	
1974 Nov 28	05:57:00	Mur	n/a		–	–	2.8-Lg	
1975 Jul 15	02:57:00	Deg	49.7914	78.0944	–	–	3.3-Lg	
1975 Oct 05	04:27:00	Deg	49.7831	78.0867	10.7	4.28	4.0-Nor	SR-1
1976 Mar 17	02:57:00	Deg	49.7556	78.0992	–	–	2.2-Lg	
1976 Apr 10	05:03:00	Deg	49.7550	78.0475	–	–	3.0-Lg	
1976 Aug 04	02:57:00	Mur	49.87	77.7	10.5	4.20	3.8-Nor	SR-2
1977 Nov 12	05:11:00	Bal	50.0522	78.8644	–	–	2.8-Lg	
1977 Nov 27	03:57:00	Deg	49.7544	78.0503	9.9	3.92	3.4-Nor	
1980 Jun 25	02:27:00	Deg	49.8258	78.0994	–	–	3.7-Lg	
1980 Oct 23	03:57:11	Deg	49.7517	78.1317	–	–	2.5-Lg	
1980 Dec 05	04:17:16	Deg	49.7517	78.1317	–	–	3.6-Lg	
1983 Nov 02	04:18:54	Deg	49.7792	78.1247	–	–	3.0-Lg	
1985 Jul 11	02:57:02	Deg	49.7506	78.0492	10.2	4.05	3.5-Nor	SR-3
1985 Jul 19	04:00:08	Deg	49.8011	78.0686	–	–	2.5-Lg	
1988 Dec 28	05:28:10	Deg	49.8011	78.0686	9.5	3.74	3.6-Nor	C

Notes:

$m_b(K)$ — calculation of m_b from K using the relationship: $m_b(K) = 0.46\ K - 0.64$.

m(NOR) — from F. RINGDAL (pers. comm., 1994), based on teleseismic signals at NORSAR.

A — This event was obscured by many Aleutian earthquakes, up to m_b 6.4 on that day.

B — The yield of this explosion has been announced as 0.23 KT (USSR NUCLEAR TESTS, 1997)

C — This event was mentioned by RINGDAL (1990).

SR — These three events are listed by SYKES and RUGGI (1986, 1989) with the following coordinates and magnitudes m_b:

1. 55.8N and 75.1E; $m_b = 4.6$.
2. 49.9N and 77.7E; $m_b = 4.1$.
3. 50.0N and 78.0E; $m_b = 4.0$.

Table 9

Eight known UNTs for which we can now assign magnitudes.

Date	Time (to nearest s)	K	$m_b(K)$	m_b	Note
1961 Oct 11	07:40:00	11.8	4.78	–	A
1962 Feb 02	08:00:00	13.6	5.63	–	B-1
1965 Jul 29	06:00:00	10.7	4.28	4.5	B-2
1965 Oct 14	04:00:00	10.7	4.28	–	C
1968 Oct 21	03:52:00	10.2	4.05	–	C
1968 Nov 12	07:30:00	10.6	4.24	–	C
1970 May 27	04:03:00	10.3	4.10	3.8	B-3
1972 Dec 28	04:27:00	11.4	4.60	4.9	B-4

Notes:
A — the first underground nuclear explosion conducted by the USSR was not included in many western lists of USSR explosions prior to publication of BOCHAROV *et al.* (1989), but it was reported (without coordinates and magnitude) by BOLT (1976), and by SYKES and RUGGI (1986, 1989).
B — detection and approximate location reported by SYKES and RUGGI (1986, 1989): 1 – no magnitude estimation; 2 – wrong time (03:05:00); 3 – coordinates with 150 km error; 4 – coordinates with 200 km error.
C — These UNTs were not published in western lists of USSR explosions prior to publication of BOCHAROV *et al.* (1989).

Murzhik for which we do not yet know the ground-truth coordinates. Thus for 17 UNTs (magnitudes 3.8–4.6), which occurred at Degelen, we can now compare the seismically-located epicenters with ground truth recently obtained for that subarea by LEITH (1998).

For these small UNTs, regional signals were acquired at CSE stations located at distances in the range 500–1400 km from STS. We mostly used data from bulletins but did read waveforms ourselves in some cases. Thus for these 17 UNTs we had 20 records and 49 station bulletin data from stations to the south; and 37 station bulletin data from stations to the east or west. So on average for the location of one event we had about one record and about three pieces of data from station bulletins located to the south of STS, and about 2 data from stations located to the east or west. On each record, 2–3 regional phases were measured (typically *Pn*, *Sn*, *Lg*). To obtain a preliminary estimate of location and origin time, we usually used (if they were available) three values of time intervals such as $t(Lg) - t(Sn)$; $t(Lg) - (Pn)$ and $t(Sn) - t(Pn)$ from each station record or bulletin. Having estimated the origin time (t_0) in this way, the next step for location was to use time intervals such as $t(Pn) - t_0$; $t(Sn) - t_0$ and $t(Lg) - t_0$.

For event location we used travel times of regional phases as given by NERSESOV and RAUTIAN (1964), based on a Pamirs-Baikal profile, slightly adapted by Khalturin for Northeast Kazakhstan. Our locations, and the comparison with ground-truth information, are given in Table 10. On average, the seismically-

Table 10

*Comparison of seismically-determined locations based on regional phases (*Khalturin *et al., 1994) and ground-truth locations (*Leith*, 1998), for small UNTs at the Semipalatinsk Test Site.*

Date	Lat.		ΔLat. km	Long.		ΔLong. km
	Seism.	G.T.		Seism.	G.T.	
1964 Jun 06	49.79	49.774	1.8	78.00	77.988	0.9
1964 Aug 18	49.81	49.821	−1.1	78.10	78.082	1.3
1965 Feb 04	49.78	49.773	0.8	78.12	77.991	9.2
1965 Mar 27	49.82	49.775	5.0	78.00	77.988	0.9
1966 Oct 29	49.74	49.785	−5.0	78.07	78.000	5.0
1966 Nov 19	49.70	49.730	−3.3	78.20	78.058	10.2
1967 Sep 02	49.79	49.742	5.3	78.02	78.026	−0.4
1968 Oct 29	49.84	49.833	0.8	78.14	78.093	3.3
1969 Apr 13	49.70	49.736	−4.0	77.92	78.105	−13.3
1969 Nov 27	49.79	49.837	−5.2	78.20	78.060	10.0
1971 Jan 29	49.77	49.805	−3.9	78.11	78.169	−4.2
1971 Apr 09	49.88	49.832	5.3	78.02	78.039	−1.4
1973 Dec 31	49.75	49.739	1.2	78.04	78.086	−3.3
1975 Oct 05	49.81	49.783	3.0	78.10	78.087	0.9
1977 Nov 27	49.80	49.754	5.1	78.06	78.050	0.7
1985 Jul 11	49.78	49.750	3.3	77.90	78.049	−10.7
1988 Dec 28	49.80	49.801	−0.1	78.06	78.069	−0.6

determined location error was only about 5 km. The ground-truth location was found to lie within the interval specified by Khalturin *et al.* (1994) as the location uncertainty in almost all cases, and only marginally outside that interval in the few cases where it was outside. The average of absolute errors for all 17 UNTs is only 3.2 km in latitude, and 4.4 km in longitude. The average of signed errors is only 0.53 km in latitude and 0.45 km in longitude (i.e., real epicenters systematically lie 0.53 km south and 0.45 km west of our estimated locations). Since the average length of the seismic paths was 750 km, the systematic error is remarkable small – about 0.07%, corresponding to an error in velocity of about 0.005 km/s.

We have thus been able to demonstrate the utility of regional seismic waves for purposes of accurate estimation of UNT locations, even when only a few records are available per event. The location uncertainty is so small in our case, because of the availability of good information, appropriate to the region, on travel times. The strongest constraint typically came from values of the time interval between *Lg* and *Pn*.

5. Comment on Magnitude Distribution and Yields

Now that we have obtained a fairly complete set of magnitudes for the nuclear tests at Semipalatinsk, it is of interest to see how they are distributed, and how well the announced information on yield is in accord with yield estimates based on seismic magnitude.

5.1. Comparison of Seismically Determined Yields, and Announced Information on Annual Total Yield

USSR NUCLEAR TESTS (1997) gave the total yield each year at each test site, and so we can compare these announced totals at STS for the years from 1964 to 1989 with the total obtained by summing the seismically determined yields. (The announced total yield at STS for the years 1961 and 1962 included several atmospheric tests; and in 1963 there were no tests. After 1963 all STS nuclear tests were underground.) Using the relation $m_b = 4.45 + 0.75 \log Y$ advocated by MURPHY (1990) and RINGDAL *et al.* (1992), we have estimated the yield of all the STS tests for which we have a magnitude. Figure 6 compares the annual total of yields determined seismically with the officially announced yield information. The yield of tests for which we do not have a magnitude is insignificant in comparison with the well-documented tests. Note that the vertical axis for this histogram is linear in yield, rather than in logarithmic units. The agreement between seismically-determined yield totals, and announced total is remarkable. The differences are

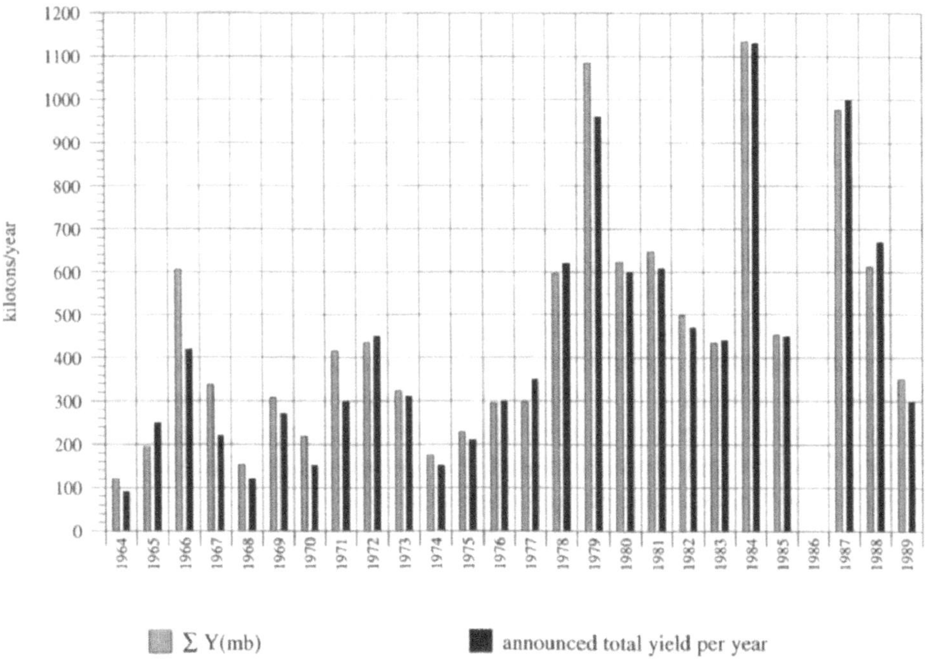

Figure 6

The announced total yield each year, of underground nuclear tests at Semipalatinsk, is compared with the total obtained by summing the estimated yield using the relationship $Y = 10\hat{}((m_b-4.45)/0.75)$, based upon the *P*-wave magnitude (AWE), for previously well-documented events plus events characterized in this paper. The aggregate announced yield for the whole period 1964–1989 is 10.8 megatons (mt), whereas the aggregate seismically-estimated yield is 11.5 mt.

somewhat greater than 10% in the earlier years, but are less than 10% for most of the last ten years of testing.

5.2. Detection Threshold and Magnitude Distribution

The small underground nuclear tests at STS provide some practical experience with certain aspects of open monitoring, albeit for years in the past, especially in the early years of underground nuclear testing, when capabilities were not as good as they are today. Thus, we can comment on three different levels of preliminary information about detected events:

(a) monitoring in the teleseismic zone (distances greater than 3000 km) without any information, other than the known position of the test site;

(b) the same as case (a) but with stations located in the regional interval of epicentral distances (700–3000 km); and

(c) the same as case (b) but when the dates of UNTs (and preferably even narrower time intervals) are known.

As our example of case (a) we can take the 271 UNTs at STS reported as seismic events by the ISC during 1961–1989 (note, not all of these events were reported at the time as nuclear in origin). However, this is monitoring of events large enough to enable good estimates of location using seismological methods. As noted above, in discussing the magnitude of these ISC events we prefer to use m_b(AWE) values rather than m_b(ISC) since the latter are known to be routinely too large, particularly for events with m_b less than about 5.0 (when the number of stations reporting magnitude is significantly lower, so that the ISC m_b is biased by reliance on such sensitive stations – see RINGDAL, 1976). Case (b) corresponds to the first stage of our analysis of regional observations (KHALTURIN *et al.*, 1994), using stations located in the 700–3000 km distance range, but with unknown origin times. Case (c) corresponds to the analysis presented in this paper, when dates of UNTs at STS were known.

It is reasonable to expect that the magnitude threshold for event reporting will decrease from case (a) to case (b) to case (c), and this expectation turns out to be correct. As a simple definition of the magnitude threshold, we chose the m_b value at which 50% of the signals are unreported. For determination of magnitude threshold, we worked with magnitudes for three sets of UNTs:

(1) m_b(AWE) for all UNTs from STS reported by the ISC;

(2) our determination of magnitudes of 27 small UNTs which were detected in the first stage of study, when days and times of explosions were not known;

(3) magnitudes of the 12 smallest UNEs which were detected later, when the date of each explosion was known. (In practice, it also proved helpful in searching for these data to know the times of day that were commonly used for UNTs at STS.)

For each of the three sets of magnitudes, we made a histogram giving the number of events in cells of width 0.2 magnitude units. We found that in the magnitude interval 4.35–4.54, AWE assigned magnitudes for 8 events and 2 were missed by ISC;

and in the magnitude interval 4.15–4.34, AWE assigned magnitudes for 4 events and 8 were missed by ISC. On this basis, the magnitude at which there was a 50% chance of an STS event being reported by the ISC corresponded to about m_b(AWE) = 4.28. For CSE regional observations, case (b), inspection of the intervals 3.15–3.34 and 2.95–3.14 indicates the magnitude corresponding to 50% chance of detection was about m_b 3.25.

The magnitude threshold for NORSAR in case (c), when the origin time is known, is very low for teleseismic signals from STS, m_b 3.3 ± 0.1. Among the 29 chemical explosions recorded regionally by CSE, NORSAR reported about 23 events using teleseismic data. NORSAR also reported all six UNTs recorded by CSE but not reported by ISC since 1973 (when NORSAR started to operate). (The NORSAR Semi-annual report for April–September 1984 indicated a detection threshold for STS in the range m_b 2.5 to 3.0.)

The study of SYKES and RUGGI (1989), published prior to the list of BOCHAROV *et al.* (1989), used numerous sources of information on teleseismic detections, including LASA and Hagfors detections of events that potentially could have been STS UNTs. An earlier and longer version of this paper is SYKES and RUGGI (1986), which gave references for the locations that were used, and stated that identification and size determination are much more questionable for events of equivalent yield less than 5 Kt. Sykes and Ruggi's success was impressive in that they found six small magnitude UNEs which had not previously been reported in the open literature. But the price for these six detections was large, in that these authors also reported 39 events as UNEs which turned out to be false alarms, namely, 17 chemical explosions at STS, together with 15 chemical explosions in the Azgir region, 6 chemical explosions in different regions of the USSR, and one earthquake. The most important point to draw from this history is that detection capability has been very good for recent decades, since all but two UNTs at STS with yields announced as greater than 1 ton have now been associated with detections. Detections have also been very good using only teleseismic data. But for accurate location and confident identification, additional data is often needed. In practice, such additional data can often be provided by regional stations.

We conclude this section with Figure 7 showing the magnitude distribution for the Balapan region, and for the Degelen and Murzhik regions. Again, it is clear that Balapan was the preferred location for the largest tests, and Degelen for the smallest. Out of the total of 340 UNTs, more than 40 UNTs had $m_b < 4.4$ and thus were probably sub-kiloton.

6. *Conclusions*

We have found and analyzed regional seismic data for underground nuclear tests (UNTs) at the Semipalatinsk Test Site (STS) that enable us here (Table 8) to report

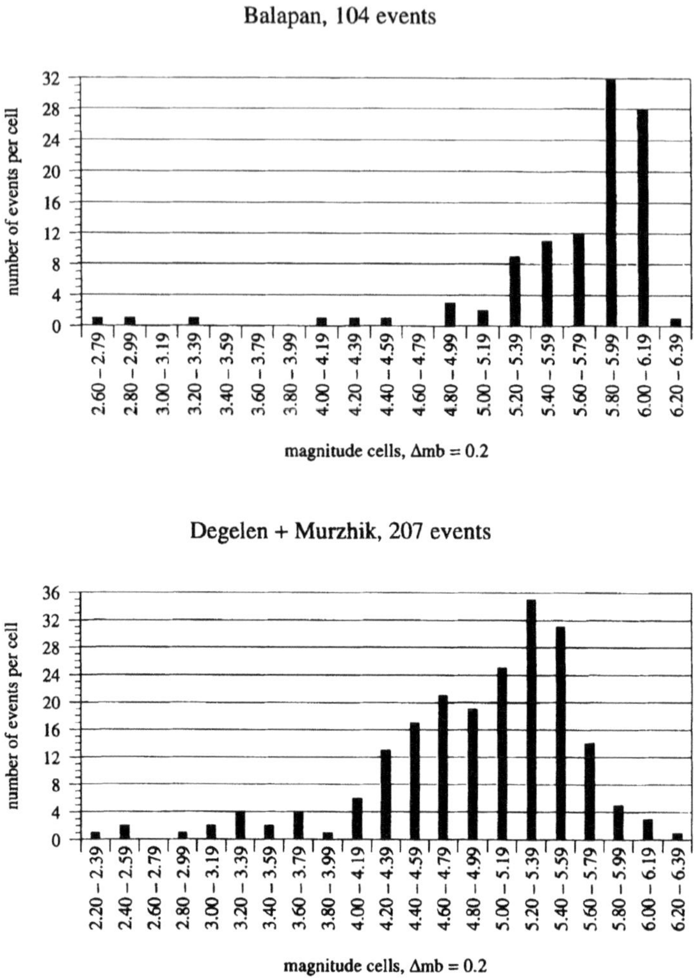

Figure 7
Magnitude distributions for Balapan (upper) and for Degelen and Murzhik subareas (lower).

the origin time and magnitude for 31 nuclear tests at this test site, that had not previously been documented in the open literature. Seismic detections for three of these UNTs were previously reported by SYKES and RUGGI (1986) and one by RINGDAL (1990), but detections for the remaining 27 were not reported in the open literature before our study.

For 19 of these events we also obtained seismically-determined estimates of location, and location uncertainty. By comparison with ground-truth location information that became available after our seismic determinations of location, we found that our location estimates were accurate to within a few km and our uncertainty estimates included the ground-truth location in most cases. We conclude that regional waves can be used to provide accurate locations even when few stations are available, provided regional travel times are well calibrated.

There are only two UNTs at STS, announced as having yields greater than one ton, for which we have been unable to find detections. Both occurred in the 1960s.

Yield estimates based upon seismic magnitudes give values for the estimated total annual yield at that differ by less than 10% from officially announced values of these yield totals, for most of the last ten years of testing at the Semipalatinsk Test Site.

The information we have been able to report here, on 67 of the 69 UNTs that were previously not well documented, can be used to assess monitoring capability for this major nuclear test site and how that capability has improved with time. The information can also be used to identify small UNTs, and chemical explosions and earthquakes near STS, suitable for evaluation of methods of discriminating between small seismic events. We recommend that efforts be supported, to build up the database of regional waveforms for small seismic events on and near STS, since these are the types of waveform that monitoring programs must be designed to detect and identify.

Acknowledgements

We thank Dr. Frode Ringdal of NORSAR and Dr. Yuri Kopnichev of CSE, who made considerable efforts to find detections and assign magnitudes for particular events reported in this paper. We also thank Dr. Bill Leith of the US Geological Survey, and Dr. Natalya Mikhailova of the National Nuclear Centre of the Republic of Kazakhstan, for assistance in obtaining ground-truth information. Our work was sponsored by contract DSWA01-97-C-0156 with the Defense Threat Reduction Agency. This is publication number 6125 of the Lamont-Doherty Earth Observatory.

References

ADUSHKIN, V., PERNIK, L., and SPIVAK, A. (1997), *Degelen Mountain Chemical Explosion Cratering Experiment*, Institute for Dynamics of the Geospheres, Technical Report No 97-3011 to DSWA, 110 pp.

BOCHAROV, V. S., ZELENTSOV, S. A., and MIKHAILOV, V. I. (1989), *Characteristics of 96 Underground Nuclear Explosions at the Semipalatinsk Test Site*, Atomnaya Energiya *67*, 210–214.

BOLT, B. A., *Nuclear Explosions and Earthquakes: The Parted Veil* (W.H. Freeman and Co., San Francisco 1976).

DOUGLAS, A. (1967), *Joint Epicentre Determination*, Nature *215*, 47–48.

GIVEN, H. K., TARASOV, N. T., ZHURAVLEV, V. V., VERNON, F. L., BERGER, J., and NERSESOV, I. L. (1990), *High-frequency Seismic Observation in Eastern Kazakhstan, USSR, with Emphasis on Chemical Explosion Experiments*, J. Geophys. Res. *95*, 295–307.

GORIN, V. V, KRASILOV, G. A., KUROCHKIN, A. I., KURKIN, A. I., MALTSEV, A. L., MATUSHENKO, A. M., ORLOV, S. L., PICHUGIN, A. V., SMAGULOV, S. G., STRUKOV, V. G., FILLIPOVSKY, V. I., KHARITONOV, K. V., and CHERNISHEV, A. K. (1994), *Semipalatinsk Test Site: Chronology of underground nuclear explosions and their primary radiation (1961–1989)*. (Bulletin of the Center of Public Information, No. 4), pp. 21–32, Moscow, Central AtomInform Institute.

HANSEN, R. A., RINGDAL, F., and RICHARDS, P. G. (1990), *The Stability of RMS Lg Measurements and their Potential for Accurate Estimation of the Yields of Soviet Underground Nuclear Explosions*, Bull. Seismol. Soc. Am. *80*, 2106–2126.

KHALTURIN, V. I., RAUTIAN, T. G., and RICHARDS, P. G. (1994), *A study of small explosions and earthquakes during 1961–1989 near the Semipalatinsk Test Site, Kazakhstan* (Technical Report for subcontract B239589 with Regents of Univ. of Calif.) 64 pp.

KHALTURIN, V. I., RAUTIAN, T. G., and RICHARDS, P. G. (1998), *The Seismic Signal Strength of Chemical Explosions*, Bull. Seismol. Soc. Am. *88*, 1511–1524.

LEITH, W. (1998), Degelen Nuclear Test and Tunnel Data, Technical Memorandum dated June 26, 1998 to the Defense Threat Reduction Agency, US Geological Survey; and Technical Reports to the Defense Threat Reduction Agency (formerly Defense Special Weapons Agency; formerly Defense Nuclear Agency), under Contract DNA001-95-C-1079.

LILWALL, R. C., MARSHALL, P. D., and RIVERS, D. W. (1988), *Body Wave Magnitudes of Some Underground Nuclear Explosions at the Nevada (USA) and Shagan River (USSR) Test Sites*, AWE Report No. O 15/88, H.M. Stationery Office, London.

LILWALL, R. C., and FARTHING, J. (1990), *Joint Epicentre Determination of Soviet Underground Nuclear Explosions 1973–1989 in Eastern Kazakhstan*, AWE Report No. O 12/90, H.M. Stationery Office, London.

MARSHALL, P. D., BACHE, T. C., and LILWALL, R. C. (1985), *Body Wave Magnitudes and Locations of Soviet Underground Nuclear Explosions at Semipalatinsk Test Site*, AWE Report No. O 16/84, H.M. Stationery Office, London.

MURPHY, J. R. (1990), *A new system for seismic yield estimation of underground nuclear explosions*. In Proceedings of the 12th annual DARPA/GL, *Seismic Research Symposium*, Key West, Florida.

MIKHAILOV, V. N. (editor) and 14 co-authors (1996), *USSR Nuclear Weapons Tests and Peaceful Nuclear explosions, 1949 through 1990*, RFNC-VNIIEF, Sarov, 96 pp.

NERSESOV, I. L., and RAUTIAN, T. G. (1964), *Kinematics and dynamics of seismic waves at epicentral distances less than 3500 km*. In Proceedings (Trudy) of the Institute of Physics of the Solid Earth, Academy of Sciences of the USSR, #32.

NNCRK (1999), Technical documentation, contributed by the National Nuclear Centre of the Republic of Kazakhstan, Oslo Workshop on IMS Location Calibration, Jan. 1999.

POOLEY, C. I., DOUGLAS, A., and PEARCE, R. G. (1983), *The Seismic Disturbance of 1976 March 20, East Kazakhstan: Earthquake or Explosion?* Geophys. J.R. Astr. Soc. *74*, 621–631.

RINGDAL, F. (1976), *Maximum-likelihood Estimation of Seismic Magnitude*, Bull. Seismol. Soc. Am. *66*, 789–802.

RINGDAL, F. (1990), *Teleseismic Event Detection Using the NORESS Array, with Special Reference to Low-yield Semipalatinsk Explosions*, Bull. Seismol. Soc. Am. *80*, 2127–2142.

RINGDAL, F., MARSHALL, P. D., and ALEWINE, R. W. (1992), *Seismic Yield Determination of Soviet Underground Nuclear Explosions at the Shagan River Test Site*, Geophys. J. Int. *109*, 65–77.

SULTANOV, J. J., ADUSHKIN, V. V., DANILOVA, T. V., KAAZIK, P. B., KUZNETSOV, O. P., NEDOSHIVIN, N. I., and RUBINSHTEIN, N. D. (1995), *Completion of Data Base of PNE and Large-scale Chemical Blasts Conducted inside the Territory of FSU and Analysis of Characteristics of Seismic Waves in Order to Improve the Methods of Seismic Event Identification*. Institute for Dynamics of Geospheres, *Report to IRIS, subcontract 201, 144 pp.*

SYKES, L., and RUGGI, S. (1986), *Soviet Nuclear Testing*, Natural Resources Defense Council, Working Paper NWD 86-4, November 1986.

SYKES, L., and RUGGI, S. (1989), *Soviet Nuclear Testing*, Chapter 10, pp. 332–375 in *Nuclear Weapons Databook*, Vol. IV, (eds. Cochran, T. B., Arkin, W. M., Norris, R. S., and Sandes, J. I.).

THURBER, C. H., QUIN, H. R., and RICHARDS, P. G. (1993), *Accurate Locations of Nuclear Explosions in Balapan, Kazakhstan, 1987 to 1989*, Geophys. Res. Lett. *20*, 399–402.

USSR NUCLEAR TESTS, Mikhailov, V. N. (editor) and 29 co-authors (UZDAT Publishing House, Moscow 1997), 248 pp.

VERGINO, EILEEN S. (1989), *Soviet Test Yields*, EOS *70*, 1511–1513 (with corrections and additions, EOS *70*, 1569).

(Received June 30, 1999, revised February 2, 2000, accepted February 9, 2000)

Pure appl. geophys. 158 (2001) 173–192
0033-4553/01/020173-20 $ 1.50 + 0.20/0

Pure and Applied Geophysics

Precise Relative Location of 25-ton Chemical Explosions at Balapan Using IMS Stations

W. SCOTT PHILLIPS,[1] HANS E. HARTSE[1] and LEE K. STECK[1]

Abstract—We test how well low-magnitude (m_{bLg} 1.8 to 2.6), 25-ton chemical explosions at Balapan, Kazakhstan, can be located using IMS stations and standard earth models, relying on precisely determined relative arrival times of nearly similar, regional and teleseismic waveforms. Three 1997 Balapan explosions were recorded by a number of currently reporting and surrogate IMS stations. Three regional stations and two teleseismic arrays yielded consistent waveforms appropriate for relative picking. Master-event locations based on the AK135 model and ground-truth information from the first, shallowest and best-recorded explosion, fell under 1 km from known locations, for depths constrained to that of the master event. The resulting 90% confidence ellipses covered 12–13 km^2 and contained the true locations; however, results for depth constrained to true depth were slightly less satisfactory. From predictions based on ground truth, we found a P_g-coda phase at Makanchi, Kazakhstan to be misidentified and poorly modeled. After accounting for this, 90% ellipses shrank to 2–3 km^2 and true-depth mislocation vectors became more consistent with confidence-ellipse orientations. These results suggest that a high level of precision could be provided by a tripartite array of calibration shots in cases where models are poorly known. We hope that the successful relocation of these small Balapan shots will support the role of calibration explosions in verification monitoring and special event studies, including on-site inspection.

Key words: Seismic verification, location, detection, CTBT, IMS.

Introduction

Because the comprehensive Test-Ban-Treaty (CTBT) prohibits all nuclear explosions, it is important to explore detection, location and event identification techniques for very small seismic sources. Here, we present a case study focusing on arrival-time determination and relative location of chemical explosions of known location and origin time (ground truth), using stations of the planned, International Monitoring System (IMS).

Ground-truth data provide travel-time path calibration, allowing high-precision location of nearby seismic sources relative to the ground-truth event. Using relative location methods, nuclear explosions have been successfully matched with satellite imagery (THURBER *et al.*, 1993, 1994; SYKES *et al.*, 1993; GUPTA, 1995; GUPTA and

[1] Seismic Research Center, Los Alamos National Laboratory, Los Alamos, New Mexico 87544, U.S.A. E-mail: wsp@lanl.gov, hartse@lanl.gov, lsteck@lanl.gov.

RICH, 1996). These explosions were of high magnitude (m_b 4.8 to 6.1 for the Balapan tests studied by THURBER *et al.*, 1993) and studies employed data from many stations. Calibration using ground truth is especially important for small sources because, in general, few data are available and results are heavily influenced by path-sensitive regional arrival times.

The often-stated goal of CTBT location efforts is a 1000 km^2 confidence ellipse because of treaty restrictions on the extent of an on-site inspection (OSI). However, it is easy to imagine situations where much higher precision might be called for, such as correlating locations with overhead imagery, or with potential shot points in the course of an OSI. Fortunately, ground-truth sources provide more than travel-time path corrections, they also provide waveforms that can be used to obtain precise relative arrival times for use in special event studies. Waveform similarity and the precise determination of relative arrival times has been intensively investigated in verification work (e.g., BUNGUM and HUSEBYE, 1971; ISRAELSSON, 1990), and used in high-precision relative location of nuclear tests (SHEARER and ASTIZ, 1997). Successful demonstrations of such processing and relocation techniques, especially for small sources, will support the use of calibration shots as an effective component of treaty verification and OSI.

The recent hole-closure program at Balapan, Kazakhstan, included a series of 25-ton shots, which provided the opportunity to test data processing and relative location procedures for small events. Seismic data were recorded by stations currently reporting to the prototype International Data Centre (pIDC) as well as by stations deployed at future IMS sites by Lamont Dougherty (LDEO, KIM *et al.*, 1996), and by the U.C. San Diego IDA group along with the USGS Albuquerque Seismological Laboratory as part of the Incorporated Research Institutions for Seismology (IRIS) Global Seismic Network. Our approach will be to use ground-truth information from one event to obtain relative locations of others in a blind manner, using IMS and surrogate stations, and a standard earth model. The ground-truth information will allow independent verification of results and calibration of data error, as well as help us to evaluate sources of error in the relative location procedure.

Data

The depth-of-burial series of chemical explosions at Balapan consisted of a 5-ton test shot, followed by three, 25-ton shots at depths of 50, 300 and 550 m in the summer of 1997 and another 50-m shot in the summer of 1998 (Fig. 1). These shots were part of a hole and tunnel closure program at the former Soviet Union test sites at Balapan and Degelen Mountain. Shots were carried out by the US Defense Special Weapons Agency in collaboration with the National Nuclear Center of Kazakhstan. The 1997, 50-m shot occurred in sedimentary rock and the remainder occurred in crystalline rock. Ground-truth location, origin time and other shot data can be found in Table 1.

Local stations were deployed during the hole-closure program to help verify origin times and study seismic radiation. Analysis of P and R_g recorded at these stations allowed the determination of laterally varying velocity structure (PEARSON

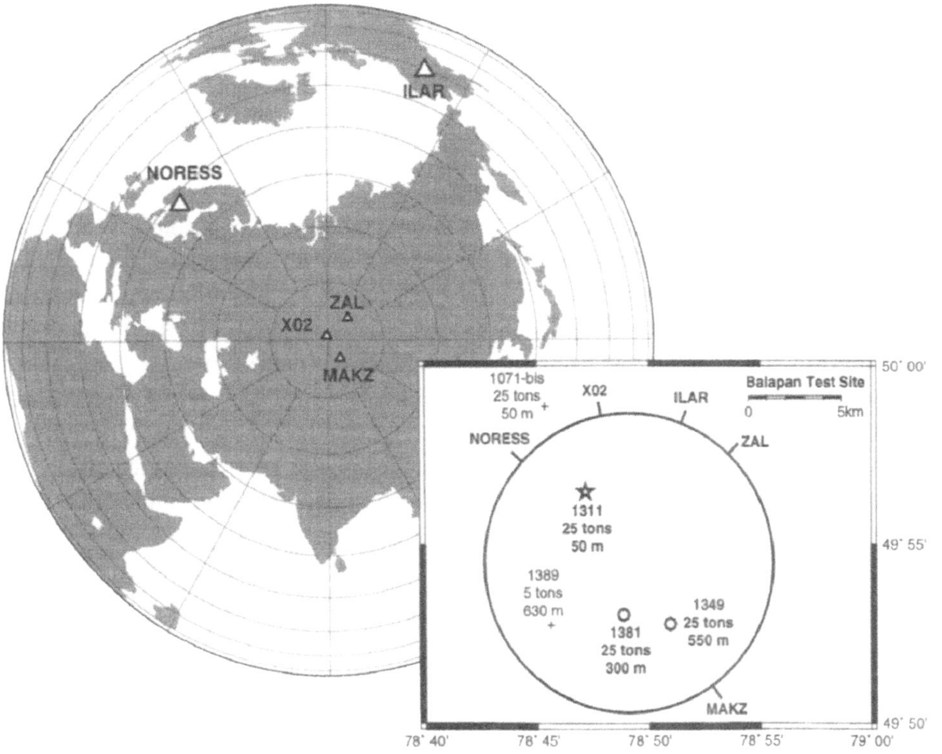

Figure 1
Balapan shots and IMS stations used in the study. The master-event shot is designated by a star, relocated shots by open circles and other shots mentioned in the text by crosses. Hole number, shot yield and depth are also indicated. Azimuths to IMS stations are shown around the encompassing circle. The background map is an orthographic projection centered on Balapan, with triangles indicating station locations. Concentric grid lines represent distance from Balapan in 10° intervals, while radial grid lines represent azimuth from Balapan in 30° intervals.

Table 1

Depth of Burial shots at Balapan

Month	Day	Year	Time	Hole	Latitude	Longitude	Depth	Yield	Geology
JUL	13	1997	08:11:04.304	1389	49.8790	78.7600	630 m	5 t	NA
*AUG	03	1997	08:07:20.040	1311	49.9412	78.7860	50 m	25 t	Shale
*AUG	31	1997	07:08:39.179	1381	49.8837	78.8147	300 m	25 t	Granosyenite
*SEP	28	1997	07:30:15.126	1349	49.8794	78.8493	550 m	25 t	Granosyenite
SEP	17	1998	07:19:40.551	1071	49.9810	78.7559	50 m	25 t	Granosyenite

Stars (*) indicate the three shots used in this study.

et al., 1998; BONNER *et al.*, 2000), which will help us later evaluate sources of relative arrival-time error.

We chose to restrict our study to IMS station data. In the future, these data will be available on the most timely basis and we need to test the ability of this sparse network to perform high-precision location. Stations used in this study are shown in Figure 1. Station data, including IMS status are included in Table 2, while geometrical relationships with Balapan are listed in Table 3.

The 1997, the 50-m shot was the best recorded of the Balapan series. This shot was included in the Reviewed Event Bulletin (REB) and assigned an m_b of 4.0. Burial effects reduced the amplitudes of the 300- and 550-m shots and these were not included in the REB. However, these shots were recorded at regional and teleseismic distances. The 1998, 50-ton shot occurred at a point in time when the station at Zalesovo (ZAL) was down. Furthermore, regional recordings from the Kazakhstan network had not yet reached the IRIS database at the time of this study. The 5-ton test shot was recorded at ZAL, but with lower signal to noise. Therefore, we chose to analyze the three, 1997, 25-ton shots at Balapan, because they provided the largest amount of data.

The m_{bLg} (NUTTLI, 1986) from the three shots have been measured using data from Makanchi, 421-km distant (HOWARD PATTON, personal communication, 1999), after applying a Q_0 of 800 and a frequency dependence of 0.8 based on L_g coda-Q maps of Eurasia (MITCHELL *et al.*, 1997). The m_{bLg} were determined to be 2.6, 2.2

Table 2

Stations used in this study

Station	Latitude (°)	Longitude (°)	Elevation (m)	Study Configuration	IMS Station Type (Code)	Planned Configuration
X02	50.7016	78.5567	240	Short-Period V	Auxiliary (KUR)	Array
MAKZ	46.8080	81.9770	590	Broadband 3C	Primary (MAK)	Array
ZAL	53.6170	84.7920	100	Short-Period 3C	Primary (ZAL)	Array
NORES	60.7353	11.5414	302	Array and 3C	Primary (NOA)	Array
ILAR	64.7716	−146.8861	418	Array and 3C	Primary (ELAK)	Array

Table 3

Geometrical relationships between stations and Balapan (50-m shot, hole 1311)

Station	Distance (km)	Azimuth (°)	Back Azimuth (°)
X02	86	349	169
MAKZ	421	145	327
ZAL	582	43	228
NORES	4225	313	75
ILAR	6689	21	328

and 1.8 for the 50-, 300- and 550-m shots, respectively. For the 50-m shot, the m_{bLg} is 1.5 magnitude units smaller than the REB m_b. The REB m_b was based on one reading (NORESS) and should be expected to be biased high because of censoring effects. In fact, RINGDAL (1990) has shown that NORESS m_b is generally 1.0 units higher than worldwide measurements for Balapan explosions and attributes this to focusing effects beneath the source and receiver regions.

We verified that the small m_{bLg} measurements are reasonable by using them to estimate yield. Extrapolation based on the data PATTON (2000) used to obtain his unified moment-magnitude (M_w) – m_{bLg} model and application of the M_w-moment (M_0) scale of HANKS and KANAMORI (1979) lead to a moment estimate of 7.3×10^{11} to 1.9×10^{12} N-m for the 50-m shot. The range reflects the one-sigma prediction error from the $M_w - m_{bLg}$ regression. Comparing to moment-yield relations of nuclear and chemical explosions (DENNY and JOHNSON, 1991) gives a yield of 9 to 22 tons, which is reasonably close to the actual value of 25 tons. We conclude that the Balapan shots studied here are two orders of magnitude smaller than the smallest nuclear tests analyzed using relative location techniques (e.g., THURBER *et al.*, 1993) and will provide an excellent case study of the low-magnitude relative location capability of a sparse global array such as the IMS.

Using a series of filters, we searched for repeatable waveforms from the three shots recorded at regional distances by IMS stations and their surrogates. We found repeatable P_g at Kurchatov, Kazakhstan (X02, 86 km), a P_g-coda phase at Makanchi, Kazakhstan (MAKZ, 421 km) and P_n at Zalesovo, Russia (ZAL, 582 km). Seismograms are aligned on their relative arrivals in Figures 2a–c. At MAKZ, the repeatable arrivals follow high frequency P_n and P_g by 10 s and 4 s, respectively. We proceed as if this were a P_g phase for purposes of the blind test; however, it will become apparent that the identification and modeling of this phase is a source of error. Any of the arrivals at these three stations should be sufficiently large to be detected by an automatic system and would serve to guide a search for poor signal-to-noise arrivals elsewhere, including at teleseismic distances. We also checked data from the slated IMS auxiliary station at Borovoye, which exhibited strong regional phases for these shots, but we found no obviously similar waveforms. Finally, we searched for, but failed to identify, repeatable S phases at regional stations. Arrival time data, including dominant periods and filter bands used to identify the similar waveforms, are listed in Table 4.

Arguably, the best and most accurately modeled relative arrivals were the P_n obtained at ZAL. However, significant clock drift has been demonstrated for this station (HANS ISRAELSSON, personal communication, 1999), based on residuals of well-constrained events that include high signal-to-noise ZAL P arrivals. In our case, arrivals from the 300-m shot reach ZAL more quickly than do those of the 550-m shot, the closer of the two, after aligning on ground-truth origin times (not shown). We downweight arrivals from ZAL by superposing an uncorrelated, 400-ms clock error on the data error when locating. The 400 ms is a rough guess based on the scatter of well constrained residuals, as defined above. This error severely reduces the influence of ZAL data on the locations, which is unfortunate because so few data constrain this problem.

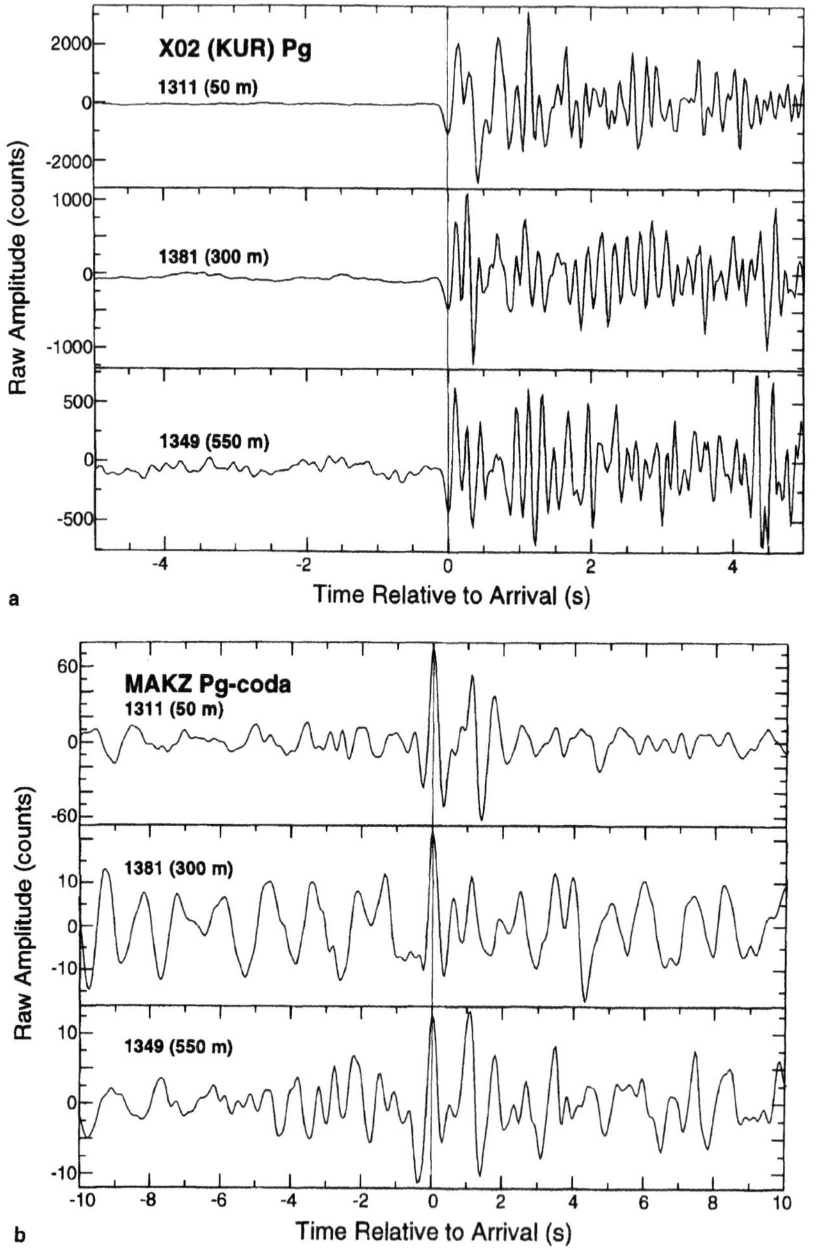

Figures 2a, b

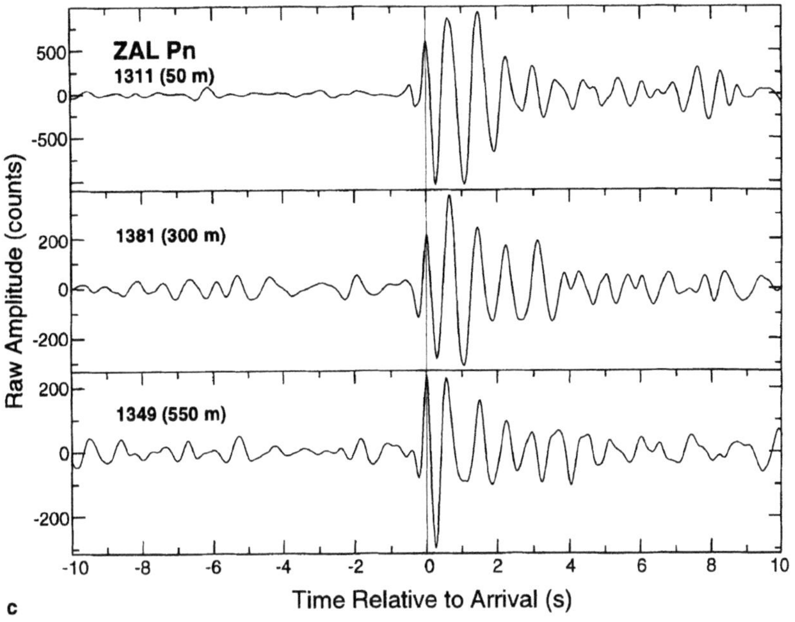

ZAL Pn
1311 (50 m)
1381 (300 m)
1349 (550 m)
Raw Amplitude (counts)
500
0
-500
200
0
-200
200
0
-200
-10
-8
-6
-4
-2
0
2
4
6
8
10
Time Relative to Arrival (s)
c

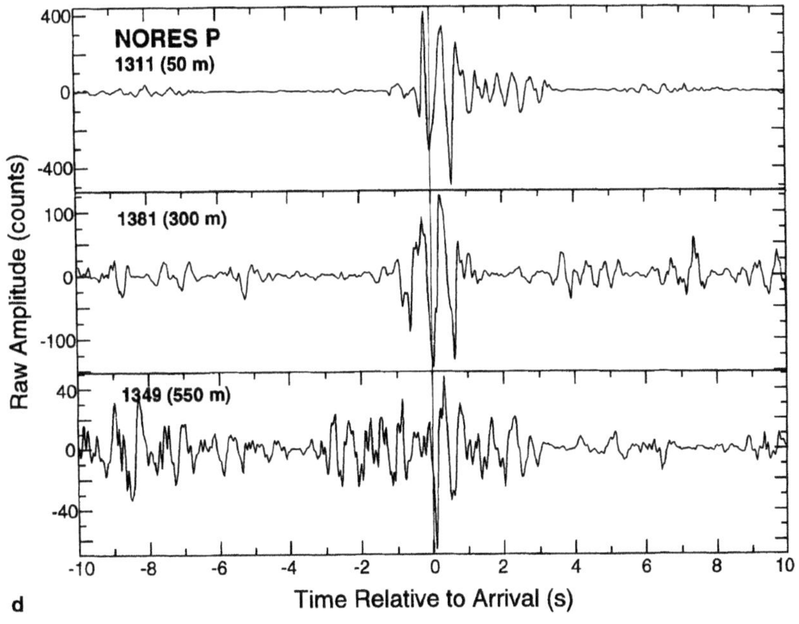

NORES P
1311 (50 m)
1381 (300 m)
1349 (550 m)
Raw Amplitude (counts)
400
0
-400
100
0
-100
40
0
-40
-10
-8
-6
-4
-2
0
2
4
6
8
10
Time Relative to Arrival (s)
d

Figures 2c, d

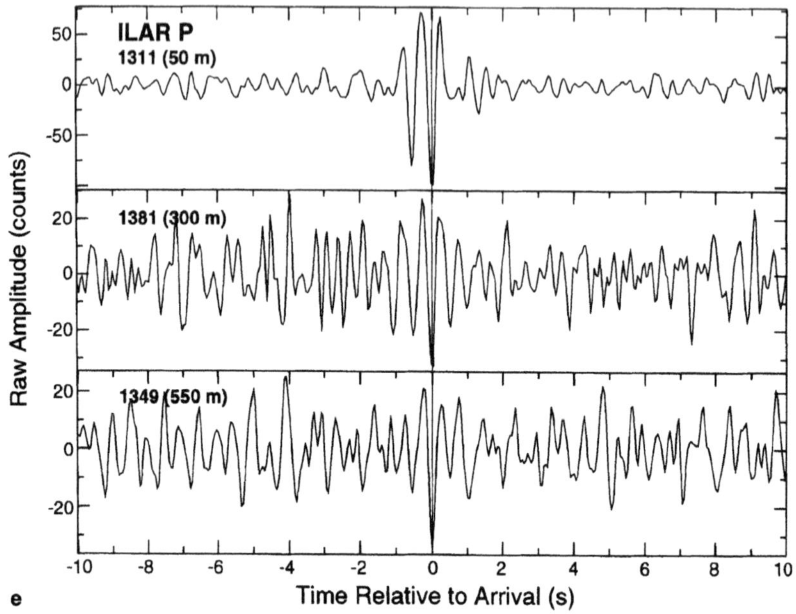

Figure 2
Vertical component seismograms, aligned on relative arrivals for a) P_g recorded by the Kurchatov cross array element 2 (X02), Kazakhstan; b) P_g–coda recorded at Makanchi (MAKZ) Kazakhstan and band-pass filtered between 0.75 and 1.5 Hz (note the high noise level in this band for shot 1381); c) P_n recorded at Zalesovo (ZAL), Russia and band-pass filtered between 0.75 and 1.5 Hz; d) P beams (bearing 74°, velocity 18 km/s) recorded at NORESS, high-pass filtered above 1.5 Hz and modified by a polarization filter; and e) P beams (bearing 321°, velocity 20 km/s) recorded at ILAR, Alaska and high-pass filtered above 1.5 Hz. Shot holes and depths are indicated above and to the left of each seismogram.

Table 4

Travel-time data

Station	Phase	1311 (s)	1381 (s)	1349 (s)	Filter (Hz)	Dominant Period (s)
X02	Pg	14.703	15.760	15.931	none	0.3
MAKZ	Pg/PmPPmP	71.322	70.471	70.317	0.75–1.5	1.0
MAKZ	Pn	62.501	61.682	NA	1.0–	0.15
MAKZ	Pb	67.323	NA	66.083	3–6	0.2
ZAL	Pn	84.757	85.251	85.403	0.75–1.5	1.0
NORESS	P	439.080	439.535	439.755	1.5–	0.6
ILAR	P	609.980	610.240	610.220	1.5–	0.6

Note: Filtering (4-pole Butterworth) and beamforming were performed using the Seismic Analysis Code (SAC).

We searched for Balapan shot arrivals at teleseismic distances using standard beamforming techniques. The NORESS and ILAR arrays yielded repeatable, waveforms appropriate for study (Fig. 2d–e). In keeping with our blind test approach, beam parameters were optimized using data from the 50-m shot, then were fixed for the remaining two. For these records, signal-to-noise was less than ideal for the two deeper shots, which lead to some difficulty in aligning waveforms. In both cases, the alignment was identified by cross correlation (using 4 s windows centered on the wave packets). A polarization filter (MONTALBETTI and KANASEWICH, 1970) emphasized arrivals at NORESS (Fig. 2d) and gave a more easily identifiable cross-correlation peak. The polarization filter was based on data from the central, three-component element and was constructed using a 2-s moving window, emphasizing linear motion trending along the back azimuth and plunging 60°. Misalignment of these waveforms would introduce errors of 600 ms. The result of misalignment on location quality will be discussed later.

Beamforming yielded a poor signal-to-noise waveform for the shallow shot at FINESS and no recognizable signal at CMAR, KSAR, EKA, GERESS and ESDC, representing all arrays within 60° for which we were able to obtain data for all three shots.

In summary, we found repeatable waveforms at three regional distance stations and two teleseismic distance arrays. These stations are distributed unevenly, with four along northerly azimuths from Balapan and only one along a southerly azimuth (Fig. 1). Large azimuthal gaps exist in east and southwest directions. The gaps are made worse by the virtual exclusion of ZAL data because of clock error. Station X02 is at 86 km, which is very close for verification monitoring; however, repeatable waveforms were also observed at arrays out to 6689 km. Signal detections at the select set of teleseismic distance stations may result from favorable noise conditions, as well as focusing in the upper mantle beneath the source or receiver sites (RINGDAL, 1990).

Relative Location Results

We used the master-event technique (EVERNDEN, 1969) to obtain relative locations of the Balapan shots. Holding the hypocenter and origin time of the 50-m shot fixed to known ground truth, we calculated travel-time residuals to the AK135 earth model (KENNETT *et al.*, 1995). These residuals quantify differences between the true earth and the model and were applied as station corrections to locate the remaining shots. Following our blind test approach, we fixed source depths to the depth of the master event.

Arrival-time residuals calculated for the 50-m shot and used as station corrections are listed in Table 5. Residuals for X02, NORESS and ILAR fell within a half-second of the model predictions. Arrivals at ZAL and MAKZ were late, 5.2 s and 2.2 s, respectively. A minor fraction of the residual comes from choosing a later pulse as the arrival, rather than the first break. Studies of crust and upper mantle in this area indicate faster lower crustal velocities and thicker crust than the AK135 model

(Leith 1987; Priestley *et al.*, 1988; Quin and Thurber, 1992; Tables 6a,b). The thick crust (40–50 km versus 35 km for AK135) must be responsible for the late P_n arrival at ZAL. Given the large difference between models, neither should we be surprised to see the 2.2-s discrepancy at MAKZ for the P_g-coda phase, which we have modeled as P_g.

We fixed data errors proportional to dominant periods in the picked traces (i.e., after filtering, Table 4). With few constraining data, the distribution of data errors can heavily influence the locations, so we adhered to this simple rule in keeping with the blind test approach. These data errors do not include modeling errors, which we might expect to be large for the crustal phases at X02 and MAKZ. We set degrees of freedom (or K-value, Jordan and Sverdrup, 1981; Bratt and Bache, 1988) to be

Table 5

Travel-time residuals used as station corrections

Station	Phase	Mode	Model	Time Residual (ms)
X02	Pg	Blind	AK135	556
MAKZ	Pg	Blind	AK135	2231
MAKZ	PmPPmP	Guided	Regional	−1035
MAKZ	Pn	Guided	AK135	2816
MAKZ	Pb	Guided	AK135	−451
ZAL	Pn	Blind	AK135	5202
NORES	P	Blind	AK135	−496
ILAR	P	Blind	AK135	445

In the guided test mode, we substituted *PmPPmP* for P_g at MAKZ and added MAKZ Pn and P_b. The regional model was taken from Quin and Thurber (1992).

Table 6a

AK135 crustal model

Depth (km)	V_p (km/s)
0	5.8
20	5.8
20	6.5
35	6.5

Table 6b

Regional crustal model (after Quin and Thurber, 1992)

Depth (km)	V_p (km/s)
0	4.5
15	6.5
50	7.05

large, allowing the input data error to control the confidence ellipses. Absolute data error will be estimated later, based on residuals from ground-truth predictions.

Locations fell 590 and 960 m from the ground-truth epicenters, while origin times were delayed 45 and 115 ms from ground-truth shot times (Run 1, Tables 7a and b, Fig. 3). Mislocations increased slightly when locations were constrained to true depths (Run 2), indicating that the extremely small mismatch in the blind test could be fortuitous.

We calibrated data errors using ground-truth information. Relative arrival-time residuals with respect to ground-truth predictions are given in Table 8 and will be referred to as ground-truth residuals. A proportionality factor was estimated such that the RMS of the data-error-weighted, ground-truth residuals from both shots was unity. Data from ZAL was excluded because of the known clock error. This procedure gave a data error of $T/5.6$, where T is the dominant period. This error represents the combined arrival-time error in master and slave events, as well as modeling error. The error estimate is large, given the waveform similarity we observe, and will be the focus of later discussion.

Table 7a

Master-event location results, shot 1381 (300 m)

Run	Description			Location			GT Difference		90% Confidence Ellipses			
	Mode	Depth (m)	Error	Latitude (°)	Longitude (°)	OT (s)	Distance (km)	Time (ms)	Major (km)	Minor (km)	Trend (°)	Area (km^2)
1	Blind	50	T/5.6	49.889	78.814	39.224	0.59	45	3.6	1.1	69	12
2	Blind	300	T/5.6	49.891	78.818	39.255	0.84	76	3.6	1.1	69	12
3	Guided	50	T/8.5	49.886	78.823	39.167	0.65	−12	2.4	0.3	67	2
4	Guided	300	T/8.5	49.888	78.827	39.196	1.01	17	2.4	0.3	67	2

Mode differentiates between blind and guided tests as discussed in the text. T represents the dominant period in the picked traces for each station as given in Table 4. Location origin time is given relative to shot hour and minute from Table 1. Origin time difference is location OT minus ground-truth OT. Origin time error is listed as 100 ms for all runs, but these values have been rounded off by the location routine.

Table 7b

Master-event location results, shot 1349 (550 m)

Run	Description			Location			GT Difference		90% Confidence Ellipses			
	Mode	Depth (m)	Error	Latitude (°)	Longitude (°)	OT (s)	Distance (km)	Time (ms)	Major (km)	Minor (km)	Trend (°)	Area (km^2)
1	Blind	50	T/5.6	49.888	78.848	15.241	0.96	115	3.7	1.1	68	13
2	Blind	550	T/5.6	49.892	78.854	15.301	1.44	175	3.7	1.1	68	13
3	Guided	50	T/8.5	49.885	78.863	15.160	1.17	34	2.5	0.3	65	3
4	Guided	550	T/8.5	49.888	78.871	15.209	1.83	83	2.5	0.3	65	3

See Table 7a comments.

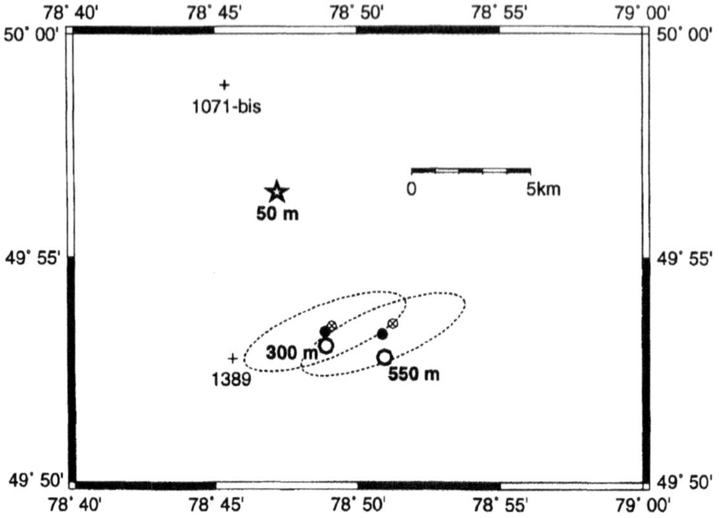

Figure 3
Blind test, master-event locations and 90% confidence ellipses for the 300- and 550-m shots at Balapan. The 50-m master event is denoted by a star, while actual locations of 300- and 550-m shots are represented by open circles. Locations and errors for depths constrained to the depth of the master event are represented by closed circles and dashed lines, respectively. Locations for depths constrained to true depths are represented by circle-x symbols. Locations of depth-of-burial shots not used in the study are denoted by crosses.

Table 8

Relative arrival-time residuals with respect to ground-truth predictions

Station	Phase	Mode	Model	Shot 1381 (ms)	Shot 1349 (ms)
X02	Pg	Blind	AK135	−39	−30
MAKZ	Pg	Blind	AK135	200	347
MAKZ	PmPPmP	Guided	Regional	59	196
MAKZ	Pn	Guided	AK135	3	NA
MAKZ	Pb	Guided	AK135	NA	63
ZAL	Pn	Blind	AK135	117	462
NORES	P	Blind	AK135	49	146
ILAR	P	Blind	AK135	−24	24

The 90% confidence ellipsoids based on $T/5.6$ data error covered 12 and 13 km^2 and included the shot epicenters; however, the 550-m shot fell close to the ellipse edge (Fig. 3). Both shots fell close to or off the edge when depths were constrained to true depths. The true-depth test indicates errors which have not been accounted for adequately, even though the location match with ground truth was still quite good. Ellipses for true-depth locations are not included in Figure 3; but, other than geographical translation, they are nearly identical to those shown.

Discussion

We find small mislocation vectors and confidence ellipses for the 25-ton shots at Balapan. However, the small ellipses have difficulty covering the known locations, a problem noted by THURBER *et al.* (1993) in a relative location study of Balapan nuclear explosions using ISC data and SPOT imagery. Furthermore, residuals based on ground-truth predictions (Table 8) can be a significant fraction of the dominant periods of the similar waveforms (Table 4), indicating sources of error other than picking are important. For example, relative arrivals at MAKZ are very late (200 and 347 ms for 300- and 550-m shot depths, respectively). The 347-ms delay is especially surprising because waveforms from the 50-m (master event) and 550-m shots overlay well (Fig. 2b). We also observe a 146-ms residual for the NORESS arrival from the 550-m shot. This is roughly 1/4 of the period and cannot be picking error alone, even considering the poor signal-to-noise we observe. In the following, we will discuss sources of relative arrival-time error.

Our relative location scheme assumes that the AK135 model yields accurate apparent velocity across the array of shots (in the reciprocal sense), even though absolute times may be poorly predicted. This works well for a phase such as P_n, which displays small (5%), but significant, fluctuations worldwide. However, this may not hold for crustal phases, because crustal structure variations are more dramatic and could lead to large differences in apparent velocity between a standard model and the true earth. For MAKZ arrivals, we find an on-azimuth apparent velocity of nearly 8 km/s based on a plane-wave fit to the ground-truth travel times, while the AK135 P_g model predicts an apparent velocity of 6.1 km/s. This mismatch will lead to relative arrival errors similar to those observed for MAKZ. Of course, 8 km/s is extremely fast for a crustal phase, but off-azimuth propagation (15° counterclockwise) allows an apparent velocity of 7.0 km/s. This suggests lateral refraction could be another source of relative arrival-time error.

The extremely high apparent velocity for the MAKZ P_g-coda phase, along with its 10-s delay relative to P_n, suggests that we have made an error in identifying the observed phase in our blind test. For example, mantle reflections (*PmP*) have been identified in near-regional data in this area (QUIN and THURBER, 1992; LANGSTON, 1995). Approximating a model developed for location studies by QUIN and THURBER (1992) with upper and lower crustal gradients (Table 6b), we check apparent velocities and delays for various phases to identify our P_g-coda phase. This model does not allow a *PmP* phase to propagate to MAKZ distances, presumably because of the strong gradient in the upper crust; however, a double-bounce, *PmPPmP* phase is permitted, moves out at 7.3 km/s and is delayed 10 s relative to P_n at MAKZ distances. We are unsure that the dominant *P* arrival at MAKZ could be such an unusual phase. However, this adjustment cuts MAKZ ground-truth residuals by 150 ms for both shots (Table 8). The importance of this type of error indicates that relative arrival errors might be more appropriately modeled by considering distance and phase type (e.g., BONDÁR and NORTH, 1999) as well as frequency content.

Concern over relying on a multiple Moho reflection prompted us to re-examine MAKZ data for other repeatable P arrivals for use in location. We found a high-frequency P_n phase that matched well between the 50- and 300-m shots (Fig. 4a). The P_n wavetrains consist of multiple arrivals; extra complexity in the 300-m record may result from depth phases. The dominant frequency is over 6 Hz. P_n from the 550-m shot matched poorly. We also found an early P_g arrival that matched well between 50- and 550-m shot records (Fig. 4b). The dominant frequency is roughly 5 Hz. We need a lower crustal ray path to describe these arrivals, but, at these distances, the

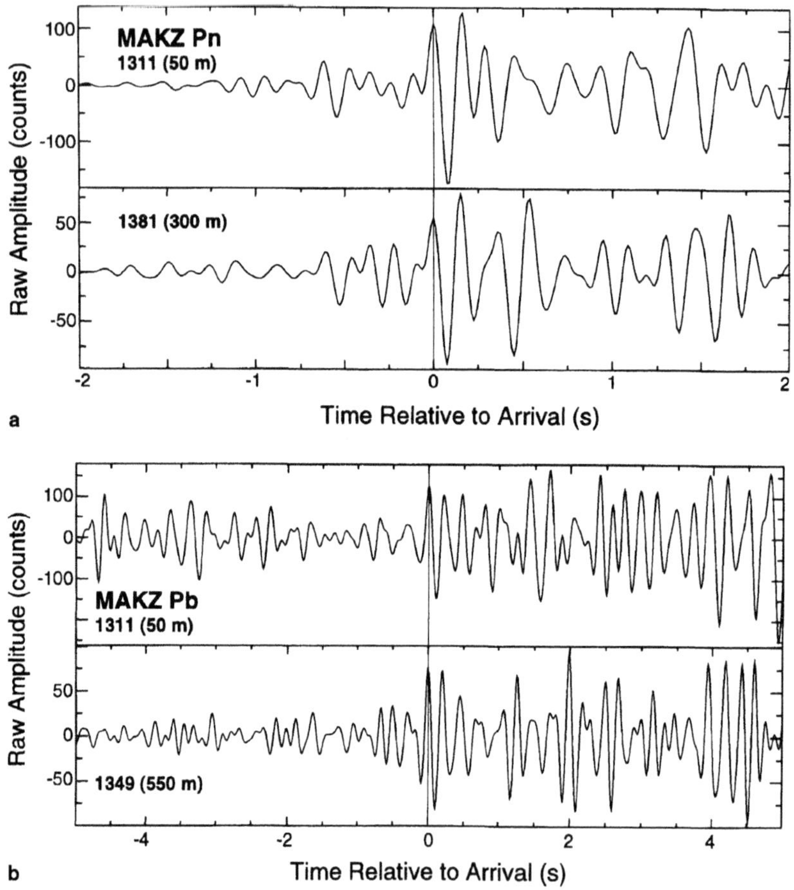

Figure 4
Vertical component data from Makanchi (MAKZ), Kazakhstan, for a) the 50- and 300-m shots, high-pass filtered at 1 Hz and aligned on chosen P_n arrival times, and b) the 50- and 550-m shots, band-pass filtered between 3 and 6 Hz, and aligned on chosen P_b arrival times. Shot holes and depths are indicated on each seismogram.

Kazakhstan model (Table 6b) will not allow such a path. We elected to use a lower crust refraction (P_b) from the AK135 model instead. We acknowledge that the ground-truth data are guiding our identification and modeling of these crustal phases, a tactic that would not be possible in a blind test.

Locations based on the original, but re-identified phase, and new P_n and P_b phases at MAKZ are offset from the ground truth by distances similar to those of the blind test (Runs 3 and 4, Tables 7a,b; Fig. 5). The low error P_n and P_b MAKZ phases exert more control than the Moho reflected phase in these runs. Excluding the MAKZ P_b, which has been modeled arbitrarily, we estimate relative arrival-time errors of T/8.5 by requiring the RMS weighted residual to be unity, as done earlier. Confidence ellipses are extremely small (2–3 km^2), owing to the addition of the low-error MAKZ data. The ellipses are also very eccentric, pointing towards the station gaps. Locations are offset from ground truth in directions parallel to the eccentric ellipses.

We investigated three specific sources of error that are found to be secondary, with respect to crustal phase modeling errors. First, as pointed out by THURBER *et al.* (1993), lateral variations in source-region velocity can produce error in relative relocation studies. Lateral velocity variations have been studied at Balapan using data from a local array deployed during the hole-closure series of shots. R_g from these shots was analyzed for a network of paths and tomographically imaged for bands from 0.5 to 2 s (BONNER *et al.*, 2000). Lateral changes are dramatic, with group velocities varying between 1.8 and 2.8 km/s at 1-s period. The 50-m

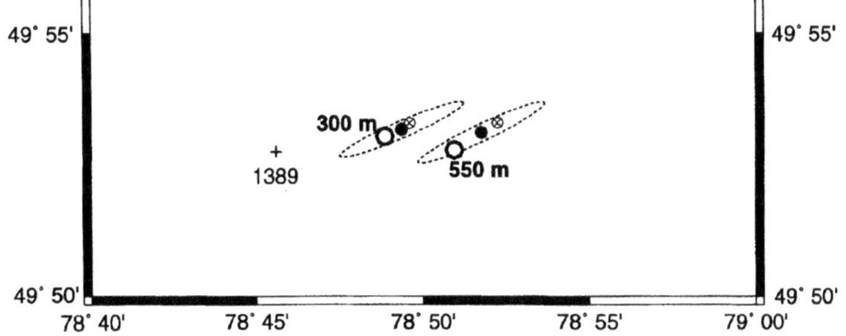

Figure 5
Master-event locations and confidence ellipses as in Figure 3 after adjusting MAKZ arrivals as guided by the ground-truth data.

shot took place in a lower velocity area of the test site, while the deeper shots took place in a high velocity area. This should produce early relative arrivals (residuals less than 0) for the deeper shots, which is opposite to what we observe at most stations (Table 8). We conclude that lateral variations contribute little to modeling error.

Second, because rays from shallow shots traverse near-surface material that is not sampled by rays from deeper shots, the velocity mismatch at shallow depths between the AK135 model and the true earth can also introduce error for turning ray paths. The upper crust compressional velocity is 5.8 km/s from the AK135 model, while apparent P velocities calculated from the local recordings range from 5.1 to 5.9 km/s (PEARSON *et al.*, 1998). The lower true velocity will delay arrivals from the 50-m shot, producing early relative arrivals from the deeper shots, again opposite to what we observe.

Finally, we have noted a pulse narrowing effect with depth for these data (Fig. 2). Such a source variation could introduce error because we pick or cross-correlate waveform peaks rather than first breaks. In this case, the narrowing with depth will result in earlier relative arrivals from the deep shots, once again opposite to what we observe.

As noted earlier, the ground-truth residual at NORESS for the 550-m shot exceeds picking error. Upper mantle anomalies have been proposed to account for spatially varying, teleseismic P amplitudes from neighboring Balapan explosions (BARKER and MURPHY, 1992; MCLAUGHLIN *et al.*, 1992). These authors show that the anomaly causes little travel-time variation to a resolution of 100 ms. However, the teleseismic arrival-time errors we observe fall at or below this level.

The last point we wish to address is the misassociation of waveforms, or cycle skipping. Misassociation could be a problem in a blind test situation, especially for poorer signal-to-noise arrivals such as the array arrivals we observe for the deeper shots (Figs. 2d,e). In a blind test, one could form a number of associations in the hope that consistent data from other stations would reject the poor associations. We tested this using our data set by adjusting NORESS and ILAR arrival times by a misassociation error of ± 600 ms, one at a time, and relocating. We examine raw and data-error weighted residuals to identify inconsistent associations. The raw residuals increased in all misassociation tests but one, while weighted residuals increased in all tests. After replacing the clock-error fraught ZAL data with predicted values, residuals degraded dramatically in all cases tested.

Conclusions

We tested the ability of the proposed IMS network to accurately locate small explosive sources (25 tons, m_{bLg} 1.8 to 2.6) after path calibration based on ground-truth information from one shot. Explosions were part of the depth-of-burial sequence of hole-closure shots at the former Soviet test site at Balapan, Kazakhstan. The study was carried out in a blind manner, after which, the full suite of ground truth was employed to evaluate errors. We hope the successful demonstration of relative relocation techniques will support the use of chemical explosions to calibrate the IMS and for special event studies, including OSI-related work.

Locations were based on a sparse set of stations where repeatable waveforms could be observed. These were the regional distance stations at Kurchatov, Makanchi and Zalesovo, as well as the teleseismic distance arrays NORESS and ILAR. Filtering helped to identify the repeatable waveforms. Array arrivals were determined on identical beams and a polarization filter helped emphasize NORESS arrivals.

We set data errors proportional to dominant periods in the waveforms that were used to determine arrival times. Results obtained using the master-event technique fell within 1 km of known ground truth. Comparing to ground-truth predictions, we estimated relative arrival-time errors to be $T/5.6$ (excluding Zalesovo data, which has known clock drift), where T is the dominant period in the picked waveform. The 90% confidence ellipses covered 12–13 km^2, far less than the standard required by OSI (1000 km^2), and approach the precision needed to associate events with overhead imagery.

In spite of the remarkably small mislocations, 90% confidence ellipses for true-depth locations covered the ground-truth location poorly, indicating errors were not well accounted for. Relative arrivals tended to be late when compared to ground-truth predictions and were too large to result from picking error, given the waveform similarity we observed. Obvious sources of error include lateral velocity variations at Balapan, poor shallow-depth model velocity, and pulse narrowing with shot depth. These mechanisms act in the wrong direction to account for the discrepancies we see. We believe misidentified and poorly modeled crustal phases are the main sources of error in this study. Error estimates fell to T/8.5 after adjusting for modeling errors along the Makanchi path. The resulting 90% confidence ellipses covered 2–3 km^2 and were highly eccentric, but consistent with mislocation vectors. The success of the original blind test in the presence of these errors proves the robust nature of the relative location procedure.

One could argue that this test benefited from a number of favorable conditions, and thus portrays an optimistic view of the ability of the IMS to perform calibrated location. Favorable conditions include the similar source type and sizes, good propagation characteristics in the vicinity of the source region, and the abundance of regional stations, beginning at 86 km, around the Balapan test site. These factors have helped us obtain high-quality, similar waveforms of sufficient numbers to constrain locations well. This should be kept in mind before extending results of this case study to other areas and to natural events. However, the study suffered from a number of factors, including the use of surrogate, three-component stations instead of the planned arrays, resulting in poor gap and noise levels that degraded waveform similarity in a few cases, clock error that eliminated otherwise high-quality data from station ZAL, as well as the aforementioned modeling errors for crustal phases. In the future, improvements to the IMS network, such as uniform timing and upgrades to arrays will increase the quality of available data and further lower magnitude thresholds for accurate relative location.

The Balapan case study suggests a new method of calibration. A tripartite array of shots would allow the calibration of travel times and apparent velocities, enabling precise relative relocations to be performed without models and model error effects. This type of calibration would be especially effective in the reciprocal sense, where the tripartite array of shots is placed around IMS stations and currently operating or temporary stations cover the surrounding regions. Obviously, the expense of additional shots is a drawback and whether or not the improved location ability justifies that expense would be a matter for discussion. However, the demonstration of such techniques in a well-covered and politically favorable region could lend support to arguments concerning the verifiability of a CTBT.

Acknowledgments

We thank the NNC of Kazakhstan, the pIDC, the IRIS DMC and W.Y. Kim at LDEO for making data available for this study. Hans Israelsson and Keith McLaughlin provided invaluable help with the data from Zalesovo. Howard Patton, Craig Pearson and George Randall contributed useful ideas to the study. We used the LOCSAT routine (Nagy, 1996) as well as GMT and SAC software. Frode Ringdal and two anonymous reviewers provided helpful suggestions for improving the manuscript. This work was performed under the auspices of the U.S. Department of Energy by Los Alamos National Laboratory under Contract W-7405-ENG-36.

References

Barker, B. W., and Murphy, J. R. (1992), *A Lithospheric Velocity Anomaly beneath the Shagan River Test Site: Detection and Location with Network Magnitude Residuals*, Bull. Seismol. Soc. Am. *82*, 980–998.

Bondár, I., and North, R. G. (1999), *Development of Calibration Techniques for the Comprehensive Nuclear-Test-Ban Treaty (CTBT) International Monitoring System*, Phys. Earth Planet. Int. *113*, 11–24.

Bonner, J. L., Pearson, D. C., Phillips, W. S., and Taylor, S. R. (2000), *Shallow Velocity Structure at the Shagan River Test Site in Kazakhstan*, Pure appl. geophys., submitted.

Bungum, H., and Husebye, E. S. (1971), *Errors in Time Delay Measurements*, Pure appl. geophys. *91*, 56–70.

Bratt, S. R., and Bache, T. C. (1988), *Locating Events with a Sparse Network of Regional Arrays*, Bull. Seismol. Soc. Am. *78*, 780–798.

DENNY, M. D., and JOHNSON, L. R. (1991), *The explosion seismic source function: Models and scaling laws reviewed.* In *Explosion Source Phenomenology* (eds. Taylor, S. R., Patton, H. J. and Richards, P. G.) (American Geophysical Union USA 1991) 268 pp.

EVERNDEN, J. F. (1969), *Identification of Earthquakes and Explosions by Use of Teleseismic Data*, J. Geophys. Res. *74*, 3828–3856.

GUPTA, V. (1995), *Locating Nuclear Explosions at the Chinese Test Site near Lop Nor*, Science and Global Security *5*, 205–244.

GUPTA, V., and RICH, D. (1996), *Locating the Detonation Point of China's First Nuclear Explosive Test on 16 October, 1964*, Int. J. Remote Sensing *17*, 1969–1974.

HANKS, T. C., and KANAMORI, H. (1979), *Moment Magnitude Scale*, J. Geophys. Res. *84*, 2348–2350.

ISRAELSSON, H. (1990), *Correlation of Waveforms from Closely Space Regional Events*, Bull. Seismol. Soc. Am. *80*, 2177–2193.

JORDAN, T. H., and SVERDRUP, K. A. (1981), *Teleseismic Location Techniques and their Application to Earthquake Clusters in the South-central Pacific*, Bull. Seismol. Soc. Am. *71*, 1105–1130.

KENNETT, B. L. N., ENGDAHL, E. R., and BULAND, R. (1995), *Constraints on Seismic Velocities in the Earth from Travel Times*, Geophys. J. Int. *122*, 108–124.

KIM, W. Y, KAZAKOV, V. V., VANCHUGOV, A. G., and SIMPSON, D. W. (1996), *Broadband and array observations at low noise sites in Kazakhstan: Opportunities for seismic monitoring of a Comprehensive Test-Ban-Treaty*. In *Monitoring a Comprehensive Test-Ban-Treaty* (eds. Husebye, E.S., and Dainty, A.M.) (Kluwer Academic Publishers Netherlands 1996) 836 pp.

LANGSTON, C. A. (1995), *Anatomy of Regional Phases and Source Characterization of the Soviet Joint Verification Experiment, Underground Nuclear Explosion*, Bull. Seismol. Soc. Am. *85*, 1416–1431.

LEITH, W. (1987), *Geology of NRDC Seismic Station Sites in Eastern Kazakhstan, USSR*, U.S. Geol. Surv. Open-File Rept. 87–597.

MCLAUGHLIN, K. L., MURPHY, J. R., and BARKER, B. W. (1992), *A Lithospheric Velocity Anomaly Beneath the Shagan River Test Site: Imaging and Inversion with Amplitude Transmission Tomography*, Bull. Seismol. Soc. Am. *82*, 999–1017.

MITCHELL, B. J., PAN, Y., XIE, J., and CONG, L. (1997), *Lg Coda Q Variation across Eurasia and its Relation to Crustal Evolution*, J. Geophys. Res. *102*, 22,767–22,780.

MONTALBETTI, J. F., and KANASEWICH, E. R. (1970), *Enhancement of Teleseismic Body Waves with a Polarization Filter*, Geophys. J. R. Astr. Soc. *21*, 119–129.

NAGY, W. (1996), *New Region-dependent Travel-time Handling Facilities at the IDC; Functionality, Testing and Implementation Details*, SAIC Tech. Rep. 96/1179, 57 pp.

NUTTLI, O. W. (1986), *Lg magnitudes of Selected East Kazakhstan Underground Explosions*, Bull. Seismol. Soc. Am. *76*, 1241–1251.

PATTON, H. (2000), *Regional Magnitude Scaling, Transportability, and* M_s:m_b Discrimination at Small Magnitudes, Pure appl. geophys., submitted.

PEARSON, D. C, BONNER, J. L., GLENN, L. A., DENNY, M. D., and PHILLIPS, W. S. (1998), *The Source Depth of Burial Experiment at Shagan River, Kazakhstan*, 20th Ann. Seismol. Res. Symp. Mon. CTBT, 86–95.

PRIESTLEY, K. F., ZANDT, G., and RANDALL, G. E. (1988), *Crustal Structure in Eastern Kazakhstan, USSR, from Teleseismic Receiver Functions*, Geophys. Res. Lett. *15*, 613–616.

QUIN, H. R., and THURBER, C. H. (1992), *Seismic Velocity Structure and Event Relocation in Kazakhstan from Secondary P-phases*, Bull. Seismol. Soc. Am. *82*, 2494–2510.

RINGDAL, F. (1990), *Teleseismic Event Detection Using the NORESS Array, with Special Reference to Low-yield Semipalatinsk Explosions*, Bull. Seismol. Soc. Am. *80*, 2127–2142.

SHEARER, P. M., and ASTIZ, L. (1997), *Locating Nuclear Explosions Using Waveform Cross-correlation*, Proc. 19th Ann. Seis. Res. Symp. Mon. CTBT, 301–309.

SYKES, L. R., DENG, J. S., and LYUBOMIRSKIY, P. (1993), *Accurate Location of Nuclear Explosions at Azgir, Kazakhstan, from Satellite Images and Seismic Data: Implications for Monitoring Decoupled Explosions*, Geophys. Res. Lett. *20*, 1919–1922.

THURBER, C. H., QUIN, H. R., and RICHARDS, P. G. (1993), *Accurate Locations of Nuclear Explosions in Balapan, Kazakhstan, 1987 to 1989*, Geophys. Res. Lett. *20*, 399–402.

THURBER, C. H., QUIN, H. R., and SALEH, R. (1994), *Catalog of Locations of Nuclear Explosions at Balapan, Kazakhstan, 1965 to 1985*, Bull. Seismol. Soc. Am. *84*, 458–461.

(Received July 2, 1999, revised January 24, 2000, accepted February 9, 2000)

Pure appl. geophys. 158 (2001) 193–209
0033-4553/01/020193-17 $ 1.50 + 0.20/0

Pure and Applied Geophysics

On the Use of Calibration Explosions at the Former Semipalatinsk Test Site for Compiling a Travel-time Model of the Crust and Upper Mantle

NADEZHDA N. BELYASHOVA,[1] VLADIMIR I. SHACILOV,[2] NATALIA N. MIKHAILOVA,[1] IGOR I. KOMAROV,[1] ZLATA I. SINYOVA,[1] ANDREY V. BELYASHOV,[1] and MARINA N. MALAKHOVA[1]

Abstract — Two chemical calibration explosions, conducted at the former Semipalatinsk nuclear test site in 1998 with charges of 25 tons and 100 tons TNT, have been used for developing travel-time curves and generalized one-dimensional velocity models of the crust and upper mantle of the platform region of Kazakhstan. The explosions were recorded by a number of digital seismic stations, located in Kazakhstan at distances ranging from 0 to 720 km. The travel-time tables developed in this paper cover the phases *P*, *Pn*, *Pg*, *S*, *Sn*, *Lg* in a range of 0–740 km and the velocity models apply to the crust down to 44 km depth and to the mantle down to 120 km. A comparison of the compiled travel-time tables with existing travel-time tables of CSE and IASPEI91 is presented.

Key words: Chemical calibration explosions, 1-D velocity model, Semipalatinsk test site.

Introduction

In 1997 and 1998, a number of chemical explosions with explosive charges up to 100 tons TNT were carried out at Balapan and Degelen of the former Semipalatinsk nuclear test site. The explosions were conducted under a joint Kazakhstan – United States agreement and the main objective of the explosions was liquidation of nuclear weapons infrastructure.

At the same time, these explosions are useful for seismic calibration purposes, since the seismic waves generated during explosions of this size generate important information about the velocity of the earth's crust and upper mantle in the region surrounding the test site. This type of study is important for monitoring the

[1] Institute of Geophysical Research, National Nuclear Center, Republic of Kazakhstan. E-mail: igr@nnc.kz
[2] Institute of Seismology, Department of Science, Republic of Kazakhstan.

Comprehensive Nuclear-Test-Ban Treaty, since it will contribute to improving the location accuracy of seismic sources in the region.

A generalized travel-time curve for distances up to 400 km and a first approximation of a 1-D crustal velocity model were previously computed by SHACILOV *et al.* (1998), using data from the 25-ton calibration explosions conducted at Semipalatinsk in 1997. The present paper is devoted to refinement and expansion to larger distances of the travel-time curve mentioned above, using data from two explosions conducted in 1998: one 25-ton explosion at the Balapan site, and one 100-ton explosion at the Degelen site. The parameters of the explosions are given in Table 1.

Station Network

A special-purpose network of temporary seismic stations, located at distances from 100 m to 15 km (see Fig. 1) was installed to register these calibration explosions. Moreover, seismic stations were installed near the Baianaul and Karkaralinsk settlements (seismic stations were operated at these sites during the joint Soviet-American experiment in 1988). Also, data from the following permanent seismic stations, belonging to the National Nuclear Center of the Kazakhstan Republic were used to determine a new travel-time curve: Kurchatov, Makanchi, Vostochnoye, Borovoye, Chkalovo and Zerenda (see Fig. 2). All of the data were recorded in digital form on PASCAL stations. The recording systems provide wide frequency and dynamic range of the recorded data and allow the determination of arrival times with high accuracy. Table 2 gives the distances between the stations and explosions.

Data Analysis

The recorded waveforms of the explosions were analyzed in detail. *P* waves, *S* waves and surface waves were observed for the closest stations S3, S4, S5, S6 (see Fig. 3). The Kurchatov station is located somewhat further from the explosion. Two

Table 1

Parameters of calibration explosions conducted during 1998

Borehole no.	Date	Time of explosion, GMT	Explosive charge (tons TNT)	Depth of placement of the charge, meters	Explosion coordinates	
					Latitude	Longitude
214	22.08.1998	05.00.18,904	100	146	49°46′00.0″	77°59′27.0″
1071-bis	17.09.1998	07.19.40,55	25	28	49°58′51.7″	78°45′21.2″

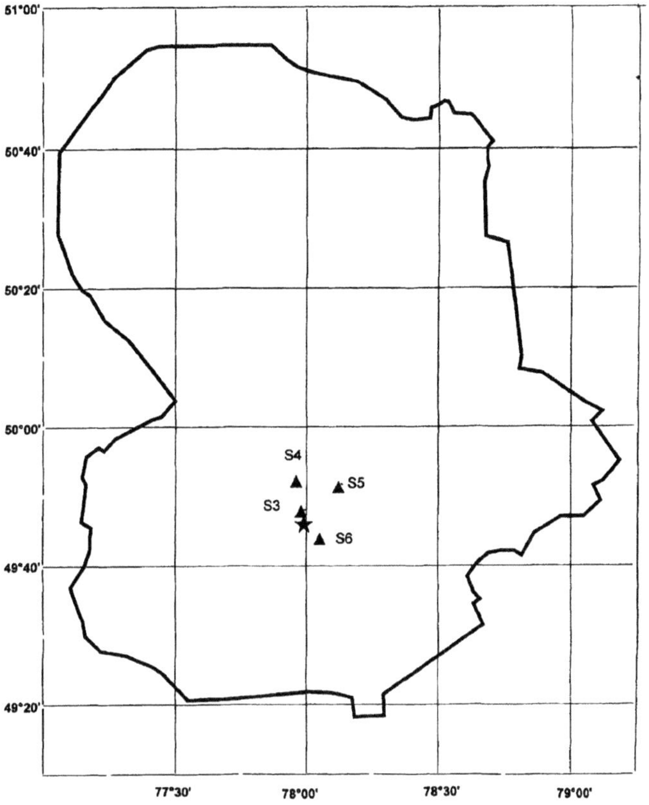

Figure 1
Deployment of nearby temporary seismic station, installed to record the 100-ton calibration explosion. The solid line shows the borders of the Semipalatinsk test site.

P arrivals, with a time difference of about 3 s, are clearly visible on the waveforms of the 100-ton explosion at this station (Fig. 4); and in this case the amplitude of the second *P* wave is larger than the amplitude of the first one. Two *S* waves could also be identified on the waveforms of this explosion. The surface waves, arriving after *S* phase, are very pronounced for the 100-ton explosion. In this connection, it should be noted that on the waveforms of Kurchatov for the 25-ton explosion, the second *P* and *S* phases were not found.

The Karkaralinsk and Baianaul stations are located at distances reflecting the most interesting part of the seismic travel-time curve (180–250 km). In this distance range the type of *P* phase which is arriving first, changes. We note (Fig. 5) that the *P* waveforms have a very complex structure. At the very beginning of the *P* arrival for the 100-ton explosion on these stations, very weak oscillations, with periods of 0.6–0.8 s are observed. Thereafter, the much stronger *Pn* phase arrives. The maximum amplitude of the *P*-wave group, exceeding by several times the amplitude of *Pn*

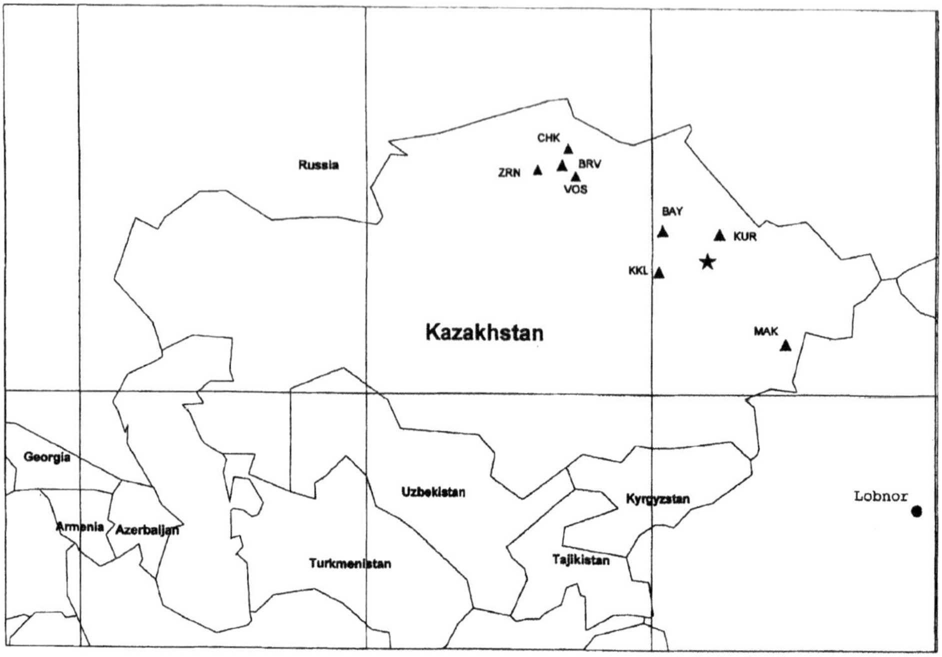

Figure 2
Map showing the location of remote stations recording the calibration explosions.

Table 2

Epicentral distances for temporary and stationary seismic stations

Station	Code	Δ, km
	100-ton explosion, Degelen	
S_3		3,3
S_4		11,5
S_5		13,6
S_6		5,9
Kurchatov	KUR	102
Karkaralinsk	KKL	194
Baianaul	BAY	210
Makanchi	MAK	442
Vostochnoye	VOS	589
Borovoye	BRV	652
Chkalovo	CHK	670
Zerenda	ZRN	719
	25-ton explosion, Balapan	
Karkaralinsk	KKL	254
Baianaul	BAY	246

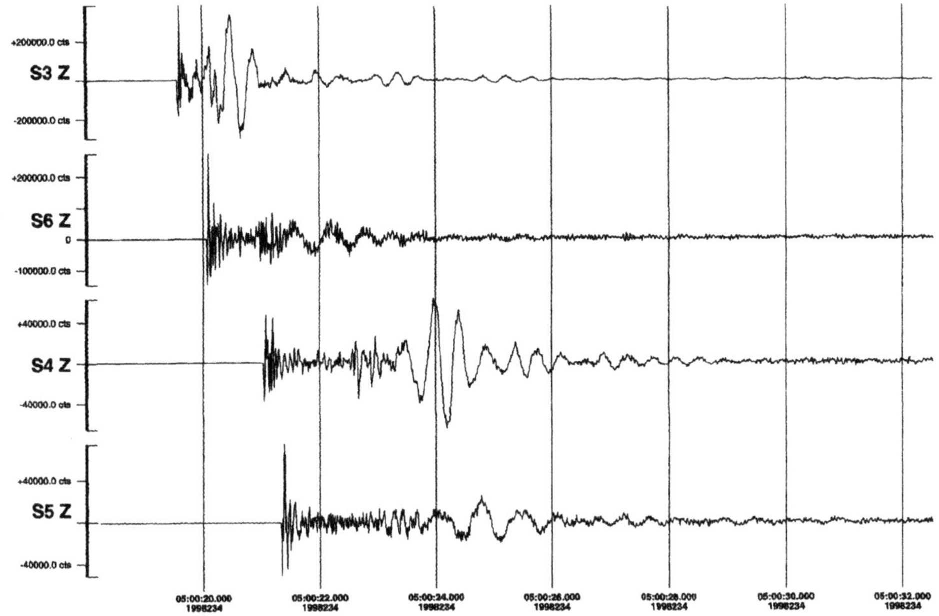

Figure 3
Waveforms of the 100-ton calibration explosion, recorded by local temporary stations (vertical component).

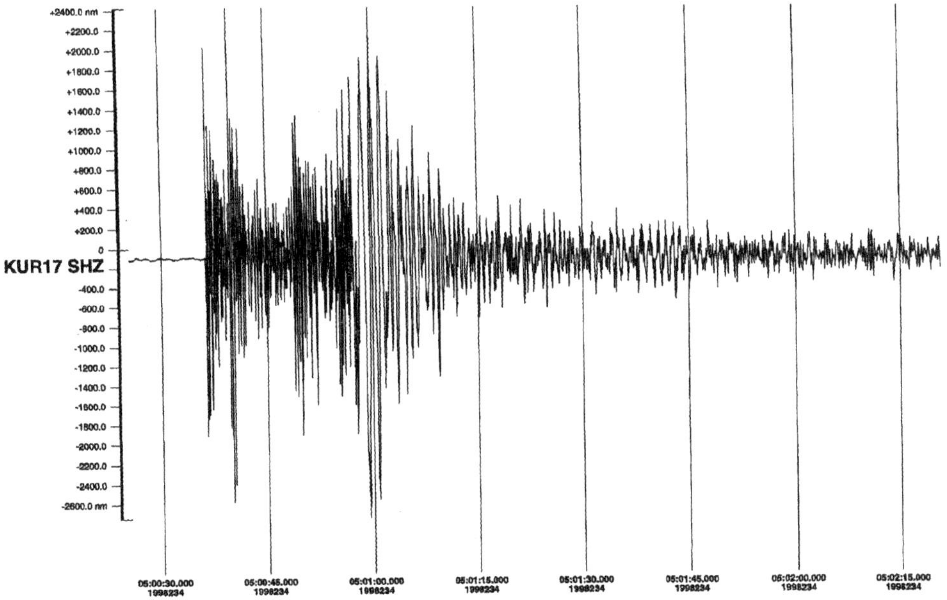

Figure 4
Waveforms of the 100-ton calibration explosion, recorded by the Kurchatov station (vertical component).

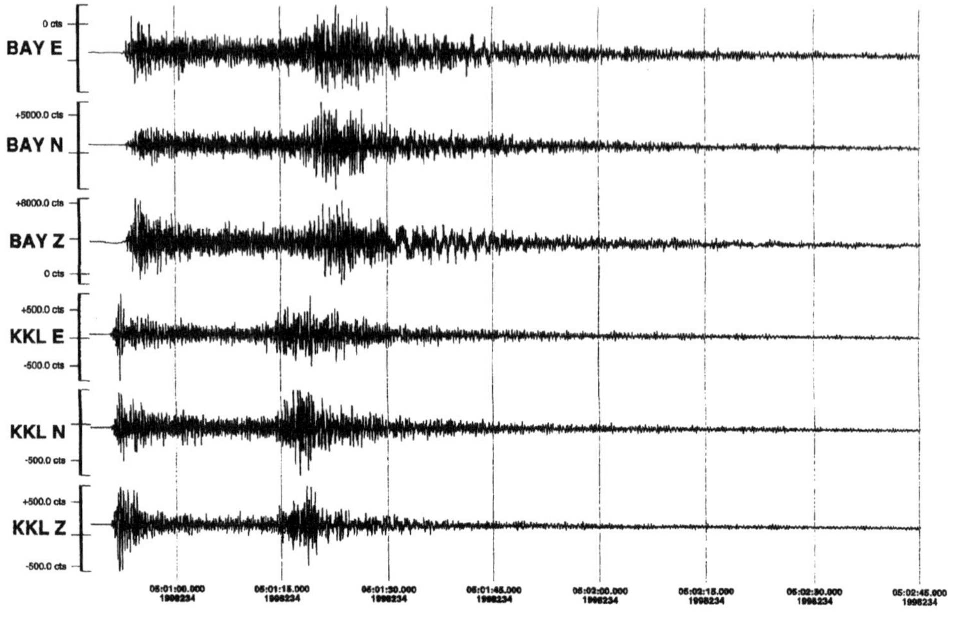

Figure 5
Waveforms of the 100-ton calibration explosion, recorded by the Baianaul and Karkaralinsk seismic stations.

phase, is observed for the *Pg* phase. In the group of *S* waves, the *Sn* phase can be observed with the dominant period shifted towards longer periods; and the amplitudes of the *Sn* phase slightly exceeds the amplitudes of the *P* coda. The *Lg* wave, arriving after *Sn*, has the maximum amplitudes for this part of the records.

The waveforms of *P* waves for the Makanchi station (Fig. 6) are also rather complicated. There are weak low-frequency oscillations during the first second after the *P*-wave arrival. The high-frequency *Pn* phase, arriving after about one second, is clearly visible. The *Pg* phase, the most intensive among the *P* waves, arrives 8.8 s later. *Sn* and *Lg* phases were also observed in the group of *S* waves.

The Borovoye large aperture array is farthest from the explosions among the stations in the network. The waveforms of the 100-ton explosion recorded at individual Borovoye sensors (Fig. 7), are quite similar to each other. Three arrivals can be noted for the group of *P* waves: the very first *P* arrival, which is very weak and of low frequency; the high-frequency *Pn* phase arriving approximately one second later; and the *Pg* wave which arrives after an additional 10–12 s. In this case the *Pn* wave is more intensive than the *Pg* wave. Also, *Sn* and *Lg* phases were observed in the group of *S* waves, with the *Lg* phase having the largest amplitude.

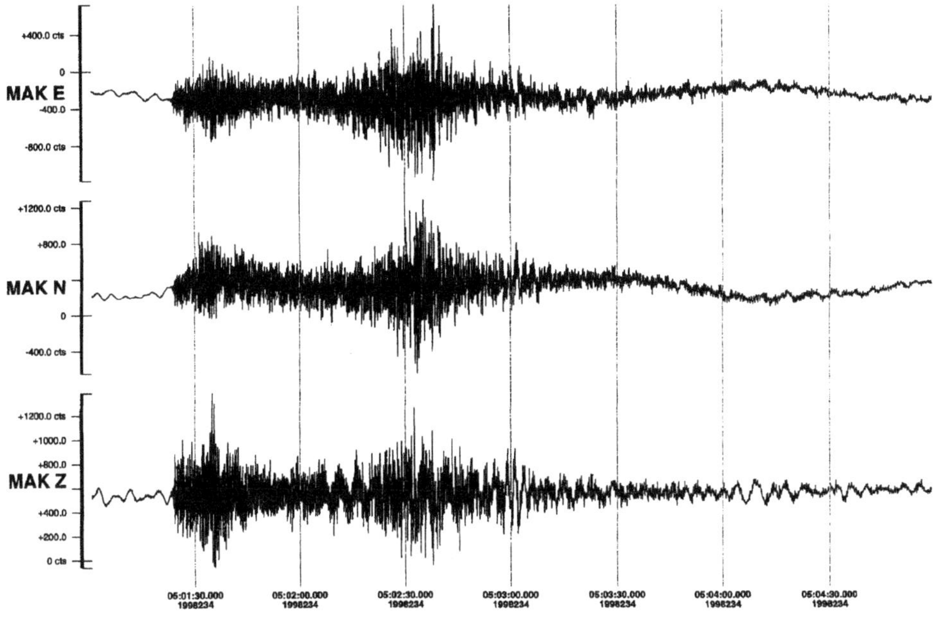

Figure 6

Waveforms of the 100-ton calibration explosion, recorded by the Makanchi station.

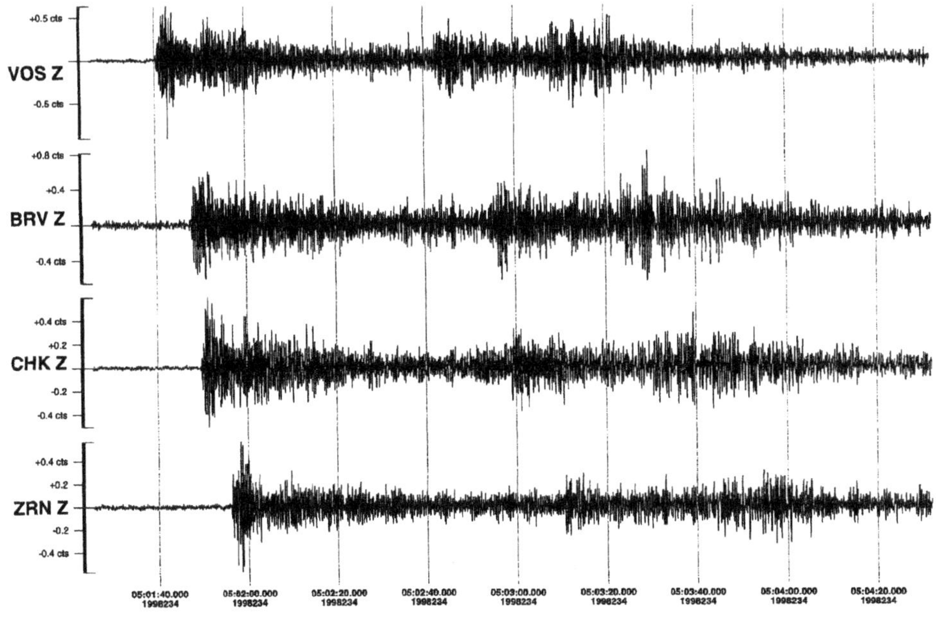

Figure 7

Waveforms of the 100-ton calibration explosion, recorded by stations of the Borovoye large aperture array (vertical component).

Travel-time Curves

The seismic recordings from all stations were used to supplement the waveforms compiled in 1997 (Fig. 8). The new data allowed the uncertainties to be eliminated in the most important part of the previously developed travel-time curve, which is the cross-over distance, i.e., where the *Pn* phase becomes the first arriving phase. The travel-time curve was extended from 400–740 km, with the *Pn* phase being the first arrival in this distance interval, using approximately regular distance step (see Fig. 9). The discrete arrival-time values of all types of *P* and *S* waves were transferred into travel times using graphical interpolation of the corresponding seismic phases.

Information regarding 'pre-critical reflections' of P_m^{ref} and S_m^{ref} waves, unique for the given region, was obtained for the first part of the travel-time curve (distance 0–160 km, using stations S3–S6 and the Kurchatov station). This is near the so-called 'starting point' which divides the travel-time curve of reflected waves into pre-critical and post-critical parts and which controls the beginning of *Pn* and *Sn* phases as secondary arrivals. The uniqueness of the information lies in the fact that, because of the large explosion power and the high quality of the seismological equipment, pre-critical phases reflected from the Moho boundary could be detected. Previously, it has rarely been possible to detect such pre-critical

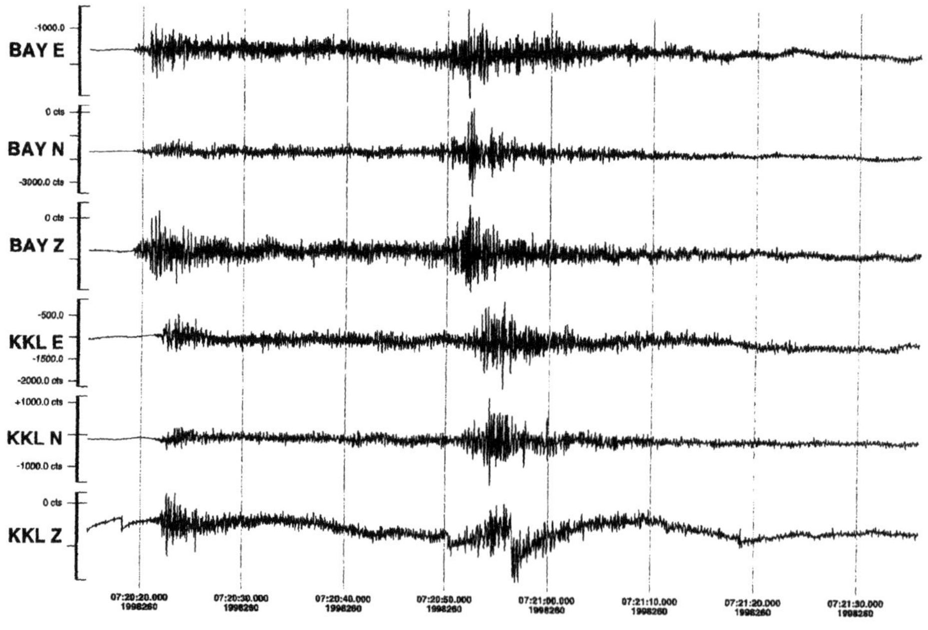

Figure 8
Waveforms of the 25-ton calibration explosion, recorded by the stations Bayanaul and Karkaralinsk.

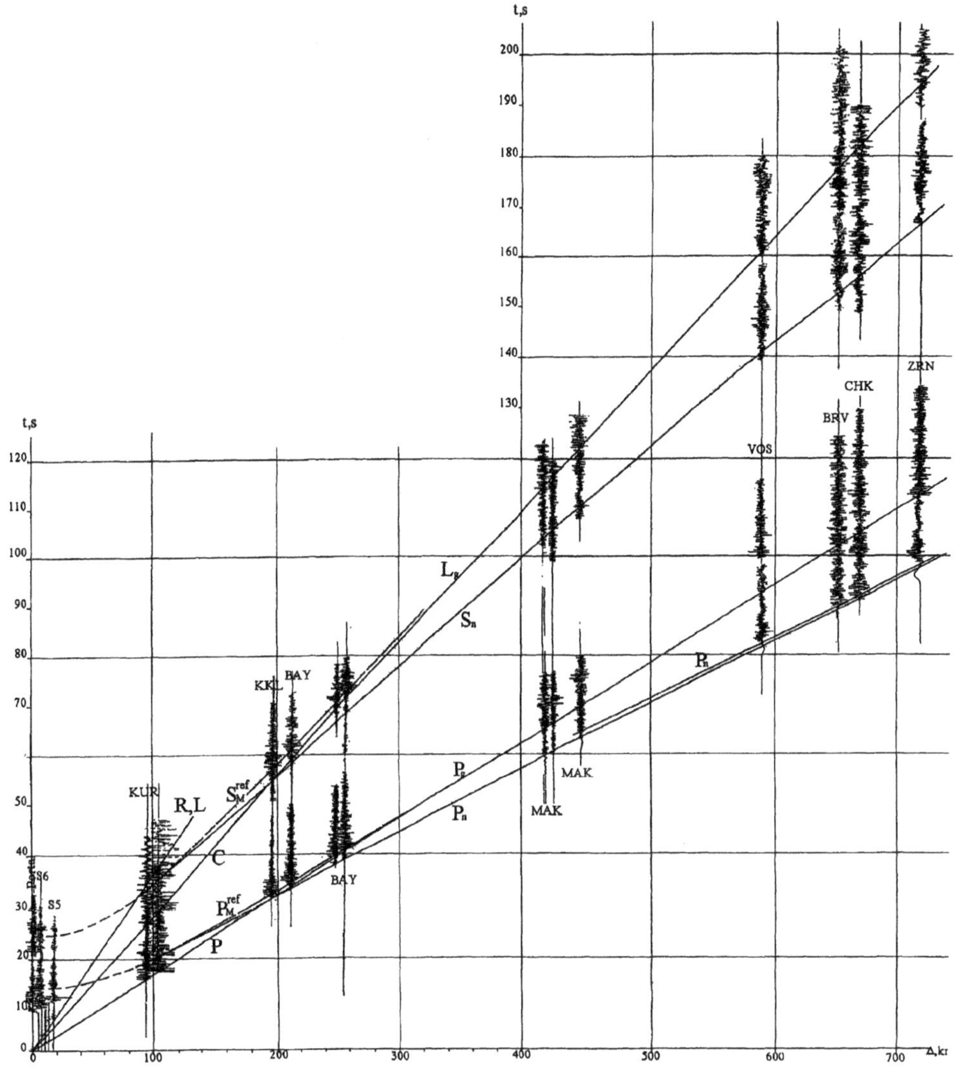

Figure 9
Generalized travel-time curves of longitudinal and transverse waves from Central Kazakhstan from calibration explosion records.

reflected phases during standard deep seismological sounding surveys, because of insufficient characteristics of the recording channels (small dynamic range and narrow frequency band).

Table 3 shows new computed travel-time tables of the main seismic phases for Central Kazakhstan (denoted CK-98) for distances 0–740 km. The travel times of *P*, *Pn*, *Pg*, *S*, *Sn*, *Lg* phases are given with 10–20 km intervals.

Table 3

Generalized travel-time table, compiled using data of the calibration explosions (CK-98)

Δ, km	*P, Pg*	*Pn*	*S, Lg*	*Sn*
0	0,0		0,0	
10	1,8		3,3	
20	3,5		6,3	
30	5,2		9,3	
40	7,0		12,1	
60	10,4		17,4	
80	13,6		22,9	
100	16,7	20,2	28,3	35,6
120	20,0	22,3	34,0	39,1
140	23,2	24,8	39,6	43,2
160	26,4	27,3	45,4	47,5
180	29,6	29,9	51,0	51,9
200	32,8	32,6	56,8	56,4
220	35,8	35,2	62,2	60,9
240	39,0	37,9	68,0	65,6
260	42,2	40,5	73,4	70,1
280	45,4	43,1	78,8	74,5
300	48,5	45,6	84,2	79,0
320	51,7	48,2	89,5	83,5
340	54,7	50,9	94,9	87,9
360	57,8	53,5	100,3	92,4
380	60,8	56,2	105,6	97,0
400	63,8	58,8	110,9	101,3
420	66,8	61,4	116,2	105,6
440	69,9	63,8	121,6	109,9
460	73,0	66,3	126,8	114,0
480	76,0	68,6	132,0	118,2
500	79,0	71,1	137,4	122,5
520	82,0	73,5	142,8	126,6
540	85,0	75,9	148,1	130,9
560	88,0	78,3	153,3	135,0
580	91,1	80,7	158,5	139,0
600	94,1	83,0	163,7	142,8
620	97,1	85,4	168,9	146,5
640	100,2	87,6	174,3	150,2
660	103,3	90,0	179,6	154,1
680	106,2	92,4	184,6	158,3
700	109,2	95,0	189,4	162,4
720	112,2	97,5	194,1	166,6
740	115,3	100,0		

Comparison with Previous Studies

A comparison of the newly created travel-time tables (CK-98) with travel-time tables compiled for this region by the Complex Seismological Expedition (CSE) of the Institute of Earth Physics of the Academy of Sciences of USSR in 1964

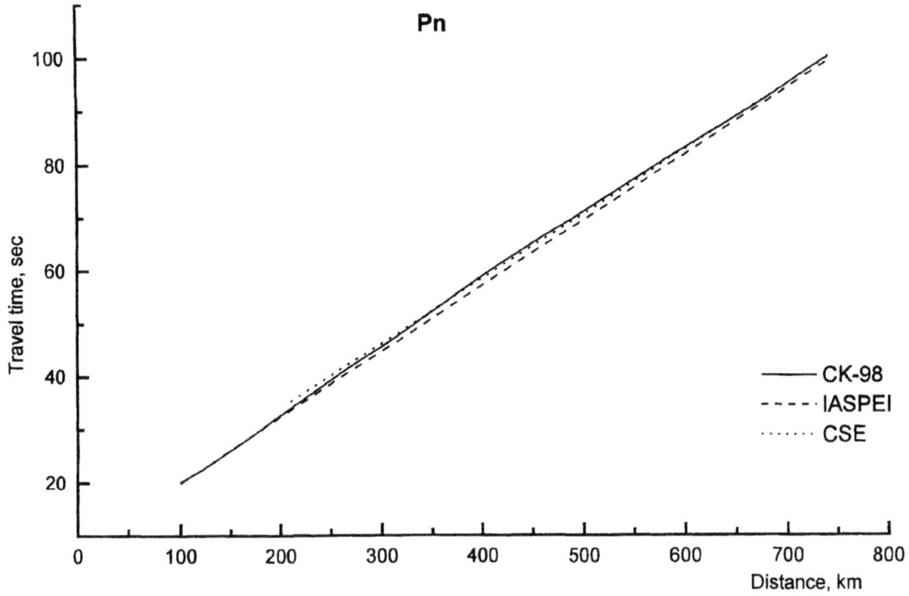

Figure 10
Travel-time comparison for *Pn* waves for CK-98, IASPEI and CSE.

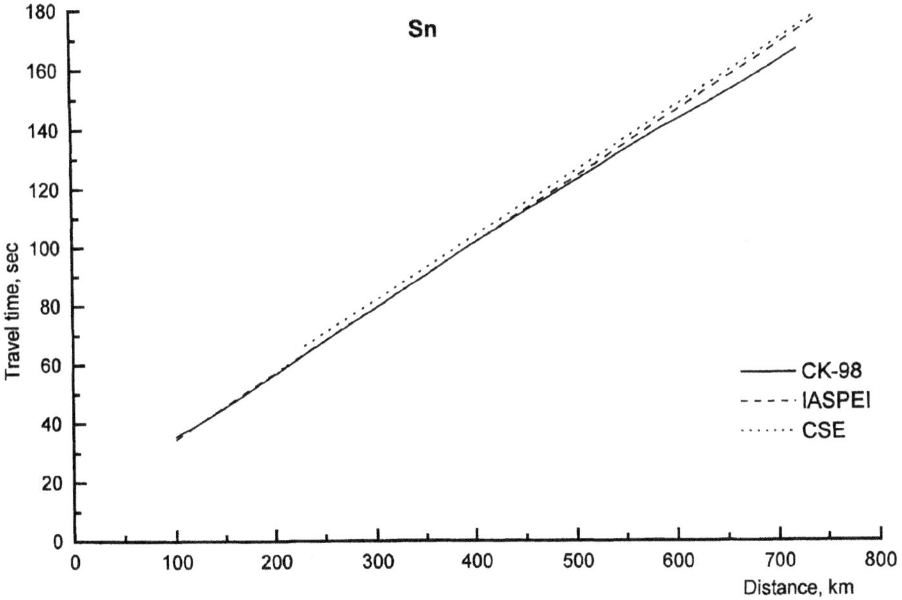

Figure 11
Travel-time comparison for *Pg* waves for CK-98, IASPEI and CSE.

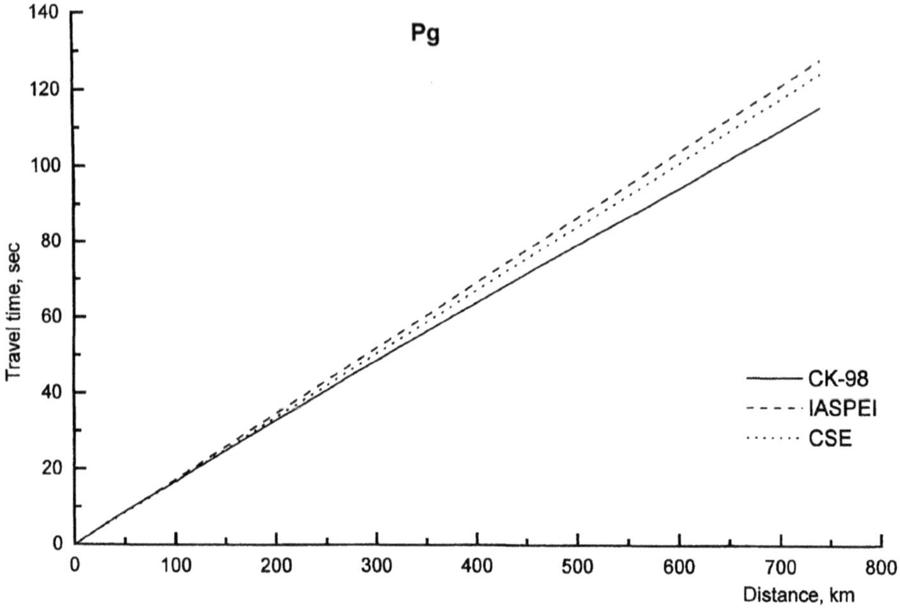

Figure 12
Travel-time comparison for *Sn* waves for CK-98, IASPEI and CSE.

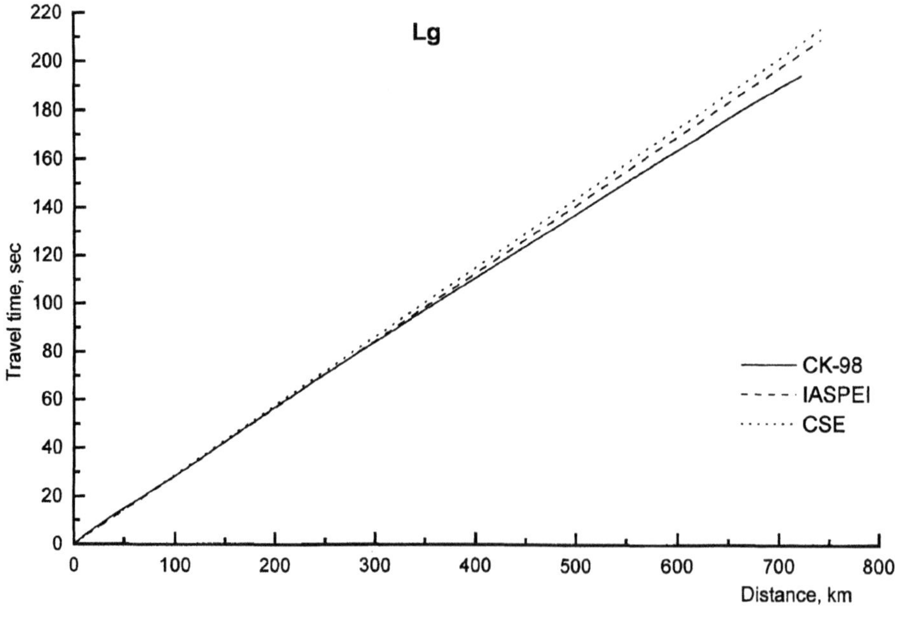

Figure 13
Travel-time comparison for *Lg* waves for CK-98, IASPEI and CSE.

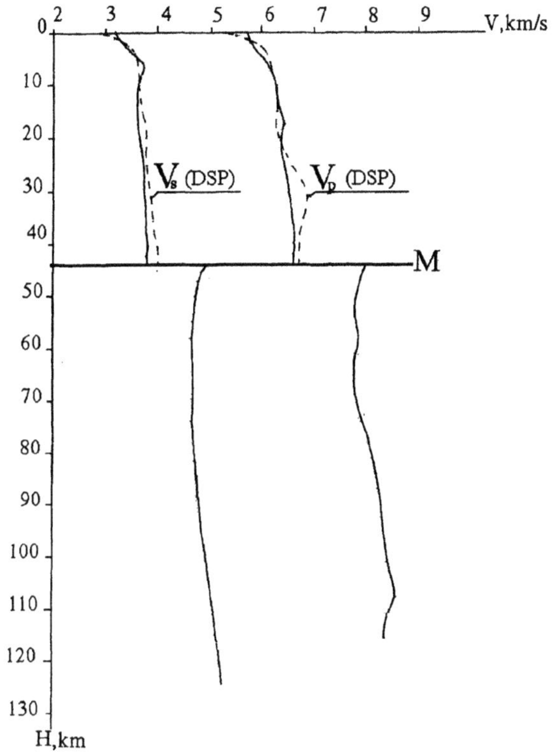

Figure 14
Generalized velocity model for longitudinal and transverse waves for the earth's crust and upper mantle for Central Kazakhstan according to the present study (solid line) and previous DSP data (dashed line).

Table 4

Generalized velocity model, compiled from data of deep seismic sounding survey

H	40	37.5	35	32.5	30	27.5	25	22.5	20	17.5	15	12.5	10	7.5	5	2.5	0	H, km
V_s	3.99	3.93	3.91	3.89	3.85	3.82	3.79	3.77	3.78	3.77	3.70	3.67	3.64	3.63	3.60	3.45	2.90	V_s, km/s
V_p	6.74	6.75	6.80	6.86	6.87	6.75	6.57	6.45	6.35	6.28	6.30	6.30	6.30	6.20	6.15	6.00	5.25	V_p, km/s

(NERSESOV and RAUTIAN, 1964) and also with the IASPEI91 travel-time tables (KENNETT, 1991) was made. Figures 10 through 13 compare travel times of main seismic *Pn*, *Pg*, *Sn*, and *Lg* waves. For the *Pn* wave all 3 models show similar travel-time values within the studied range of distances (Fig. 10). For the *Sn* wave all 3 models also show similar travel time for distances reaching 600 km (Fig. 11).

Table 5

Velocity model of the earth's crust, computed from calibration explosion travel-time curves

V_s		V_p		V_s		V_p	
H, km	V, km/s	H, km	V, km/s	H, km	V, km/s	H, km	V, km/s
0.1	3.18	0.1	5.72	32.0	3.76	37.8	6.64
2.6	3.39	1.8	5.78	36.7	3.77	39.7	6.62
6.0	3.73	5.2	6.06	40.4	3.82	41.9	6.62
8.2	3.67	8.6	6.24	42.4	3.78	44.0	6.66
11.0	3.61	10.9	6.29	44.0	3.79		
14.4	3.60	14.3	6.31				
17.3	3.61	17.3	6.45				
21.9	3.68	18.8	6.36				
26.0	3.73	23.6	6.40				
28.6	3.73	32.1	6.52				

For the *Pg* wave there are major deviations in the new calculated travel times as compared to IASPEI91 and CSE (Fig. 12). The velocity of this wave from CK-98 significantly exceeds the corresponding value from the other two models. This is also typical for the *Lg* wave but to a lesser extent (Fig. 13).

Velocity Model

A generalized 1-D velocity model for each of the travel-time tables for the earth's crust and upper mantle was computed separately, using the'INVERS' (inverse problem of refracted waves) algorithm (NERSESOV and RAUTIAN, 1964; MISHENKINA *et al.*, 1983). The algorithm allows calculation of the parts of the cross section with positive and negative values of the $\partial V/\partial z$ gradient. In our case the medium for obtaining the one-dimensional cross section was represented in the form of plane parallel velocity layers. This allowed use of a two-dimensional inverse problem solution algorithm for velocity model calculation. The computed velocity models are given in Figure 14. This figure also shows generalized models for the crust, obtained previously by DSP. The computed velocity models are also shown in Tables 4, 5, 6.

Discussion

The difference between the travel-time curves obtained in the present study (CK-98) and the IASPEI91 curves becomes quite understandable if one takes into account that travel-time curves compiled on the basis of earthquake data (like IASPEI91) reflect the orogenic lower part of the earth's crust, where the V_p and V_s have lower

Table 6

Velocity model of the earth's mantle, computed from calibration explosion travel-time curves

V_s	H, km depth from the ground	44.1	44.4	47.2	52.6	58.4	63.9	68.7	73.9	78.6	82.0	85.1	88.2	99.7	116.2	122.2		
	H, km from the Moho	0.1	0.4	3.2	8.6	14.4	19.9	24.7	29.9	34.6	38.0	41.1	44.2	55.7	72.2	80.2		
	V_s	4.88	4.88	4.77	4.70	4.65	4.67	4.67	4.65	4.70	4.74	4.75	4.77	4.93	5.16	5.25		
V_P	H, km depth from the ground	44.1	44.4	45.9	49.7	54.3	57.6	60.9	65.1	68.8	76.2	84.6	88.4	92.6	101.2	107.6	110.9	115.4
	H, km from the Moho	0.1	0.4	1.9	5.7	10.3	13.6	16.9	21.1	24.8	32.2	40.6	44.4	48.6	57.2	63.6	66.9	71.4
	V_P	8.0	8.0	7.94	7.84	7.81	7.88	7.82	7.80	7.82	8.05	8.24	8.31	8.35	8.47	8.62	8.46	8.41

values as compared with the platform values. The CK-98 travel-time curve, which is compiled using explosion data, reflects exactly the platform crust. The agreement between the travel-time curves for the Pn phase for CK-98 and IASPEI91 suggests approximately equal values of V_p at depths between 45–120 km for platform and orogenic mantle.

The length (740 km) of the travel-time curve developed in this paper is sufficient to study the velocity model of the earth's crust to its entire extent, using the *P* + *Pg* and *S* + *Lg* phase travel-time curves, which describe 'direct' refraction waves, penetrating to the very bottom of the earth's crust. According to generalized data obtained by previous geophysical surveys (using converted waves of earthquakes and deep seismic sounding), the average depth of the Moho layer in this area is 44 km. In this case a sufficient length of the travel-time curve would be about 600 km. Beyond that, we see 'wing-like' in-phase axes, corresponding to the *Pg* and *Lg* waves. Thus, it becomes possible to study the velocity characteristics of platform crust, and, equally important, of the thick layer of the upper mantle (80 km). This has been impossible using methods of deep seismic sounding, which supply a travel-time curve with a range generally not exceeding 250–300 km, or by using ordinary seismological tools, because of the absence of strong earthquakes in this region.

Acknowledgement

The equipment used for local seismic recording of the calibration explosions was provided by Los Alamos National Laboratory and Lamont-Doherty Earth Observatory, and the data acquisition for both the temporary and permanent station networks was carried out in cooperation with scientists from these institutions. We acknowledge in particular the contributions by Dr. Won-Young Kim and Dr. Michael Haggerty, who participated in this work under an agreement between the IRIS Consortium and the National Nuclear Center of the Republic of Kazakhstan.

References

Kennett, B. L. N. (1991), *IASPEI 1991 Seismological Tables*, Research School of Earth Sciences, Australian National University, 167 pp.

Mishenkina, Z. R., Sheludko, I. F., and Krylov, S. V., *The usage of linearized model for inverse kinematics problem for two-dimensional time fields t(x,l) of refracted waves* (Nauka, Novosibirsk 1983). In *Numerical Methods in Seismic Research* (Nauka Publisher, Siberian Division, Novosibirsk, 1983) pp. 140–152.

Nersesov, I. L., and Rautian, T. G., *Kinematics and dynamics of seismic waves at distance up to 3500 km*, Proceedings of the Institute of Earth Physics (Moscow, 1964) *32*, 63–87.

SHACILOV, V. I., BELYASHOV, A. V., and MALAKHOVA, M. N., *Utilization of calibration explosions for verification of velocity structure of the earth's crust of a former nuclear test site.* Reports on the Second International Conference on Non-proliferation Problems, Kurchatov, September 14–17, 1998. p. 58.

(Received July 21, 1999, revised January 20, 2000, accepted February 9, 2000)

Pure appl. geophys. 158 (2001) 211–240
0033–4553/01/020211–30 $ 1.50 + 0.20/0

Pure and Applied Geophysics

Improving Regional Seismic Event Location in China

LEE K. STECK,[1] AARON A. VELASCO,[1] ALLEN H. COGBILL,[1] and HOWARD J. PATTON[1]

Abstract—In an effort to improve our ability to locate seismic events in China using only regional data, we have developed empirical propagation path corrections and applied such corrections using traditional location routines. Thus far, we have concentrated on corrections to observed *P* arrival times for crustal events using travel-time observations available from the USGS Earthquake Data Reports, the International Seismic Centre Bulletin, the preliminary International Data Center Reviewed Event Bulletin, and our own travel-time picks from regional data. Location ground truth for events used in this study ranges from 25 km for well-located teleseimic events, down to 2 km for nuclear explosions located using satellite imagery. We also use eight events for which depth is constrained using several waveform methods. We relocate events using the EvLoc algorithm from a region encompassing much of China (latitude 20°–55°N; longitude 65°–115°E). We observe that travel-time residuals exhibit a distance-dependent bias using IASPEI91 as our base model. To remedy this bias, we have developed a new 1-D model for China, which removes a significant portion of the distance bias. For individual stations having sufficient *P*-wave residual data, we produce a map of the regional travel-time residuals from all well-located teleseismic events. Residuals are used only if they are smaller than 10 s in absolute value and if the seismic event is located with accuracy better than 25 km. From the residual data, correction surfaces are constructed using modified Bayesian kriging. Modified Bayesian kriging offers us the advantage of providing well-behaved interpolants and their errors, but requires that we have adequate error estimates associated with the travel-time residuals from which they are constructed. For our *P*-wave residual error estimate, we use the sum of measurement and modeling errors, where measurement error is based on signal-to-noise ratios when available, and on the published catalog estimate otherwise. Our modeling error originates from the variance of travel-time residuals for our 1-D China model. We calculate propagation path correction surfaces for 74 stations in and around China, including six stations from the International Monitoring System. The statistical significance of each correction surface is evaluated using a cross-validation technique. We show relocation results for nuclear tests from the Balapan and Lop Nor test sites, and for earthquakes located using interferometric synthetic aperture radar. These examples show that the use of propagation path correction surfaces in regional relocations eliminates distance bias in the residual curves and significantly improves the accuracy and precision of seismic event locations.

Key words: Seismic, location, China, kriging, regional.

[1] Los Alamos Seismic Research Center, EES-3 MS D408, Los Alamos National Laboratory, Los Alamos, NM 87545, USA. E-mail: lsteck@lanl.gov

1. Introduction

Seismic detection and location of small events remain a key issue for effectively monitoring a Comprehensive Nuclear-Test-Ban Treaty (CTBT), which has an implicit goal of locating seismic events within a contiguous area smaller than 1000 km^2. In China, where stations reporting to the international bulletins are sparse and crustal structure is strongly heterogeneous, this goal presents a significant challenge. China is host to some of the most dramatic terrain in the world, including the Tibetan Plateau (the largest and highest of its kind in the world), and the Tarim Basin. Structures such as these can have profound effects on the travel times and amplitudes of seismic waves propagating through them, and these effects, if not accounted for, may cause significant errors in the associated event locations. This is particularly true for seismic events in the regional context, where the number of observing stations is small and the signal-to-noise ratios (SNRs) of seismic phases can be low.

When a seismic event occurs near an existing nuclear test site, significant political problems can arise when the event is poorly recorded and located. A good example of this is the 16 August 1997 seismic event near the Russian nuclear test site at Novaya Zemlya (Israelsson *et al.*, 1997; Richards and Kim, 1998; Hartse, 1998). This $m_b = 3.9$ event created concern within the United States government due both to the event's occurrence in a region of low seismicity and to its proximity to the test site. The event was only observable at stations within 2500 km of the test site, and the seismic location was ambiguous at first because of low SNRs for P and S arrivals at many stations. The political fallout of this event resulted in the U.S. government requesting an explanation from Russian government officials, with Russian officials responding that the event was a small earthquake that occurred in the Kara Sea. Eventually, analysis of waveforms by United States researchers confirmed that pronouncement (Israelsson *et al.*, 1997; Hartse, 1998). Small events such as these will continue to create controversy until they can be characterized quickly and properly.

To improve regional seismic location in China, we follow an empirical approach by developing spatially varying propagation path corrections (PPCs) to P-wave travel-time observations for shallow events. For this study we have used bulletin data from the United States Geological Survey's (USGS) Earthquake Data Reports (EDRs), the International Seismic Centre (ISC) bulletin, the prototype International Data Center's (pIDC) Reviewed Event Bulletin (REB), and our own travel-time picks from regional data. We have chosen to locate events within a region defined from 20–55°N latitude and from 65–115°E longitude, using the EvLoc algorithm (Bratt and Bache, 1988; Nagy, 1996). While our intent is to calibrate all regional seismic stations in Asia, many of the stations of interest, particularly those that will be part of the International Monitoring System (IMS), have not been in operation long enough to acquire sufficient data for our empirical approach. To circumvent this problem we use data from stations reporting to the EDR and ISC as surrogates for those of the IMS.

The paper is organized in the following fashion. We first describe our data sets, velocity models, and sources of ground truth. For distances $<20°$, we find that our travel-time residual surfaces show some distance dependence when global travel-time tables are used, suggesting an inadequate model of regional crustal and upper mantle structure for China. To properly construct the travel-time residual surfaces, we must attempt to remove this trend. For this reason, we develop and examine some 1-D regional velocity models for China prior to surface construction. Next, we discuss event relocations based on four velocity models. We follow this with a section on the development of the PPC surfaces. COGBILL and STECK (1997) found that by using PPC surfaces, they were able to obtain stable regional locations when as few as 3 to 5 stations were available, even if the closest stations were $\sim 10°$ distant. Furthermore, the regional location bias due to lateral structure variation was reduced. Grid-search locations (KENNETT, 1991) also improved with the use of PPCs, but not as much as with the EvLoc algorithm (COGBILL and STECK, 1997). Here, we enhance these PPC surfaces by using more regional data and adding more ground-truth information. We develop the PPC surfaces using modified Bayesian kriging (SCHULTZ *et al.*, 1998) for 68 EDR/ISC and 6 IMS stations. We then validate and test these PPC surfaces using cross-validation and available ground-truth information. Following the section on cross-validation and testing of our surfaces, we present several case studies to illustrate the impact of applying PPCs to event location. Our principal sources of ground truth for verifying results are: (1) accurate locations of a number of nuclear tests, both from the Chinese nuclear test site near Lop Nor in western China (GUPTA, 1995) and from the former Soviet Union test site in east Kazakhstan (THURBER *et al.*, 1993), (2) information from earthquake surface ruptures inferred from interferometric synthetic aperture radar (InSAR) data (PELTZER *et al.* 1999), and (3) depth-contrained events developed in this study. We have been able to construct the empirical PPCs largely due to the relatively high seismicity in China, especially western China. In aseismic regions, it may be prudent to apply other methods of predicting PPCs. We conclude that by developing and applying PPCs, we can reduce travel-time bias and mislocation using only regional data.

2. *Method and Data*

2.1. Method

Unless one has a very dense network of stations and events, it is difficult to develop three-dimensional velocity models close enough to reality to accurately construct regional travel-time curves. However, it is straightforward to try to develop corrections to 1-D travel-time curves, if one has a sufficient number of well-located events. The drawback to such an approach is that a relatively high level of seismicity is required in order to garner enough data to form good statistics. Much of China and the surrounding areas have a relatively high level of seismicity, with some notable exceptions (e.g., the interior of the Tarim Basin). Thus, our general philosophy to improve event location has been to develop PPCs to one-dimensional velocity models, rather than trying to develop three-dimensional velocity models. There are additional, pragmatic reasons for this approach. First, the location codes in use at both the pIDC and the U.S. National Data Center use the interpolation of standard travel-time curves as the primary method for calculating travel times from an event to a station. In addition, facilities exist in these location codes that permit spatially varying corrections to be easily utilized. In effect, the empirical travel-time corrections provided by the PPC surfaces, when applied to the reference velocity model from which they were developed, provide travel-time curves that account for the heterogeneous nature of the crust and upper mantle near a given station.

The propagation path corrections that we develop are source-to-station corrections. Thus, each station has its own set of corrections; for a given station, the corrections vary depending upon geographic position of the source. In principle, the corrections should vary as a function of source depth, as well. However, because we are most interested in surface sources, we have restricted our investigations to crustal events, and therefore the corrections we develop are not necessarily applicable to deeper sources.

We base our analysis on seismic events occurring between 20°–55° latitude and 65°–115°E longitude, using regional and teleseismic phase data from a variety of sources. For the years between 1986 and 1989, we use data from the ISC bulletin. Between 1990 and January 1998, we use the USGS EDRs. *P*-wave residual means and variances at stations reporting to both catalogs are similar, which suggests that we are not introducing any additional biases by combining these data sets. We supplement these data with travel-time picks from the pIDC's Reviewed Event Bulletin (REB), and our own travel-time picks from regional data. Figure 1 shows seismicity for the region in this study.

A critical element of our location effort lies in improving our ability to accurately estimate the errors in travel-time measurements. This is important both in determining the location error ellipse (or confidence region), and in providing accurate estimates of travel time for interpolation. To aid in this, we have refined estimates of modeling and measurement errors, the sum of which form the total error assigned to a given travel time residual. The modeling error is defined as the standard deviation of travel time of a particular model as a function of distance, while the measurement error is defined as the picking error associated with each phase. Optimally, signal-to-noise ratio (SNR) should be used to determine the measurement error, but this is not reported in most catalogs. The seismic event location algorithm (EvLoc) that we use allows for the inclusion of modeling and measurement error. Furthermore, for each station, the algorithm can utilize different velocity models,

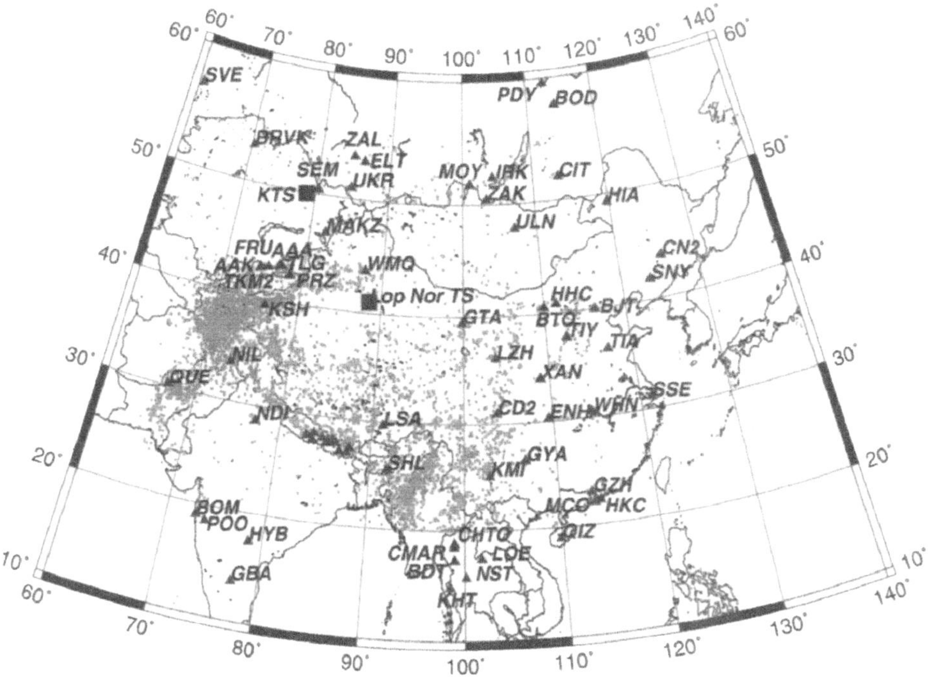

Figure 1
Seismicity (shaded cirlces) and seismic stations (triangles) used in this study. The Lop Nor and Kazakhstan nucelar test sites are also plotted (shaded squares).

bulk station corrections, and spatially varying travel-time corrections. Another option of the algorithm is to discard arrivals that have large residuals. Thus, we define all *P* phases to be used in the relocations (including the depth phase *pP*), but allow the algorithm to remove large residual phases.

2.2. *Measurement Errors, Regional Picks, and Other Arrival Information*

We have little information about measurement errors in the global EDR and ISC catalogs (i.e., SNR). We use two approaches to solve this problem. First, when possible, we substitute our regional picks (for which we have SNR), for global catalog picks. Secondly, if no regional picks are available, we down-weight the arrivals by assigning the picks a uniform measurement error of 2 s, as is done in the EDRs. We have found that many stations of the Chinese Digital Seismic Network (CDSN) have phase picks that are rounded or truncated to the nearest second. Our second approach discussed above is sufficient to handle this problem. Figure 2 shows the inconsistency of some of the regional picks from the global catalogs for waveforms at CDSN stations LZH and KMI. The travel times are truncated and/or rounded off to the nearest second for the EDR catalog, while using the digital data

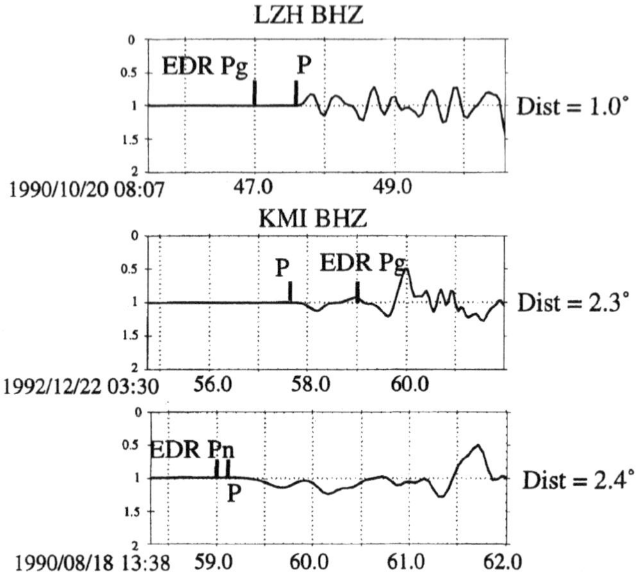

Figure 2
Waveforms and arrival time picks at LZH and KMI. Note that for these examples, the travel times are truncated to the nearest second for the EDR catalog, while using the digital data will result in precise travel-time estimates. The measurement error for the EDR times can be greater than 1.0 sec.

will result in precise arrival time estimates. Thus, when possible, we use our regional picks. This example also illustrates that the measurement error for the EDR arrival times can be greater than one second.

Our waveform segments were retrieved from the IRIS database using the event origin information from the global catalogs mentioned above, and are the result of a large regionalization effort (HARTSE *et al.*, 1997). Stations that we use are located in China, India, Pakistan, Khyrghyzstan, Kazakhstan, Mongolia, Russia, South Korea, and Thailand. Data from the 1980s are mostly from the Chinese Digital Seismic Network (CDSN). Travel-time picks have been made on broadband channels when available; otherwise, they were made on short-period channels (HARTSE *et al.*, 1997). The waveforms are high-pass filtered at 1 Hz, and when visible, the arrivals *Pn*, *Pg*, *Sn*, and *Lg* have been hand-picked. For the locations, we use only the first-arriving *P* travel times, and have added about 6800 of these picks to our data set. We base the measurement error on the signal-to-noise ratio of each phase, using the convention of NAGY (1996). The measurement error is defined as 0.1 s for phases with SNR > 50, 1.0 s for SNR < 5, and 1.0 (0.9 log 10(SNR/5.0)) s for SNR between 5 and 50. These same criteria are used to define measurement error for REB picks.

We check if we are introducing any biases by combining our waveform picks along with the EDR and REB catalogs. However, of the 1984 *P* phases in our catalog that match the EDR catalog, we find that the mean difference in travel times is only

−0.02 ± 0.69 s. For the 321 *P* phases that match the REB catalog, the mean difference in travel times is 0.06 ± 0.62 s. For all 2513 regional phases that overlap with the EDR, the travel-time difference reduces to 0.002 ± 0.77 s. This highlights two points. First, we are not introducing travel-time bias (i.e., a skewed estimate of travel times) by using the REB phases or the truncated travel times for the CDSN EDR phases. Second, our picks seem to be at least consistent with the global catalogs.

For many IMS arrays, azimuth and slowness values are reported. We utilize this information, but check that we are not biasing our results by including this information. For approximately 400 events with 30 or more defining phases, we find virtually no difference in origin time, latitude, longitude, and depth when azimuth and slowness are incorporated. For almost 900 events with less than 30 defining phases, we find that the mean differences to still be virtually zero, but the standard deviations increase to 0.98 s for origin time, 0.25° in latitude, 0.60° in longitude, and 14 km in depth. Thus, if we choose events with a high number of defining phases, using array azimuth and slowness data does not significantly effect our results.

Caution must be applied when merging information from different data sources, because inconsistencies can exist for station information. For example, station names can be confused depending on the network, as is the case for station MAKZ, an IRIS/USGS Global Seismograph Network (GSN) digital station located in Makanchi, Kazakhstan. Before this station became a GSN station in 1996, it was named MAK (Kim *et al.*, 1996). However, MAK is the name of a USGS station in Makhachkala, Dagestan. Because we have retrieved data from both stations, we renamed the USGS MAK station to MAKU to avoid confusion. Inconsistencies also exist for latitude, longitude, and elevations of various stations depending on the reporting agency. Ruppert *et al.* (1999) has systematically documented the most accurate information about many digital stations available by obtaining information from individual network operators. We use this information when possible, and default to the USGS when this information is unavailable.

2.3. Velocity Models and Modeling Errors

We use several velocity models for this analysis. We begin by using the global velocity models IASPEI91 (Kennett and Engdahl, 1991) and *ak135* (Kennett *et al.*, 1995). In addition to these, we have developed travel-time tables for 1-D models that may be more representative of the China region. We compute travel times for *P*, *Pn*, and *pP* for regional 1-D velocity models that are based on results published in Li and Mooney (1998), Kosarev *et al.* (1993), and Jih (1998), which we call *China_LM*, *Asia*, and *China_SSB*, respectively. For these regional 1-D models, we place their crustal structure on top of the IASPEI91 mantle structure.

The *China_LM* model is developed by averaging seven crustal models based on deep seismic sounding results for China (LI and MOONEY, 1998). Our strategy was to assume a three-layer crust with each layer having the average crustal thickness and velocity of the three thickest layers of each of the seven models. Thin sedimentary layers and anomalous mid-crustal zones were eliminated. The *China_SSB* model is a 1-D model developed for China by the Chinese State Seismological Bureau during the middle 1970s (JIH, 1998). The KOSAREV *et al.* (1993) model for the Tien Shan is included primarily for the purpose of comparison. Figure 3 shows all *P*-wave models used in this study. The major differences in the models are the layer and crustal thicknesses. However, the *Asia* model has a lower average crustal velocity than the others and a positive gradient throughout the crust.

Modeling error is the more difficult of the two types of error to estimate. We begin by using estimates of the modeling error developed from arrivals compiled by the pIDC for the global velocity model IASPEI91 (SWANGER, personal communication, 1998). This estimate of modeling error does not account for intrinsic

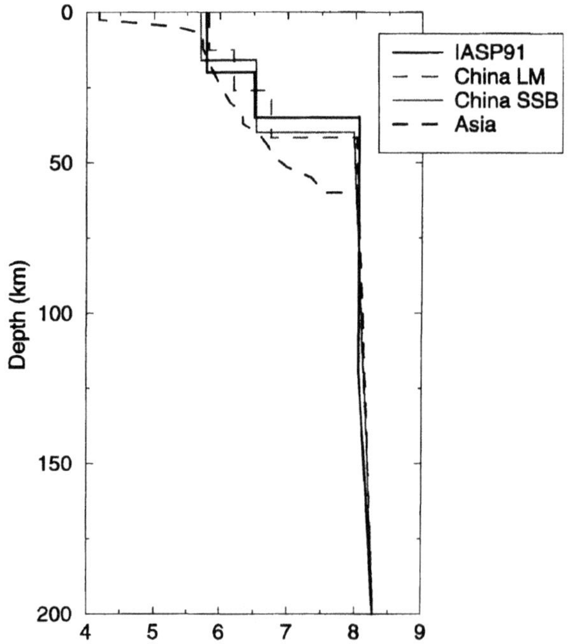

Figure 3

Four *P*-wave velocity models investigated in this study. We use the global model of IASPEI91 (KENNETT and ENGDAHL, 1991), and regional models derived from LI and MOONEY (1998), JIH (1998), KOSAREV *et al.* (1993), which we call *China_LM*, *China_SSB*, and *Asia*, respectively. The major differences in the models are the layer and crustal thicknesses. However, the *Asia* model has a lower average crustal velocity than the others.

biases present in our locations, arising from differences in network geometry and from using a 1-D base model when the true structure is laterally variable. These modeling errors vary as a function of distance and are largest in the regional distance range (see next section for more details). This suggests that although IASPEI91 provides an excellent average crustal model for teleseismic phases, at regional distances better 1-D travel-time models are needed. We use these modeling errors at first, but later develop our own modeling errors based on regional 1-D velocity modeling in China. To accurately interpolate travel-time corrections, distance trends must be eliminated.

2.4. Ground Truth

To test and validate our travel-time corrections and relocation algorithms, ground-truth information is needed. We include several types of ground-truth events in this study, and we describe these below.

2.4.1. Nuclear explosions

The best available-constrained ground-truth information available for the region is that from nuclear explosions. We include explosions from two test sites: the Chinese test site at Lop Nor and the former Soviet Union test site at Balapan, Kazakhstan. GUPTA (1995) relocated many Lop Nor nuclear explosions using satellite imagery and a relative location algorithm. THURBER *et al.* (1993) obtained locations for 20 explosions at Balapan, Kazakhstan, also using a joint epicentral determination (JED) technique in conjunction with SPOT satellite images. Table 1 lists the nuclear explosions used in this study. Because the origin times for these events is not known, these events are considered to be accurate within 2 km, although the reported location error is typically less than 2 km.

2.4.2. Earthquake surface ruptures

We also include ground rupture information obtained for a large earthquake (M_w = 7.5) that occurred on Nov. 8, 1997, approximately 750 km south of the Lop Nor nuclear test site. The mainshock is a shallow, strike-slip, intraplate event (VELASCO *et al.*, 1999). VELASCO *et al.* (1999) obtain a consistent source model over a wide range of frequencies, from a wide variety of seismic data. However, mapping of the actual ground rupture was obtained from InSAR data for the mainshock (PELTZER *et al.*, 1999). The surface rupture is approximately 170 km long, showing a very linear fault (Fig. 5). We can use this ground rupture information as ground truth, as long as we can identify and locate the event responsible for the surface break. Since the mainshock is an east-west striking fault, we can use the mainshock rupture as a guide for validating our path correction surfaces since the depth and latitude are well constrained.

Table 1

Nuclear events relocated in this study

Date	Origin Time	Lat. (_N)	Lon. (_E)	Depth	m_b	M_a	Source
87/03/12	01:57:17	49.9358	78.8261	0	5.6	3.9	Thurber
87/04/03	01:17:08	49.9186	78.7794	0	6.2	4.8	Thurber
87/04/17	01:03:05	49.8831	78.6708	0	6.0	4.7	Thurber
87/06/20	00:53:05	49.9367	78.7464	0	6.1	4.5	Thurber
87/08/02	00:58:07	49.8806	78.8750	0	5.9	4.3	Thurber
87/11/15	03:31:07	49.8981	78.7575	0	6.1	4.8	Thurber
87/12/13	03:21:05	49.9614	78.7933	0	6.1	4.7	Thurber
87/12/27	03:05:05	49.8789	78.7253	0	6.1	4.5	Thurber
88/02/13	03:05:06	49.9322	78.8681	0	6.1	4.6	Thurber
88/04/03	01:33:06	49.9069	78.9064	0	6.0	4.8	Thurber
88/05/04	00:57:07	49.9500	78.7494	0	6.1	4.9	Thurber
88/06/14	02:27:06	50.0364	78.9675	0	5.1	4.1	Thurber
88/09/14	03:59:58	49.8781	78.8239	0	6.1	4.8	Thurber
88/11/12	03:30:04	50.0467	78.9689	0	5.4	4.3	Thurber
88/12/17	04:18:07	49.8797	78.9236	0	5.9	4.7	Thurber
89/01/22	03:57:07	49.9411	78.7869	0	6.0	4.7	Thurber
89/02/12	04:15:07	49.9167	78.7142	0	5.8	4.8	Thurber
89/07/08	03:46:58	49.8675	78.7792	0	5.6	4.3	Thurber
89/09/02	04:16:57	50.0094	78.9856	0	5.1	3.9	Thurber
89/10/19	09:49:57	49.9300	78.9456	0	5.9	4.8	Thurber
90/05/26	07:59:57.9	41.569	88.701	0	5.4	–	Gupta
90/08/16	04:59:57.7	41.514	88.739	0	6.2	–	Gupta
92/05/21	04:59:57.5	41.513	88.774	0	6.5	5.0	Gupta
92/09/25	07:59:58.5	41.716	88.336	0	5.0	–	Gupta
93/10/05	01:59:56.6	41.667	88.695	0	5.9	4.7	EDR
94/06/10	06:25:57.8	41.527	88.710	0	5.8	–	EDR
94/10/07	03:25:57.8	41.574	88.730	0	5.9	–	EDR
95/08/17	00:59:57.0	41.570	88.730	0	6.1	–	EDR
96/06/08	02:55:58.0	41.657	88.690	0	5.9	4.3	EDR
96/07/29	01:48:57.8	41.824	88.420	0	4.9	–	EDR

2.4.3. Depth constrained events

We also include 26 events that have independent depth constraints as ground truth. While perhaps not as valuable as epicentrally-constrained ground truth for calibration purposes, these data nevertheless offer an improvement over events for which the depth is poorly constrained. It is important to emphasize that, in the absence of realistic strategies for gaining ground-truth information, this approach of establishing ground truth may be the best means available to verification seismologists for calibrating a region. Furthermore, much can be learned about the performance of seismic techniques from comparative analysis, and such knowledge can be used to design improvements in techniques and/or to identify new data requirements.

Recently, PATTON (1998) and WOODS *et al.* (1998) have reported focal depths for 26 central Asian earthquakes from a variety of methods and sources. Focal depths are obtained from the ISC hypocenter determination, *pP-P* travel-time delays as reported by the ISC, hypocenter determination in the PDE and EDR issued by the National Earthquake Information Center (NEIC), simultaneous focal depth and moment tensor (M_{ij}) determination reported by the NEIC based on Sipkin's method (SIPKIN, 1982), Harvard centroid moment tensor (CMT) catalog, *pP-P* modeling from WOODS *et al.* (1998), broadband modeling of regional seismograms in the time domain from WOODS *et al.* (1998), modeling regional surface-wave spectra (PATTON, 1998), and depth phase determinations from ENGDAHL *et al.* (1998). Origin time, epicentral location, magnitude, focal depth, and root-mean-squared (RMS) residual from the median depth are listed for all 26 earthquakes in Table 2. The focal depth is

Table 2

Central Asian Earthquakes

Ref. #	yrmmddhrmn	Lat. N	Lon. E	M_b(Lg)	Focal depth (km)	Uncertainty (km)	Total # depth est.
01	8701052252	41.95	81.32	6.1	17.8	2.5	9
02	8701240809	41.48	79.33	6.6	18.0	2.5	9
85	8702251956	38.08	91.16	6.2	16.6	3.4	8
10	8703030941	41.34	79.29	5.3	5.8	4.6	5
14	8709182158	47.11	89.53	5.3	14.0	4.2	6
18	8712220016	41.32	89.62	6.1	16.0	2.5	8
27	8806171330	42.92	77.49	5.8	18.4	3.3	8
29	8807230738	48.70	90.56	6.4	15.6	3.4	7
31	8812150640	46.53	95.59	5.5	21.0	3.3	7
42	9006141247	47.85	85.06	6.4	47.8	3.8	8
43	9008030915	47.97	84.97	6.2	33.9	2.8	8
44	9010242338	44.11	83.88	5.6	19.1	3.3	8
45	9011121228	42.94	78.07	6.3	19.5	3.5	8
46	9012011809	40.89	73.59	5.6	29.6	3.3	7
47	9102251430	40.36	78.98	6.3	21.9	3.3	8
48	9108190605	47.00	85.31	5.3	36.9	3.6	7
49	9212240509	42.15	72.24	5.6	35.0	3.5	8
86	9310020842	38.17	88.66	6.6	13.1	2.7	9
50	9312301424	44.74	78.80	5.6	15.0	3.3	7
51	9405090914	40.22	78.90	4.8	32.6	3.4	8
52	9408231418	40.09	78.79	4.8	14.7[h]	10.5[h]	4
80	9509260439	41.75	81.58	5.5	22.8	3.9	6
81	9511011229	42.92	80.31	5.2	18.4	2.5	8
82	9601180933	41.81	77.50	5.4	25.3	4.9	5
83	9603121843	48.44	88.14	5.6	17.1	2.7	5
84	9603191500	39.99	76.70	6.8	22.2	2.6	6

[h] Mean and standard deviation.

computed by taking the weighted average of depth estimates available for each event, excluding those from the CMT (that appear to be bounded at 15 km).

Uncertainties on estimates are given if they are available. ISC uncertainties are standard deviations (one sigma), as are uncertainties given in Woods *et al.* (1998) for *pP* modeling. Depth uncertainties from modeling regional broadband seismograms are believed to be within 5–10 km. The NEIC states only that depth estimates in the PDE have been found to be accurate within 25 km. No error estimates are provided for depth by Sipkin's method or the CMT, although notes about the treatment of depth in the CMT inversion are available and posted in Table 3. (Notes about the treatment of PDE depth are also posted.) Depths from surface-wave spectra were initially believed to be accurate to within 5 km at the one sigma level.

The largest variance from the weighted mean focal depth is found for regional surface-wave estimates (RSS), while the depth estimate from the NEIC Moment tensor has the smallest variance. Their variances translate into standard deviations of 14.0 km and 2.0 km, respectively. Thus, surface-wave spectral modeling shows a standard deviation that is almost three times what was anticipated based on the stated accuracy of 5 km. The other standard deviations are: 11.5 km for the ISC Location, 5.5 km for the ISC *pP*, 7.7 km for the NEIC PDE, 2.0 km for the NEIC Moment tensor (M_{ij}), 8.7 km from the HVD CMT, 12.0 km for the Woods *et al.* (1998) *pP*, 11.0 km for the Woods *et al.* (1998) regional body wave (RBB), and 4.9 for the depth phase solutions of Engdahl *et al.* (1998).

Estimates of depth from depth phases appear most stable, and the global moment tensor solutions also appear to provide fairly good depth constraint. The regional surface and body-wave inversions have a larger standard deviation than originally stated. However, as mentioned above, the CMT catalog constrains the depth to be at least 15 km (this is not the case for the NEIC moment tensor inversions). Since both regional methods, RBB and RSS, rely on surface-wave amplitudes, one might anticipate some correlation between the two. However, the RBB method models additional energy on the regional record, such as *Pnl* wave trains, and for many events, was applied to data recorded on multiple stations. The RSS method was applied to surface waves only recorded by a single station, WMQ. Presumably, the methods are uncorrelated because of the contributions of *Pn* waves. Events with good depth control may be beneficial for improving location estimates in this critical region. However, to use these events as ground truth requires a different strategy than employed for well-constrained epicentral events. We describe this approach in our path-correction surface development section.

3. Relocations

We perform our relocations using 9500 events with over 300,000 *P*-wave arrivals and spanning 11 years. We relocate all events using the four velocity models

Table 3

Estimates of focal depth, H, in km

	ISC				NEIC		HVD	Woodward-Clyde			HJP	EHB
	Location		$pP - P$		PDE	M_{ij}	CMT	$pP - P$ modeling		RBB	RSS	
Ref. #	H	ΔH	H	ΔH	H	H	H	H	ΔH	H	H	H^{s}
01	15	NA	14	0.6	17	19	15	15	4.0	25	20	14.9
02	29	NA	26	2.0	29	14	15^{i}	12	4.0	10	5	6.6
85	26	NA	23	1.7	26	NA	15^{i}	4	4.0	5	25	3.5^{t}
10	10	NA	9	3.8	33	NA	33	15	8.0	NA	40	2.2^{t}
14	25	NA	24	3.4	33^{j}	NA	15	6	5.1	15	30	4.6
18	21	NA	18	0.3	21^{h}	11	19	20	4.0	NA	20	18.6
27	18	5.7	20	2.3	24	NA	15^{i}	19	7.7	18	30	13.1^{t}
29	19	NA	15	0.7	18	NA	18^{k}	16	6.0	20	20	14.1
31	1	17.	19	1.7	24	NA	24^{k}	16	5.1	15	20	21.5
42	NA	NA	54	1.1	58^{h}	46	36	5	4.0	18^{m}	25	36.7
43	16	8.7	35	1.2	33^{h}	31	32	33	4.0	NA	35	36.6
44	20	8.9	19	1.4	20	NA	15^{l}	19	6.0	25	35	19.6
45	9	5.7	17	1.3	19^{h}	19	15^{i}	18	4.0	NA	5	19.1
46	30	9.6	33	2.4	29	NA	32^{l}	23	6.0	NA	20	27.6^{t}
47	21	NA	21	1.8	21	NA	15^{i}	20	5.1	18^{m}	5	24.2
48	25	9.5	33	1.1	30	NA	55	42	5.1	NA	35	41.0
49^{n}	NA	NA	36	1.9	38	NA	26	8	6.0	10	1	35.3
86°	14	NA	16	1.2	14^{h}	12	15^{i}	15	4.0	15	15	8.5
50	1	14.	13	2.2	15^{h}	NA	21^{l}	13	4.0	NA	25	18.6
51	4	19.	15	5.3	46	NA	33^{l}	5	7.7	5	35	35.7
52	33	NA	NA	NA	33^{j}	NA	15^{l}	4	5.1	15	25	NA
80	23	NA	20	3.9	33^{j}	NA	43	10	5.1	15	10	19.4
81	20	NA	21	1.3	20^{h}	14	15^{l}	15	5.1	15	15	21.0
82^{p}	–	–	–	–	33^{j}	NA	20	4	4.0	15	25	26.1^{t}
83	–	–	–	–	17	17	17^{k}	14	5.1	15	15	12.4
84	–	–	–	–	28	19	22	32	4.0	25	10	21.3

ISC: International Seismological Centre; HVD: Harvard; CMT: Centroid Moment Tensor; RBB: Regional broadband seismogram modeling; RSS: Regional surface-wave spectra modeling; HJP: PATTON (1998); EHB: depth phase determination of ENGDAHL *et al.* (1998).

^{h} fixed by geophysicists.

^{i} could be < 15 km.

^{j} fixed at "normal" depth.

^{k} fixed to PDE depth.

^{l} fixed by educated guess.

^{m} mid-point of a range of depths.

^{n} regional waveform inversion yields H = 8 km (GHOSE *et al.*, 1988).

^{o} teleseismic *P*-wave waveform inversion yields H = 15 km (WOODS *et al.*, 1998).

^{p} regional waveform inversion yields H = 16 km (GHOSE *et al.*, 1998).

^{r} teleseismic *P*-wave waveform inversion yields H = 30 km (WOODS *et al.*, 1998).

^{s} 10 or more depth phases used.

^{t} 5 to 10 depth phases used.

described above, letting the algorithm remove outliers based on travel-time residual size. To ensure high quality locations to be used in our 2-D correction surfaces, we require that events must have been recorded by at least 30 stations with an azimuthal gap <180°, and have a sum of depth plus depth error less than the Moho depth of the model. Applying these criteria reduces the number of useable events to 705, 915, 907, and 1074 for the IASPEI91, *China_LM*, *China_SSB*, and *Asia* models, respectively. From these sets of events we then gather residuals by station, discarding residuals larger than 10 s in absolute value. Thus, our data set is teleseismically constrained.

3.1. Location Comparisons

We first compare our results using the global velocity model IASPEI91 to locations given in the ISC and EDR catalog. For the ISC catalog, we find for 296 events an origin time difference of 1.51 ± 2.14 s, a latitude difference of −0.36 ± 8.47 km, a longitude difference of 0.65 ± 9.55 km, and a depth difference of −3.16 ± 15.37 km. The origin time difference reflects the difference in the travel times for the different models used, where the JEFFREYS and BULLEN (1940) *P*-wave travel-table used by the ISC differs from the IASPEI91 *P*-wave travel time tables on the order of 1.8–1.9 s (KENNETT and ENGDAHL, 1991). For the EDR catalog, we find for 516 events an origin time difference of 0.54 ± 2.26 s, a latitude difference of 0.10 ± 11.63 km, a longitude difference of 0.18 ± 10.84 km, and a depth difference of −11.06 ± 14.72 km. The main difference between our results and the EDRs is the depth. The EDRs will set the depth to 33 km if the depth is poorly constrained. We set the depth to 0 km if the depth is not constrained. Where both methods solved for depth, the depth difference is reduced to 5 km ± 16.50 km for 89 events, although the standard deviation of the difference does not decrease.

We next compare location results using the different velocity models. For the IASPEI91-*China_LM* models, we find for 786 events an origin time difference of 0.07 ± 1.28 s, a latitude difference of 0.09 ± 4.86 km, a longitude difference of −0.25 ± 4.68 km, and a depth difference of 0.55 ± 8.5 km. For the IASPEI91-*China_SSB* models, we find for 785 events an origin time difference of 0.39 ± 1.39 s, a latitude difference of 0.68 ± 4.97 km, a longitude difference of −0.21 ± 5.05 km, and a depth difference of 1.47 ± 9.45 km. For the IASPEI91 – *Asia* models, we find for 623 events an origin time difference of 3.15 ± 1.88 s, a latitude difference of 2.34 ± 7.99 km, a longitude difference of −1.21 ± 7.65 km, and a depth difference of 5.93 ± 11.57 km. Thus, we are not obtaining significantly different location results using these regional 1-D models of *China_LM* and *China_SSB* compared with the IASPEI91 global model. The *Asia* model, however, is slightly shifting the hypocentral locations and origin times. To further explore the effects of these models on location, we must analyze the travel-time residual bias for the region.

3.2. *Travel-time Residuals and Modeling Errors*

To explore 1-D travel-time bias in our data set, we bin *P*-wave travel-time residuals as a function of epicentral distance over all the stations. The IASPEI91 velocity model produces bias in our travel-time residuals at regional distances (Fig. 4a), which suggests that the IASPEI91 global model is too fast for China, particularly for *Pn*. The residual biases have been slightly reduced by using our 1-D models (Fig. 4a). In particular, the *China_LM* model, which features a somewhat thicker crust and faster upper mantle than IASPEI91 (Fig. 3), removes some of the bias that we see with the global model. Figure 4b shows the standard deviation of the travel-time residuals versus distance, which represents the root of the modeling error for each of the models. Also plotted is the original estimate of modeling error from IASPEI91 *P* waves (SWANGER, personal communication, 1998). The global and regional 1-D model errors appear to be consistent at regional distances, where modeling errors are largest. At teleseismic distances, all of our models are consistent with each other, but the errors are on the order 0.4 s larger than those provided by the pIDC. The primary impact of applying these new modeling errors will be to change the error ellipse sizes. When we compare our IASPEI91 locations with the new and old modeling error differences, we find that that the error ellipses grow on average by only 0.4 ± 0.7 km for the semi-major axis and 0.3 ± 0.5 km for the semi-minor axis.

3.3. *Ground-truth Locations*

For nuclear explosions that were located by GUPTA (1995), we constrain the latitude, longitude, and depth and invert only for origin time. We do this for all velocity models, so that we can construct travel-time residual surfaces for all velocity models used. For our depth-constrained events, we set the depth and let the program invert for latitude, longitude, and origin time. While the main effect of constraining depth is to more accurately determine origin time, we expect some improvement in epicentral location, as well as more consistent residual estimates. To use our 26 depth-contrained events, we check the difference betwen the free- and fixed-depth solutions for hypocentral location. For models IASPEI91, *China_LM*, *China_SSB*, we find 8 events for which depths for fixed and free depth solutions are within 5 km of each other. The latitudes and longitudes are also within 0.1 km of each other for these solutions. Thus, we use these eight events for our 2-D surfaces below. For the *Asia* model, we find 11 events that pass our criteria.

4. *Correction Surface Development*

As mentioned above, we ensure high-quality locations to be used for our 2-D surface corrections by requiring that events must have been recorded by at least 30 stations and have an azimuthal gap of less than 180 degrees. By employing the previous two criteria, we restrict ourselves to residuals from events located with an accuracy of about 25 km (E.R. ENGDAHL, personal communication). An additional criterion we impose is that the event has a depth plus depth error of less than the

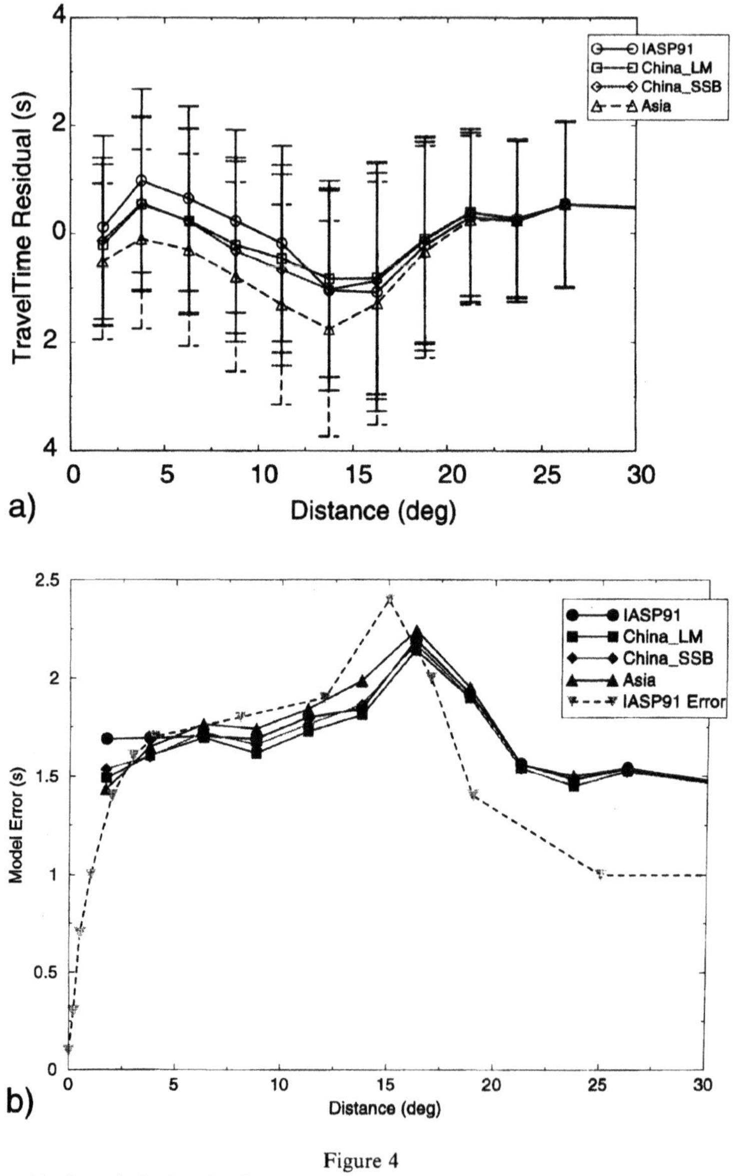

Figure 4
Travel-time residual analysis for the four velocity models used in this study. a) Plotting travel-time residuals versus distance for the four velocity models shows a distance dependence for regional distances. In general, the distance dependence is slightly less for the regional models than the global model. b) The standard deviation of travel-time residuals versus distance for the same models also shows distance dependence.

Moho depth of the model. Furthermore, only travel-time residuals less than 10 s in absolute value are used. Following the relocations, *P*-wave residuals were gathered by a station to be used for the correction surfaces.

Propagation path correction surfaces are created by interpolating the residuals onto 2-D surfaces using modified Bayesian kriging (SCHULTZ *et al.*, 1998) for 68 stations reporting to the ISC and/or EDR catalogs, plus 6 of the pIDC stations. Unfortunately, the limited phase data available from the pIDC stations, due to their more recent installation, make construction of correction surfaces for these stations of reduced value at this stage. Using modified Bayesian kriging, the value of the PPC surface goes to zero at distances greater than a specified correlation length away from the data points, and an estimate of the variance associated with the surface is calculated based on user-supplied residual variances and an estimate of the background variance. We use 5° for both the model and data correlation lengths and a background variance of 7 s^2. For station ZAL, Figure 6 shows the kriged *P*-residual surface and its associated variance surface.

4.1. Surface Validation

There are two principal questions we wish to answer here: 1) How well do the kriged surfaces describe the data, and 2) how well can kriging predict surface values at data points that have been left out? We use a method called cross-validation to explore the latter question. Our approach is to perform a leave-one-out scenario, whereby for each surface we sequentially rekrig, eliminating a different data point each time for all points in that surface. We then determine the mean and variance of

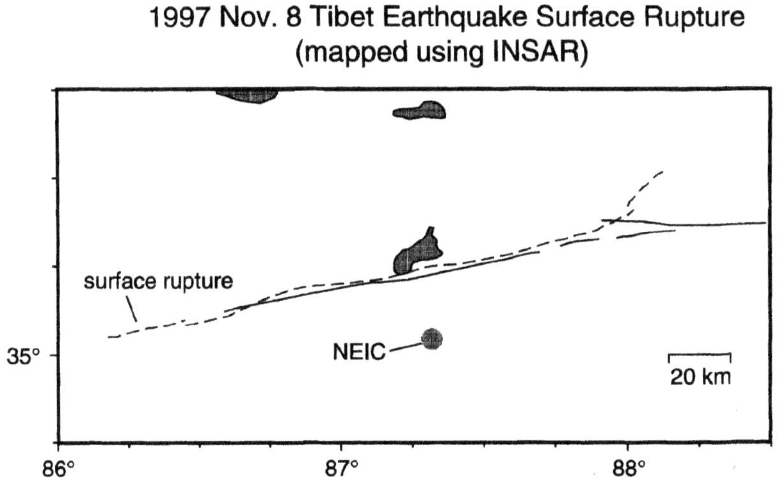

Figure 5
Ground-truth information for the Nov. 8, 1997 (M_W = 7.5) Tibet earthquake surface rupture (dashed line) derived from InSAR (provided by G. Peltzer) and the NEIC epicenter. The NEIC teleseismic location is approximately 20 km south of the surface rupture, even though this is a well-recorded event.

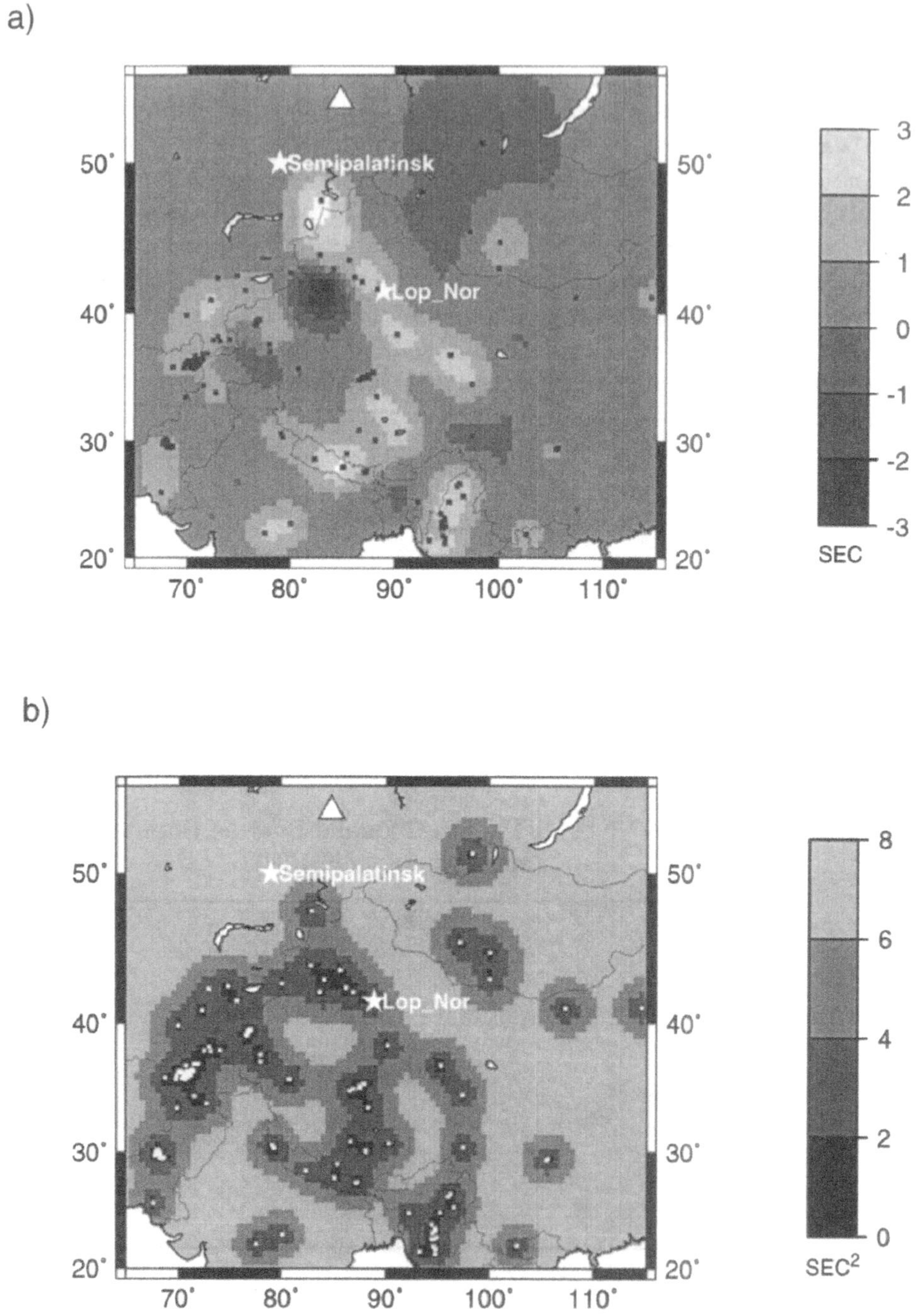

Figure 6
Propagation path corrections (PPCs) derived for IMS station ZAL. a) Kriged PPC surface (event locations: small black squares), and b) variance surface (event locations: small white squares).

the differences between the value of the cross-validated surface at the eliminated point and the value at that point for the original surface. The mean and variance are 0.1 s and 1 s^2, respectively, suggesting that the surfaces are robust. For comparison, the average variance of the input residual data is about 5 s^2. The average variance reduction of the surfaces themselves from the data is about 70%. Figure 7 compares a cross-validated surface to the original kriged surface. Note that in areas where data are plentiful, there is good correlation between the two. Where points are isolated, there is little correlation as is expected. Eliminating stations where a cross-validated surface has low correlation with the original surface is unnecessary, because the kriging process itself accounts for the relative accuracy of the input data points.

A second way to qualitatively test the kriging procedure and the robustness of the residual data going into it is to look at kriged surfaces for stations that are close to one another. We do this for stations AAA, FRU, PRZ, and TLG. Figure 8 illustrates the surfaces for each station along with the mean surface and its standard deviation. With the exception of a few outliers, the surfaces are similar, with standard deviations from the mean typically less than 1 s. This simple test suggests that our data and the kriging process itself are robust for the stations shown.

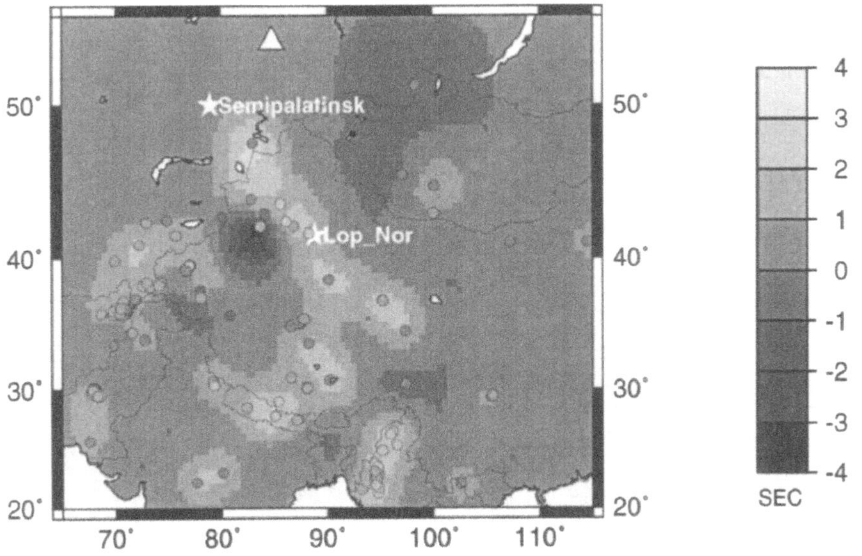

Figure 7
Cross-validated PPC data for station ZAL. Grayscale-filled circles are the cross-validated points, and these overlay the kriged surface from Figure 6a. Where grayscales within the circles match those on the background surface, kriging was able to successfully predict the data value at that point.

5. Relocation Using Propagation Path Correction Surfaces

To test the effect of the PPCs on epicentral location, the events are again relocated in EvLoc using both teleseismic and regional phases. Figure 9 compares the

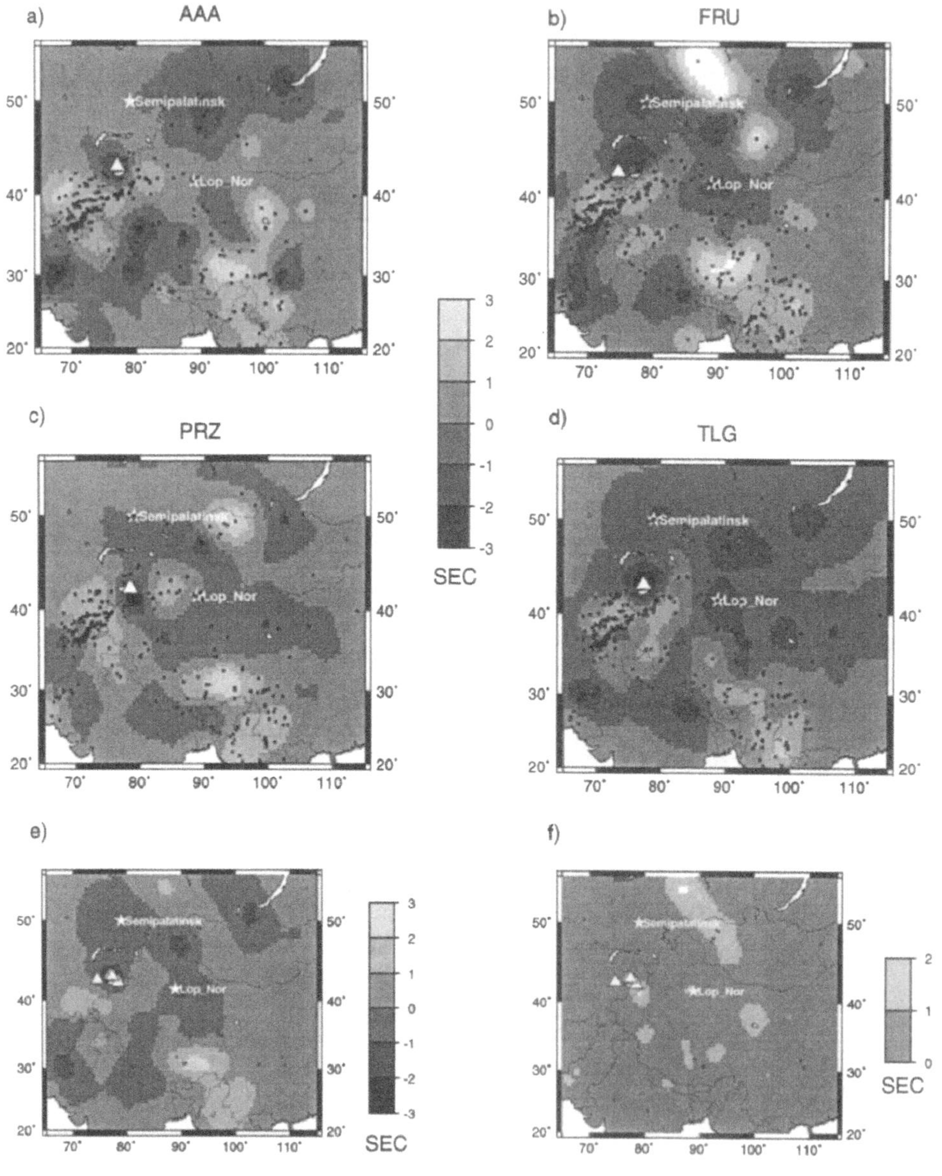

Figure 8
PPC surfaces for stations a) AAA, b) FRU, c) PRZ, and d) TLG. The average and standard deviations of the four surfaces are plotted in e) and f), respectively.

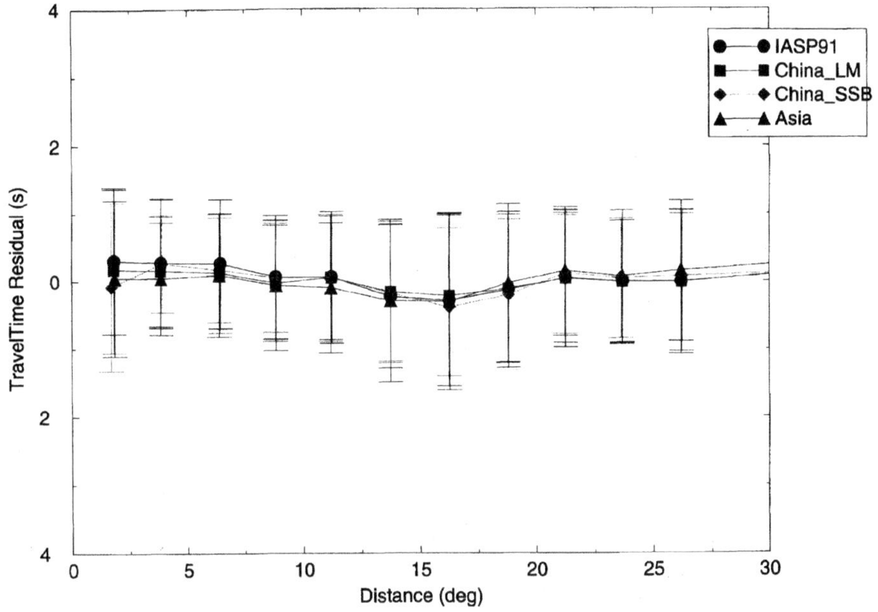

Figure 9
Travel-time residuals for the four velocity models when PPCs are applied. Note that all regional dependence is removed independent of the velocity model.

residuals from relocations using PPCs to those of the earlier locations and shows that we successfully remove the remaining distance biases with the 2-D corrections. We use our ground truth at Lop Nor, Balapan, and Tibet in order to assess improvements in location resulting from using PPCs.

5.1. Lop Nor Explosions

Figure 10a (top) gives the teleseismic relocations in the *China_LM* model for 6 nuclear explosions at Lop Nor, both with (black lines) and without (gray lines) PPCs. In both cases location errors are about the same, about 2 km, though when PPCs are used the location procedure is more robust (all 6 of the events are located, versus only 5 for the non-PPC case). The northernmost event is problematic due to poor regional station coverage, with a location error of 10 km. Note that teleseismic locations are relatively unaffected by the use of PPCs. This is due to the resultant down-weighting of the regional data in this context: available data numbers in the several hundreds, with only 74 stations at regional and far-regional distances. When we look at regional relocations using only the stations for which we have PPCs, the effect of the PPCs is more profound (Fig. 10a, bottom), though in this case, still not dramatic. There are two distinct testing locations in this figure, and nuclear tests at the southwestern location are all located more accurately and with greater precision with the use of PPCs than without. With PPCs, all explosions (except the problematic event) are

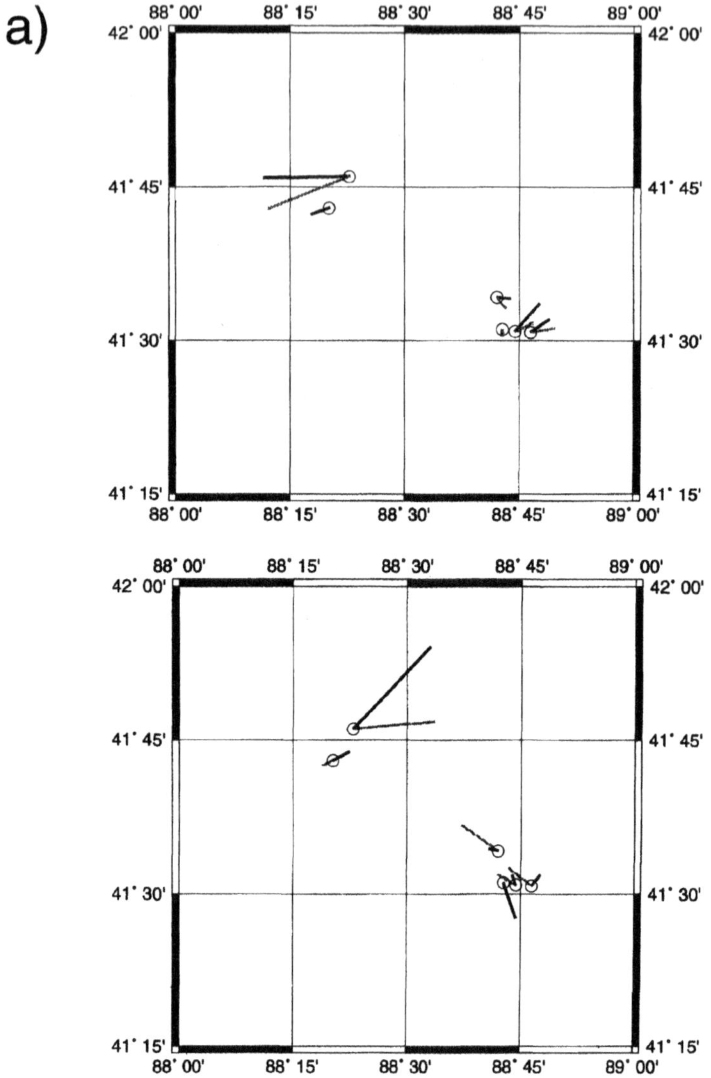

Figure 10

Ground-truth comparisons at a) Lop Nor, b) Balapan, and c) the 1997 $M_w = 7.5$ Tibetan earthquake. In a) and b), upper plots show relocations using all available stations, while the lower plots show relocations using only stations for which PPCs have been determined. Also, in a) and b), black represents runs using PPCs, while gray represents locations without PPCs. Note that at Lop Nor, the use of PPCs increases the accuracy and precision of regional relocations for the cluster of events to the southeast. One event in this cluster was not locatable without PPCs. The northernmost Lop Nor event has a large mislocation which biases the statistical results presented in Table 4. In c), although we have little to no eastwest information on the rupture initiation, implementing the PPCs has removed the northsouth teleseismic bias by locating the event on the fault.

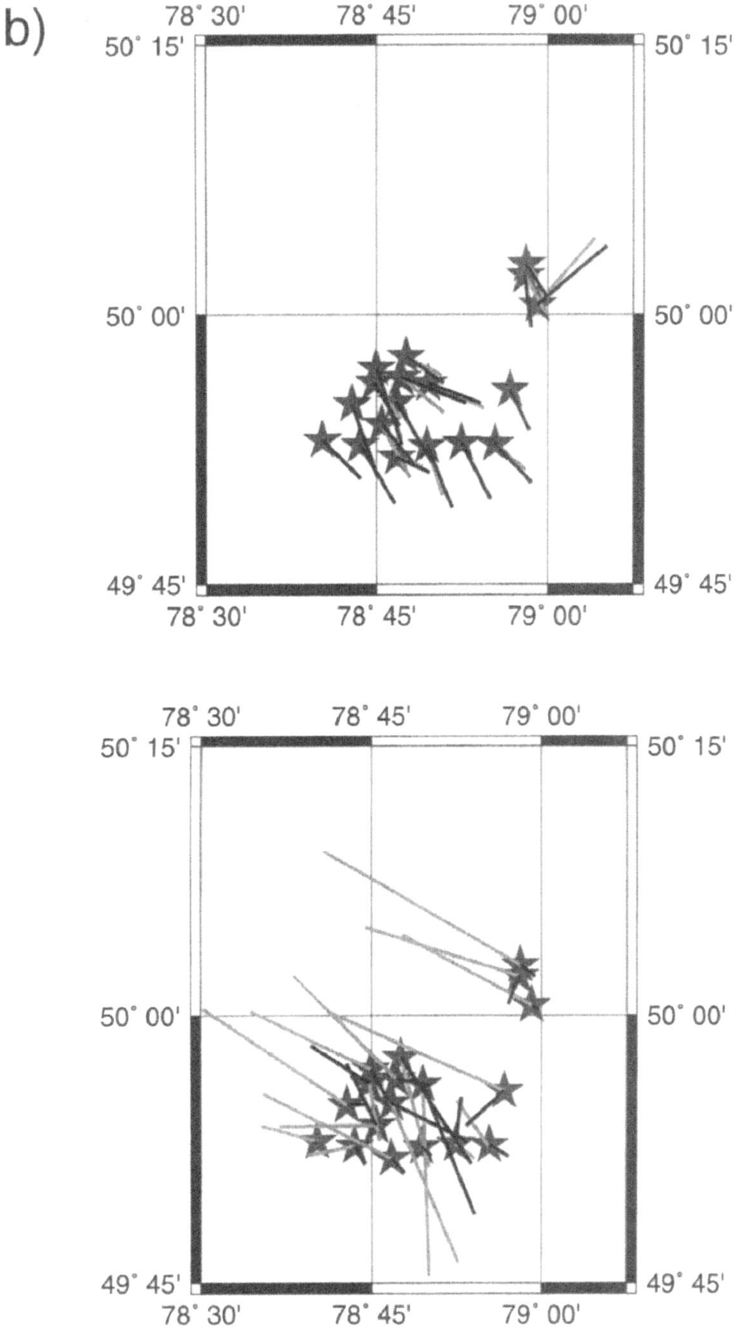

Figure 10 (Continued)

c)

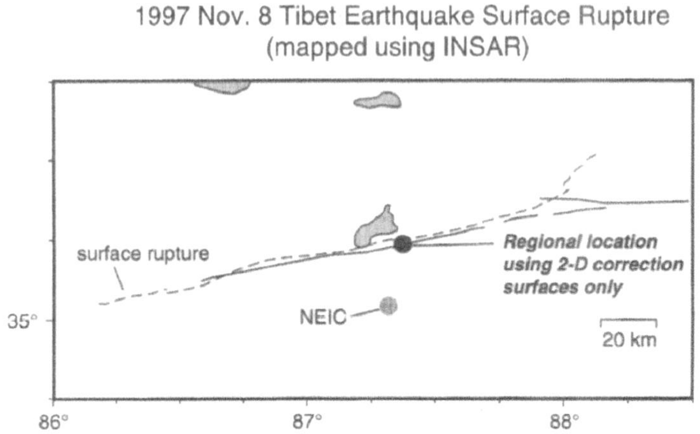

Figure 10 (Continued)

located to an accuracy equivalent to the teleseismic case, and again the location algorithm is able to converge for all 6 of the test events, while without PPCs one event is missing due to algorithm divergence. The northernmost event, with large mislocations, is relatively small for a nuclear test, with a magnitude of 4.6, and consequently it was not recorded on as many stations as other nuclear tests. For this event, all regional stations are located to the south of Lop Nor, with one exception (WMQ). Furthermore, a cluster of stations 15° distant and to the southwest has residuals that differ by more than 4 seconds. When this cluster is removed, the regional mislocation is reduced to half its previous value (from 14 to 7 km). Average mislocation errors for the four velocity models studied are given in Table 4 for Lop Nor. From this table one might infer that PPCs actually create a location bias. However, this is being driven by the large mislocation of one event (the northernmost), and caution must always be used in drawing conclusions based on statistical properties from such a small sample.

5.2. Balapan Explosions

For explosions at the former Balapan nuclear test site, the improvement in location using PPCs is much more obvious. Figure 10b (top) shows the teleseismic cases with and without PPCs, and as expected, the results are similar, with mislocations averaging about 4 km. Using only the regional stations and no PPCs, mislocations are on the order of 20 km (Fig. 10b, bottom). With PPCs, the mislocation error drops to about 4 km. Average mislocation errors for the four velocity models studied are given in Table 5 for Balapan. For Balapan, we see an

Table 4

Comparing relocations with ground truth at Lop Nor

	China_LM			China_SSB			IASPEI91			Asia		
	lat.	lon.	n	lat.	lon.	n	lat.	lon.	n	lat.	lon.	n
R + P	2.4 ± 6.9	4.2 ± 7.7	6	2.2 ± 6.3	4.9 ± 7.8	6	−2.4 ± 6.6	6.3 ± 9.0	6	14.1 ± 0	23.3 ± 0	1
R	2.0 ± 2.4	−0.2 ± 11.7	5	−1.3 ± 5.8	−1.9 ± 11.7	6	−22.8 ± 48.1	25.9 ± 52.7	5	−1.5 ± 2.5	−8.9 ± 0.9	2
T	−1.3 ± 3.0	−1.4 ± 10.6	5	−2.1 ± 3.4	−1.7 ± 10.9	5	−2.4 ± 4.6	−2.2 ± 12.0	5	−3.3 ± 2.6	2.1 ± 3.2	4
T + P	0.8 ± 2.5	−1.9 ± 10.1	6	1.7 ± 4.0	−2.6 ± 12.1	6	−1.0 ± 4.0	−1.9 ± 11.1	6	−2.5 ± 8.7	0.9 ± 7.2	2

Table 5

Comparing relocations with ground truth at Balapan

	China_LM			China_SSB			IASPEI91			Asia		
	lat.	lon.	n	lat.	lon.	n	lat.	lon.	n	lat.	lon.	n
R + P	−1.7 ± 4.8	1.8 ± 5.5	18	−2.1 ± 5.2	2.3 ± 6.9	18	0.3 ± 7.4	1.7 ± 6.2	18	−10.5 ± 13.8	3.6 ± 11.3	6
R	0.3 ± 9.5	−11.4 ± 13.6	17	0.1 ± 10.5	−11.7 ± 14.0	17	2.7 ± 10.6	−12.4 ± 11.9	17	−0.6 ± 14.5	−12.5 ± 21.0	14
T	−3.4 ± 3.2	−5.0 ± 2.0	17	−3.8 ± 3.0	5.2 ± 2.0	18	−2.8 ± 3.0	5.3 ± 2.1	20	−6.2 ± 2.9	3.9 ± 2.1	18
T + P	−4.0 ± 3.1	5.5 ± 2.5	18	−4.0 ± 3.0	4.7 ± 2.5	20	−3.3 ± 3.0	6.4 ± 2.3	20	−7.3 ± 2.9	4.5 ± 2.4	19

increase in both precision and accuracy with the use of PPCs in the regional context, and in contrast to Lop Nor, these results are statistically significant.

5.3. *Tibet Surface Rupture*

Figure 10c compares the NEIC location to the regional location using *China_LM* and PPCs. The epicenter now sits squarely on the surface rupture identified using InSAR. Because the rupture runs E-W, this illustrates that we have good constraints on latitude. There is some indication from source modeling that this rupture was bilateral (VELASCO *et al.*, 1999), however, this has not been positively identified, and our ground-truth constraints on longitude remain weak. This illustrates a potential problem with using InSAR for ground-truth locations of large events: such events will have large rupture planes. However, even some m_b 5 earthquakes can cause enough surface deformation to be detectable with InSAR and this offers some hope that the method can be used to generate sufficiently accurate ground for calibration.

6. *Discussion*

In general, without using corrections, teleseismic locations using hundreds of stations perform fairly well at Lop Nor, with mislocations less than the 1000 km^2 prescribed by the on-site inspection provisions of the CTBT. To explore the problem of regional event location, we examined the nature of event location at Lop Nor using only the 10 closest stations. By systematically removing the most distant of these stations and relocating with and without the propagation path corrections, a sense of the likely regional performance was obtained (the primary caveat being that the arrival times we used had signal-to-noise ratios higher than one would expect for a truly regional problem). Our results show that the provided corrections improve the stability and accuracy of seismic event location at Lop Nor (COGBILL and STECK, 1997), even when data are available from only a few stations. It is important to note, however, that the routine application of the corrections should be checked manually when few stations are available, as damping levels in the EvLoc algorithm may need to be adjusted to ensure convergence. In the above tests, all occurrences of algorithm divergence or larger-than-desired mislocations can be attributed to one or both of two cases: 1) no (or few) reporting stations within regional distances, and 2) poor station geometry. While the results at Lop Nor show that propagation path corrections can greatly reduce mislocation errors, we have very little other ground truth in the China region. Therefore, the applicability of the provided correction surfaces cannot be judged in any other geographical province.

Crustal thickness variations in the study region are quite large, ranging from 30–70 km (LI and MOONEY, 1998). Inasmuch as our PPCs have been developed empirically, we investigated whether the values of the PPCs were at all predictable from changes in crustal thickness. We used the depth-to-Moho data available from the Cornell University database (FIELDING *et al.*, 1993) to estimate the crustal thickness at the locations of all the events used to construct PPC surfaces. The results were surprisingly simple: the travel-time residuals seem to be completely uncorrelated with crustal thickness (STECK *et al.*, 1998).

An important question is whether or not a "best" 1-D model exists for locating seismic events in western China. We have performed relocations of nuclear tests from the Lop Nor region using a number of different 1-D models, including those for western China based on deep seismic sounding, and none is able to eliminate the regional location bias we see with respect to the relocations of GUPTA (1995). Furthermore, the crustal structure in this region is extremely varied, and using a 2-D approach, as developed in this study, improves our ability to account for the extreme lateral effects in the region. Independent of the velocity model chosen, it is possible to remove that location bias using propagation path corrections (COGBILL and STECK, 1997). However, for kriging to be most effective, it is important to remove all trends in the data sets, and better 1-D models can remove some trends. To continue to improve regional location, we must consider 3-D approaches, including the development of 3-D correction surfaces, construction of detailed 3-D models, and tomographic velocity imaging. A significant body of work on estimating regional velocity models currently exists for China, however the bulk of it is published in the Chinese literature. JIH (1998) has translated some of the regional 1-D velocity model results, and the reader is referred to that reference for additional information.

Ground-truth information is vital for improving location, and our approach allows for new ground-truth data to be easily incorporated into a PPC surface. All that is required is to rekrig our data set with the new point. However, developing ground truth in remote regions of the world remains a significant challenge and requires innovations and new approaches. InSAR information appears to be an excellent means to gather shallow event locations. Large, shallow events can create large ruptures, which can provide valuable information, as shown above. However, the large lateral extent of these ruptures make this information limiting for ground truth. An interesting aspect of InSAR is that smaller events can also be identified. PELTZER *et al.* (1999) have identified a secondary rupture approximately 50 km south of the 1997 Tibet mainshock, and estimated its rupture length at 5 km. We are currently trying to identify the event that caused this rupture, but this event could be classified as GT5 at worst. Mining information can also provide ground-truth information. Thus far, we have been unable to gather significant mining information in the region. Calibration explosions can also provide excellent ground-truth information if they are recorded on regional networks. The recent (1997) calibration explosions at Kazakhstan are valuable, but are not well recorded at regional distances (PHILLIPS *et al.*, 1999). A new type of ground-truth information that we have presented in this study is that of the depth-constrained events. Although the location accuracy of these events is difficult to assess using our 2-D correction approach, these events will be vital for developing 3-D velocity models and travel-time correction surfaces.

The data used in this study are a mixture of global catalog information and regional information gathered from digital data from a significant regionalization effort (HARTSE *et al.*, 1997; AGUILAR-CHANG *et al.*, 1999). As demonstrated above, the global catalog arrivals are of varied quality, and obtaining accurate time picks is critical for improving regional location. The quantity of regional arrivals in the global catalogs is also limited, and we find it necessary to gather independent arrival times from regional seismograms to obtain the best results possible. Furthermore, measurement errors are difficult to assign for the global picks given the varied data quality, and measurement error estimates are critical to obtaining improved error or confidence ellipse estimates. For our regional picks, we believe that the signal-to-noise (SNR) criteria used for our measurement error estimate needs further study for regional seismograms. The reason is that regional seismograms are highly varied in nature, and not as stable as teleseismic *P*-waves, for which the SNR relationship was developed.

7. *Conclusion*

The goal of this research is to develop an accurate and precise location algorithm that can be universally and easily applied over large portions of the globe while maintaining regional location capability. To this end, we have developed 2-D empirical *P*-wave propagation path corrections for 74 seismic stations in and around China. We have shown that these corrections improve the precision and accuracy of regional event locations at the Lop Nor and Balapan nuclear test sites, as well as for a Tibetan earthquake located with InSAR. In the teleseismic case, our relocations are dominated by the more abundant distant stations, and the regionalized PPCs have minimal impact. In any seismic event location, accuracy depends on a number of factors, including adequate station coverage, correct phase identification, accurate phase timing, and knowledge of the regional velocity structure. For small events recorded by few stations, all of these factors should be presumed suspect, and if the event is of great interest these factors should be vigorously investigated.

Some distance dependence is observed in our correction surfaces when locations are performed with the IASPEI91 velocity model; for accurate kriging this bias should be eliminated beforehand. To remove this bias we have developed a simple 1-D model for China based on Deep Seismic Sounding results (LI and MOONEY, 1998). Relocations with this model compare favorably to those using a 1-D China model developed by the Chinese State Seismological Bureau. Application of PPCs based on our average China model successfully eliminates all remaining distance bias in the PPCs, except at distances less than 2.5°. Because of sparse seismicity in some areas of China, and a lack of data at some critical stations, we have begun looking at alternative methods of predicting PPCs. Initial research on using Moho depth to predict PPCs shows no correlation between event Moho depth and *P*-wave residual. On the other hand, *P*-wave tomography reveals some coherent lateral variations in *Pn* velocities that may be used to predict PPCs (STECK *et al.*, 1998). Our next step will be to look at developing correction surfaces for detailed 3-D velocity models derived from all available velocity information in the China region.

Acknowledgments

We thank IRIS/DMC for the data used in this study, and G. Peltzer for providing us with his InSAR results. We appreciate Kevin Anderson's help regarding the statistics of the depth-constrained events, and our colleagues in the Geophysics Group (EES3) at LANL for help with the travel-time picks. Critical reviews by an anonymous referee helped clarify the text. This research was supported by US DOE contract No. W-7405-ENG-36. This is contribution # 2 of the Los Alamos Seismic Research Center.

REFERENCES

AGUILAR-CHANG, J., VELASCO, A. A., and HARTSE, H. E. (1999), *Seismic database in support of regional monitoring research in Asia.* In *Proceedings for the 21st Annual Seismic Research Symposium*, 6 pp.

BRATT, S. R., and BACHE, T. C. (1988), *Locating Events with a Sparse Network of Regional Arrays*, Bull. Seismol. Soc. Am. *78*, 780–798.

COGBILL, A., and STECK, L. K. (1997), *The Use of Propagation Path Corrections to Improve Seismic Event Location in Western China*, Los Alamos National Laboratory Report, LA-UR-97-4876.

ENGDAHL, E. R., VAN DER HILST, R., and BULAND, R. (1998), *Global Teleseismic Earthquake Relocation with Improved Travel Times and Procedures for Depth Determination*, Bull. Seismol. Soc. Am. *88*, 722–743.

FIELDING, E. J., BARAZANGI, M., and ISACKS, B. L. (1993), *A Geological and Geophysical Data Base for Eurasia*, Final Technical Report, ARPA NMRO F29601-91-K-DB08.

GHOSE, S., HAMBURGER, M. W., and AMMON, C. J. (1998), *Source Parameters of Moderate-size Earthquakes in the Tien Shan from Regional Moment Tensor Inversion*, Geophys. Res. Lett. *25*, 3181–3184.

GUPTA, V. (1995), *Locating Nuclear Explosions at the Chinese Test Site near Lop Nor*, Science and Global Security *5*, 205–244.

HARTSE, H. E. (1998), *The August 16, 1997 Novaya Zemlya Seismic Event as Viewed from GSN Stations KEV and KBS*, Seismol. Res. Lett. *69*, 206–215.

HARTSE, H. E., TAYLOR, S. R., PHILLIPS, W. S., and RANDALL, G. E. (1997), *A Preliminary Study of Regional Seismic Discrimination in Central Asia with Emphasis on Western China*, Bull. Seismoc. Soc. Am. *87*, 551–568.

ISRAELSSON, H., FISK, M. D., YANG, X., and NORTH, R. G. (1997), *The August 16, 1997 Event in the Kara Sea*, Tech. Rep. CMR-97/38, 47 pp.

JEFFREYS, H., and BULLEN, K. E., *Seismological Tables* (British Association for the Advancement of Science, London 1940).

JIH, R.-S. (1998), Location calibration efforts in China, In *Proceedings for the 20th Annual Seismic Research Symposium*, 12 pp.

KENNETT, B. L. N. (1991), *Locating Oceanic Earthquakes – the Influence of Regional Models and Location Criteria*, Geophys. J. Int. *108*, 848–845.

KENNETT, B. L. N., and ENGDAHL, E. R. (1991), *Travel Times for Global Earthquake Location and Phase Identification*, Geophys. J. Int. *105*, 429–465.

KENNETT, B. L. N., ENGDAHL, E. R., and BULAND, R. (1995), *Constraints on Seismic Velocities in the Earth from Travel-Times*, Geophys. J. Int. *122*, 108–124.

KIM, W-Y., KAZAKOV, V. V., VANCHUGOV, A. G., and SIMPSON, D. W. (1996), *Broadband and array observations at low noise sites in Kazakhstan: opportunities for seismic monitoring of a comprehensive test-ban treaty* Monitoring a Comprehensive Test Ban Treaty (E. S. Husebye and A. M. Dainty, eds.) pp. 467–482.

KOSAREV, G. L., PETERSEN, N. V., VINNIK, L. P., and ROECKER, S. W. (1993), *Receiver Functions for the Tien Shan Analog Network: Contrasts in the Evolution of Structure Across Talasso-Fergana Fault*, J. Geophys. Res. *98*, 4437–4448.

LI, S., and MOONEY, W. D. (1998), *Crustal Structure of China from Deep Seismic Sounding Profiles*, Tectonophysics *288*, 105–113.

MCNAMARA, D. E., WALTER, W. R., OWENS, T. J., and AMMON, C. J. (1997), *Upper Mantle Velocity Structure beneath the Tibetan Plateau from Pn Travel-time Tomography*, J. Geophys. Res. *102*, 493–505.

NAGY, W. (1996), *New Region-dependent Travel-time Handling Facilities at the IDC: Functionality, Testing, and Implementation Details*, Tech. Rep. SAIC-96/1179 57 pp.

PATTON, H. J. (1998), *Bias in the Centroid Moment Tensor for Central Asian Earthquakes: Evidence from Regional Surface Wave Data*, J. Geophys. Res. *103*, 26963–26974.

PELTZER, G., CRAMPE, F., and KING, G. (1999), *Evidence of Nonlinear Elasticity of the Crust from the M_w 7.6 Manyi (Tibet) Earthquake*, Science *286*, 272–275.

PHILLIPS, W. S., HARTSE, H. E., and STECK, L. K. (1999), *Precise Relative Location of 25 Ton Chemical Explosions at Balapan Using IMS Stations*, Pure appl. geophys., submitted.

PRIESTLEY, K. F., and PATTON, H. J. (1997), *Calibration of $m_b(Pn)$, $m_b(Lg)$ scales and transportability of the M_s : m_b Discriminant to New Tectonic Regions*, Bull. Seismol. Soc. Am. *87*, 1083–1099.

RICHARDS, P. G., and KIM, W-Y. (1998), *Testing the Nuclear Test Ban Treaty*, Nature, *389*, 565–573.

RUPPERT, S. D., HAUK, T. F., O'BOYLE, J. L. DODGE, D. MOORE, M., and DAVELER, S. (1999), *Lawrence Livermore National Laboratory's Middle East and North Africa Database*, In *Proceedings for the 21st Annual Seismic Research Symposium*, 9 pp.

SCHULTZ, C. A., MEYERS, S. C., HIPP, J., and YOUNG, C. J. (1998), *Nonstationary Bayesian Kriging: A Predictive Technique to Generate Spatial Corrections for Seismic Detections, Location, and Identification*, Bull. Seismol. Soc. Am. *88*, 1275–1288.

SIPKIN, S. A. (1982), *Estimation of Earthquake Source Parameters by the Inversion of Waveform Data: Synthetic Waveforms*, Phys. Earth Planet. Int. *30*, 242–259.

SMITH, W. H. F., and WESSELL, P. (1990), *Gridding with Continuous Curvature Splines under Tension*, Geophysics *55*, 293–305.

STECK, L. K., COGBILL, A. H., and VELASCO, A. A. (1998), *The use of propagation path corrections to improve regional seismic event location in western China*, In *Proceedings for the 20th Annual Seismic Research Symposium*, 10 pp.

THURBER, C. H., Quin, H. R., and Richards, P. G. (1993), *Accurate Locations of Nuclear Explosions in Balapan, Kazakhstan, 1987 to 1989*, Geophys. Res. Lett. *20*, 399–402.

VELASCO, A. A., RANDALL, G. E., STECK, L. K., and COGBILL, A. H. (1998), *Seismic Event Location in China: Sensitivity to Modeling Errors and Travel-time Residual Bias*, EOS Trans. AGU *79*, S210.

VELASCO, A. A., AMMON, C. J., and BECK S. L. (1999), *Broadband Source Modeling of the November 8, 1997 Tibet (M_w = 7.5) Earthquake and its Tectonic Implications*, in J. Geophys. Res., preparation.

WOODS, B. B., SAIKIA, C. K., and THIO, H. K. (1998), *Focal Depths and Source Parameters for Earthquakes in Northwest China*, Final Scientific Report, Woodward-Clyde Federal Services, Pasadena, CA.

ZHAO, L.-S., and XIE, J. (1993), *Lateral Variations in Compressional Velocities beneath the Tibetan Plateau from Pn Travel-time Tomography*, Geophys. J. Int. *115*, 1070–1084.

(Received August 5, 1999, revised November 22, 1999, accepted February 9, 2000)

Pure appl. geophys. 158 (2001) 241–257
0033-4553/01/020241-17 $ 1.50 + 0.20/0

Pure and Applied Geophysics

Seismic Event Location: Nonlinear Inversion Using a Neighbourhood Algorithm

M. S. SAMBRIDGE[1] and B. L. N. KENNETT[1]

Abstract—A recently developed direct search method for inversion, known as a neighbourhood algorithm (NA), is applied to the hypocentre location problem. Like some previous methods the algorithm uses randomised, or stochastic, sampling of a four-dimensional hypocentral parameter space, to search for solutions with acceptable data fit. Considerable flexibility is allowed in the choice of misfit measure.

At each stage the hypocentral parameter space is partitioned into a series of convex polygons called Voronoi cells. Each cell surrounds a previously generated hypocentre for which the fit to the data has been determined. As the algorithm proceeds new hypocentres are randomly generated in the neighbourhood of those hypocentres with smaller data misfit. In this way all previous hypocentres guide the search, and the more promising regions of parameter space are preferentially sampled.

The NA procedure makes use of just two tuning parameters. It is possible to choose their values so that the behaviour of the algorithm is similar to that of a contracting irregular grid in 4-D. This is the feature of the algorithm that we exploit for hypocentre location. In experiments with different events and data sources, the NA approach is able to achieve comparable or better levels of data fit than a range of alternative methods; linearised least-squares, genetic algorithms, simulated annealing and a contracting grid scheme. Moreover, convergence was achieved with a substantially reduced number of travel-time/slowness calculations compared with other nonlinear inversion techniques. Even when initial parameter bounds are very loose, the NA procedure produced robust convergence with acceptable levels of data fit.

Key words: Hypocentre, event location, neighbourhood algorithm.

1. Introduction

The location of seismic events in space and time is an important part of seismological practice with a wide range of applications, but has particularly significance in the context of monitoring a comprehensive test-ban treaty. A primary discriminant between man-made events and the majority of natural earthquakes is provided by the depth of the source. Drilling techniques for the emplacement of devices preclude a source depth of more than a few kilometres. Where the depth of an event can be established to be deeper than 15 km, the event is very unlikely to be

---

[1] Research School of Earth Sciences, Institute of Advanced Studies, Australian National University, Canberra ACT 0200, Australia. E-mail: malcolm@rses.anu.edu.au, brian@rses.anu.edu.au

man-made, even allowing for location errors. A significant role for location procedures in monitoring compliance with the treaty is therefore to act as a winnowing device to segregate the very large number of intermediate and deep-focus events occurring in subduction zones, especially around the Pacific Rim. For shallower events there is a need to obtain the best possible location by combining data from regional and more distant stations, and then to try to characterise the source using additional information.

One of the difficult tasks for a global monitoring system with a limited number of stations is the association of arrivals and waveform segments from a widely separated seismic stations with a single event. Across the network, it is quite likely that seismic phases from more than one event will be being recorded simultaneously. At least in the initial stages of location estimation, the procedures used should cope with the possibility that some phase readings may be misassociated, and not belong to the event of interest.

2. *Hypocentre Location*

In order to describe the hypocentre of a seismic event we have to specify the spatial location and the origin time of the event. These four pieces of information, the latitude, longitude, depth and the time of initiation of seismic energy have to be inferred indirectly. The primary information is the times of arrival of seismic phases at different stations; when seismic arrays are used these times can be supplemented with vector slowness information for phases, comprising the azimuth of the wavefront crossing the array and the horizontal slowness of the wavefront.

We can formulate the task of locating an event in the following terms: Given the arrival times [and slownesses] of seismic waves at a number of different receivers, deduce the best estimate of the origin time and spatial location of the hypocentre of the event (i.e., the point at which radiation is initiated). This task depends on us having a clear identification of the seismic phases whose arrival time [or slowness] is measured, and a good representation of the propagation processes associated with the measured passage times [or slownesses] for the different phases.

The task of phase association is not entirely straightforward but can be aided by the use of array information (see e.g., KVAERNA and DOORNBOS, 1986; RINGDAL and KVAERNA, 1989) or by using multiple attribute analysis for three-component broadband records (CHRISTOFFERSSON *et al.*, 1988; TONG and KENNETT, 1995, 1996). In this work we will assume that the identity of each observed phase is known but allow for the possibility of using different classes of phases and types of information.

The approach used to determine an event location should not be limited by the representation of the propagation process. Ideally we would like to take account of the full three-dimensional structure of the earth, or at the very least allow for regional structure in the neighbourhood of the source. A practical first step is to determine a location with a spherically symmetric earth model, such as the *iasp91* model of KENNETT and ENGDAHL (1991) which is currently employed in the International Data Center for the CTBT, or *ak135* (KENNETT *et al.*, 1995) which has an improved representation for S and core phases. Once an approximate location is established, region specific corrections can be applied or alternatively a more complex model representation used.

The specification of the location of the event requires the determination of four parameters
t_h — the origin time of the seismic disturbance
$\mathbf{x}_h = (x_h, y_h, z_h)$ — the spatial coordinates of the hypocentre
which together define the 4-vector $\hat{\mathbf{h}}$ of hypocentral parameters.

We will assume that we have N_T observations t_i of the arrival times of identified phases at a set of seismic receivers with associated errors σ_i; these arrival times may be associated with different ray paths for a single wave type or different wave types. The time information may also be supplemented with N_S observations of the vector slowness of arrivals (derived from seismic arrays).

For the i-th seismic phase we calculate the travel time $t_{ri}(\mathbf{x}_s)$ for passage between a source at $\mathbf{x}_s = (x_s, y_s, z_s)$ and the requisite receiver by ray tracing in an earth model. From these travel times we construct estimates t_{ci} of the arrival times of the phases for an assumed origin time t_s as

$$t_{ci}(\mathbf{x}_s, t_s) = t_s + t_{ri}(\mathbf{x}_s), \tag{1}$$

and these arrival times have to be compared with the observations t_i. We note the clear separation between the dependence of the estimated arrival time t_{ci} on the spatial and temporal components of the estimated location.

In a similar way we construct estimates for the j-th vector slowness $\mathbf{s}_{cj}$ to be compared with the measured vector slownesses $\mathbf{s}_j$.

We now need to establish suitable measures of the extent to which the calculated arrival times and slowness match the observations so that we use these misfit measure to guide the inversion for the hypocentre $\hat{\mathbf{h}}$.

For arrival times the statistics of the residuals can often be well represented by a distribution suggested by JEFFREYS (1932) which consists of a Gaussian superimposed on a slowly varying pedestal function. Commonly, the narrower Gaussian provides a representation of the effect of picking errors and the broader background arises mostly from the differences between the real earth and the times predicted from a simplified model. The net effect is that the probability distribution has a substantially longer 'tail' than a Gaussian (L_2 norm).

A simpler form with a similar property is to use L_p measures of misfit with $p < 2$. For many purposes, satisfactory results can be produced with $p = 1.25$ which is tolerant to large outliers, even though this may not represent a close approximation to the actual probability distribution for the residuals (BILLINGS *et al.*, 1994). In particular, when many different types of seismic phases are employed, the residual distributions can differ significantly between phases. An advantage of the L_p measures with $p < 1.5$ is that they are robust and much less affected by large outliers than the conventional squared residual measures.

We construct an L_p measure for the misfit between observed and calculated arrival times as

$$C_T = \left[\frac{1}{N_T}\sum_{i=1}^{N_T} |t_i - t_{ci}|^p / \sigma_i^p\right]^{1/p}, \tag{2}$$

where σ_i is the estimate of the error in the i-th arrival time observation.

Array information can be included by the inclusion of an extra term in the misfit function (see e.g., BRATT and BACHE, 1988). For each vector slowness measurement $\mathbf{s}_j$, we construct corresponding theoretical slownesses $\mathbf{s}_{cj}$ for the current source location and calculate

$$C_S = \left[\frac{1}{N_S}\sum_{j=1}^{N_S} |\mathbf{s}_j - \mathbf{s}_{cj}|^p / \sigma_{sj}^p\right]^{1/p} \tag{3}$$

where N_S is the number of slowness measurements and σ_{sj} is the error in the j-th slowness observation. Although the residual distribution for vector slowness is likely to be somewhat different from that for arrival times, we have achieved good results by using the same value of p for both the time and slowness misfit terms C_T and C_S.

We use the composite misfit function

$$C = C_T + C_S \tag{4}$$

as the measure of the quality of the location estimate.

2.1. Nonlinear Inversion

The conventional approach to event location is to use L_2 misfit representations in iterative inversion with linearisation around successive location estimates, which requires calculation of the derivatives of the times and slownesses with respect to the hypocentral parameters. The calculation of such derivatives limits the class of velocity models that can be conveniently employed.

However, with the advent of faster computers it is now feasible to calculate the travel times afresh for each postulated source location rather than rely on linearisation and so avoid any differentiation. This more flexible formulation allows us to use the more general representations of the misfit functions such as L_p norms introduced above. For teleseismic arrivals, the tau-spline procedure of BULAND and CHAPMAN (1983) gives an efficient scheme for multi-phase travel time or slowness calculation for a given epicentral distance in a spherical model and has been adopted for the IASPEI91 travel-time tables (KENNETT and ENGDAHL, 1991).

We have noted earlier the separation of the dependence of the arrival times for the seismic phases on the estimate of the origin time t_s and the spatial location of the event x_s. This has been used as the basis of a number of different schemes in which a four-dimensional search procedure isolates the regions of best fitting hypocentres. For example, SAMBRIDGE and KENNETT (1986) used a directed spatial grid search with bracketing of the origin time and KENNETT (1992) introduced a contracting grid scheme which exploits the fact that the arrival times for different origin times do not require recalculation of travel times for a fixed source location. Such directed search procedures cannot guarantee that a global minimum for the misfit function will be found by the discrete search. However, the results obtained with different starting meshes are generally very consistent.

Alternative classes of nonlinear procedures based on direct calculation of travel times and different classes of optimisation algorithm have also been successfully used to find hypocentral estimates. Such methods include genetic algorithms (KENNETT and SAMBRIDGE, 1992; SAMBRIDGE and GALLAGHER, 1993; BILLINGS *et al.*, 1994), simulated annealing (BILLINGS, 1994) and evolutionary programming (MINSTER *et al.*, 1995). Even so, grid methods retain the attraction of simplicity of implementation and good performance even with poor network geometry.

In this paper we introduce the neighbourhood algorithm (NA) for location. The NA procedure exploits all previous information in the inversion procedure to focus in on those parts of parameter space for which the current misfit measure is least (SAMBRIDGE, 1999a). For the location problem the neighbourhood algorithm, as currently used, has approximately the character of a contracting irregular grid in four-dimensional hypocentre space. The procedure is very efficient and able to work with multiple local minima (SAMBRIDGE, 1998).

2.1.1. The use of a neighbourhood algorithm

The neighbourhood algorithm is a derivative-free method for searching a multi-dimensional parameter space for models (hypocentres) of acceptable data fit. It is based on some simple geometrical concepts, involves two tunable parameters, and makes use of randomised, or stochastic, sampling of parameter space. We outline the approach below. Full details can be found in SAMBRIDGE (1999a).

The central idea behind the neighbourhood algorithm is the use of Voronoi cells to guide the search in parameter space. Voronoi cells are simply nearest neighbour regions, as defined by a suitable distance norm (VORONOI, 1908; DELAUNAY, 1934). Here we use the L_2-norm. Figure 1 shows examples of Voronoi cells, defined about an irregular set of points in the plane. Each cell is a convex polygon about its defining point. Note that the sides of each cell are formed from the perpendicular bisectors between neighbouring pairs of points. Voronoi cells have many useful properties and have found applications in a number of fields (see OKABE *et al.*, 1992, for a review). From Figure 1 we see that the size and shape of Voronoi cells automatically adapt to the density and distribution of the defining points. This is the property which is

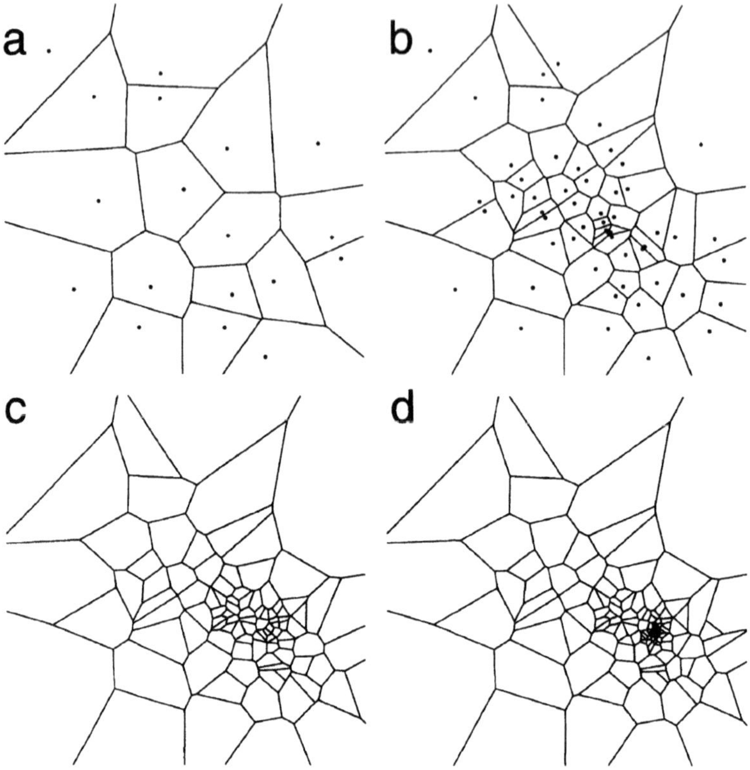

Figure 1
Voronoi cells produced by a neighbourhood algorithm after a) 1, b) 5, c) 10 and d) 20 iterations. At each iteration 9 samples are added into the cells about the 2 epicentres with the least data misfit.

exploited in the search algorithm. Figure 1 shows that the Voronoi cell defines a neighbourhood about each point.

The neighbourhood hypocentre location algorithm can be summarised as follows: First, an initial population of n_s hypocentres is generated. Usually these follow a uniform random distribution within the parameter space bounds. At each iteration, the current n_r hypocentres with lowest data misfit are determined and a new set of n_s hypocentres is uniformly distributed inside their Voronoi cells, that is, n_s/n_r in each cell. Subsequent iterations are repeated in a similar manner, with the n_r best hypocentres and their Voronoi cells being updated at each iteration. The number of hypocentres per iteration n_s and the number of Voronoi cells to be occupied n_r are the two tuning parameters of the NA procedure. For each hypocentre tested, we calculate the associated arrival time and slowness information and use this to generate the appropriate misfit measure.

Note that the algorithm only uses the misfit measure to rank the goodness of fit of the hypocentres; the actual value of the misfit is not used directly. The use of a rank measure to drive the algorithm allows considerable freedom in controlling the way in which the data influence the NA search. For example, one may choose a single misfit measure combining arrival time and array measurements as in (4), with the relative weighting depending on the choice of assigned errors. Alternatively, one might separately rank hypocentral estimates according to the misfit for arrival time and array data; an overall rank could then be constructed from some combination of the two.

Figure 1, shows an example of the neighbourhood algorithm (used for epicentral location) at iterations 1, 5, 10, and 20, with $n_s = 9$ and $n_r = 2$. Notice how the Voronoi cells adapt to the sampling and quickly focus in the southeastern part of the domain. The precise details of the sampling method are described in SAMBRIDGE (1999a). It turns out that the bookkeeping calculations are surprisingly straightforward and may be handled very efficiently.

The behaviour of the search algorithm depends on the two control parameters, n_s and n_r. As can be seen in Figure 1, for small values of n_r it takes on the character of a contracting irregular grid, and can efficiently optimise a function. The power of the approach is due to the fact that at each iteration the sampling is driven by the size and distribution of the Voronoi cells, and these are controlled automatically by all previous samples.

2.1.2. Location assessment

In the absence of external information about the seismic source, the accuracy of the hypocentral location has to be estimated from the information available in the inversion. The implied precision of the estimates is therefore model-based and can be distinctly misleading as to the error in the location of the true hypocentre, e.g., through neglect of the influence of 3-D earth structure.

Commonly confidence intervals for the hypocentral location can be deduced by examining the distribution of

$$\Gamma(h) = C(h) - C(\hat{h}), \tag{5}$$

where C is the misfit criterion and $\hat{h}$ is the best estimate of the location of the hypocentre.

The results from the neighbourhood algorithm inversion can be used *a posteriori* to build up an estimate of the probability distribution for the hypocentral estimates (SAMBRIDGE, 1999b). Bayesian measures of information (for example the posterior model covariance matrix, resolution matrix and marginal density distributions) may be useful in assessing the degree of constraint and trade-off between parameters. A fully nonlinear calculation of Bayesian measures involves the evaluation of various integrals of the posterior probability density distribution (PPD) over the hypocentral parameter space. SAMBRIDGE (1999b) describes a 'resampling' technique whereby estimates of these integrals may be obtained from the finite ensemble of hypocentres generated during the search stage. The accuracy of these measures depends in part on the size and distribution of the available ensemble.

It is common for covariance and resolution matrices to be determined using local linearisation, for example, by approximating the contours of $\Gamma(h)$ by ellipsoids about the best fit hypocentre (for example, see BULAND, 1976). If the nonlinearity is severe, e.g., with a sparse network, then the linearised estimates are likely to be both inaccurate and biased. In such circumstances, we anticipate that the fully nonlinear approach would produce more accurate results. However, in either case knowledge of the statistics of all noise process in the data is required and this may be difficult in some cases.

2.2. Illustrations of Event Location

We illustrate the NA location scheme with two examples using different types of data. The first example is for a test event in the Marianas including arrival time and array information, and the second example is for P arrivals from a nuclear blast at the E. Kazakh test site with known location. These examples have previously been used with other nonlinear procedures (KENNETT, 1992, 1996) and so provide a good base for comparative performance.

From a wide variety of tests we have chosen to use the NA procedure with an initial population of 9 random estimates spanning the search region and then 20 iterations in which 9 further estimates are produced at each stage. We have allowed resampling of up to 2 cells at each stage (i.e., $n_s = 9$ and $n_r = 2$). With these settings we have found that we achieve good convergence on a location estimate with a modest total of 189 sets of travel-time/slowness calculations.

2.2.1. Arrival time and array slowness information

The first example is based on an event in the Marianas. The arrival times have been supplemented with simulated array information, and so we use the extended definition of the misfit function.

The phase information used for this trial is presented in Table 1. Apart from two close stations, the data are from teleseismic distances with a limited range of azimuths. We have undertaken a number of different inversions for this data set using different styles of algorithms and different assumptions for constraints on the location.

For the first group of tests we have taken a region 2° across in both latitude and longitude, with a depth interval of 60 km and a time interval of 40 s centred on the ISC location. We have then used a variety of different nonlinear algorithms with the IASPEI91 reference model. In each case we have used the same L_1-norm measure of misfit and we compare the results in Table 2. All these nonlinear methods give some shift from the ISC location derived using the Jeffreys-Bullen travel-time tables (without the array information). The smallest variations are obtained using the simulated annealing method [SA] (BILLINGS, 1994), and a similar result with a shallower depth is also found from the use of the contracting grid scheme [CG] (KENNETT, 1992). By contrast the application of genetic algorithms [GA] (KENNETT

Table 1

Marianas Test Event 20:17:49.4 14.5 N 146.9 E 33.00 km

Code	Phase	Hour Min Sec	Range [deg]	Back Azimuth
GUA	Pn	20 18 24.20	2.05	242.4
GUMO	Pn	20 18 24.20	2.07	244.1
WB2	P	20 24 52.20	36.35	200.0
BJI	P	20 24 57.00	36.92	319.4
PKI	P	20 27 43.60	58.23	293.6
KKN	P	20 27 44.50	58.34	293.8
DMN	P	20 27 45.80	58.50	293.6
COL	P	20 28 38.20	66.61	25.0
INK	P	20 29 17.00	72.82	22.5
YKA	P	20 30 04.40	81.20	27.7
SOD	P	20 30 35.00	87.64	340.4
KJF	P	20 30 41.00	88.98	337.5
SUF	P	20 30 48.00	90.37	336.6
NUR	P	20 30 56.00	92.21	335.2
HFS	P	20 31 17.30	96.64	338.4
Code	**Phase**	**Azimuth**	**Range [deg]**	**Back Azimuth**
WB2	P	200.000	36.35	200.0
YKA	P	104.250	81.20	27.7
Code	**Phase**	**Slowness**	**Range [deg]**	**Back Azimuth**
WB2	P	0.080	36.35	200.0
YKA	P	0.050	81.20	27.7

Table 2

Comparison of location estimates

Method	Lat.	Long.	Depth [km]	Time	L_1 Misfit
ISC	14.50 N	146.90 E	33.00	49.40	1.380
SA	14.51 N	146.92 E	23.10	49.44	0.755
CG	14.52 N	146.90 E	22.80	49.40	0.748
GA 1	14.49 N	146.67 E	49.97	53.35	0.732
GA 2	14.50 N	146.72 E	44.10	52.70	0.720
NA	14.50 N	146.78 E	41.64	52.23	0.711

and SAMBRIDGE, 1992) produce a better fit with a deeper source; two different GA results are displayed based on the use of different seed for the random processes. For each of these GA inversions 1320 sets of travel-time evaluations were undertaken to arrive at the location estimate. The neighbourhood algorithm produces a result with the smallest misfit of any of the methods. The NA method employs only 189 travel-time and slowness evaluations which is less than a quarter of that for any of the other techniques.

The NA procedure does not depend on having a good initial estimate of the location. We illustrate this for the Marianas event event by taking a starting model displaced by more than 100 km from the likely location and with a significant time offset. We employ a grid with a ±2° range in latitude and longitude, ±40 km range in depth and ±30 s in origin time about 13.52 N, 147.5 E, 39.0 km, 47.9 s for which the L_1 misfit is 1.916. We show in Figure 2 the progress of the NA scheme with an L_1 measure of misfit towards the best fitting location by means of projections onto the spatial axes and a time bar. The display covers almost the entire allowed ranges on the parameters.

The neighbourhood algorithm initially explores much of the allowed domain but quite quickly shifts attention to a zone with improved misfit. The convergence towards a cluster of well-fitting location estimates is rapid even though the starting point is well displaced from the best estimate of the location. Each of the samples tested in an individual iteration is indicated by the same symbol. As the iteration number increases the number of sides is also increased and then solid symbols are introduced for the last few iterations. The best fit location is very close to that reached with the domain centred at the ISC solution [14.484 N, 146.761 E, 43.156 km, 52.30 s] with an L_1 misfit of 0.716.

The progress of the NA procedure for this case is illustrated in Figure 3. The initial model estimate used as the centre of the bounds on the parameter distribution provides a moderately good fit. It requires several iterations of the exploration of the model space to find a better model. Exploration continues and from time to time a model is found which has lower misfit than all its predecessors. As the 20th iteration is approached, all models are tightly clustered around the best fit (as can also be seen in Fig. 2).

As a final test for this same data set we have undertaken an inversion with a very broad set of initial bounds [±5° in latitude and longitude, ±60 km in depth and ±60 s in origin time] about the same starting point as in the previous example. If we run the NA procedure for 20 iterations as in the previous example we obtain a very satisfactory location given the weak imposed constraints: [14.441 N, 146.625 E, 59.03 km, 52.30 s] with a misfit of 0.780. This solution is not far from that found with much smaller bounds but has higher misfit. This shows that with the control parameters used in this case, it is possible for the NA search to get distracted by local minima and therefore not reach the best possible solution.

With a low cost approach such as the neighbourhood algorithm it is possible to construct solutions from a number of different starting points and accept that which gives the best fit. Thus with the same coarse grid around the centre point [14.0 N, 146.0 E, 50.0 km, 58.0 s] we reach an L_1 misfit of 0.712 at [14.496 N, 146.734 E, 45.189 km, 52.64 s] in 189 sets of travel-time calculation.

Once a reasonable solution has been found it is also effective to restart the NA search with a smaller zone in order to refine the hypocentre. In our experience the neighbourhood algorithm is the most robust method for assessing likely locations when the bounds on the allowed solution are very large, as is likely to be the case

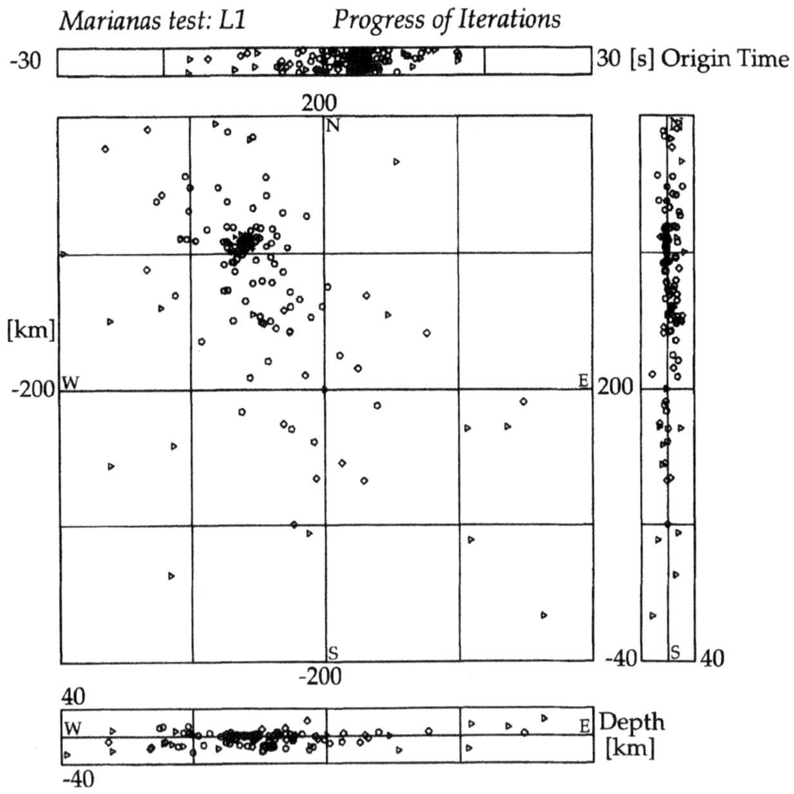

Figure 2
Representation of the progress of the NA technique towards convergence on a location estimate via projections onto the different spatial planes. The centre is the reference point about which the parameter bounds were specified. The group of models generated at each iteration are indicated by the same style of polygons. The number of sides increases as the iterations proceed, but for clarity solid symbols are introduced for the last few iterations.

with the sparse global IMS network operating under the Comprehensive Nuclear-Test-Ban Treaty.

2.2.2. East Kazakh event

In the second example we consider a nuclear test at the Semipalatinsk test site for which precise coordinates are available from Soviet sources (BOCHAROV *et al.*, 1989). We have chosen one of the smaller events and used the set of arrival times reported by the Bulletin of the International Seismological Centre. The station distribution relative to the event is illustrated in Figure 4, and displays a high concentration of stations in Europe with a scatter across the rest of the globe and a broad open azimuth sector towards the Pacific. There are no close stations so that the resolution in depth is expected to be rather poor.

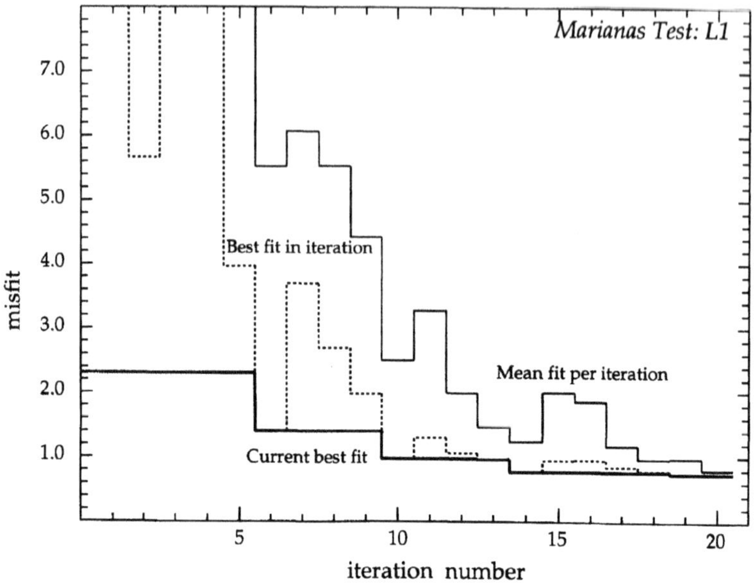

Figure 3
The progress of convergence in the NA procedure for the Marianas test. The best fitting model per iteration, the mean misfit across all the models in each iteration and the current best fit are shown. All converge over the 20 iterations of the neighbourhood algorithm.

We illustrate the progress of the hypocentral inversion in Figure 5 with an $L_{1.25}$ measure of misfit using a range of $\pm 1°$ in latitude and longitude, ± 10 km in depth and ± 10 s in origin time, centred at the Soviet location. Each of the symbols in Figure 5 represents one of the location estimates which has been assessed in the progress of the NA inversion. The symbols are coded in greytone by the level of misfit with darker tones indicating better fit. The display in Figure 5 shows only a portion of the original search region and therefore displays the nature of the misfit function in the neighbourhood of the best fitting model. The lightest symbols have a misfit approximately twice that of the best. The immediate neighbourhood of the best locations indicates relatively slow variation but away from this region the misfit in arrival times grows quite rapidly. As indicated above the sampling of the misfit function in the course of the NA inversion can be put to use in a retrospective assessment of the probability distribution for the hypocentre parameters.

The Soviet location (at the centre of the main grid) is one of the models tested in the NA process but does not provide the best fit because of the influence of both 3-D heterogeneity in the earth and the geometry of the available network. In all the different inversions we have tried, the best fitting location is displaced about 6 km to the east and 1 km to the south of the true location. With the IASPEI91 travel times the origin time lies within 0.6 s of the Soviet information. Because of the limited

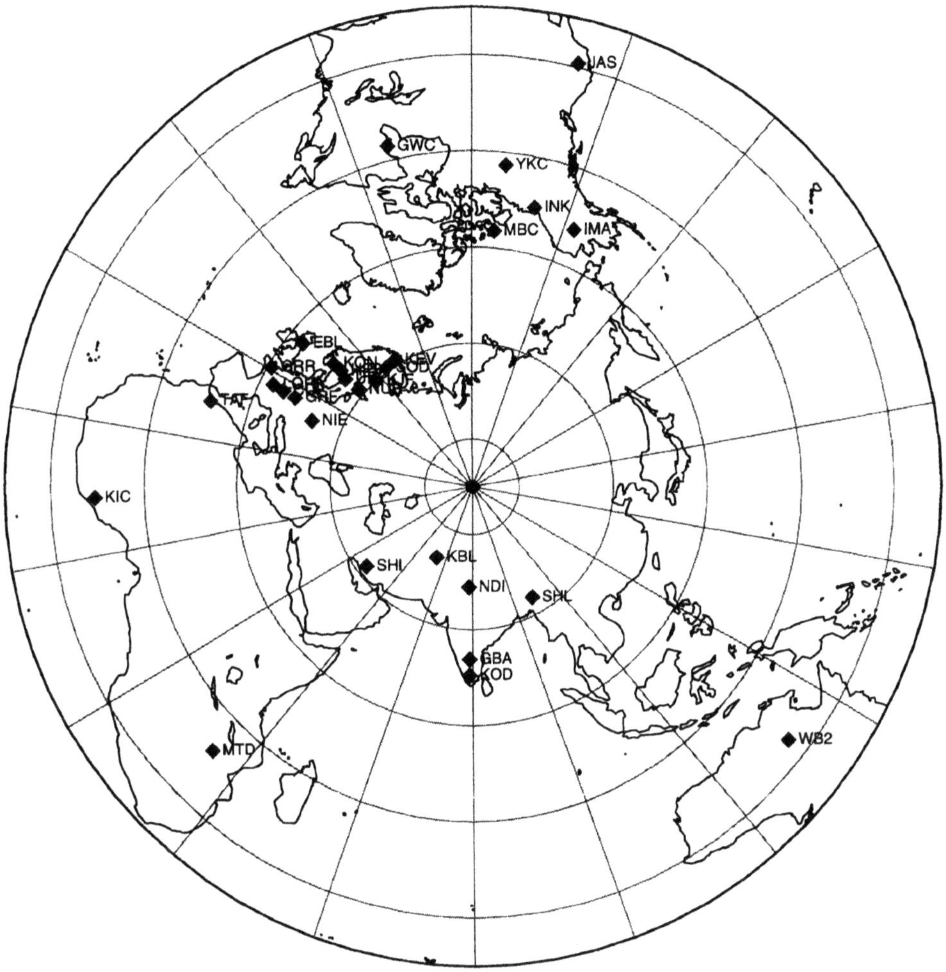

Figure 4
The distribution of stations used in the hypocentral estimates for the East Kazakh event. Polar projection centred on the announced hypocentre.

depth constraints the depth estimate depends more strongly on the range of allowed depths. With an allowed depth range from 0–50 km we find a depth of 7 km and tighter limits on the depth sampling produce shallower solutions with a misfit that decreases slowly with decreasing depth. The best fit is achieved for a constrained solution at the Soviet depth of 0.139 km but with only a very small shift in horizontal position (about 0.25 km).

As discussed in section 2.1, it is possible to make use of the ensemble of hypocentral estimates generated in the NA search to generate an *a posteriori* probability distribution for the hypocentral parameters.

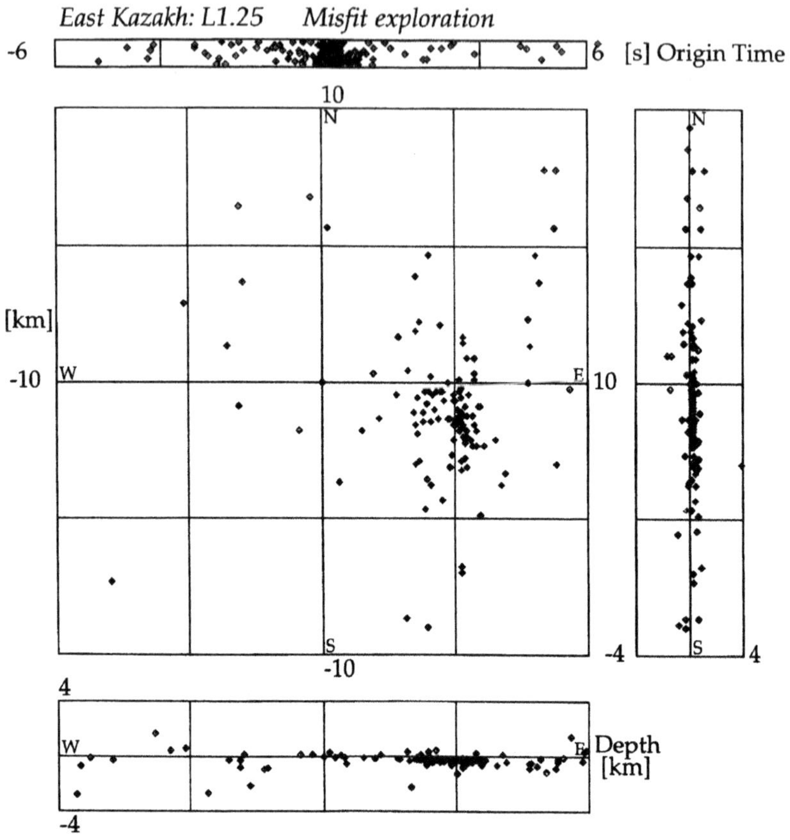

Figure 5
The central portion of the parameter space showing the sampling of model misfit in the NA procedure. The darker symbols indicate better fit and their distribution indicates the reliability of the estimated hypocentre. The environs of the best fitting model are well explored and this information is available for assessment of the probability distribution of the hypocentral parameters.

The distribution of misfit in hypocentral space can also be used directly as an indication of the reliability of the postulated hypocentre. If we can prescribe a satisfactory level of fit, then the 'cloud' of hypocentres whose misfit lies below this level outline a nonlinear error domain for the solution. The process can be illustrated with Figure 5 where the cluster of the darkest symbols (indicating least misfit) occupy a zone about 3 km E/S by 5 km N/S.

The quality of the location estimates we have achieved using the IASPEI91 travel times is very good when we allow for the limited data set and the influence of lateral heterogeneity. Inclusion of site specific corrections would readily shift the solution to the 'ground truth'.

2.2.3. Phase rejection during location

For any global network, a significant aspect of the location of seismic events is the separation of seismic phase information for a particular time period into sets of information associated with separate events. In an initial scan it may not be apparent whether certain observations should be associated with others in the definition of an event.

KENNETT (1996) has shown how a contracting grid procedure with a robust measure of misfit such as an L_1 or $L_{1.25}$ norm can be used to recognise observations which do not match with the main group of phase readings, so that these can be removed from further consideration. The approach is to undertake a preliminary inversion over a wide area (e.g., 20° across and 200 km deep) with a coarse initial grid and look for a location estimate with the least misfit even though the variation in the misfit may be small because of the major outliers. This approximate location is then used to identify those observations which are regarded as having too large residuals to be reliably associated with the event. The data set is then reduced to include only the consistent information and then a finer grid search, over a smaller region, is used to improve the location estimate.

This style of procedure is readily adapted to use with the neighbourhood algorithm. Once again the initial model scan is over a broad region and the algorithm is run for a limited number of iterations and then the best fitting location estimate is used to screen the observations. We recall that we depend only on the ranking of the misfit of the different location estimates and the algorithm is not affected by a large bias from the discordant observations. Once the data set is reduced the neighbourhood algorithm is used over a smaller zone such as those employed above to improve the estimate of the event location.

3. *Discussion*

The neighbourhood algorithm is able to achieve good quality locations with far fewer travel-time evaluations than any previous nonlinear location procedure we have tried. In the examples we have shown we have been able to get good results with a total of 189 travel-time computations. In contrast the contracting grid scheme of KENNETT (1992) requires at least 500 travel-time computations to achieve comparable levels of misfit. Note that trial bounds can be specified independently on each parameter so that it is possible to impose strong constraints on the depth of an event if required.

The particular advantage of the neighbourhood algorithm approach is that the computational effort is concentrated on those portions of hypocentral space where the misfit is least, and further the scheme can follow multiple minima (SAMBRIDGE,

1998) and therefore provide information on alternative location possibilities. When combined with the retrospective assessment of the probability distribution of the location parameters, the scheme provides both an effective location estimation and assessment procedure which can be used with any class of velocity model for which ray-tracing can be undertaken including full 3-D structure.

REFERENCES

BILLINGS, S. (1994), *Simulated Annealing for Earthquake Location*, Geophys. J. Int. *118*, 680–692.

BILLINGS, S., KENNETT, B. L. N., and SAMBRIDGE, M. S. (1994), *Hypocentre Location: Genetic Algorithms Incorporating Problem-Specific Information*, Geophys. J. Int. *118*, 693–706.

BOCHAROV, V. S., ZELENTSEV, A., and MIKHAILOV, V. N. (1989), *Characteristics of 96 Underground Nuclear Explosions at the Semipalatinsk Test Site (in Russian)*, Atomnaya Energiya *67* (3).

BRATT, S. R., and BACHE, T. C. (1988), *Locating Events with a Sparse Network of Regional Arrays*, Bull. Seismol. Soc. Am. *78*, 780–798.

BULAND, R. (1976), *The Mechanics of Locating Earthquakes*, Bull. Seismol. Soc. Am. *66*, 173–187.

BULAND and CHAPMAN (1983) *The Computation of Seismic Travel Times*, Bull. Seismol. Soc. Am., *73*, 1271–1302.

CHRISTOFFERSSON, A. E., HUSEBYE, E. S., and INGATE, S. F. (1988), *Wavefield Decomposition Using ML Probabilities in Modelling Single Site 3-component Records*, Geophys. J. Int. *93*, 197–213.

DELAUNAY, B. N. (1934), *Sur la sphere vide*, Bull. Acad. Science USSR VII: Class. Sci. Math., 793–800.

JEFFREYS, H. (1932), *An Alternative to the Rejection of Observations*, Proc. R. Soc. Lond. *137A*, 78–87.

KENNETT, B. L. N. (1992), *Locating Oceanic Earthquakes — The Influence of Regional Models and Location Criteria*, Geophys. J. Int. *108*, 945–854.

KENNETT, B. L. N. *Event location and source characterisation*. In *Monitoring a Comprehensive Test-Ban-Treaty* (eds. HUSEBYE, E. S., and DAINTY, A.) (Kluwer, Dordrecht 1996) pp. 501–520.

KENNETT, B. L. N., and ENGDAHL, E. R. (1991), *Travel Times for Global Earthquake Location and Phase Identification*, Geophys. J. Int. *105*, 429–465.

KENNETT, B. L. N., and SAMBRIDGE, M. S. (1992), *Earthquake Location: Genetic Algorithms for Teleseisms*, Phys. Earth. Planet. Int. *75*, 103–110.

KENNETT, B. L. N., ENGDAHL, E. R., and BULAND, R. (1995), *Constraints on Seismic Velocities in the Earth from Travel Times*, Geophys. J. Int. *122*, 108–124.

KVAERNA, T., and DOORNBÖS, D. J. (1986), *An integrated approach to slowness analysis with arrays and three-component stations*, Semiannual Technical Summary, 1 October 1985–31 March 1986, NORSAR Sci. Report No. 2-85/86, Kjeller, Norway.

MINSTER, J. -B. H., WILLIAMS, N. P., MASTERS, T. G., GILBERT, J. F., and HAASE, J. S., *Application of evolutionary programming to earthquake hypocenter determination*. In *Proc. 4th Annual. Conf. of Evol. Prog.* (eds. McDonnell, J. R., Reynolds, R. C., and Fogel, D. B.) (MIT Press, Cambridge 1995) pp. 3–17.

OKABE, A., BOOTS, B., and SUGIHARA, K., *Spatial Tessellations Concepts and Applications of Voronoi Diagrams* (John Wiley & Sons, New York 1992).

RINGDAL, F., and KVAERNA, T. (1989), *A multi-channel Processing Approach to Real-time Network Detection, Phase Association and Threshold Monitoring*, Bull. Seismol. Soc. Am. *79*, 1927–1940.

SAMBRIDGE, M. S. (1998), *Exploring Multidimensional Landscapes without a Map*, Inverse Problems *14*, 427–440.

SAMBRIDGE, M. S. (1999a), *Geophysical Inversion with a Neighbourhood Algorithm – I. Searching a Parameter Space*, Geophys. J. Int. *138*, 479–494.

SAMBRIDGE, M. S. (1999b), *Geophysical Inversion with a Neighbourhood Algorithm – II. Appraising the Ensemble*, Geophys. J. Int. *138*, 727–746.

SAMBRIDGE, M. S., and KENNETT, B. L. N. (1986), *A Novel Method of Hypocentre Location*, Geophys. J. R. Astr. Soc. *87*, 313–331.

SAMBRIDGE, M. S., and GALLAGHER, K. (1993), *Earthquake Hypocenter Location Using Genetic Algorithms*, Bull. Seismol. Soc. Am. *83*, 1467–1491.

TONG, C., and KENNETT, B. L. N. (1995), *Towards the Identification of Later Seismic Phases*, Geophys. J. Int. *123*, 948–958.

TONG, C., and KENNETT, B. L. N. (1996), *Automatic Seismic Event Recognition and Later Phase Identification for Broad-band Seismograms*, Bull. Seismol. Soc. Am. *86*, 1896–1909.

VORONOI, M. G. (1908), *Nouvelles applications des paramètres continus à la théorie des formes quadratiques*, J. reine angew. Math. *134*, 198–287.

(Received June 30, 1999, revised October 6, 1999, accepted February 9, 2000)

Pure appl. geophys. 158 (2001) 259–276
0033–4553/01/000000–00 $ 1.50 + 0.20/0

Pure and Applied Geophysics

Best Practice in Earthquake Location Using Broadband Three-component Seismic Waveform Data

ROBERT A. UHRHAMMER,[1] DOUGLAS DREGER[1] and BARBARA ROMANOWICZ[1]

Abstract—We present an earthquake location algorithm, the Broadband Waveform Regional Earthquake Location Program (*BW_RELP*), which utilizes phase onset times and wave azimuths recorded by three-component broadband seismic stations and an adaptive migrating grid search algorithm to find the global minimum in an arbitrary normed misfit parameter. The performance of *BW_RELP* is demonstrated using regional (300–800 km distant) broadband recordings to locate events in the 1995 Ridgecrest, California earthquake sequence. The purpose of this study is to introduce the *BW_RELP* algorithm in detail and to expand on the previous paper by DREGER *et al.* (BSSA, *88*, 1353–1362, 1998), using one Berkeley Digital Seismic Network (BDSN) station (YBH) and two USNSN stations (ELK and MNV) which span 300–800 km in distance and 55 degrees in azimuth, to further investigate the capability of a sparse broadband network of three-component stations at monitoring a region located outside of the network, as will be the case in the monitoring of the Comprehensive Test-Ban-Treaty (CTBT) for low magnitude seismic events. We assess the capability of this sparse three-station broadband network and we compare locations estimated from phase onset time and wave azimuth measurements to a ground-truth catalog of high-quality earthquake locations derived from data recorded by the Southern California Seismic Network (SCSN). The results indicate that in the regional distance range it is possible, when an appropriate calibration event is available, to obtain absolute event locations to within 18 km as is prescribed by the CTBT.

Key words: Broadband network, three-component stations, earthquake location algorithm, CTBT.

Introduction

Most conventional earthquake location algorithms in widespread use by the seismological community (i.e., LEE and LAHR, 1972) use phase onset time data and some variant of Geiger's method (GEIGER, 1910) in which a linearized inversion process is iteratively solved until a convergence criteria is satisfied. These linearization algorithms generally suffer from two fundamental limitations: they make limited use of the information inherent in seismic wavefield recordings obtained from modern three-component broadband seismometers coupled to high-resolution digital data loggers, and they implicitly assume that the error surface is ellipsoidal (GEIGER, 1910; FLINN, 1965). We take advantage of the additional information inherent in

[1] Berkeley Seismological Laboratory, University of California, Berkeley, CA 94720-4760, U.S.A.

these broadband recordings by measuring the wave azimuth as well as the phase onset time for use in locating the seismic source. The wave azimuth data are complimentary to the phase onset data and their combined use results in more robust solutions. Direct mapping of the iso-misfit surface shows that it is generally asymmetrical and not ellipsoidal in shape (BULAND, 1976). The inherent assumption that the error surface is ellipsoidal can lead to significant systematic errors in the estimated uncertainties. We introduce a recently developed earthquake location algorithm, the *Broadband Waveform Regional Earthquake Location Program* (*BW_RELP*), which does not suffer from the above limitations and which was specifically designed for the problem of locating earthquakes when the available data are from a sparse broadband network and when the source region is outside of the network. Our motivation was, in part, to develop a robust algorithm for locating earthquakes in California and vicinity when the only available data was from a few Berkeley Digital Seismic Network (BDSN) broadband stations. *BW_RELP* uses both phase onset time and wave azimuth data and an adaptive migrating grid search algorithm to find the minimum in an arbitrarily normed misfit parameter space. The corresponding uncertainties are determined by direct mapping in the misfit parameter space.

The recent proliferation of broadband seismic station networks, such as BDSN, offers the opportunity to analyze seismic events over a wide range of seismic moment release. This will likely be the case in the monitoring of the Comprehensive Test-Ban-Treaty (CTBT) using the International Seismic Monitoring System (ISMS) for small magnitude events, and for studying earthquakes in the United States using the National Seismic Network (NSN). Sparse broadband networks have been deployed in northern and southern California (BDSN and TERRAscope, respectively), in the Mediterranean (MedNet), in Japan (FREESIA NET), in the continental US (USGS NSN), and on a global scale (IRIS GSN). Stations from these networks contribute to the International Seismic Monitoring System (ISMS). A key design consideration for the Primary and Auxiliary seismic networks of the ISMS was that these networks be capable of locating $M \geq 4$ events with an uncertainty of less than 1000 km^2, i.e., within a circle of radius less than 18 km. For low-yield explosions or evasively tested nuclear devices the resulting small magnitudes will preclude the recording of signals from IMS stations at teleseismic distances, and will necessarily focus the analysis effort on distances of a few hundred to perhaps 2000 km, where lateral heterogeneity in earth structure significantly complicates the problem.

The purpose of this paper is to describe the *BW_RELP* algorithm and to expand on our study (DREGER *et al.*, 1997) which evaluated the effectiveness of a sparse broadband network for monitoring a region. The 1995 M_L 5.8 Ridgecrest, California earthquake sequence (Fig. 1; Table 1) is ideally suited for this study because the events were located employing a local seismic network, because the energetic sequence produced numerous aftershocks $M_L \geq 3.5$, and owing to very compact the source region. The Ridgecrest sequence events studied here are given in Table 1

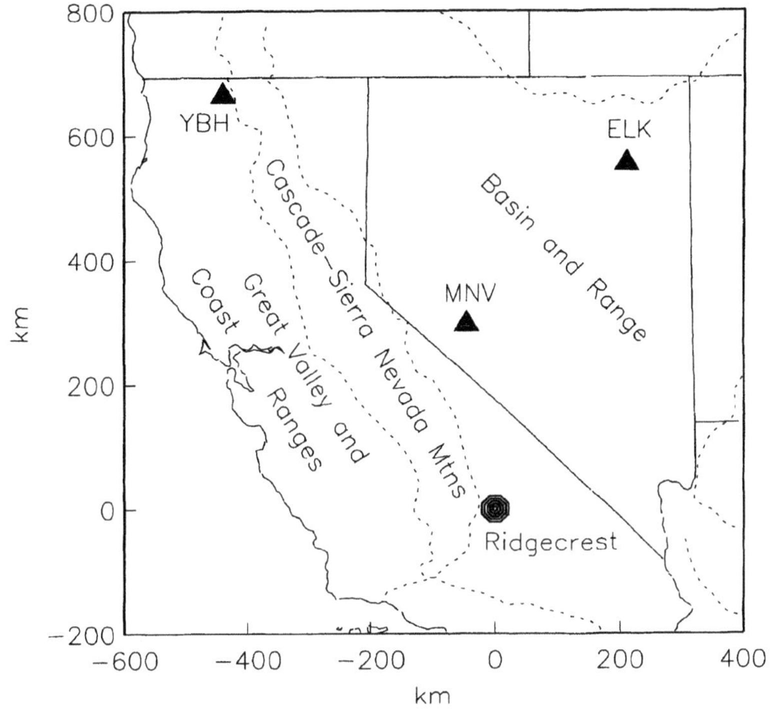

Figure 1
Map showing the locations of BDSN station at Yreka, California (YBH), the USNSN stations in Mina, Nevada (MNV) and Elko, Nevada (ELK), the location of the 1995 Ridgecrest sequence, and the Physiographic Provinces. The Ridgecrest sequence "ground-truth" locations all plot within a circle one-fifth the diameter of the octagon. There is considerable variation in the physiological structure along the propagation paths to the three stations as indicated by the dashed lines that outline the Great Valley and Coast Ranges, the Cascade-Sierra Nevada Mtns, and the Basin and Range Provinces.

where the locations, adopted as "ground truth" for comparison purposes, were determined using data from the local Southern California Seismic Network (SCSN). All the hypocenters in Table 1 fit within a 4-km-radius sphere so the source region for the M_L 5.8 sequence is unusually compact. A compact source region is advantageous when comparing the broadband waveforms and picking phase onset times because the absolute travel times to any one station vary only by approximately ± 1 second between events.

We also discuss modifications to standard processing techniques which can improve performance. The location analysis is evaluated using one Berkeley Digital Seismic Network (BDSN) station located in northern California (YBH, an auxiliary IMS station) and two USNSN stations (ELK and MNV) located in Nevada (see Fig. 1), which span 300–800 km in distance and 55 degrees in azimuth, to further investigate the capability of a sparse broadband network of three-component stations to monitor a region located outside of the network.

Table 1

1995 Ridgecrest sequence locations

Event #	Date	Time	Lat.	Long.	Depth	M_L
1	1995/08/17	22:39:58.99	35.776	−117.662	5.55	5.4
2	1995/08/30	15:29:54.62	35.791	−117.641	4.96	4.2
3	1995/08/30	15:54:22.46	35.796	−117.640	3.36	4.0
4	1995/08/31	01:58:58.78	35.793	−117.644	5.15	3.9
5	1995/09/11	18:37:23.75	35.788	−117.663	6.61	4.2
6	1995/09/20	23:27:36.27	35.761	−117.638	5.42	5.8
7	1995/09/20	23:56:58.50	35.781	−117.666	5.32	3.9
8	1995/09/21	07:46:53.65	35.759	−117.637	5.09	3.9
9	1995/09/21	07:57:41.34	35.756	−117.633	5.38	4.0
10	1995/09/21	23:48:39.16	35.761	−117.643	5.44	4.0
11	1995/09/24	13:15:30.04	35.791	−117.660	6.34	3.7
12	1995/10/06	19:15:04.26	35.758	−117.636	5.40	3.6
13	1995/10/11	14:45:35.14	35.799	−117.633	5.13	3.5
14	1996/01/07	14:32:53.06	35.766	−117.649	5.90	5.2

Earthquake Location Algorithm

To improve the sparse network locations, a new algorithm, herein referred to as *BW_RELP*, was developed to make use of multiple phase onset times and wave azimuth information (DREGER *et al.*, 1996, 1997). The algorithm was specifically designed with the goal of providing robust earthquake locations under adverse conditions during which conventional algorithms fail. As such, we drew on a combination of methodologies to avoid the known limitations of conventional algorithms, to utilize more of the information inherent in broadband seismic records, and to develop a "best practice" algorithm. The major components of the *BW_RELP* algorithm are described in the following paragraphs. The reader is referred to the Berkeley Seismological Laboratory (BSL) web page (URL given below) for more detail.

BW_RELP finds the location of the event by searching for the global minimum in a misfit function in which the misfit is a measure of the difference between the observed and the calculated data. The algorithm is designed to utilize both phase onset time and wave azimuth data to locate the event. Since phase onset time (in seconds) and wave azimuth (in degrees) are not in the same units, we cast both the time residuals and the azimuth residuals in terms of a distance metric. The misfit function that is used in *BW_RELP* is:

$$P = \sum_{i=1}^{n} \left(\sqrt{(1-\alpha)(\delta r)^2/n_t + \alpha(\Delta\delta\varphi)^2/n_\varphi} \right)^{\text{norm}}, \tag{1}$$

where P is the misfit value; α is a parameter which provides relative weighting between travel time and wave azimuth data; δr is the travel-time residual which is cast in terms of a distance metric where $\delta r = \delta t^* v$ (where δt is the travel-time residual and v is the wave velocity); $\Delta\delta\varphi$ is the azimuth residual which is also cast in terms of a distance metric in the form of an arc length (where $\delta\varphi$ is the azimuth residual and Δ is the distance to a station); and n_t and n_φ are the number of time picks and azimuth picks, respectively. The misfit function utilizes only travel time information if $\alpha = 0$ and only azimuth information if $\alpha = 1$ and a typical value for α is $\sim$0.1. The wave velocity (v) is set to an average wave velocity, typically $\sim$6 km/s. The norm value can range between 0 and ∞. However, there is no compelling reason for using norm values larger than the least-squares value of 2. In practice, the preferred norm value is one which closely matches the observed distribution of the residuals. Empirically, we find that a value of 1.25, as suggested by KENNETT (1996), provides a good match to the residual distribution and also robust solutions. The summation in equation (1) is performed for n phase readings where the phase readings can be either wave arrival time(s) or wave azimuth(s). The algorithm can support any number of phase readings from a single station. Empirically, we have found that using both travel time and wave azimuth data, and 1-D velocity models in the forward calculations, has the added advantage that the misfit function (P) is very robust in that it has only a global minimum and no local minima. One might expect this to be the case since the time and azimuth distance metrics used in constructing the misfit parameter are approximately orthogonal to each other.

In lieu of an inverse approach, a grid search technique (SAMBRIDGE and KENNETT, 1986) is used to locate the minimum in the misfit function. This approach provides a level of flexibility that would be difficult to achieve with a standard inverse approach and does not suffer from the numerical instabilities found in inverse algorithms (BULAND, 1976). In *BW_RELP*, the grid searching algorithm is both adaptive and migratory. The grid search parameters are set by the user to control the starting grid extent, the mesh size, the migration rate, the shrinkage rate of the four-dimensional grid (origin time, latitude, longitude, and depth), and the convergence criteria. The grid search algorithm is adaptive in the sense that the grid migrates and shrinks as it converges on the minimum in the misfit parameter in the four-dimensional parameter space. Figure 2 shows an example of how the adaptive grid search procedure converges. The grid migrates if the minimum in the misfit function is near the edge of the grid and shrinks if it is near the center of the grid. The process continues until both the successive iteration step size and the grid mesh size are both smaller than the size of the convergence parameter. This approach has been found to be both robust and efficient. The adaptive grid search is also more amenable to the incorporation of 3-D velocity models because it is not necessary to compute complicated partial derivatives.

In inverse location algorithms, the quality of the solution is typically indicated by an error ellipsoid whose size is determined using a first-order Taylor series expansion approach to map the solution variance from the data variance (GEIGER, 1910; FLINN, 1965). This approach is not ideally suited for analyzing sparse network data using *BW_RELP* because it requires the calculation of derivatives and for the reason that it implicitly assumes that the error surface is an ellipsoid and that the number of observations is large. The problem at hand is to determine the size and shape of the iso-misfit surface in the misfit parameter space which represents a desired level of confidence (one standard deviation or the 95% confidence level, namely) in the location of the event. BULAND (1976) investigated mapping of the error surface and demonstrated that when the volume of the parameter space in which the linear approximation is valid is sufficiently large, the uncertainty corresponding to given iso-misfit value may be simply calculated using a χ^2 statistic, and his numerical example using 4 stations and 8 travel-time observations indicated that a 99% confidence error could be calculated reliably. TARANTOLA and VALETTE (1982) also provide a detailed description of mapping uncertainties in inherently nonlinear problems. The approach used in *BW_RELP*, to estimate the size and shape of the iso-misfit error surface at a desired confidence level (one standard deviation, say), is to determine (in sequential order) the standard error of the residuals (δr and $\Delta\delta\varphi$ in Equation (1)), the equivalent origin time standard error, the average of the iso-misfit

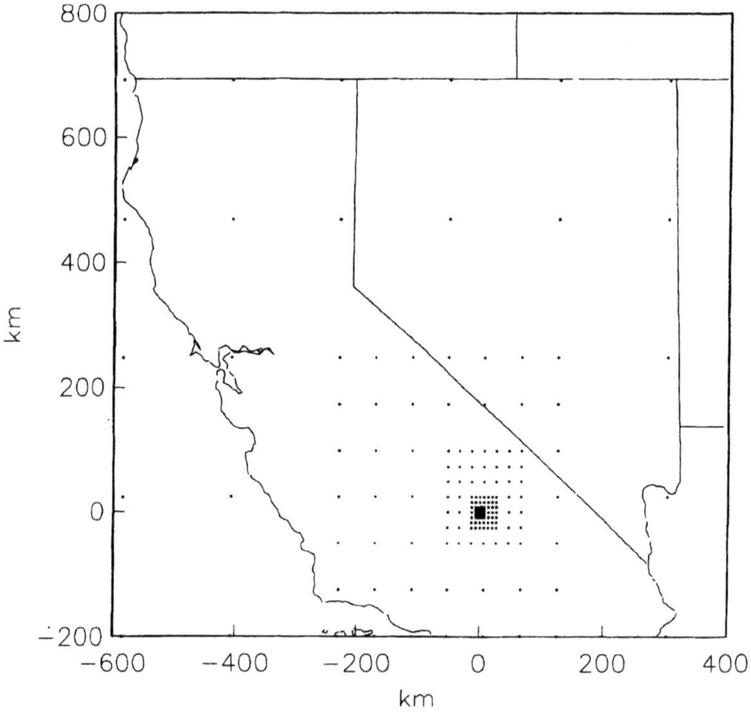

Figure 2
Example of an adaptive grid search. The dots designate the specific locations that were tested. The grid shifts and contracts as the method converges. The "ground-truth" location is at 0 km on both axes. The grid converged to within a km of the final solution in 5 iterations.

value corresponding to perturbations of ±1 standard deviation in the origin time around the global minimum solution, and finally to determine the corresponding uncertainties in the latitude, longitude, and depth. Actually the confidence level, at which the uncertainties in the solution are calculated in *BW_RELP*, is a parameter that is set by the user. Several tests have been performed comparing the size and shape of the error surfaces calculated by *BW_RELP* and by a convention inversion algorithm (for equivalent and well-determined least-squares problems using only travel-time data), and they produced uncertainty estimates that differed by less than 5 percent. An example of the horizontal projection of error surfaces corresponding to various levels of confidence is shown in Figure 3.

The *BW_RELP* algorithm, man pages, and representative examples are available online via the BSL home page via URL http://www.seismo.*berkeley.edu/seismo/algorithms*. The default model for the calculation of travel times and azimuths is a gradient layer over a half space where the model parameters are user-supplied. For a more complex model, the user must supply the appropriate travel time and azimuth calculation algorithms. *BW_RELP* is written in FORTRAN, however any

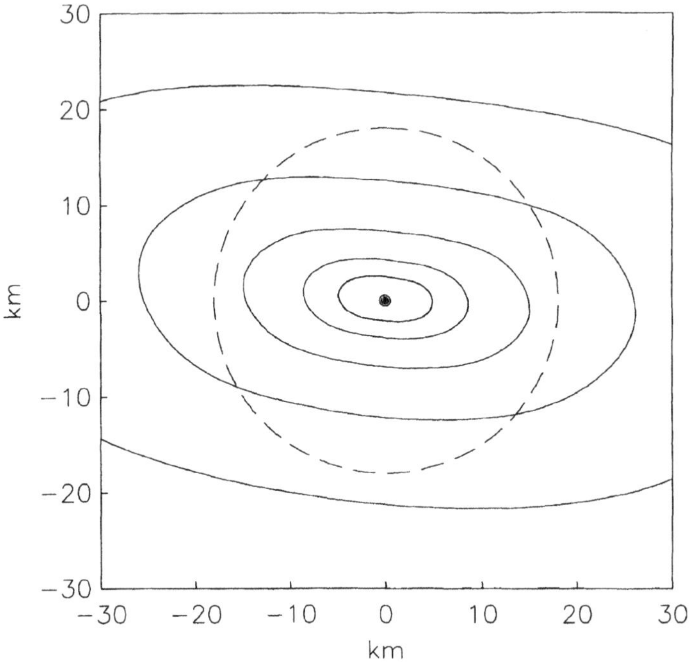

Figure 3
Example of mapping the P_{95} region for Event #1 in Table 1. The misfit parameter was determined every 1 km on a grid, and the resulting values were contoured. The innermost contour represents the P_{95} region (2 standard deviations) and the remaining contours represent 4, 8, 16, and 32 standard deviations. The general shape of the contours reflects the sparse broadband network geometry. The circular dashed line is the prescribed CTBT 1000 km^2 area (18 km radius). The solid circle is the ground-truth location.

user-supplied forward model calculation algorithms need not be restricted to that programming language. There are no default parameters in the algorithm and that all parameters must be set by the user. This was intentionally done for flexibility in the algorithm and because appropriate parameter values are not necessarily known *a priori*. A local event is used to facilitate a comparison of the solutions obtained from the *BW_RELP* algorithm and from standard inversion algorithms. A comparison of the uncertainty volumes determined using *BW_RELP* and a conventional inversion algorithm is included.

Analysis of Sparse Broadband Station Event Locations

The locations of the three regional broadband stations and the Ridgecrest events used in this study are shown in Figure 1. To demonstrate the capabilities of the *BW_RELP* algorithm at far-regional distances, we purposely chose the BDSN station YBH in northern California, as it has been designated as an Auxiliary International Monitoring System (IMS) station, and the two USNSN broadband stations in Nevada (ELK and MNV) to provide azimuth coverage of the source region. These three stations span ~300–800 km in distance and ~55 degrees in azimuth relative to the earthquakes.

There are two fundamental impediments to determining accurate absolute epicenter locations when analyzing events that are at regional distances and external to a sparse broadband seismic network. First, the propagation paths can be quite complex if a variety of physiographic provinces are traversed by the seismic waves. The propagation paths from Ridgecrest to the sparse three-station broadband network happens to be rather complex, as shown in Figure 1, and the use of inadequate velocity models can easily lead to systematic location errors in excess of 100 km (DREGER *et al.*, 1997). Second, consistent picking of the phase onset times and wave azimuths become problematic at large distances and at small magnitude, owing to the degradation in the broadband signal-to-noise ratio (SNR). Appropriate passband filtration of the broadband signals to maximize the SNR, combined with cross-correlation methodologies, significantly enhances the determination of relative phase arrival times and wave azimuths. If an appropriate calibration event with an accurately known location is available, these impediments can be abated since the calibration event can be used both as a template for phase arrival and wave azimuth measurements and as a model calibrator to provide high accuracy relative locations. If no appropriate calibration event is available, one must resort to the daunting task of determining a velocity model with sufficient accuracy to reduce the systematic errors to an acceptable level. In the following sections we use the largest event in the Ridgecrest sequence (M_L 5.8; event #6 in Table 1) as a calibration event for determining the locations of the remaining events in the sequence.

Broadband Waveform SNR

As an example of the large SNR variation with earthquake magnitude that is encountered at regional distances, we display the raw broadband waveforms recorded at YBH (800 km from Ridgecrest) in Figure 4. Note that many of the events are not easily discernible to the eye when viewing the raw broadband data. Figure 5 shows the P-wave and S-wave SNR for the largest event in the Ridgecrest sequence (M_L 5.8) recorded at YBH. The P-wave SNR is highest in the 0.3–4 Hz

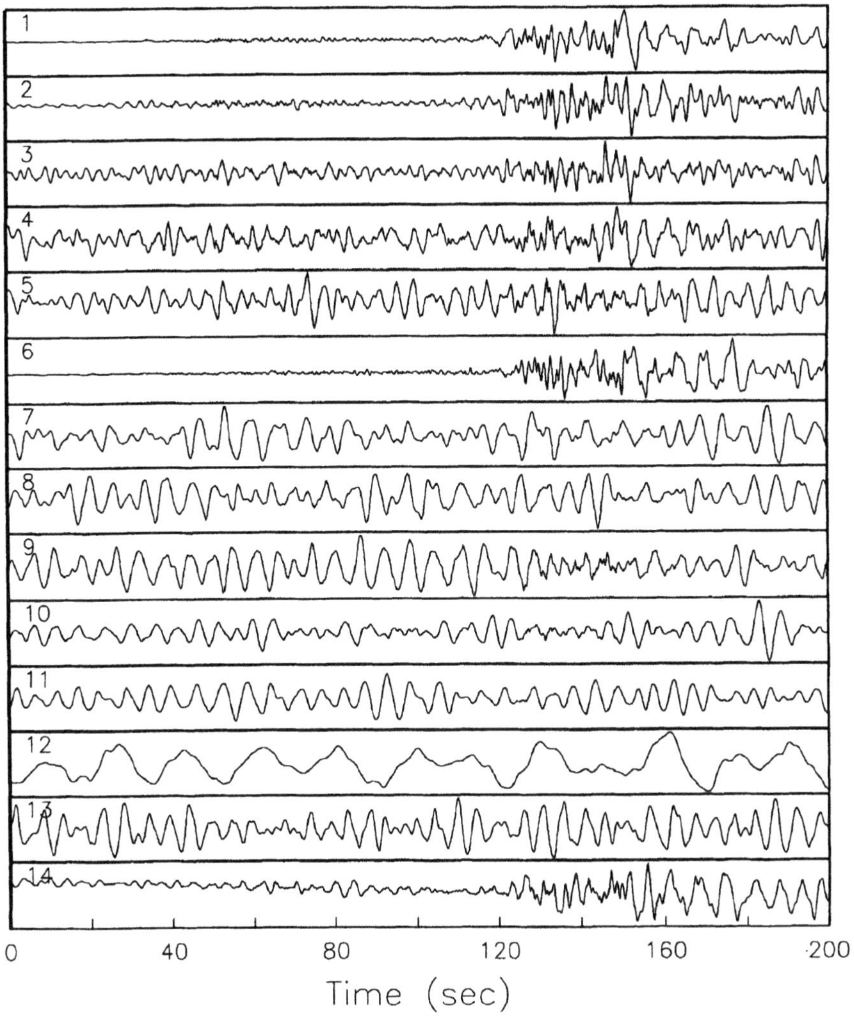

Figure 4
The raw BHZ waveforms recorded at YBH for the 23 Ridgecrest events in Table 1. Note that the smaller events are not readily discernible and filtration is required prior to reading the seismic phases onsets and wave azimuths. The trace numbers correspond to the event numbers in Table 1.

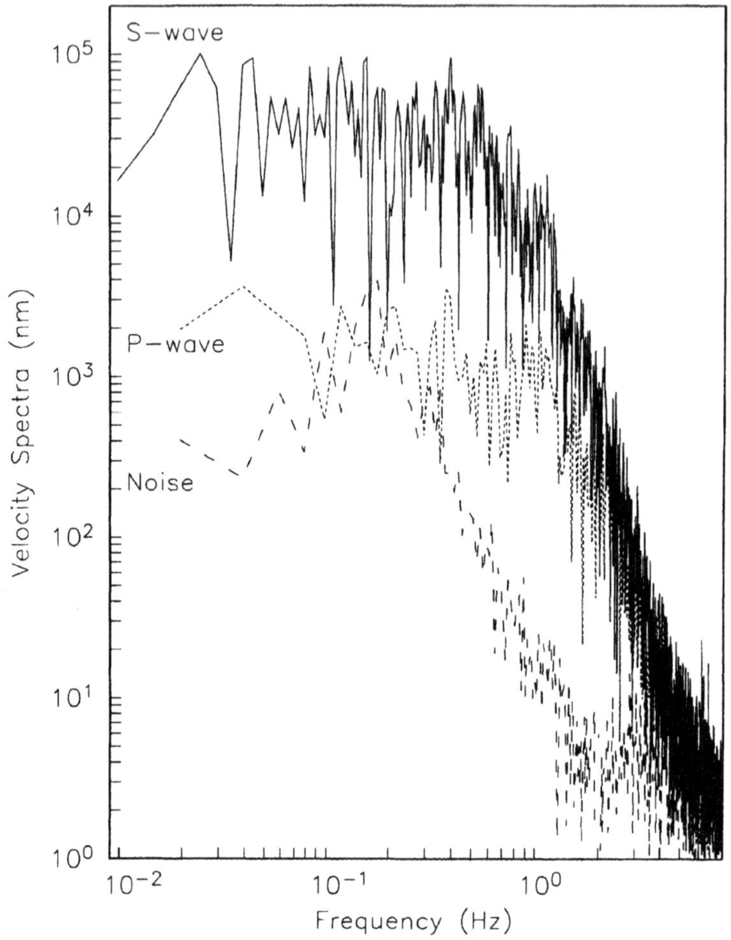

Figure 5

Velocity spectra of the P wave and S wave for the largest Ridgecrest earthquake (M_L 5.8). At 800 km from the source, the signal is below the noise level for frequencies above 4 Hz and the S-wave signal is about 30 dB above the P-wave signal at frequencies below ~0.5 Hz.

passband, it is ~40+ dB in the 1–2 Hz passband and it degrades to ~0 dB at approximately M_L 3.8. The S-wave SNR is 30+ dB in the 0.02–4 Hz passband and 40+ dB in the 0.6–2 Hz passband; the SNR degrades to ~0 dB below 0.3 Hz at approximately M_L 4.3 and it degrades to ~0 dB in the 0.6–2 Hz passband at approximately M_L 3.8. Assuming that a minimum SNR of ~6 dB is required to consistently pick phases, we will be unable to reliably pick the P- and S-wave phase data at YBH for Ridgecrest events smaller than M_L ~4.1. Reliably determining wave azimuths to a precision of order 1 degree requires an SNR of approximately 12 dB or better within an appropriate passband on the radial component relative to the transverse component (assuming that a 5 second window is used to determine the azimuth). This implies that the wave azimuth recorded at YBH will not be well resolved for Ridgecrest earthquakes smaller than approximately M_L 4.5. Similar analyses for ELK (~600 km from Ridgecrest) and MNV (~300 km from Ridgecrest) indicate that corresponding M_L thresholds are ~0.4 and ~1.4 lower than the values for YBH, respectively. The active tectonic region of the Western US has relatively high seismic wave attenuation and in other regions of the world, where the seismic wave attenuation is lower, the M_L threshold will be correspondingly reduced.

Determination of Phase Arrival Times

We show in detail the process of picking the P-wave phase arrival times for only YBH as it is the most distant of the three broadband stations and it has the lowest SNR. The other two broadband stations (ELK and MNV) were analyzed in a similar fashion. Owing to the large SNR variation observed in the YBH three-component broadband waveforms, and based on the spectral analysis conducted above, we passband filtered the P-wave portion of the waveforms using a six-pole Butterworth filter with a 0.6–2 Hz passband, resulting as shown in Figure 6. Note that the P-wave onset is not easily discernible for many of the events and that there are secondary P-wave arrivals that are readily apparent for all except the smallest events. Consequently, we chose to determine the times of the secondary P-wave arrival for use in determining the location of the events. We used the maximum in the cross-correlation over a 5 second time window to determine the relative arrival time of this phase for each event to an accuracy of approximately ±25 ms for all except the smallest events. Some of the cross-correlation maxima were negative, owing to differences in the source mechanisms (DREGER *et al.*, 1997). The arrival times of the S wave at YBH were calculated in a like fashion. The arrival time results for all three broadband stations are shown in Table 2.

Determination of Wave Azimuth

As was the case with travel time, we show in detail the process for determining the wave azimuth for only YBH as it is the most distant of the three broadband stations and it has the lowest SNR. The other two broadband stations were analyzed in a similar fashion. We used the same filtered waveforms as in the above section. To determine the wave azimuth we search for the azimuth θ which maximizes the correlation function:

$$c(\theta) = \frac{\sum_{j=1}^{m}(Z_j(N_j \cos(\theta) + E_j \sin(\theta))}{\sum_{j=1}^{m}(E_j \cos(\theta) - N_j \sin(\theta))^2} \qquad (2)$$

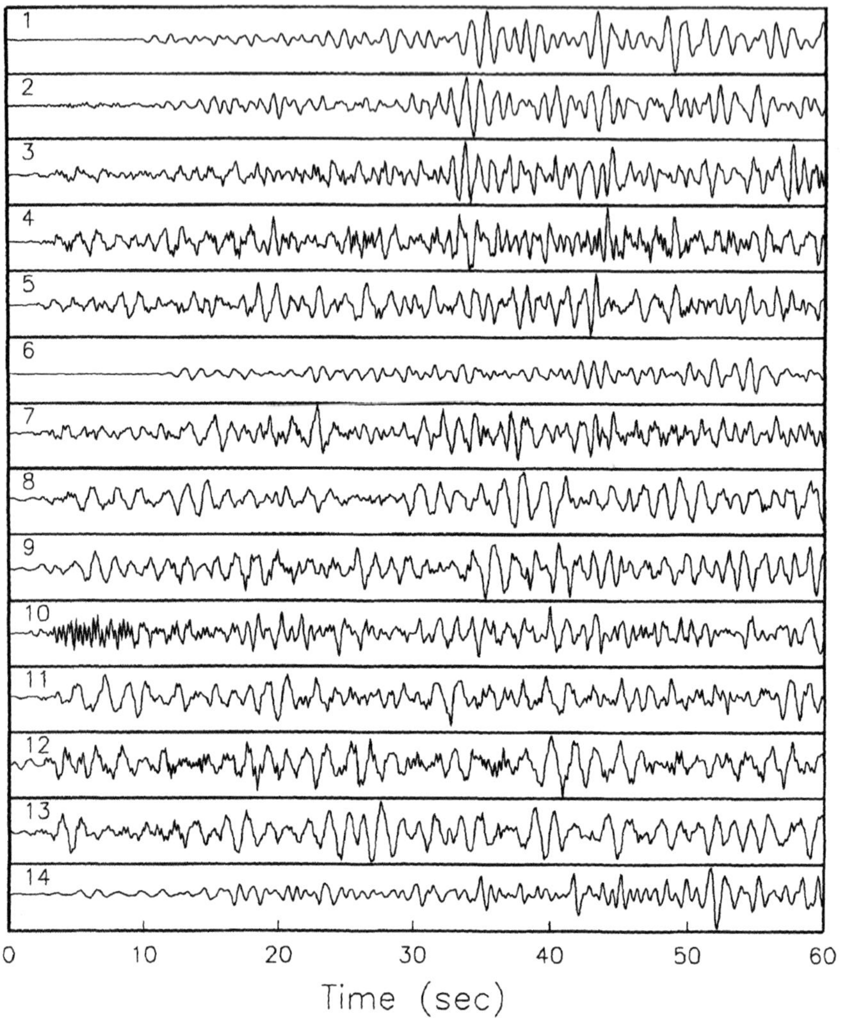

Figure 6
P-wave filtered (6PBP at 0.3–4 Hz) Z-component waveforms recorded at YBH. The secondary P-wave arrival at circa 42–44 seconds was chosen for reading as it was visible to all except the smallest M_L events. The numbers on the traces correspond to the event numbers in Table 1.

where Z_i, N_i, and E_i are the vertical, north, and east components, respectively, of the filtered broadband ground velocity at time i. The uncertainty in θ is approximately 0.5 degree for the larger earthquakes and it degrades rapidly when the SNR is below approximately 12 dB as estimated in the SNR section above. The wave azimuth results for all three broadband stations are shown in Table 2.

Locating Ridgecrest events using only phase picks and a standard travel-time inverse procedure without adjustments were found to have large absolute misloca-

Table 2

Observed phase arrival times and wave azimuths

Station	MNV	ELK	YBH
Distance (km)	300.1	591.4	796.4
Azimuth (deg)	351.19	19.98	327.87
P Travel time (sec)	50.464	110.393	148.295
P Time residual (sec)	0.038	0.025	0.251
P Velocity (km/sec)	5.947	5.357	5.371
P Azimuth (deg)	342.0	48.0	324.0
P Az residual (deg)	−9.2	27.2	−2.4
S Travel time (sec)	89.080	179.393	198.445
S Time residual (sec)	0.118	0.271	0.459
S Velocity (km/sec)	3.369	3.297	4.103
S Azimuth (deg)	349.0	49.0	335.0
S Az residual (deg)	−2.2	28.2	8.6

tions, in some cases exceeding 100 km (DREGER *et al.*, 1997). Table 1 lists the event locations, adopted as "ground truth" for comparison purposes, which were determined applying local data recorded by the Southern California Seismic Network (SCSN).

Calibration Event Analysis

Event #6 in Table 1, the largest event in the Ridgecrest sequence, was chosen as the calibration event for the purpose of calibrating the location procedure. For this analysis the SCSN location, given in Table 1, is adopted as "ground truth." The secondary P-wave and S-wave onset times for this event at the three broadband stations were measured and the results are given in Table 2. The corresponding wave azimuths were determined with the aid of Equation (2). Table 2 also supplies the absolute travel times, distances, great circle azimuths, and the inferred wave velocities for comparison. The wave velocities and the wave azimuths, relative to the great circle azimuth, vary considerably among the three stations. This variation is expected given the structural differences (see Fig. 1) along the propagation paths from Ridgecrest to the three stations. Interestingly, the largest wave propagation direction deviation is seen at ELK where the apparent propagation direction is $\sim$28 degrees clockwise (CW) of the great circle path (of course this assumes that the ELK horizontal seismometers were properly aligned N and E). Upon closer inspection, the CW deviation from the great circle path is real and it is attributed to the lateral refraction of the wavefield along the high contrast boundary between the Sierra Nevada and the Basin and Range physiographic provinces (see Fig. 1) where the Sierra Nevada has the higher velocities. One might have expected the deviations to be the largest at YBH since the propagation path traverses three physiographic provinces as shown in Figure 1.

The "ground-truth" calibration event location and origin time were adopted and the phase arrival times at the three stations were determined and used to construct a 1-D linear gradient layer over a half-space velocity model, using a forward modeling approach, for each of the propagation paths and the onset time residuals were determined using these path-dependent models. Azimuth residuals relative to a great circle propagation path were also determined. These phase time and azimuth residuals were adopted as the corresponding station phase onset times and wave azimuth adjustments for locating the events. The path-dependent velocities, phase times and azimuth, and the corresponding phase and azimuth adjustments are given in Table 2. As a check on the location procedure, the calibration event was relocated using the path-dependent 1-D velocity models and the corresponding station phase arrival time and wave azimuth adjustments and the relocated epicenter were within 160 meters of the adopted "ground-truth" location as shown in Table 1. To determine the sensitivity of the calibration event solution to missing and or erroneous phase time and azimuth data, a series of tests were performed in which the phase times and azimuths were randomly omitted, the phase times were randomly perturbed by ± 1 second, and the wave azimuths were randomly perturbed by ± 1 degree. The resulting solutions were all within 8 km of the "ground-truth" location, and the corresponding P_{95} contour shape changed slightly as the data were omitted, and its size increased as the "mislocation" distance increased. This implies that the location procedure is quite robust and capable of locating regional events within the CTBT criteria. As a final test, the entire process was repeated using a single 1-D velocity model with corresponding phase and azimuth adjustments for each station. When the phase times and azimuths were randomly omitted, the phase times were randomly perturbed by ± 1 second, and the wave azimuths were randomly perturbed by ± 1 degree, the scatter in the solutions increased by about 15 percent over the path-dependent model results. The geological map shows both Quaternary volcanic flow rocks and Mesozoic granite in the vicinity of the Ridgecrest source region consequently the differences in the observed scatter between the two velocity models may be due to lateral structure in the source region.

Ridgecrest Sequence Analysis

The Ridgecrest sequence events in Table 1 were analyzed and located, systematically proceeding from the largest to the smallest M_L events. The path-dependent velocity models and phase time and wave azimuth adjustments derived from the calibration event, given in Table 2, were used in the location procedure and the results portrayed in terms of the "mislocation" distance from the "ground-truth" location are shown in Table 3 and Figure 7. The largest events, $M_L \geq 5$, generated the shortest mislocation distances (less than 3.6 km) and the "mislocation" distance increased as the event M_L decreased. This is as expected since the SNR decreases

Table 3

Event mislocation distances

Event #	M_L	Mislocation distance (km)	P_{95} radius (km)
1	5.4	0.3	3
2	4.2	7.2	4
3	4	6.9	4
4	3.9	13.7	11
5	4.2	1.9	4
6	5.8	0.16	0.045
7	3.9	9	15
8	3.9	11.9	16
9	4	3.6	6
10	4	6.2	7
11	3.7	9.5	18
12	3.6	9.8	20
13	3.5	16.4	24
14	5.2	3.6	4

proportionally with decreasing M_L. Note that the P_{95} average radius increases significantly below $M_L \sim 4$. This is because the phase time and wave azimuth data recorded at YBH and the wave azimuth data recorded at ELK were not well resolved, owing to degradation in the SNR. As discussed in the section on broadband waveform SNR, we expected the P-wave and S-wave SNR at YBH (800 km from Ridgecrest) to degrade to 0 dB at $M_L \sim 3.8$ and the wave azimuth data to be only moderately resolved below $M_L \sim 4.5$. Thus it is not surprising that the threshold M_L for reliably reading any body-wave phase data at YBH is approximately M_L 4. One inference is that the minimum SNR necessary to read phase times is ~4 dB. Also from the broadband waveform SNR analysis, the phase azimuth data at ELK

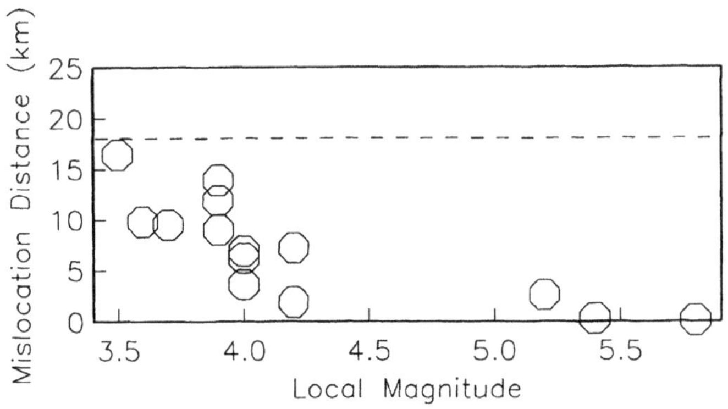

Figure 7

Event mislocation distance relative to the adopted "ground-truth" locations given in Table 1. The dashed line is the 18 km CTBT threshold. Note that the mislocation distance increases significantly below $M_L \sim 4$.

(600 km from Ridgecrest) will not be resolvable below a $M_L \sim 4$ threshold. This is consistent with the general increase in the "mislocation" distance and the P_{95} average radius below M_L 4, as shown in Table 3, and it is directly attributable to the omission of unreliable phase time and wave azimuth when determining the event location. Note that the "mislocation" distance (Table 3 and Figure 7) for all events is below the 18-km CTBT criteria. However, this does not insure that all the events are reliably located within the 18-km radius area. Note that the average P_{95} threshold radius is larger than ~15 km for most of the $M_L < 4$ events in Table 3 therefore the probability that these events are within the 18-km radius CTBT criteria can be considerably less than 95 percent. For example, event #13 has a "mislocation" distance of 16.4 km and a P_{95} average radius of 24 km consequently there is only a 43 percent probability that it is actually located within the 18-km radius region.

Conclusions

We have introduced the *BW_RELP* location algorithm and demonstrated its performance using data from a sparse three-station regional network of three-component broadband seismic stations, to locate events in the 1995 Ridgecrest earthquake sequence. *BW_RELP* employs an adaptive migrating grid search algorithm to locate the hypocenter by finding the global minimum in a misfit parameter. The algorithm was specifically designed to utilize more of the information inherent in broadband seismic recordings, by using both phase onset times and wave azimuth data to locate events. It is very flexible in that it allows for path-dependent velocity models, station adjustments, and azimuth adjustments. Also the algorithm has no default parameter values, and the user must supply all the required parameter values. This was done deliberately as appropriate parameter values are not necessarily obvious *a priori*. The adaptive grid search procedure is also more amenable to the incorporation of 3-D velocity models because it is not necessary to compute complicated partial derivatives. Unlike conventional linearized least-squares algorithms, *BW_RELP* also makes no assumptions regarding the shape of the error surface and it provides more robust parameter uncertainty estimates, especially when using data from a sparse network to locate distant events.

The results of our analysis of the Ridgecrest sequence reveal that it is indeed possible to resolve event locations with an accuracy bettering 18 km, as prescribed by the CTBT, when using a sparse three-station network of regional distance (300–800 km) broadband, three-component seismic stations. Achievement of this degree of accuracy requires the availability of an appropriate event to calibrate the location procedure. Without a calibration event, the "mislocation" distances can be quite large, often exceeding 100 km. The primary advantages of a calibration event are that it provides a template for phase selection and cross-correlation and a "ground-truth" location to determine appropriate path-dependent velocity models and phase arrival time and wave azimuth adjustments. With appropriate selection of the seismic phases and wave azimuths to measure for use in the location algorithm, we find that all three broadband stations in the sparse network provide useful data for locating the Ridgecrest events down to a $M_L \sim 4$ threshold. The limiting factors are the SNR and the intrinsic attenuation/scattering along the seismic wave propagation path.

Below $M_L \sim 4$, the events can be located with reduced accuracy, by omitting phase arrival time and wave azimuth from the farthest (YBH, 800 km distant) and also wave azimuth data from the next most distant station (ELK, 600 km distant), down to $M_L \sim 3.5$. Consistency in the determination of the phase arrival time and wave azimuth picks is paramount to achieving the most reliable relative event locations. We find that secondary P-wave arrivals (Pg, for example) are consistently observed over a wider M_L range, than are the primary P-wave arrivals, and with appropriate filtration of the broadband seismic signals they can be reliably measured as long as the SNR is more than ~4 dB for measuring the arrival time and exceeds ~12 dB for measuring wave azimuth. Below these thresholds the accuracy degrades rapidly, and it is better to omit the data than to chance biasing the solution with unreliable observations.

Acknowledgements

This work was partially supported by the Lawerence Livermore National Laboratory, through the Department of Energy's Comprehensive Test-Ban-Treaty Research and Development (CTBT R & D) Program, under the Inter-University (IU) Agreement No. B331608. The data used in this work were acquired from the Northern California Earthquake Data Center (NECDE; http://quake.geo.berkeley.edu), the Southern California Earthquake Data Center (SCEC_DC; http://scecdc.scec.org), and the IRIS DMC (http://www.iris.washington.edu). This is Berkeley Seismological Laboratory contribution 00-03.

References

Buland, R. (1976), *The Mechanics of Locating Earthquakes*, Bull. Seismol. Soc. Am. *66*, 173–187.

Dreger, D., and Romanowicz, B. (1994), *Source Characteristics of Events in the San Francisco Bay Region*, USGS Open-File Report *94–176*, 301–309.

Dreger, D., Pasyanos, M., Uhrhammer, R., Romanowicz, B., and Ryall, A. (1996), *Evaluation of the performance of broadband networks and short-period arrays in global monitoring*, Proceedings of the 18th Annual Seismic Research Symposium on Monitoring a CTBT, *PL-TR-96-2153*, 704–713.

Dreger, D., Pasyanos, M., Uhrhammer, R., Franck, J., and Romanowicz, B., (1997), *Evaluation of the Performance of Broadband Stations and Regional Arrays in Global Monitoring – Phase II*. Final Report, submitted July 1997, 10 pp.

Flinn, E. A. (1965), *Confidence Regions and Error Determinations for Seismic Event Location*, Rev. Geophys. *3*, 157–185.

GEIGER, L. (1910), *Herdbestimmung bei Erdbeden aus den Ankunftzeiten*, K. Gessel. Wiss. Goett. *4*, 331–349.

KENNETT, B. L. N., *Event location and source characterization.* In *Monitoring a Comprehensive Test Ban Treaty* (E. S. Husebye and A. M.Dainty, eds.) (Kluwer Academic Publishers, 1996) pp. 501–520.

LEE, W. H. K., and LAHR, J. C. (1972), *HYPO71: A Computer Program for Determining Hypocenter, Magnitude, and First Motion Pattern of Local Earthquakes*, U.S. Geol. Survey, Open File Report 72–0224.

SAMBRIDGE, M. S., and KENNETT, B. L. N. (1986), *A Novel Method of Hypocenter Location*, Geophys. J. Royal astr. Soc. *87*, 679–697.

TARANTOLA, A., and VALETTE, B. (1982), *Generalized Nonlinear Inverse Problems Solved Using the Least-squares Criterion*, Rev. Geophys. and Space Phys. *20*, 219–232.

(Received July 20, 1999, revised August 3, 2000, accepted September 3, 2000)

Pure appl. geophys. 158 (2001) 277–289
0033–4553/01/020277–13 $ 1.50 + 0.20/0

Pure and Applied Geophysics

HYPOSAT – An Enhanced Routine to Locate Seismic Events

JOHANNES SCHWEITZER[1]

Abstract—A program package, called HYPOSAT, has been under development that attempts to use the maximum information possible to estimate the hypocenter of a seismic source. The standard input parameters can be used: arrival times of first and later onsets with backazimuths and ray parameters (or apparent velocities). In addition, travel-time differences between different phases observed at the same station can be optionally used. The observed standard deviations are used to weight all input parameters and the inversion is done with a generalized matrix inversion code.

A starting solution with *a priori* uncertainties can be calculated as the intersection of all backazimuth observations. If S observations are also available, a preliminary origin time is estimated using Wadati's approach to estimate a source time.

Global earth models and user-defined horizontally layered local or regional models can be used alone or together to locate seismic events. To gain the best result from all input data, observations of all seismic phases as defined in the IASPEI91 tables can be inverted. Station corrections and corrections for phases with reflection points at the earth's surface can be applied by using local velocity structures.

Key word: Hypocenter inversion.

Introduction

Since the early days of seismology, seismologists have tried to locate the source of observed seismic waves in space and time. The precision of the estimated locations has always depended strongly on our knowledge of the distribution of seismic velocities inside the earth. That is, seismic event locations are only true locations within the framework of the applied velocity model, and all estimated uncertainties must be considered in relation to other hypocenter solutions using the same model.

Although many modern location routines are based on nonlinear inversion techniques, the approach of GEIGER (1910, 1912) is still the most frequently applied algorithm today. He solved the nonlinear problem of hypocenter determination with a step-wise linearized least-squares algorithm. Limitations in computing capabilities were the reason for concentrating on the inversion of first P onsets for the next 50 years after Geiger. As computer facilities became more readily available, it became possible to include larger data sets and to develop new ideas. Location programs like

[1] NORSAR, P. O. Box 51, N-2027 KJELLER, Norway. E-mail: johannes@norsar.no

HYPO71 (LEE and LAHR, 1972), HYPOINVERSE (KLEIN, 1978), and HYPOCENTER (LIENERT *et al.*, 1986) came into use worldwide for local and regional events. Another important step was the development of the seismic array concept which created the possibility of measuring more accurately the slowness vector of an observed seismic phase, and algorithms to locate events with this parameter became available (e.g., LILWALL and DOUGLAS, 1968; GJØYSTDAL *et al.*, 1973). BRATT and BACHE (1988) combined the array location procedures with the classical travel-time inversion in the program TTAZLOC. However, this program was based on ideas already published at the beginning of the 20th century, when ABT (1907) observed the propagation of seismic phases over a net of seismic stations by assuming a plane-wave propagation. He located earthquakes by taking the backazimuth from the observed slowness vector and calculated the epicentral distance from the travel-time difference between the different onsets. A newer version of TTAZLOC, called LOCSAT, is now in use at the prototype International Data Center (pIDC) being developed to monitor the Comprehensive Nuclear-Test-Ban Treaty (CTBT). LIENERT and HAVSKOV (1995) published a new version of HYPOCENTER, which was also able to invert for later phases, backazimuth, slowness observations, and explicitly given travel-time differences between different phases observed at one station.

However, in 1996 at the Ruhr-University Bochum, I started to develop my own program, called HYPOSAT, for the purpose of utilizing the largest possible set of available information for locating events. This program has similarities with other location programs but also newer ideas or newly combined features which will be explained in this contribution. The algorithm can invert travel times of all P- and S-type onsets, for which travel-time information is available, backazimuth observations, ray parameters (or apparent velocities), and if desired travel-time differences between phases observed at the same station. Using the ray parameter in an inversion gives a relatively weak indication of the epicentral distance, however the ray parameter is a good criterion to identify (especially later) phases, and a large ray-parameter residual for an otherwise defined phase may also indicate problems with corresponding backazimuth estimation.

The Concept of HYPOSAT

The Starting Solution

For every hypocenter determination it is necessary to define the data used and the kind of corrections applied to increase the quality of the determination. For this purpose, a location program is needed, which is open for all input parameters and for various corrections.

Event location procedures are defined as iterative processes. The first step in determining a hypocenter is the definition of a starting solution for the iterations. If no preliminary hypocenter information is available, the starting solution must be derived from the input data. Because the installation of seismic arrays and the polarization analysis of signals at 3C stations has increased in recent years, measuring the backazimuth of a seismic onset is an increasingly common feature. These backazimuth observations are very useful for determining a starting solution. Taking all backazimuth observations from the different phases at the different stations, a first estimate of the epicenter can be derived by calculating the mean point of all intersections of the backazimuth directions. From the scatter of all these intersections and the uncertainties of the single backazimuth observations, an uncertainty for this starting epicenter can also be derived. If insufficient data are available, other methods must be used for such a determination, e.g., single array locations.

The next parameter of a starting solution is the source time. If P-type and S-type onsets are both reported, a simple method to obtain a source time is Wadati's approach (WADATI, 1933). Searching for S-P, Sn-Pn, Sb-Pb, and Sg-Pg travel-time differences at all stations, and sorting these travel-time differences for phase types, one can easily calculate a source time using Wadati's formula for each phase type separately. The source time of the starting solution and its standard deviation can now be calculated as the mean value of all source times after weighting with the uncertainties of the single phase readings.

The depth of an event should be either *a priori* fixed, inverted from the beginning or the inversion should be able to start with a fixed depth which will be freed for the final iterations, after a stop criterion has been reached.

Theoretical Travel Times

A location routine should capably handle observed data from all epicentral distances. Therefore different velocity models are needed: for teleseismic observations the standard radially symmetric earth models should be available. In the program, the models JEFFREYS-BULLEN (1940), PREM (DZIEWONSKI and ANDERSON, 1981), IASPEI91 (KENNETT and ENGDAHL, 1991), SP6 (MORELLI and DZIEWONSKI, 1993), and AK135 (KENNETT *et al.*, 1995) are implemented. To calculate the travel time and all derivatives with respect to distance and source depth, these models were prepared to be used by the tau-spline interpolation software of BULAND and CHAPMAN (1983).

However, these global models are usually not the right choice at local or regional distances. Here different models should be usable together with the global one. Therefore regional models of horizontal layers can be defined for which travel time and all derivatives can be calculated. In addition to the phase list of the tau-spline software, it should also be possible to calculate reflections from the Conrad and/or the Mohorovicic discontinuity. The implemented algorithm for regional models includes the corrections needed to transform a horizontally layered model into layers of the spherical earth (MÜLLER, 1977). If desired, a local or regional velocity model should be derivable from published crustal models. The possibility of deriving by bilinear interpolation, velocity models of source regions from the published model CRUST 5.1 (MOONEY *et al.*, 1998) was implemented.

Corrections

Velocity models contain neither the effect of topographic differences between seismic stations nor the differences in the local seismic velocities below the seismic stations. Adding station-dependent local P- and S-velocities, the observed onset times should be corrected for these effects with respect to phase type and incidence angle of the actual phase.

Following the ideas of Engdahl *et al.* (1998) a location routine should be able to correct surface reflections (e.g., pP, sP, sS, PP, P'P', ...) for the local velocity structure at the reflection points. The corrections should be calculated by using the actual ray parameter (i.e., incidence angle) of the reflection and estimating the travel time and distance effect for the velocity model at the reflection point, both in the global model and the local model. The difference between these two estimates then gives the corresponding correction. The model CRUST5.1 was also implemented here to interpolate for crustal velocities and elevations (or sea depth) at the actual reflection points.

Because all velocity models are derived for the earth as a sphere and not as an ellipsoid, this must be corrected. The program uses ellipticity corrections to calculate theoretical travel times. Therefore, all internal calculations must be done in geocentric instead of geographic latitudes (GUTENBERG and RICHTER, 1933), and the travel times must be corrected for the ellipticity of the earth. Ellipticity corrections either provided by KENNETT and GUDMUNDSSON (1996) or listed in the IASPEI91 tables are used.

Observations

As observations, the onset times of all seismic phases as defined in the IASPEI91 tables or by the local/regional model should be available for the inversion. Usually, seismic events are well defined by inverting for the first P onsets and some S observations. However, the possibility of inverting for all later phases is important in all cases in which only a small amount of observed data is available. The same is valid for slowness observations: to invert for backazimuth and ray parameter is far more important for events observed only at some stations than for well observed ones. The usage of slowness observations may become even more important when slowness corrections will be available for more seismic arrays and 3C stations (e.g., SCHWEITZER, 2001). In addition, the standard deviations of all observations should be given so that all input parameters can be adequately weighted.

As an option, the travel-time differences between phases arriving at the same station should be internally calculable and usable during the inversion. In the case of ideal, error-free data, these travel-time differences are a linear combination of the directly measured onset times and they cannot contribute new information to the inversion. However the situation changes in the case of erroneous and incomplete data (see the examples), which is the general case in all location problems. All travel-time differences are dependent on the epicentral distance but not on the source time or systematic timing errors; the influence of source-depth errors and velocity anomalies below the stations is also reduced. In the case of reflections (e.g., pP, sP, pS, sS, PmP, SmP, PcP, PcS, ScP, ScS) the travel-time difference between these phases and a direct phase is strongly influenced by the source depth and less dependent on the epicentral distance. The usage of travel-time differences can decrease the influence of model uncertainties, because the travel-time differences are less sensitive to base-line shifts between different models. However, because the systematic errors and the quality of the travel-time difference measurements are unknown, the best method for calculating standard deviations of travel-time differences is to derive them from the standard deviations of the single phases.

Intuitively, utilizing all this information for locating events should present a possibility of obtaining better location estimates (origin time, latitude, longitude, and depth).

The Inversion

Most frequently, the location process of a seismic event is formulated as an iterative inversion of a linearized system of normal equations (GEIGER, 1910, 1912). In this program this equation system is solved with the Generalized-Matrix Inversion (GMI) technique (e.g., MENKE, 1978) using the Single-Value Decomposition algorithm (SVD) as published in PRESS *et al.* (1992). This inversion technique was chosen because subsequently further information can also be retrieved pertaining to the quality of the inversion, i.e., the information density matrix can aid removal of unimportant input data, the resolution matrix explains the quality of the inverted parameter, and the covariance matrix can be used to analyze the trade-off between the modeled parameter. In addition, it is easy to weight the equation system with the *a priori* uncertainties of the hypocenter parameters to be modeled.

In the location routine presented here, all partial derivatives are calculated during the inversion process and the Jacobi matrix is recalculated and reconstructed for each iteration. The standard deviations of the observed data (independently given for every onset, backazimuth, and ray-parameter observation) are used respectively to weight the corresponding equation in the equation system. The given (or calculated) uncertainties of the parameters to be modeled (i.e., the source parameters) are initially used to weight the inversion. For a new iteration the equation system is always reweighted with the standard deviations of the modeled parameters as calculated during the former iteration, now used as *a priori* information. This will keep relatively well-defined model parameters mostly unchanged in the next iteration. For example, if the epicenter is well defined by the data, the remaining observed residuals are then used mainly to resolve source time and depth. The final standard deviations of the modeled parameters are given as the uncertainties of the estimated solution.

The system of equations to be solved has the following principle form:

$$\begin{bmatrix} 1 & \frac{\partial t_i}{\partial \text{lat}} & \frac{\partial t_i}{\partial \text{lon}} & \frac{\partial t_i}{\partial z_o} \\ 0 & \frac{\partial dt_j}{\partial \text{lat}} & \frac{\partial dt_j}{\partial \text{lon}} & \frac{\partial dt_j}{\partial z_o} \\ 0 & \frac{\partial p_k}{\partial \text{lat}} & \frac{\partial p_k}{\partial \text{lon}} & \frac{\partial p_k}{\partial z_o} \\ 0 & \frac{\partial azi_l}{\partial \text{lat}} & \frac{\partial azi_l}{\partial \text{lon}} & 0 \end{bmatrix} \begin{bmatrix} \delta t_o \\ \delta \text{lat} \\ \delta \text{lon} \\ \delta z_o \end{bmatrix} = \begin{bmatrix} \Delta t_i \\ \Delta dt_j \\ \Delta p_k \\ \Delta azi_l \end{bmatrix} \tag{1}$$

where

t_i – rows with travel times and their residuals Δt_i

dt_j – rows with travel-time differences between two phases observed at the same station and their residuals Δdt_j

p_k – rows with observed ray parameter (or apparent velocity) observations and their residuals Δp_k

azi_l – rows with observed backazimuth (from station to epicenter) observations and their residuals Δazi_l

δt_o – the calculated change in the source time for one iteration

δlat – the calculated change in the latitude for one iteration

δlon – the calculated change in the longitude for one iteration

δz_o – the calculated change in the source depth for one iteration (if not fixed)

One known problem of inverting for hypocenter solutions is the occurrence of oscillations between a set of hypocenter solutions. In this case, the iteration process is automatically interrupted by calculating the mean values for all hypocenter parameters as new starting solutions for the next iteration, and this helps to resolve the global minimum of all solutions. If only the depth is oscillating (together with the source time), the depth will be automatically fixed at the mean value of all solutions or at the earth's surface.

Test Examples

A Synthetic Test

The following examples are meant to illustrate the advantages of using travel-time differences as additional parameters in the inversion. In the case of error-free onset observations, the travel-time differences are linear combinations of the absolute travel times and therefore they do not change the inversion results. But in the case of erroneous or insufficient data, the usage of travel-time differences can improve the result.

To demonstrate this, a synthetic example was chosen. The coordinates of the event are listed in the first row of Table 1. The travel times calculated for model AK135 (KENNETT *et al.*, 1995) to the stations ARCES, FINES, and NORES are listed in Table 2, and Figure 1 shows a map with the seismic stations and the hypothetical event. These data were inverted to reestimate the theoretical source using different approaches. The results of these inversions are also listed in Table 1. The solution and especially the depth estimation of this example depends on the initial epicenter because of the disadvantageous geometry of source and observing stations. The initial epicenter for all further inversions was set to latitude 54.5° and longitude 21.5°; backazimuth or ray parameter values were not used for this test. In the first two inversions the original data were inverted once with, and once without the usage of travel-time differences (TTD). The solution in both cases is, within numerical limits, the same. The differences between the two solutions and the differences from the theoretical location can be partly explained by the truncation of the input onset times to 1/100 s, partly by the usage of a finishing convergence criterion for defining a solution, and partly by the disadvantageous geometry. In the next step, the absolute onset times at FINES were disturbed by adding 1 s for both phases (Pn and Sn) to simulate a systematic timing error. Because the source depth was no longer resolvable in this case, it was fixed at 10 km (S1). In the next simulation (S2) the theoretical travel times were unaltered at FINES and NORES, but a 3 s delay was added for all onsets at ARCES. This was done to simulate a station at a larger distance with a weak onset leading to late picks for both Pn and Sn. In the last theoretical test (S3) a combination of such effects was introduced: the onsets at ARCES were 3 s delayed, for FINES Sn was 1 s delayed and Pn was made to come 1 s too early, and both onsets at NORES were 1 s too early.

In all cases with erroneous data (S1–S3) the inversion with travel-time differences gives a solution closer to the 'true' source and the corresponding quality parameters (i.e., standard deviations and the RMS values) are smaller.

A Case Study: The November 11, 1999 Explosion in the Dead Sea

Finally, the new program HYPOSAT was used to locate an explosion in the Dead Sea. On November 11, 1999 the Geophysical Institute of Israel (GII) blasted a charge of 5 ton TNT in a water depth of ca. 70 m in the Dead Sea as a calibration shot for the CTBT (see Table 4). GII published the exact coordinates of this explosion and it was observed by some of the seismic stations contributing data to the pIDC. Table 3 lists all parameters of the onsets associated by the pIDC to this event and used in

Table 1

Theoretical (first line) and inverted source coordinates with and without travel-time differences (TTD). The cases S1–S3 have more or less biased onsets; for further details see text.

Time	Latitude [°]	Longitude [°]	Depth [km]	Location Error [km]	RMS [s]	Remarks
00:00:00.000	55.0000	22.0000	10.00			theoretical source
23:59:59.988 ± 0.015	55.0022 ± 0.0026	21.9990 ± 0.0011	9.67 ± 0.39	0.41	0.002	with TTD
23:59:59.985 ± 0.018	55.0027 ± 0.0030	21.9989 ± 0.0012	9.60 ± 0.46	0.51	0.002	without TTD
00:00:00.417 ± 0.416	55.0016 ± 0.0265	21.9244 ± 0.0390	10.0 fixed	4.85	0.363	S1, with TTD
00:00:00.500 ± 0.781	55.0069 ± 0.0518	21.9171 ± 0.0573	10.0 fixed	5.37	0.367	S1, without TTD
00:00:00.684 ± 1.518	54.9728 ± 0.0967	21.9053 ± 0.1424	10.0 fixed	6.78	1.341	S2, with TTD
00:00:00.378 ± 2.902	54.9516 ± 0.1909	21.9063 ± 0.2132	10.0 fixed	8.07	1.348	S2, without TTD
23:59:59.148 ± 1.875	54.8996 ± 0.1194	21.8362 ± 0.1766	10.0 fixed	15.35	1.439	S3, with TTD
23:59:58.785 ± 3.542	54.8752 ± 0.2328	21.8489 ± 0.2608	10.0 fixed	16.95	1.447	S3, without TTD

Table 2

The theoretically calculated onset times for the inversion tests of Table 1.

Station	Distance [°]	Phase	Onset Time
NORES	8.003	Pn	00:01:56.15
NORES	8.003	Sn	00:03:26.58
FINES	6.810	Pn	00:01:39.80
FINES	6.810	Sn	00:02:57.27
ARCES	14.676	Pn	00:03:27.28
ARCES	14.676	Sn	00:06:09.74

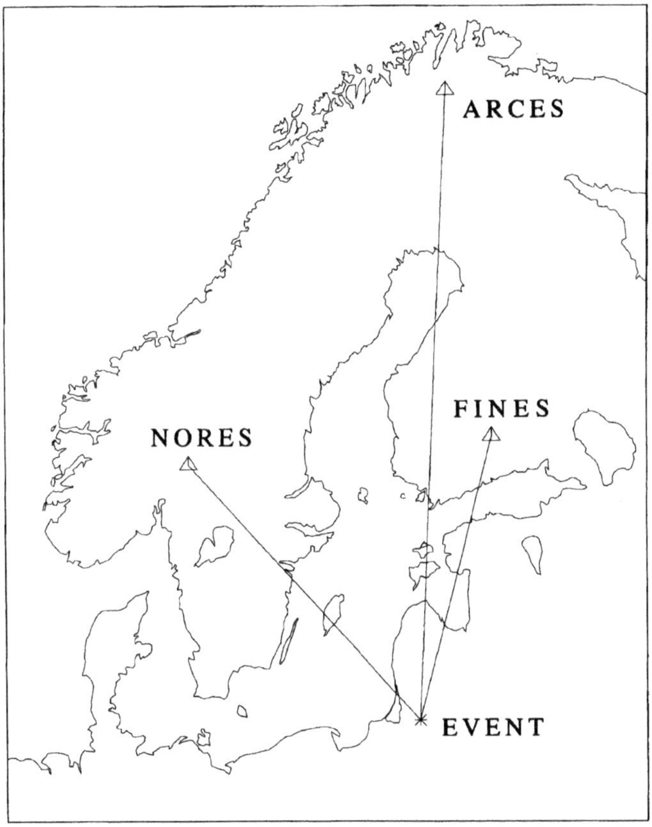

Figure 1
Map showing the locations of the seismic arrays (triangles) and the theoretical test event (star) to demonstrate the advantage of including travel-time differences in the inversion process (for details see text).

their location as published in their bulletin (REB): the pIDC location is quite precise with only a small location error of 2.45 km. These pIDC data were used as input for HYPOSAT and the event was relocated with several options. In all relocations with

HYPOSAT, the IASPEI91 model was used as the theoretical model. Without any further information about the local structure below the different stations, all elevation effects were corrected with the same velocities (V_P = 5 km/s, V_s = 2.89 km/s), which are slightly lower than the uppermost velocities in IASPEI91. However, no other model modifications were applied.

Using all intersections of the listed backazimuth observations (43 combinations could be used), a starting epicenter (latitude 37.7 ± 13.7°, longitude 34.9 ± 4.8°) could be estimated. Comparing this epicenter with the true one, the discrepancy for the latitude of about 690 km too far to the north is obvious, although the error in the longitude of approximately 50 km is relatively small. However, the known location is inside the estimated uncertainties of this starting solution. The Wadati analysis of the S-P onset times gave a preliminary source time of 15:00:07.665 ± 1.404 s which is about 8 s too late with respect to the true origin time. Consequently, this starting solution is not the best choice with respect to the two regional stations, however with respect to the teleseismic observations this epicenter is a usable solution as a first guess.

After trial inversions it became clear that ray parameter and backazimuth data contribute very little to the final inversion because of their large uncertainties (Table 3). Therefore they were not used further. Also the inversion for the source depth showed that this parameter has a very low resolution in this case and was therefore fixed at 0 km.

To demonstrate with real data the different effects in using travel-time differences as an additional condition, the following relocations were tried: "HYPOSAT 1" is the result using all observed travel times but no travel-time differences. If one looks at the absolute location error of 3.04 km (see Table 4), the location routine works quite well. "HYPOSAT 2" uses travel-time differences but not the absolute travel times of the corresponding phases. That is, the two regional stations contributed in this case only with their travel-time differences. Not surprisingly, the location error is now larger (5.32 km). In "HYPOSAT 3" the travel-time differences were used and in addition the absolute travel times of the corresponding P onsets to fix the absolute timing. The error of 2.76 km is now smaller with respect to the first tests. "HYPOSAT 4" displays the opposite configuration: the usage of the absolute travel times of the S onsets and the travel-time differences. In this case the error becomes larger again. In a final test ("HYPOSAT 5") all absolute travel times and all possible travel-time differences were used. The differences in the location errors are relatively small, but again, the result including both the absolute travel times and travel-time differences exhibits the smallest absolute location error of only 2.39 km. This demonstrates the advantage of including travel-time differences as additional data in the inversion process to find the best solutions.

Table 3

Observed onsets of the November 11, 1999 Dead Sea explosion as measured at the pIDC including their standard deviations (STD). The phase identification in parentheses is from HYPOSAT and the distances are the theoretical ones.

Station	Phase	Distance [deg]	Onset Time ± STD	Azimuth [deg] ± STD	slowness [s/deg] ± STD
MRNI	Pg(Pn)	1.475	15:00:28.34 ± 0.120	348.52 ± 11.5	15.68 ± 3.15
MRNI	Lg(Sg)	1.475	15:00:48.491 ± 0.424	300.14 ± 07.4	19.93 ± 2.57
EIL	Pn	1.905	15:00:34.626 ± 0.120	25.86 ± 09.4	12.56 ± 2.07
EIL	Lg(Sn)	1.905	15:01:00.901 ± 1.002	20.39 ± 09.2	14.20 ± 2.28
MLR	Pn	15.777	15:03:45.880 ± 1.070	23.65 ± 15.6	11.22 ± 3.06
GERES	P	23.839	15:05:16.325 ± 0.838	127.41 ± 23.8	11.06 ± 4.12
ARU	P	29.656	15:06:06.085 ± 1.049	202.23 ± 37.8	10.67 ± 6.91
BGCA	P	30.699	15:06:18.063 ± 1.070	355.36 ± 17.5	13.49 ± 4.11
ESDC	S	32.807	15:06:35.350 ± 0.765	98.58 ± 08.6	7.29 ± 1.09
PDYAR	P	56.913	15:09:48.200 ± 0.966	267.82 ± 20.0	8.48 ± 2.94

Table 4

Hypocenters for the November 11, 1999 Dead Sea explosion. Listed are the hypocenter as announced by the Geophysical Institute of Israel (GII), the epicenter published by the pIDC (REB), and the results of several relocations using the HYPOSAT routine (for details see text). The given uncertainties are, for the pIDC solution, the 90% confidence limits, and for the HYPOSAT solutions, the standard deviations of the calculated source parameters. Additionally given is the number of defining data (#), the RMS values for the absolute arrival times used, and the absolute horizontal location error. The number of definings contains both the number of absolute travel times and the number of travel-time differences used for the inversion. For details of the different HYPOSAT solutions, see text. The source depth was fixed for all inversions at 0. km.

Model	Origin Time	Latitude [deg]	Longitude [deg]	#	RMS [s]	Error [km]
GII	15:00:00.795	31.5336	35.4413	–	–	–
pIDC (REB)	15:00:00.78 ± 1.21	31.5199 ± 17.0 km	35.4616 ± 10.3 km	10	0.89	2.45
HYPOSAT 1	15:00:00.595 ± 0.760	31.5520 ± 0.0091	35.4651 ± 0.0926	10	0.883	3.04
HYPOSAT 2	15:00:01.034 ± 0.555	31.5744 ± 0.0253	35.4709 ± 0.1072	8	1.067	5.32
HYPOSAT 3	15:00:00.615 ± 0.181	31.5508 ± 0.0096	35.4623 ± 0.0918	10	0.962	2.76
HYPOSAT 4	15:00:00.893 ± 0.427	31.5733 ± 0.0136	35.4678 ± 0.0911	10	0.958	5.07
HYPOSAT 5	15:00:00.633 ± 0.161	31.5550 ± 0.0081	35.4440 ± 0.0828	12	0.903	2.39

Remarks

HYPOSAT employs the new version of the regionalization in seismic units (YOUNG *et al.*, 1996). The program will be further developed and is open to new ideas and improvements, and its use is therefore encouraged. For instance, the method of uniform reduction to handle long-tailed residual distributions (JEFFREYS, 1932) as an alternative weighting method of the observations, the usage of travel-time differences to test the plausibility of phase identifications, and the calculation of confidence ellipsoids for the determined hypocenter are planned for the next implementation of the program. The program HYPOSAT is available including all necessary data files, examples, a manual, and the source code. The newest version can be found on the ftp server of NORSAR (ftp.norsar.no) under/pub/outgoing/ johannes/hyposat.

Acknowledgments

Part of this work was accomplished while I worked at the Ruhr-University Bochum, Germany and this research was partly sponsored by the Nuclear Treaty Programs Office of the U.S. Department of Defense and monitored by AFTAC, Patrick AFB, FL32925, under contract no. F08650-96-C-0001. Two anonymous reviewers strongly influenced this contribution with their constructive criticism. This is NORSAR contribution No. 675.

REFERENCES

ABT, A., *Vergleichung seismischer Registrierungen von Göttingen und Essen (Ruhr)*, Ph.D. Thesis (Georg-August-Universtät, Göttingen 1907).

BRATT, S. R., and BACHE, T. C. (1988), *Locating Events with Sparse Network of Regional Arrays*, Bull. Seismol Soc. Am. *78*, 780–798.

BULAND, R., and CHAPMAN, C. H. (1983), *The Computation of Seismic Travel Times*, Bull. Seismol. Soc. Am. *73*, 1271–1302.

DZIEWONSKI, A. M., and ANDERSON, D. L. (1981), *Preliminary Reference Earth Model*, Phys. Earth Planet. Inter. *25*, 297–356.

ENGDAHL, E. R., VAN DER HILST, R., and BULAND, R. (1998), *Global Teleseismic Earthquake Relocation with Improved Travel Times and Procedures for Depth Determination*, Bull. Seismol. Soc. Am. *88*, 722–743.

GEIGER, L. (1910), *Herdbestimmung bei Erdbeben aus den Ankunftszeiten*, Nachrichten der Königlichen Gesellschaft der Wissenschaften zu Göttingen, mathematisch-physikalische Klasse, 331–349.

GEIGER, L. (1912), *Probability Method for the Determination of Earthquake Epicenters from the Arrival Time only*, Bulletin St. Louis University *8*, 60–71.

GUTENBERG, B., and RICHTER, C. F. (1933), *Advantages of Using Geocentric Latitude in Calculating Distances*, Gerlands Beiträge zur Geophysik *40*, 380–389.

GJØYSTDAL, H., HUSEBYE, E. S., and RIEBER-MOHN, D. (1973), *One-array and Two-array Location Capabilities*, Bull. Seismol. Soc. Am. *63*, 549–569.

JEFFREYS, H. (1932), *An Alternative to the Rejection of Observations*, Proc. Roy. Soc. London *A137*, 78–87.

JEFFREYS, H., and BULLEN, K. E., *Seismological Tables* (British Association for the Advancement of Science, London 1940).

KENNETT, B. L. N., and ENGDAHL, E. R. (1991), *Travel Times for Global Earthquake Location and Phase Identification*, Geophys. J. Int. *105*, 429–466.

KENNETT, B. L. N., ENGDAHL, E. R., and BULAND, R. (1995), *Constraints on Seismic Velocities in the Earth from Travel Times*, Geophys. J. Int. *122*, 108–124.

KENNETT, B. L. N., and GUDMUNDSSON, O. (1996), *Ellipticity Corrections for Seismic Phases*, Geophys. J. Int. *127*, 40–48.

KLEIN, R. W., *Hypocenter Location Program HYPOINVERSE* (U.S. Geol. Surv. Open-File Report. 78–694, 1978).

LEE, W. H. K., and LAHR, J. C., *HYPO71: A Computer Program for Determining Hypocenter, Magnitude and First Motion Pattern of Local Earthquakes* (U. S. Geol. Surv. Open-File Report 1972).

LIENERT, B. R., BERG, E., and FRAZIER, L. N. (1986). *HYPOCENTER: An Earthquake Location Method Using Centered, Scaled, and Adaptively Damped Least-squares*, Bull. Seismol. Soc. Am. *76*, 771–783.

LIENERT, B. R., and HAVSKOV, J. (1995), *A Computer Program for Locating Earthquakes Both Locally and Globally*, Seism. Res. Lett. *66(5)*, September–October, 26–36.

LILWALL, R. C., and DOUGLAS, A. (1968), *Epicenter Determinations by Seismic Arrays*, Nature *220*, 362–363.

MENKE, W., *Geophysical Data Analysis: Discrete Inverse Theory*, revised edition, International Geophysics series *45* (Academic Press, San Diego 1978).

MOONEY, W. D., LASKE, G., and MASTERS, T. G. (1998), *CRUST 5.1: A Global Crustal Model at 5° × 5°*, J. Geophys. Res. *103*, 727–747.

MORELLI, A., and DZIEWONSKI, A. M. (1993), *Body-wave Traveltimes and a Spherically Symmetric P- and S-wave Velocity Model*, Geophys. J. Int. *112*, 178–194.

MÜLLER, G. (1977), *Earth Flattening Approximation for Body Waves Derived from Geometric Ray Geometry – Improvements, Corrections and Range of Applicability*, J. Geophys. *42*, 429–436.

PRESS, W. H., TEUKOLSKY, S. A., VETTERLING, W. T., and FLANNERY, B. P., *Numerical Recipes in FORTRAN, the Art of Scientific Computing*, second edition (Cambridge University Press, Cambridge 1992).

SCHWEITZER, J. (2001), *Slowness Corrections - One Way to Improve IDC Products*, Pure appl. geophys., 158, 375–396.

WADATI, K. (1933), *On the Travel Time of Earthquake Waves. Part. II*, Geophys. Mag. (Tokyo) 7, 101–111.

YOUNG, J. B., PRESGRAVE, B. W., AICHELE, H., WIENS, D. A., and FLINN, E. A. (1996). *The Flinn-Engdahl Regionalization Scheme: The 1995 Revision*, Phys. Earth Planet. Int. *96*, 223–297.

(Received July 15, 1999, revised February 4, 2000, accepted February 9, 2000)

Pure appl. geophys. 158 (2001) 291–317
0033–4553/01/020291–27 $ 1.50 + 0.20/0

Pure and Applied Geophysics

Global Event Location with Full and Sparse Data Sets Using Three-dimensional Models of Mantle *P*-wave Velocity

MICHAEL ANTOLIK,[1] GÖRAN EKSTRÖM,[1] and ADAM M. DZIEWONSKI[1]

Abstract—In order to improve on the accuracy of event locations at teleseismic distances it is necessary to adequately correct for lateral variations in structure along the ray paths, either through deterministic model-based corrections, empirical path/station corrections, or a combination of both approaches. In this paper we investigate the ability of current three-dimensional models of mantle *P*-wave velocity to accurately locate teleseismic events. We test four recently published models; two are parameterized in terms of relatively long-wavelength spherical harmonic functions up to degree 12, and two are parameterized in terms of blocks of constant velocity which have a dimension of a few hundreds of km. These models, together with detailed crustal corrections, are used to locate a set of 112 global test events, consisting of both earthquakes and explosions with *P*-wave travel-time data compiled by the International Seismological Centre (ISC). The results indicate that the supposedly higher resolution block models do not improve the accuracy of teleseismic event locations over the longer wavelength spherical harmonic models. For some source locations the block models do not predict the range of observed travel-time residuals as well as the longer wavelength models. The accuracy of the locations largely varies randomly with geographic position although events in central Asia are particularly well located. We also tested the effect of reduced data sets on the locations. Multiple location iterations using 30 *P*-wave travel times indicate that teleseismic events may be located within an area of 1000 km^2 of the true location 66% of the time with only the model-based corrections, and increasing to 75% if calibration information is available. If as few as 8 phases are available then this is possible only 50% of the time. Further refinement in models and/or procedure, such as the addition of P_n phases, azimuth data, and consideration of *P*-wave anisotropy may provide further improvement in the teleseismic location of small events.

Key words: Event location, seismic tomography, mantle heterogeneity.

Introduction

One stated technical goal for monitoring of the Comprehensive Nuclear-Test-Ban Treaty (CTBT) is to locate events of $M \geq 4$ with an estimated uncertainty of 1000 km^2 or less for the purpose of on-site inspection. Because of the lateral heterogeneities present within the real earth, this goal is usually not achieved using conventional location techniques with standard one-dimensional velocity models. Two general approaches can be used to improve the quality of locations. The first is

[1] Department of Earth and Planetary Sciences, Harvard University, Hoffman Laboratory, Cambridge, MA 02138, U.S.A.

the application of empirically derived station corrections. While such corrections are only calculated once and then stored, and therefore can be applied extremely quickly in most location algorithms, they are often critically dependent on the source region. Station corrections which are regionally invariant often give little or no improvement in location. Further, the application of source-region dependent corrections depends upon previous sampling of ray paths from all possible source regions. The second approach is the use of a laterally varying earth model. While this approach does not suffer from the above disadvantage, it requires the calculation of a travel-time correction for each ray. In addition, the potential resolution of laterally heterogeneous models is limited by the quality and coverage of the data employed, and also by the computational resources available for their derivation.

Global three-dimensional (3-D) velocity models of the earth's mantle continue to evolve and become parameterized on an ever finer scale. Models of both compressional and shear-wave velocity are now commonly parameterized in terms of constant velocity blocks (e.g., VASCO and JOHNSON, 1998; GRAND *et al.*, 1997; VAN DER HILST *et al.*, 1997; BOSCHI and DZIEWONSKI, 1999a) rather than spherical harmonic functions. Such "high-resolution" models, using blocks with sizes on the order of a few hundred kilometers, have provided sharper images of coherent smaller-scale heterogeneities, in particular fast, sheet-like anomalies presumed to correspond to slabs penetrating the lower mantle. These models should lead to better characterization of *P*-wave residuals for paths through subduction zones or other areas where small-scale lateral heterogeneities are present. However, a number of factors may lead to lower resolution of large-scale, smaller amplitude anomalies in block models. Because of the higher number of unknown parameters, it is still impractical to invert combinations of very large data sets (waveforms and travel times), as is frequently done with spherical harmonic models (SU and DZIEWONSKI, 1997; LI and ROMANOWICZ, 1996). This may result in lower resolution in certain areas (particularly the shallow mantle) to which particular data sets are sensitive. In addition, the division into arbitrary, constant velocity blocks may induce an unrealistic shape in long-wavelength anomalies. A lack of correlation between new, high-resolution earth models and earlier longer wavelength models has previously been noted (GRAND *et al.*, 1997). Another possible factor in this discrepancy may be the use of regularization or damping in the solution of the inverse problem (BOSCHI and DZIEWONSKI, in press).

Recently, the work of SMITH and EKSTRÖM (1996) has shown that spherical harmonic models of *P*-wave velocity are of sufficient quality to be useful in improving teleseismic event locations. By inverting *P*-travel times from a data set of events with known or very accurately determined "ground-truth" locations, they demonstrated that the average mislocation distance was reduced by approximately 40% using the 3-D model S&P12/WM13 (hereafter referred to as SP12) (SU *et al.*, 1994) as compared to PREM or IASPEI91. However, only small improvements were obtained for earthquakes occurring in geologically complex areas along plate boundaries, presumably due to the inadequate representation in SP12 of anomalies with wavelengths of a few hundred kilometers or less. On the other hand, this may be due to the fact that most of these latter events are earthquakes with less accurate ground-truth locations.

In this paper we use essentially the same data set as that used by SMITH and EKSTRÖM (1996) (hereafter SE96), with additional explosions from the Chinese Lop Nor test site. We test the improvement which can be obtained in teleseismic event location using the newer block models of mantle *P*-wave velocity compared to spherical harmonic models and to PREM. The performance of PREM compared to other global 1-D models such as IASPEI91 or the tables of JEFFREYS and BULLEN (1958) has already been tested by SE96. We concentrate only on models of *P*-wave velocity since block models of shear velocity are as yet few in number, and previous work using *S* waves has shown them to be mostly of use in improving depths rather than epicentral locations (EKSTRÖM *et al.*, 1997) due to the larger picking errors associated with them. We test the models both with and without empirical station corrections. After a brief description of the models and of the location method, the Analysis section of the paper discusses the accuracy of locations obtained using the entire ISC *P*-wave data set for each model and with different combinations of station corrections employed.

Since the terms of the CTBT require the elimination of nuclear tests of all yields, it is also desirable to examine the utility of 3-D models for the location of small or moderate sized events (*M* 4–5). At teleseismic distances, small magnitude events may be recorded by only a few stations. We have therefore conducted experiments in which only a portion of the available *P*-wave data is used and examined the performance of each of the 3-D models relative to PREM. These results are presented in the second half of the Analysis section. Using data sets consisting of 30 phases, the 3-D models are able to locate the test events to within the accuracy goal of the CTBT on the order of 70% of the time. The results improve when using station corrections derived from the ground-truth locations of nearby test events.

3-D Models and Event Data

We test four 3-D models in this paper. In assessing the accuracy in event location obtained by the various models, it is important to keep in mind the data used in constructing each model as well as the inversion method. Model SP12 was obtained by SU and DZIEWONSKI (1993) through joint inversion of both *P*- and *S*-wave travel times compiled by the ISC and differential travel times compiled by other researchers. To improve resolution in the mid-mantle, they also employed long-period body and mantle waveform data. Their starting model was obtained by adapting and scaling an earlier shear-wave velocity model (SU *et al.*, 1994). The final model is parameterized horizontally in terms of spherical harmonics up to degree 12 and radially in terms of Chebyshev polynomials up to order 13. This provides a nominal radial resolution on the order of 200 km and a horizontal resolution of about 1,700 km. Although this resolution is rather low, SE96 obtained an average reduction in mislocation distance of 40% over standard 1-D models using SP12 for explosions with known locations.

SU and DZIEWONSKI (1997) subsequently carried out a joint inversion for shear and bulk sound velocity in the mantle using much of the same data. Approximately 40,000 waveforms and well over a million travel times were used. Their final models were obtained in two stages; first by solving for the perturbations in shear modulus only, and then by adding perturbations in the bulk modulus. They used the same parameterization as in SP12 and the same starting model. We test a *P*-velocity model derived from the bulk sound and shear velocity models (referred to as MK12).

The other two 3-D models that we test are parameterized in terms of blocks having a constant velocity. The first of these is Model BDP98, obtained by BOSCHI and DZIEWONSKI (1999a) using ISC residuals from epicenters corrected for lateral heterogeneity using model SP12. The model describes *P*-wave velocity with respect to PREM. The blocks have a dimension of $5° \times 5°$ at the equator (roughly equivalent to spherical harmonic degree 40). The area of the blocks is kept constant from the equator to the poles. In their inversion, BOSCHI and DZIEWONSKI heavily damped the roughness of the solution over its norm. The second block model is that obtained by VAN DER HILST *et al.* (1997), who used *P*-wave residuals derived from the new locations published in ENGDAHL *et al.* (1998). We refer to this model as HWE97. HWE97 describes *P*-velocity anomalies with respect to the 1-D model ak135 KENNETT *et al.* (1995). Its parameterization is in $2° \times 2°$ blocks where the area of the blocks is not preserved. BOSCHI and DZIEWONSKI (1999a) compare the spherical harmonic spectra of these two models, and note that BDP98 has higher power at lower degrees than HWE97 but a far more rapid decay in power at higher degrees.

The set of test events which we employ to test these models is the same as that used by SE96, with the addition of additional explosions at the Chinese Lop Nor test site for which accurate ground-truth information is available from satellite data (WALLACE and TINKER, 2001). Figure 3 shows the locations of these events. There are 30 explosions and 82 earthquakes. The epicentral information for the explosions has been determined by non-seismic means, while the location accuracy for the earthquakes has been reported as 5 km and the origin time uncertainty as 0.5 s (KENNETT and ENGDAHL, 1991). From these events we first use all of the *P* phases in the ISC catalog in the distance range 25° to 96°. The number of available phases ranged from about 45 to over 500, and the events range in magnitude from m_b 5.0 to 7.5. Next we investigate the effect of using a sparse data set by restricting the number of phases to first 30 and then to 8.

Procedure

Following SE96, we compute separate travel-time perturbations for each *P*-wave raypath used in the inversions corresponding to the effects of mantle structure, crustal structure, ellipticity, and station elevation, which are then added to the travel time calculated for the reference model (PREM or ak135). For the spherical harmonic models, the mantle structure corrections are obtained using the method of DZIEWONSKI (1984) which uses coordinate rotations and an integral table to achieve rapid calculation of the travel-time perturbation corresponding to each coefficient of the spherical harmonic expansion. We take advantage of Fermat's principle in assuming that the mantle corrections are stationary with respect to small changes in the raypath [SE96] (i.e., the assumed raypath is that in the reference model). For block models the mantle correction is simply

$$\delta t_M = -\sum_{i=1}^{N} dt_i \left(\frac{\delta v}{v_o}\right)_i \tag{1}$$

where dt_i is the travel time in the ith block, and the expression in parentheses is the velocity perturbation of the ith block with respect to the reference 1-D velocity (v_o).

Crustal corrections are incorporated into the relocations in two ways. The first is a degree 12 spherical harmonic expansion of the ocean-contintent function and is described in SU *et al.* (1994). The expansion is normalized to give an average crustal thickness of 24.4 km. We also employed the full CRUST5.1 model of MOONEY *et al.* (1998), which is defined in $5° \times 5°$ blocks. This correction was applied by calculating travel time for a ray through this model having the same slowness as in the reference 1-D model. Both source and receiver-side corrections were used. Corrections for crustal structure can be quite large in regions of thick crust, and are negative in the ocean basins (Fig. 1). We calculated the elevation correction by assuming a vertical raypath through crust with an average velocity of 5.8 km s^{-1}.

Station Corrections

We investigated the use of empirical station corrections with the 3-D models in an attempt to correct for unmodeled structure. Despite the high resolution of the block models, we expect that shallow, localized anomalies with wavelengths less than ~100 km may introduce significant error into locations determined from the mantle models alone, as may deeper, longer wavelength anomalies if their amplitudes are not as well determined as in lower resolution models. We calculated region-dependent corrections by dividing the earth into bins 5° wide in azimuth and either 2.5° or 5° wide in epicentral distance. No correction is made for event depth. We then use one of the 3-D models to relocate all of the events in the compilation

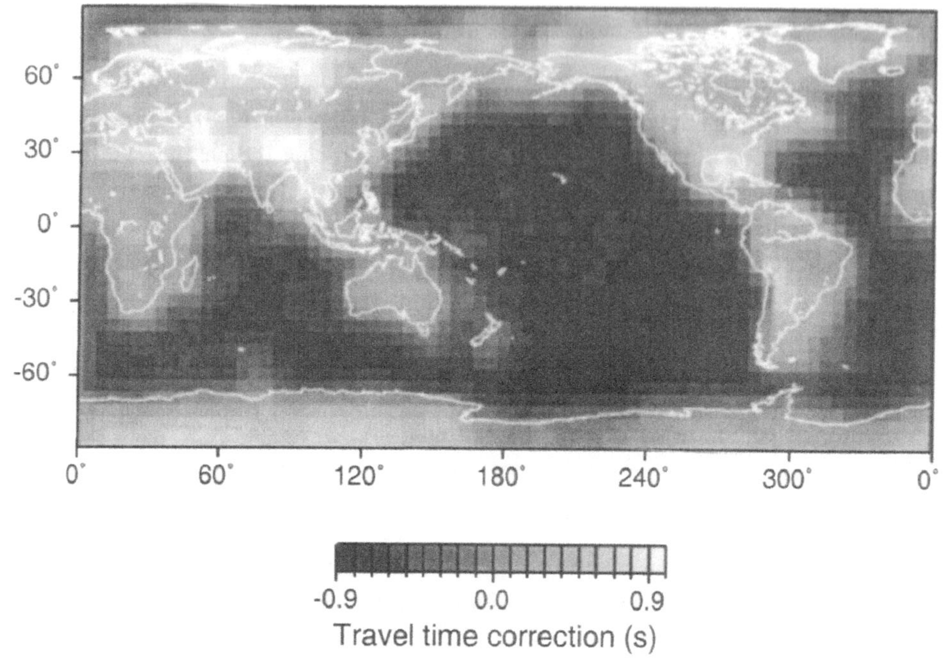

Figure 1
One-way travel-time corrections (relative to PREM crust) for a P wave in model CRUST5.1 of MOONEY *et al.* (1998). Vertical incidence is assumed at the base of the crust.

of ENGDAHL *et al.* (1998) in each of the i source bins, and the correction for each bin is then

$$S^{ij} = \frac{1}{K} \sum_{k}^{K} t_r^{ijk} \tag{2}$$

where the travel-time residual after relocation is

$$t_r^{ijk} = [t_{\text{obs}} - (t_o + \delta t_M + \delta t_C + \delta t_E + \delta t_{EL})]^{ijk}. \tag{3}$$

The superscript j refers to the station and the sum is over each k event within the source bin, t_o is the reference model travel time, δt_C is the crustal correction for the raypath, δt_E is the ellipticity correction and δt_{EL} is the elevation correction. Thus, the station corrections incorporate local site effects as well as unmodeled mantle or crustal structure which might produce similar delays at nearby stations. In theory, it would seem preferable to calculate corrections for bins which are as small as possible. However, it is unclear whether there are enough data currently available to constrain region-dependent corrections for many stations. We have arbitrarily employed a cutoff for each source bin of four observations; if a bin contains fewer events for any particular station the correction term is set to zero. All of the station corrections were calculated using CRUST5.1 as the crustal model. We used these station corrections in the analysis using the full data set.

Figure 2 compares corrections for station JAS in California and models SP12 and BDP98. Many areas exhibit strong correlation in the corrections between adjacent bins, indicating significant unmodeled structure. However, the mean value of the corrections is quite small, on the order of 0.1 s. In some areas (South America and the Caribbean) the corrections vary considerably within a short distance, and

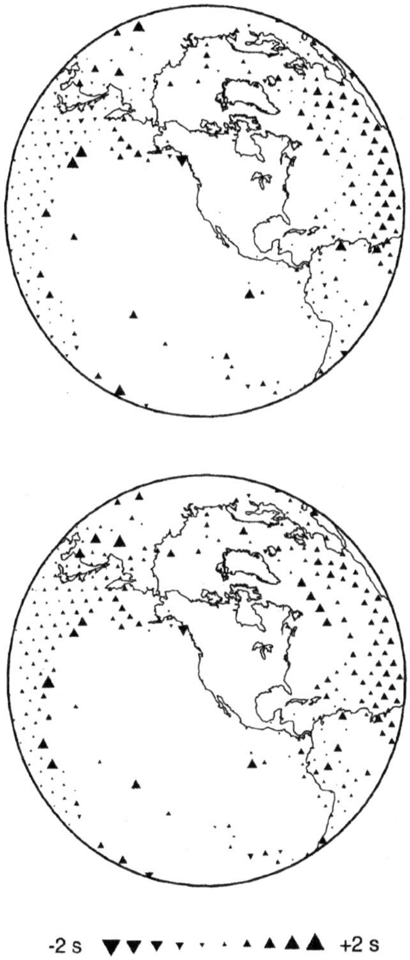

Figure 2
Station correction terms for station JAS in California plotted at the center of 5° × 5° equal-area bins. The top map shows corrections calculated for model SP12 and the lower map for model BDP98. The overall pattern and amplitude of the corrections are very similar for both models.

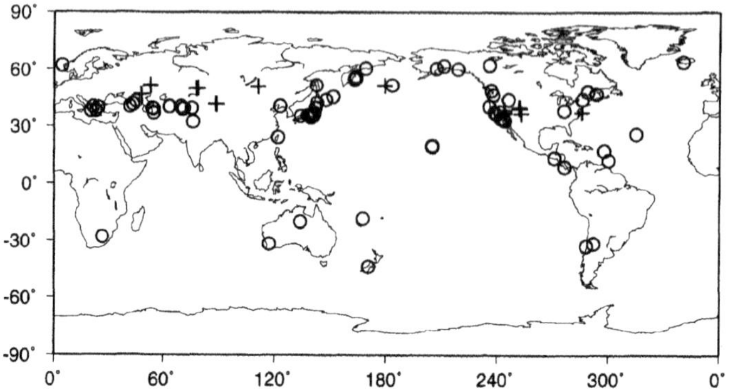

Figure 3
Locations of test events used in this study. Explosions are indicated by crosses and earthquakes by circles. Epicentral information is listed in SMITH and EKSTRÖM (1996) and KENNETT and ENGDAHL (1991).

probably are determined largely by structure near the particular source bins. The corrections are similar for both models and are dominantly positive. This indicates the presence of additional slow anomalies, probably in the upper mantle under JAS, which are not completely compensated for in either model.

Relocation Method

Each event is relocated by standard nonlinear least-squares, with the depths held fixed at the surface for explosions and at the depth reported in KENNETT and ENGDAHL (1991) for the earthquakes, owing to the lack of depth resolution of teleseismic P waves. Since the change in epicentral location is small during the inversion, the 3-D mantle correction calculated for the initial location, that of the ISC, is used throughout. When we tested the effect of varying the 3-D mantle corrections between iterations for model SP12, the average difference in location for the test events was ~0.2 km.

The inversion process continued until the root-mean-square of the weighted travel-time residuals changes by less than 0.001 s, usually in 3–5 iterations. Weighting was assigned to the residuals in the following manner:

$$\begin{aligned} w &= 1.0, \quad Z < 3.333 \\ w &= e^{-(0.3*Z)^2-1}, \quad 3.333 < Z < 20.0 \\ w &= 0.0, \quad Z > 20.0 \end{aligned} \tag{4}$$

where $Z = |\bar{t} - t_r|$. $\bar{t}$ is the mean travel-time residual in the azimuthal quadrant of the observation. This type of weighting scheme was used in order to include residuals in the inversion which are large due to a poor initial location.

Analysis

Full Data Set

The results of relocating all of the test events using the complete set of ISC teleseismic *P* waves, and without empirical station corrections, are summarized in Table 1. As was found by SE96, there is a considerably better improvement in locations of the explosions than the earthquakes relative to PREM, however for these results the improvement in the earthquake locations is greater. This is probably due to the fact that we have used a smaller distance cutoff than that used by SE96, which only inverted data recorded at distances larger than 30°. The most striking aspect of these results is that model SP12 generates the smallest mislocations for both earthquakes and explosions. The rms mislocation for this model is about 2 km smaller than that using the other 3-D models. As mentioned above, model MK12, although it has the same parameterization as SP12, was originally derived as a model of bulk sound and shear velocity. Thus, we might expect this model to provide a smaller improvement in event location when using only *P* waves. The other two models (BDP98 and VWE97), however, are block models with nominally higher resolution and were constructed using only *P* wave (also *pP* for VWE97) travel-time residuals. VWE97 contains the most free parameters of the 3-D models. The performance of these two models relative to SP12 is therefore somewhat surprising.

Table 1 suggests that appropriate correction for crustal structure is just as important as the choice of 3-D mantle model. Not surprisingly, we observe a large improvement in location quality using the full CRUST5.1 model over the rather crude crustal correction employed by Su *et al.* (1994). This is especially true for the explosion events where this improvement is equal to that of using SP12 over the other mantle models. The rather coarse parameterization of CRUST5.1 may be a factor in the small improvement observed for the earthquakes, since many of these events

Table 1

Root-mean-squares mislocation in km for earthquakes and explosions using a particular velocity model and all available teleseismic P phases.

Model	Explosions	Earthquakes
PREM + C5.1	12.92	18.83
SP12 + OCF	8.47	15.57
SP12 + C5.1	7.83	15.37
MK12 + OCF	11.14	17.92
MK12 + C5.1	9.53	17.40
BDP98 + OCF	11.43	17.22
BDP98 + C5.1	9.80	17.13
VWE97 + C5.1	10.51	17.33

OCF refers to use of the ocean-continent function crustal corrections, C5.1 to the use of the CRUST5.1 model.

occur along coastlines where the crustal thickness varies rapidly over short distances. Nevertheless, this model is quite effective in reducing much of the regional bias in *P*-wave travel times due to crustal structure.

Mislocation vectors for the test events in models SP12 and and BDP98 are compared in Figure 4. The considerably smaller location errors for explosion events, even compared with earthquakes located nearby, is obvious. The general trend for location errors using BDP98 to be larger than those using SP12 exists for all regions. Locations derived from BDP98 are generally displaced from the true location in the same direction as those derived from SP12, although by longer distances. This suggests that the amplitudes of large velocity anomalies are not as well recovered in BDP98, although the lateral positions of the anomalies are similar to SP12.

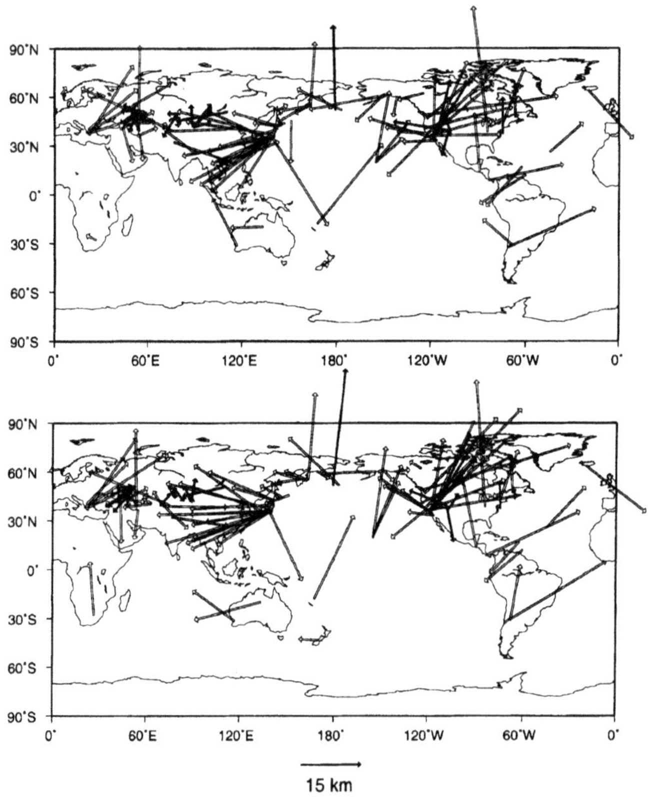

Figure 4

Mislocation vectors for the test events using the complete ISC *P*-wave travel time set, for models SP12 (*top*) and BDP98 (*bottom*) without station corrections. Length of the vectors is proportional to the magnitude of the mislocation. The base of each arrow is plotted at the ground-truth location and each vector points in the directon of the model-derived location. Explosion events are plotted as the solid vectors and earthquakes as the open vectors. Events located with BDP98 are generally mislocated in the same direction as with SP12, but with a larger error.

One area in which the location difference between BDP98 and SP12 is especially large is in the western United States. It is interesting to compare residuals predicted for these two models for events in this region. Although the areas for which the two models predict particularly large residuals are similar (Fig. 5), in general the amplitudes of the residuals predicted by SP12 are larger. The residual pattern predicted by SP12 markedly better matches the wide variation in the observed residuals (Fig. 6). In contrast, the residuals predicted by both models for events at the Semipalatinsk nuclear test site are notably more similar, although differences (notably in North Africa and Australia) do exist. As a result the locations derived from both models for events in central Asia are quite similar.

One expected result of using 3-D models containing larger numbers of free parameters is a better fit to the travel times. Figure 7 shows the rms residual obtained for the test events using each of the 3-D models. SP12 and MK12 produce similar

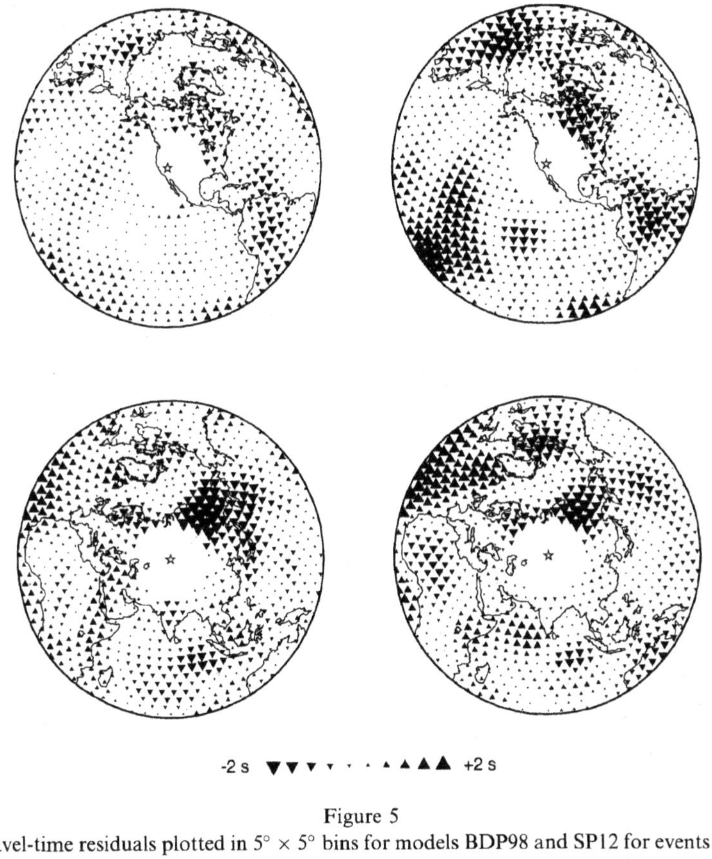

Figure 5
Predicted travel-time residuals plotted in 5° × 5° bins for models BDP98 and SP12 for events located at the Nevada and Semipalatinsk nuclear test sites. Each triangle represents the value of the mantle travel-time correction in the center of the bin after removal of the mean value over all bins. Top two maps are for an event at the Nevada test site and the bottom two maps are for an event at the Semipalatinsk test site. Maps on the left are for BDP98 and those on the right are for SP12.

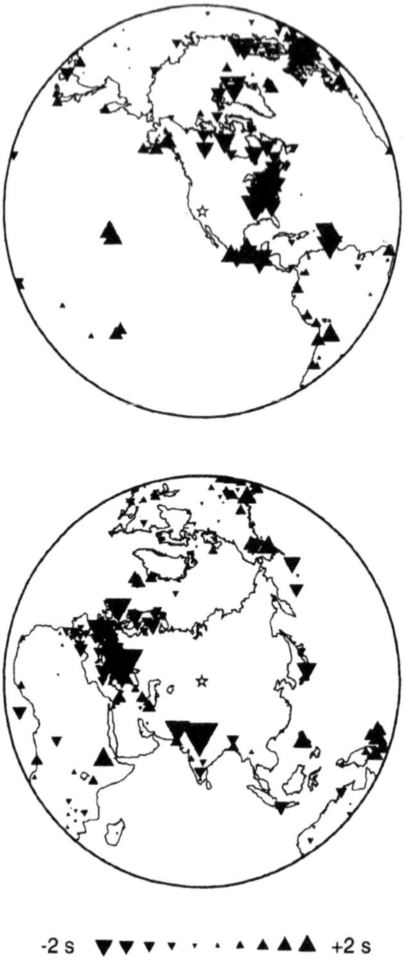

Figure 6
Mean observed P-wave travel-time residuals for explosions at the Nevada (*top*) and Semipalatinsk (*bottom*) test sites. The value plotted is the difference between the observed travel time and the value predicted for PREM plus all additional corrections except for the mantle correction. The mean of all of the values has been subtracted from each data point. Residuals with a magnitude larger than 5 s have been excluded.

magnitude residuals, whereas the rms residual for the block models is 0.3–0.4 s smaller. We also note that the origin times are on average better fits using the block models, yet the block models do not improve the fit to the location parameters over the other models. Smaller residuals alone should not be taken as an indication of the reliability of locations.

We then applied the empirical station corrections described above in the locations. Table 2 lists the average mislocation obtained for all test events, using three types of station corrections as well as no station corrections. The station

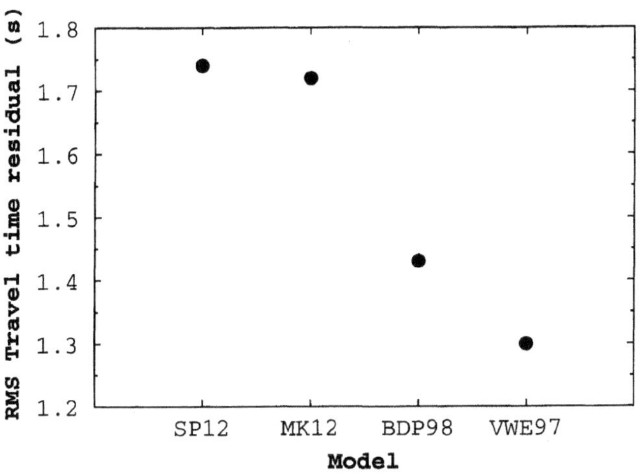

Figure 7
The rms travel-time residual remaining after relocation of the test events using each of the 3-D models and all available phase observations. The mean value was removed from each residual before calculating the rms. The rms residual values for the two models parameterized by blocks (BDP98 and VWE97) are significantly smaller than for the other two models.

Table 2

Average mislocation in km for all events using various types of station corrections.

Model	No corrections	Invariant corrections	5° bins	2.5° bins
SP12	11.71	12.22	11.75	11.73
MK12	13.11	13.58	13.12	13.09
BDP98	13.06	13.46	13.09	13.06
VWE97	13.72	14.07	13.73	13.69

Bins for source region-dependent corrections are either 5° (column 4) or 2.5° (column 5) wide in epicentral distance.

correction terms are in addition to the corrections for the 3-D model and the crust. The third column of Table 2 displays results obtained when station corrections which are not dependent on the source region of the event are used. The last two columns list the average mislocation obtained when the station corrections are calculated using source bins with a width of 5° in azimuth and either 5° or 2.5° in epicentral distance from each station. The bins are non-overlapping.

The invariant station corrections produce mislocations which are consistently larger for all 3-D models. It is quite clear that mixing correction terms for events in all source regions is not appropriate. Somewhat better results are found when using the region-dependent corrections. However, the improvement in the average mislocation is still negligible. This is most likely due to the fact that for only a fraction of the phases used for each event is there enough information to constrain the station correction term. As mentioned above, the correction term was set to zero if there are fewer than four events located within the source bin. The results listed in Table 2 were obtained utilizing all available *P* waves, regardless of whether a station correction term was available. For most of the events, there are insufficient stations with nonzero correction terms to significantly affect the location error. The lack of improvement may also result because most of the unmodeled structure is removed through the crustal correction. A third possibility is that the application of corrections which vary in such a rough manner is not appropriate, and that applying smooth function to the correction values may result in greater improvement to the locations, much in the manner of SCHULTZ *et al.* (1998).

Somewhat better results are obtained when restricting the data set to only those stations with nonzero correction terms. In this case, using the region-dependent correction terms and restricting the analysis to only those events with 20 or more such stations, the average mislocation is improved by about 0.5 km. The results are similar for all of the 3-D models.

Figure 8 suggests that significant improvement in location accuracy over that enabled by current 3-D models may only be achieved by taking into account structure very local to the source. We calculated "site-specific" station corrections for a group of explosions in the western U.S. and a group in central Asia. One event from each group was selected as a reference event which was relocated using model SP12 and the resulting residual taken as the correction for each station. Figure 8 shows the relocations obtained for both groups of events both with and without these corrections. This procedure resulted in significant improvement in location accuracy for those events located within 50 km of the reference event but degraded the accuracy for events only slightly more distant. Further reduction of the source bin sizes for the purpose of calculating empirical station corrections would thus be desirable although there is not enough phase data available in current event bulletins to make such an effort fruitful.

Sparse Data Sets

As noted above, it is important to examine the performance of 3-D models when locating events teleseismically with a limited number of phases. The primary seismic network of the International Monitoring System, for example, is to consist of only 50 3-component stations and arrays worldwide. Station coverage in many areas is therefore sparse. In this section we discuss location experiments using only subsets of the available *P*-wave data set.

For each of the test events we randomly selected 30 of the available phases and relocated the event according to the procedure described above. We repeated this

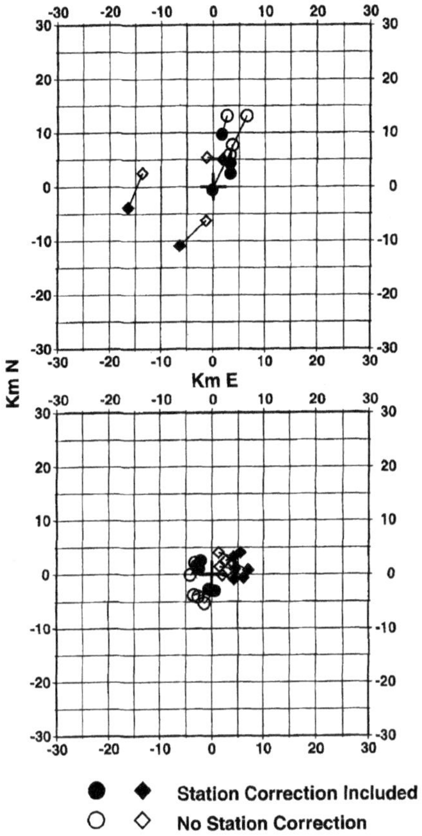

Figure 8
Map grids showing relocation of explosions using site-specific station corrections. Top map shows relocation of explosions in the western U.S. and the bottom map from the former Soviet Union. In each case, a set of station corrections was computed applying the residuals resulting from relocation of one of the events in model SP12. Test events in the same region were then relocated both with (closed symbols) and without (open symbols) these station correction terms. Circles correspond to those events located in the immediate vicinity (within 50 km) of the reference events, while the diamonds correspond to events located up to 5° distant. The ground-truth location for each event is assumed to be at the cross located in the center of the grid. In each case the location accuracy obtained for the closer events is improved while that for the more distantly located events is degraded.

procedure 100 times for each event. No consideration with regard to the azimuth or epicentral distance of the reporting station was made when selecting the phases, except that the distance was restricted to between 25° and 96° as before. A location trial is deemed to be "successful" if it results in a location within the 1000 km^2 circular area surrounding the actual location specified by the CTBT. Formal calculation of error ellipses often results in areas of less than 1000 km^2 uncertainty which may not include the ground-truth epicenter. However, for the purposes of the CTBT, the area set aside for on-site inspection would include the actual epicenter in these cases.

In addition to relocating the events in this manner using only the model travel-time corrections, we also computed new station corrections using only the set of test events. We divided the test events into groups containing three or more events, and from each group selected a reference event, usually the event with the largest number of recorded phases. Each group consisted of events located 500 km or less from the reference events. We then take as the station correction the residual remaining after relocation of the reference event in the 3-D model. These corrections are referred to as "model-based" corrections. Station corrections were also calculated using the ground-truth location for each event and are referred to as "ground-truth" corrections. This procedure was thus similar to that described in Figure 8 except that we limited the available phases to only those with a nonzero correction (i.e., to those stations recording the reference event). However, since the stations reporting nearby test events are similar, most of the available stations have a correction for a given event. This procedure allowed computation of station corrections for 69 of the total of 112 test events.

Figure 9 presents the results of this process for an explosion in the Ural mountains. Results of location trials using model SP12 are compared with those using PREM. Each closed circle is the result of one location trial using 30 *P*-wave phases, and the model used in the relocations is displayed at the top of the appropriate map. For the top two panels in Figure 9, we used no station corrections in addition to the models indicated. The two bottom grids depict results using SP12 and either the model-based or the ground-truth station corrections. Each level of complexity in the model which is employed results in better locations. For PREM the average location error is about 10 km and about 5 km for SP12. Using SP12, most of the location trials lie within the 1000 km^2 objective of the CTBT. The use of station corrections aids removal of both the scatter and the bias in the location trials.

Results of the location trials are displayed geographically in Figures 10–14 for the four 3-D models and for PREM. The top map in each figure displays the percentage of the location trials which satisfies the CTBT objective for the trials using 30 phases without station corrections. The bottom map displays the same for trials with use only 8 phases from each event (250 trials per event). As can be seen from these figures, the quality of the locations seems to vary randomly with geographic location. Events in central Asia are consistently well-located, with generally above 60% of the location trials satisfying the treaty criterion. Most of these events are explosions. The areas showing particularly less satisfying results are western North America (including the Nevada Test Site explosions) and southern Europe. Again these maps suggest a large effect resulting from structure close to the sources, as the results are often quite different for events located near to each other.

Results for three of the four 3-D models are quite similar and provide noticeable improvement over PREM in most areas. Model VWE97 gives fewer successful trials than the other 3-D models. The percentage of successful trials is, however, considerably degraded using only 8 phases. Few of the events have as many as

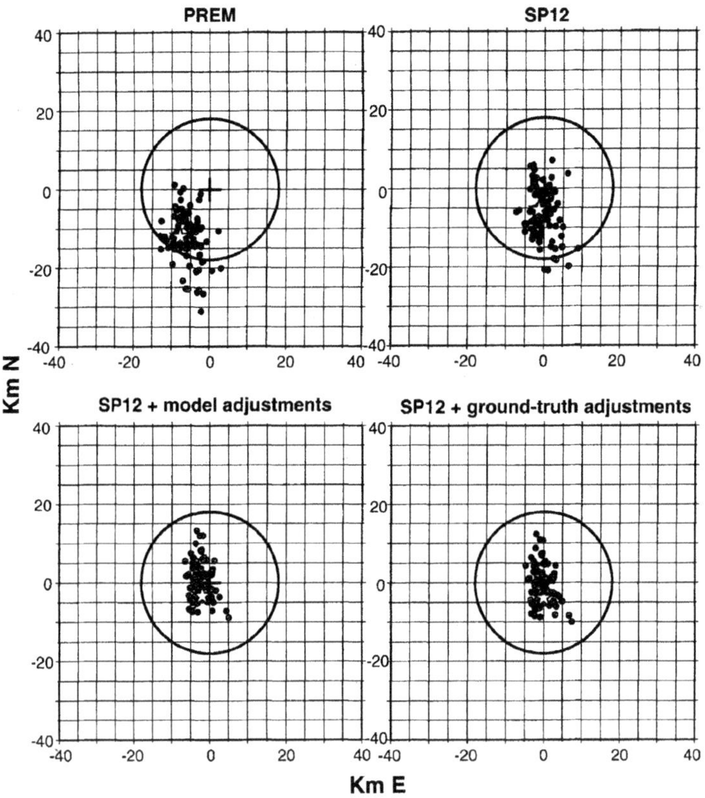

Figure 9
Results of 100 location trials, each using only 30 *P*-wave observations, for a single explosion located near the southern Ural mountains. Each grid is a map-view with the ground-truth location denoted by the black cross at the center. The velocity model used for each set of trials is shown at the top of the grid. The grid at top left is for PREM with no station corrections; that at the top right for SP12 with no station corrections; the lower-left grid is for SP12 with model-based station corrections; and the lower-right grid is for SP12 with ground-truth corrections. Each filled circle depicts the location resulting from one of the trials. The white stars indicate the locations obtained from all available phases. The black circles delineate an area of 1000 km^2 surrounding the ground-truth location.

60% of the location trials which satisfy the treaty criterion regardless of the model used.

The breakdown of the location trials is summarized by Tables 3 and 4 and Figure 15. When using no station corrections and 30 total phases, the percentage of successful trials is 65–70%, and the average mislocation is only slightly greater than that obtained using all phases (Table 2). The model-based station corrections produce improvement in the average mislocation although not in the number of successful trials. The ground-truth corrections produce significant improvement in both average location and percentage of successful trials. If ground-truth information is available, then an event can be located at a confidence level of

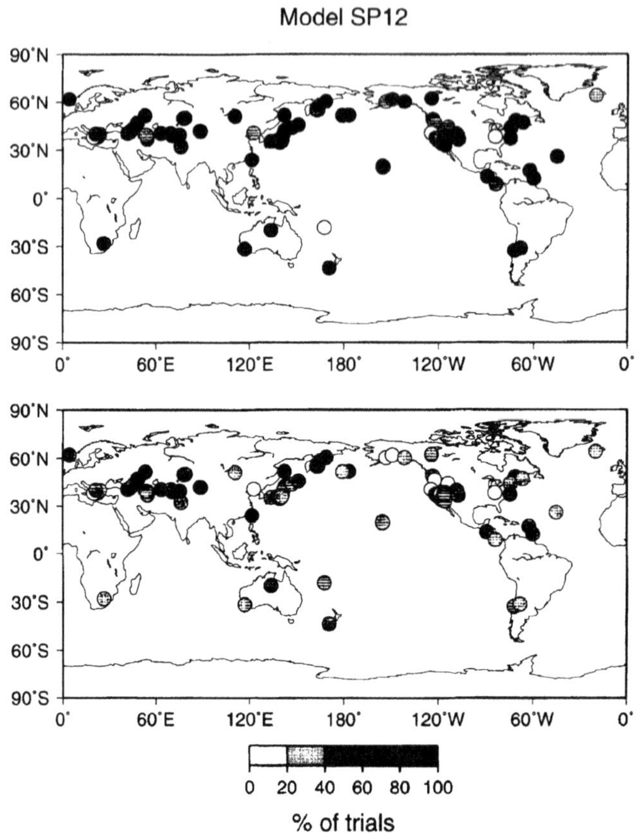

Figure 10
Maps showing the percentage of trials resulting in a location within an area of 1000 km^2 about the ground-truth location for each of the test events. The degree of gray-shade for each event is proportional to the number of successful trials. The top map is for the trials using 30 phases and the bottom map for those using 8 phases. The model used was SP12 with the CRUST5.1 crustal corrections and no empirical station corrections.

75–80% within the goal of the CTBT. This is even true when using PREM as the location model, which points to the importance of calibration information, if available. The ground-truth corrections reduce the difference between the results for the different models. If only 8 phases are used in the inversion, fewer than half of the trials are successful and the average mislocation exceeds 30 km for all the 3-D models.

Figure 15 displays the number of trials with location errors in 5-km bins as well as greater than 20 km. Again, the distribution of the location trials is similar for three of the 3-D models with VWE97 resulting in slightly larger errors. Use of ground-truth information results in ~50% of the trials yielding a location error of less than 10 km (with 30 phases). In this case, the distribution of the location errors using PREM is

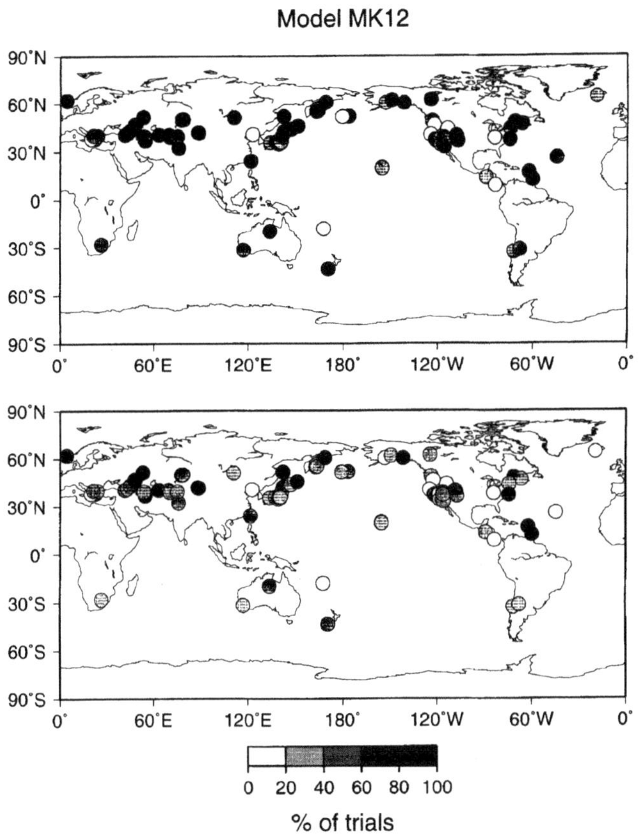

Figure 11
Similar figure to Figure 10 using model MK12.

almost exactly the same as using one of the 3-D models. In the absence of calibration information, however, the quality of the locations derived from the 3-D models is clearly superior. Without station corrections, the median mislocation is roughly 12 km using 30 phases, and 20 km using 8 phases.

Discussion

Enough data seem to be currently available to conclude that, while all 3-D models of mantle velocity seem to provide substantial improvement over 1-D models in the ability to locate teleseismic events, the degree of improvement does not necessarily increase with the complexity of the model. Since it appears that neither the parameterization chosen for the tomographic inverse problem, nor the particular inversion technique used greatly affect the solution obtained (BOSCHI

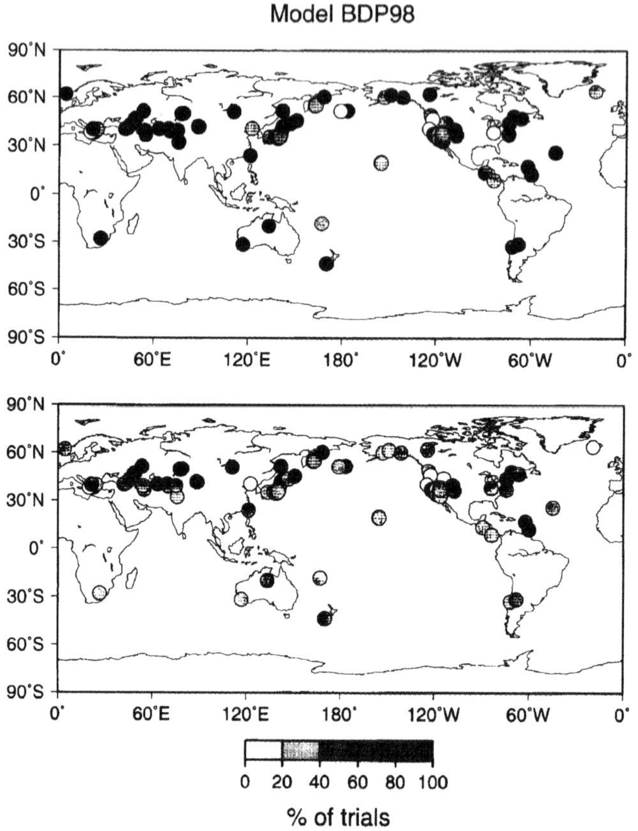

Figure 12
Similar figure to Figure 10 using model BDP98.

and DZIEWONSKI, 1999a), an alternative explanation for this result should be invoked. For example, the type of regularization and damping employed in the inversion for block models with nonuniform data coverage might result in substantially lower amplitude anomalies in regions where the coverage is relatively low. This might lead to a tendency for block models to underpredict the range of observed residuals for events in some regions, such as we observe in Figure 5.

Another possibility which requires full investigation is the effect of multiple data sets to constrain the *P*-wave velocity in the mantle. Many spherical harmonic models of mantle velocity such as SP12 and MK12 have been obtained using a variety of absolute and differential times as well as waveform data. However, present limitations of computer memory prevent the use of such large data sets when inverting for models of high resolution. *P*-wave travel times are largely sensitive to structure in the mid- and lower mantles whereas the combination of travel times and

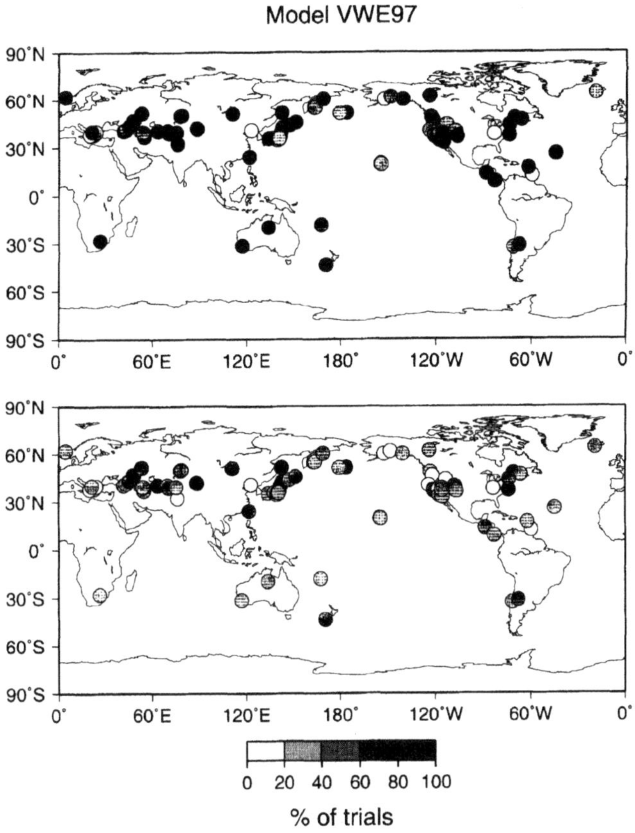

Figure 13
Similar figure to Figure 10 using model VWE97.

waveforms may lead to more even resolution with depth. Nevertheless, it would seem rather unlikely that the phases recorded for the test events used here are overly sensitive to areas of the mantle where resolution is relatively poor for the two block models. Most of the events occur in regions where nearby seismicity is relatively abundant.

The fact that the block models produce substantially better fits to the ISC residuals, while failing to improve the quality of teleseismic locations, also brings into question the quality of the data, particularly in the early years. The effect of poor phase picks may limit the ability to accurately locate events and the production of more detailed models for this purpose may prove futile. The mapping of earthquake mislocations into the production of tomographic models therefore probably does not decrease with increasing model complexity.

To some extent, the results from the location trials with sparse data sets are encouraging in that small to moderate teleseismic events may be located to within the

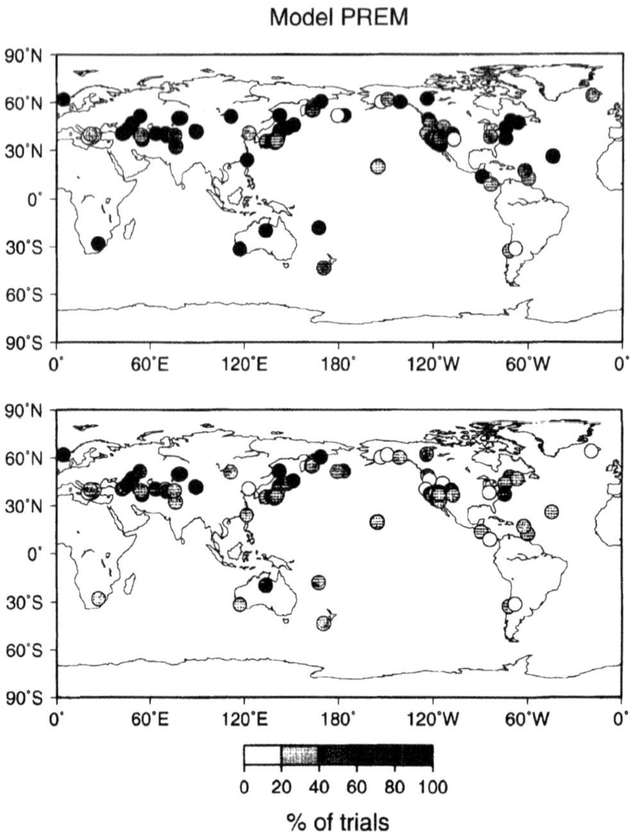

Figure 14
Similar figure to Figure 10 using PREM.

Table 3

Average mislocation in km for location trials using a random selection of 30 phases. 100 trials were computed for each event.

Model	No corrections	Model-based corrections	Ground-truth corrections
SP12	14.32 (71.5%)	13.74 (70.8%)	11.02 (79.1%)
MK12	15.24 (67.7%)	14.30 (68.3%)	10.97 (78.8%)
BDP98	14.80 (69.3%)	14.27 (68.8%)	10.79 (79.6%)
VWE97	16.49 (64.9%)	16.11 (64.2%)	11.88 (75.8%)
PREM	17.42 (61.8%)	...	11.23 (78.8%)

See text for description of station corrections. Numbers in parentheses are the percentage of location trials falling within a circular area of 1000 km^2 surrounding the ground-truth location.

specified 1000 km^2 area for the CTBT about two-thirds of the time. The median mislocation in such circumstances is around 12 km. This is true even if calibration information is not available for any of the stations. Further improvement may be

Table 4

Average mislocation in km for location trials using a random selection of 8 phases. 250 trials were computed for each event.

Model	No corrections	Model-based corrections	Ground-truth corrections
SP12	38.79 (41.3%)	33.10 (52.7%)	31.06 (56.6%)
MK12	36.90 (41.9%)	32.74 (52.4%)	32.50 (57.3%)
BDP98	33.87 (44.5%)	31.10 (52.8%)	28.62 (58.3%)
VWE97	37.81 (36.9%)	32.65 (49.2%)	29.26 (56.4%)
PREM	42.57 (34.6%)	...	29.16 (56.8%)

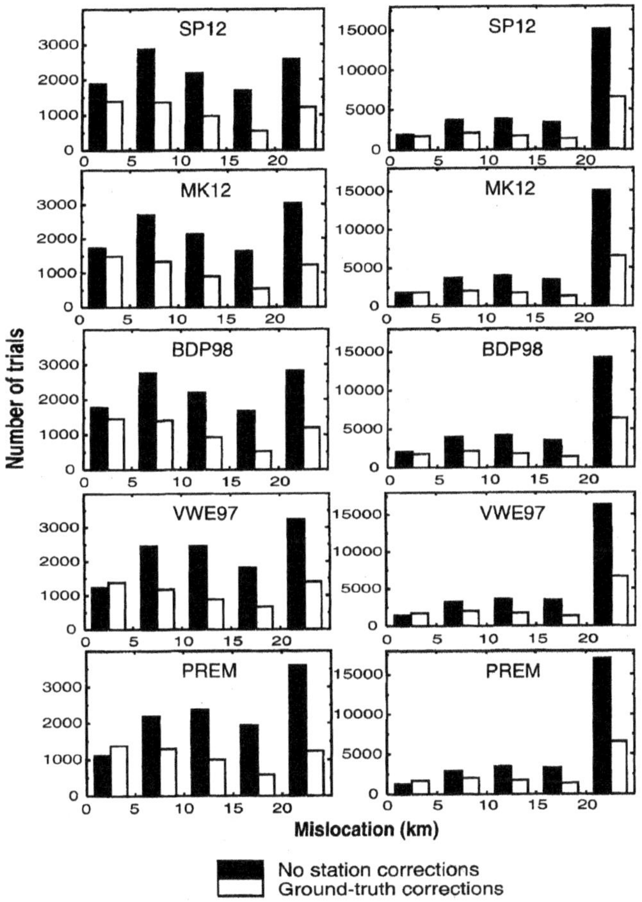

Figure 15

Histograms depicting the distribution of location errors for the trials using 30 phases (*left*) and 8 phases (*right*) for all the 3-D models and for PREM. The height of each bar represents the number of location trials resulting in a mislocation within a 5-km wide bin, except that the last bin contains all location trials with a mislocation larger than 20 km. Filled bars show results for location trials using no empirical station corrections while the open bars are for the trials using the ground-truth corrections. The open bars do not include results for the reference events used to calculate the ground-truth corrections.

possible if care is taken to ensure that sufficient observations are used from all azimuthal quadrants, which we have not done in this study. The probability of achieving this accuracy is increased to over 75% if calibration information is available.

On the other hand, for very small events only a handful of observations may be available. In such cases it probably will be impossible to achieve adequate azimuthal coverage. The results demonstrate that calibration information is essential if one is to achieve the desired accuracy with a confidence level of even 50%. For these events the addition of regional phases is likely to be essential, which places importance on accurate mapping of P_n velocities (e.g., SMITH and EKSTRÖM, 1999). Due to their frequency content and low background noise level, first-arriving *P* waves are likely to remain the most useful data for teleseismic event location, although *S* waves, *pP* and core phases provide important constraints on depths (ENGDAHL *et al.*, 1998). The use of azimuth data may hold further promise. Such data are already used by the Prototype International Data Center in producing their event bulletin (BONDÁR *et al.*, 1998) and have already been shown to substantially improve the accuracy of regional locations [DREGER *et al.*, 1998; BRATT and BACHE, 1988). Three-dimensional models may be used to provide corrections for the azimuth data in the absence of calibration information for certain stations.

Accurate corrections for crustal structure are seen to considerably improve prospects for teleseismic event location. Further improvement and refinement of existing crustal models is certainly desirable and may in the future lessen the need for empirical station corrections to teleseismic phases. Based on Figure 2, it is likely that most of the unmodeled structure resides in the crust and upper mantle directly beneath the station. In this regard, we also believe that the incorporation of anisotropy into 3-D tomographic modeling may further improve the utility of teleseismic *P* waves in event location (BOSCHI and DZIEWONSKI, 1999b). Both the upper mantle (e.g., BEGHOUL and BARAZANGI, 1990; SMITH and EKSTRÖM, 1999) and perhaps the area near the CMB likely contains *P*-wave anisotropy on the order of several percent, however its nature and how it is best parameterized in tomographic modeling is as yet unclear.

Conclusions

In summary, we have used a data set of explosions and earthquakes with well-known locations to test the accuracy of teleseismic event locations obtained using four different 3-D models of *P*-wave velocity in the mantle. The 3-D models include two parameterized horizontally in terms of spherical harmonic functions reaching degree 12 and two which are represented in terms of constant velocity blocks. Thus the latter two models provide better horizontal resolution. However, the two block models do not improve upon the accuracy of the locations. Using all the available teleseismic *P*-wave observations in the ISC catalog, the average mislocation obtained for the test events using the two block models is 2–3 km larger than the best of the spherical harmonic models, model S&P12/WM13 of SU and DZIEWONSKI (1993). This occurs despite the fact that the block models provide a better fit to the travel times. From analysis of the mantle corrections to the travel times, it appears that existing block models may underestimate the amplitudes of long-wavelength travel-time anomalies.

The largest effect on the accuracy of the locations appears to originate from structure in the crust or upper mantle near the sources. Use of the crustal model of MOONEY *et al.* (1998) to correct travel times on both the source and receiver side results in a large reduction in the mislocation over an earlier, less detailed crustal model. Source-region dependent station corrections calculated for the 3-D models have little effect on the locations unless the source bins are very small (much less than 5° in dimension). Current event bulletins probably do not contain enough data to adequately constrain empirical corrections in such small bins, so as to significantly improve upon locations derived from mantle 3-D models and crustal corrections.

In the second half of the paper we tested the ablity of the 3-D models to provide accurate locations for small to moderate events recorded by only a limited number of phases. For each of the test events we performed 100 separate location trials applying a different random selection of 30 phases and 250 separate trials using 8 phases. Again, the block models do not improve the outcome of the trials over the spherical harmonic ones. Using 30 phases, approximately two-thirds of the trials result in a location within the 1,000 km^2 area specified for on-site inspection under the CTBT. This percentage decreases to less than half using only 8 phases. If ground-truth information is available to correct the travel times, however, the percentage of trials which result in a location within the specified area is some 79% and and 57% using 30 and 8 phases, respectively. Adopting PREM to locate the test events along with ground-truth calibration is equally effective as using any of the 3-D models. At present, however, ground-truth information is available for only limited regions of the globe and for seismic stations that have remained in extended operation.

Acknowledgements

The authors wish to thank Terry Wallace for providing data on the Lop Nor explosions and Lapo Boschi and Rob van der Hilst for access to their mantle *P*-velocity models. Comments by M. Tinker and an anonymous reviewer enhanced the final manuscript. Most of the figures were produced using the Generic Mapping Tools (GMT) software. This research was supported by the U. S. Department of Defense under contract DSWA001-97-C-0124.

References

Bratt, S. R., and Brache, T. C. (1988), *Locating Events with a Sparse Network of Regional Arrays*, Bull. Seismol. Soc. Am. *78*, 780–798.

Beghoul, N., and Barazangi, M. (1990), *Azimuthal Anisotropy of Velocity in the Mantle Lid beneath the Basin and Range*, Nature *348*, 536–538.

Bondár, I., Yang, X., Wang, J., Bahavar, M., Israelsson, H., and McLaughlin, K., *Tuning and calibration activities at the PIDC*, Proc. 20th Annual Seismic Res. Symp. on Monitoring a Comprehensive Test-Ban-Treaty, Santa Fe, NM (J. Fantroy *et al.*, eds.) 1998, 1–10.

Boschi, L., and Dziewonski, A. M. (1999a), *High and Low-resolution Images of the Earth's Mantle: Implications of Different Approaches to Tomographic Modeling*, J. Geophys. Res. *104*, 25,567–25,594.

Boschi, L., and Dziewonski, A. M. (1999b), *Whole Earth Tomography from P, PcP, and PKP Delay Time Measurements: The Possibility of Laterally Varying Radial Anisotropy Throughout the Mantle*, EOS *80*, S13.

Dziewonski, A. M. (1984), *Mapping the Lower Mantle: Determination of Lateral Heterogeneity in P Velocity up to Degree and Order 6*, J. Geophys. Res. *89*, 5929–5952.

Dreger, D. S., Uhrhammer, R., Pasyanos, M., Franck, J., and Romanowicz, B. (1998), *Regional and Far-regional Earthquake Locations and Source Parameters Using Sparse Broadband Networks: A Test on the Ridgecrest Sequence*, Bull. Seismol. Soc. Am. *88*, 1353– 1362.

Ekström, G., Dziewonski, A. M., and Smith, G. P. (1997), Strategies for improving seismic event location at regional and teleseismic distances. In *Proc. 19th Annual Seismic Res. Symp. on Monitoring a Comprehensive Test-Ban-Treaty* (M. J. Shore *et al.*, eds.) 221–229.

Engdahl, E. R., van der Hilst, R. D., and Buland, R. P. (1998), *Global Teleseismic Earthquake Relocation with Improved Travel Times and Procedures for Depth Determination*, Bull. Seismol. Soc. Am. *88*, 722–743.

Grand, S. P., van der Hilst, R. D., and Widiyantoro, S. (1997), *Global Seismic Tomography: A Snapshot of Convection in the Earth*, GSA Today *7*(4), 1–7.

Jeffreys, H., and Bullen, K. E., *Seismological Tables* (British Association for the Advancement of Science, London, 1958).

Kennett, B. L. N., and Engdahl, E. R. (1991), *Travel Times for Global Earthquake Location and Phase Identification*, Geophys. J. Int. *105*, 429–465.

Kennett, B. L. N., Engdahl, E. R., and Buland, R., (1995), *Constraints on Seismic Velocities in the Earth from Travel Times*, Geophys. J. Int. *122*, 108–121.

Li, X., and Romanowicz, B. (1996), *Global Mantle Shear Velocity Model Developed Using Nonlinear Asymptotic Coupling Theory*, J. Geophys. Res. *101*, 22,245–22,272.

Mooney, W. D., Laske, G., and Masters, T. G. (1998), *CRUST 5.1: A Global Crustal Model at* 5° × 5°, J. Geophys. Res. *103*, 727–747.

Schultz, C. A., Myers, S. C., Hipp, J., and Young, C. J. (1998), *Nonstationary Bayesian Kriging: A Predictive Technique to Generate Spatial Corrections for Seismic Detection, Location, and Identification*, Bull. Seismol. Soc. Am. *88*, 1275–1288.

Smith, G. P., and Ekström, G. (1996), *Improving Teleseismic Event Locations Using a Three-dimensional Earth Model*, Bull. Seismol. Soc. Am. *86*, 788–796.

Smith, G. P., and Ekström, G. (1999), *A Global Study of* P_n *Anisotropy beneath Continents*, J. Geophys. Res. *104*, 963–980.

Su, W., and Dziewonski, A. M. (1993), *Joint 3-D Inversion for P and S Velocity in the Mantle*, EOS *74*, 557.

Su, W., Woodward, R. L., and Dziewonski, A. M. (1994), *Degree-12 Model of Shear Velocity Heterogeneity in the Mantle*, J. Geophys. Res. *99*, 6945–6980.

Su, W., and Dziewonski, A. M. (1997), *Simultaneous Inversion for 3-D Variations in Shear and Bulk Velocity in the Mantle*, Phys. Earth Planet. Inter. *100*, 135–156.

van der Hilst, R. D., Widiyantoro, S., and Engdahl, E. R. (1997), *Evidence for Deep Mantle Circulation from Global Tomography*, Nature *386*, 578–584.

VASCO, D. W., and JOHNSON, L. R. (1998), *Whole Earth Structure Estimated from Seismic Arrival Times*, J. Geophys. Res. *103*, 2633–2671.

WALLACE, T. C., and TINKER, M. A. (2001), *Locations and Yields of Lop Nor Underground Nuclear Explosions*, Pure appl. geophys., this issue.

(Received August 5, 1999, revised January 4, 2000, accepted February 9, 2000)

Pure appl. geophys. 158 (2001) 319–347
0033–4553/01/020319–24 $ 1.50 + 0.20/0

Pure and Applied Geophysics

Improving Seismic Event Location: An Alternative to Three-dimensional Structural Models

C. Piromallo[1] and A. Morelli[1]

Abstract—We devise and apply a method to account for the effect of the aspherical structure of the Earth in locating earthquakes. This technique relies upon the ability to detect the average structural signal present in the residuals between source and receiver and correct for this signal during location, using a phenomenological description that we call Empirical Heterogeneity Corrections (EHC). EHC are employed in the relocation of a large set of well-constrained teleseismic earthquakes selected among the events reported by the Bulletins of the International Seismological Centre 1964–1995. The rms length of EHC relocation vectors for these events is about 10 km. The method is also tested against a selected set of ground-truth events, both earthquakes and explosions, whose locations are independently known by nonseismic means. The rms length of the mislocation vectors for the test events, compared to their original mislocation in the reference 1-D model SP6, is reduced in the EHC relocation by 17% for explosions and 12% for earthquakes. Our technique provides a successful alternative to the use of 3-D structural models, approximately reaching the same value of effectiveness in improving event location.

Key words: Earthquake location, P waves, travel time, teleseismic, velocity anomaly.

1. Introduction

The accuracy of predicted seismic travel times is crucial to the determination of earthquake hypocentral parameters. Location procedures, both at global and regional scale, often employ one-dimensional (1-D) velocity models to compute theoretical travel times. However, lateral variations in seismic wave velocity can easily reach values of a few percent, particularly near subduction zones. The use of globally-averaged travel times, which neglect departures from spherical symmetry in wave velocity, may cause systematic location bias.

The observation of the pattern of systematic discrepancy between observed travel times and those computed for a reference Earth model, led to the development of techniques to account for geographical variations of wave propagation properties during earthquake location. Classical methods range from the use of station

[1] Istituto Nazionale di Geofisica, Rome, Italy.

correction terms of different order (CLEARY and HALES, 1966; HERRIN and TAGGART, 1968; LILWALL and DOUGLAS, 1970; DZIEWONSKI and ANDERSON, 1983; ZHOU and WANG, 1994), representing the contribution of the average upper mantle structure beneath the receiver (SOURIAU and WOODHOUSE, 1985; ROBERTSON and WOODHOUSE, 1997), to relative location algorithms, such as the Joint Hypocenter Determination (FREEDMAN, 1967; DOUGLAS, 1967) or the master event technique (e.g., FITCH, 1975).

More recent studies explicitly consider departures of the Earth from spherical symmetry when computing theoretical travel times. Some authors perform an iteration of the step-wise sequence of event location and structure determination by tomographic techniques (DZIEWONSKI, 1984), while others propose tomographic joint inversion for three-dimensional (3-D) Earth structure, hypocentral parameters and station corrections, which ideally provide the most complete method for taking lateral variation of travel times into account (SPAKMAN and NOLET, 1988; SPAKMAN *et al.*, 1993; VAN DER HILST *et al.*, 1993; BIJWAARD *et al.*, 1998). SMITH and EKSTRÖM (1996) demonstrated that 3-D *P*- and *S*-wave models offer significant improvement in event location over all the 1-D models considered. Nevertheless , in tomographic imaging the effort to locate earthquakes is closely connected to the problem of determining model parameters for 3-D Earth structure, causing a direct trade-off between the amplitudes of imaged velocity variations and the size of relocation vectors. Different data and modelling techniques produce significant differences among proposed tomographic Earth models, and the choice of such a 3-D model is therefore not unique (e.g., BIJWAARD *et al.*, 1998; BOSCHI and DZIEWONSKI, 1999).

For the sole purpose of improving event location, a tomographic 3-D Earth structural model may not be necessary. PIROMALLO and MORELLI (1998), in a study focused on the Mediterranean region, showed that a detailed statistical scheme of dealing with travel-time heterogeneity can be just as efficient as a 3-D mantle velocity model. They developed a procedure which does not require any explicit tomographic inversion, therefore avoiding most sources of model instability, and is computationally very efficient and straightforward to implement. This procedure improved source location yielding a rms misfit reduction by more than 20%, according to their estimate. A similar approach was also followed by FIRBAS *et al.* (1997, 1998), in an effort to calibrate the Comprehensive Test-Ban-Treaty (CTBT) International Monitoring System (IMS), who suggested the use of travel-time corrections from maps based on a 3-D velocity model, called Source-Station Specific Corrections (SSSC). In this study, we follow the approach of PIROMALLO and MORELLI (1998), extending our method to make it suitable for location of earthquakes on a global scale, using teleseismic observations of the data set reported by the Bulletins of the International Seismological Center (ISC). Geographically-averaged summary residuals with respect to a reference model provide the basis for empirically modeling aspherical variations in wave propagation during the subsequent relocation stage.

To estimate the improvement in location accuracy, we check the obtained locations against those given by a small data set of ground-truth test events and explosions, and compare them to those obtained with a standard 1-D location.

2. Data Processing

2.1. Event Selection and Location Procedure

We consider teleseismic P-wave arrivals of global events reported by the ISC Bulletins for the time period 1964–1995. We use all first arrivals with epicentral distance in the range 30° to 90°, to avoid upper mantle triplications of the travel-time curve and core-mantle boundary diffractions. The data selection criteria, to reasonably limit bias possibly due to uneven path distribution, are the following: we only retain shallow events (depth ≤50 km), we require each event to have a minimum of 30 teleseismic reported arrivals, distributed in at least 6 out of 8 adjacent azimuthal sectors, the largest teleseismic azimuthal gap being, at worst, 180° (which is comparable with the criterion suggested by ENGDAHL (1998)). The bottoming depths for these paths are located in the lower mantle, deeper than 700 km.

We reprocess all selected earthquakes, using the radial global Earth model SP6 (MORELLI and DZIEWONSKI, 1993), to remove the systematic bias in hypocentral parameters due to the ISC location procedure based on JB travel-time tables (JEFFREYS and BULLEN, 1940). The location is performed by means of an iterative least-squares optimization of first arrival residuals, with starting epicenters and fixed source depth given by the ISC report, including ellipticity (DZIEWONSKI and GILBERT, 1976) and station elevation corrections. Let r_{ij} be the residual of the P arrival of event i recorded by station j. For each event i the location algorithm adjusts origin time and epicentral coordinates $(\tau_i, \lambda_i, \phi_i)$ to minimize residuals r_{ij} for those stations which observed event i (i.e., for all $j \in \mathcal{S}^i$). Following the uniform reduction scheme to account for the non-normal distribution of residuals (JEFFREYS, 1932, 1939), the location minimizes the summed squared weighted residuals:

$$\sum_{j \in \mathcal{S}^i} (w_{ij} r_{ij})^2$$

where $w_{ij} = w(r_{ij})$ and

$$w(x) = \left\{ 1 + \mu \exp\left[\frac{(x-m)^2}{2\sigma^2}\right] \right\}^{-1} \tag{1}$$

is the data weighting function specified by JEFFREYS (1932). Parameters m, σ^2 and μ represent respectively, the mean and the variance of the Gaussian distribution, and a parameter accounting for the outliers. For σ^2 and μ, slowly varying, we use average distribution parameters found by ROBERTSON and WOODHOUSE (1995) for teleseismic P residuals.

Each first arriving phase is associated to the theoretical first arrival at that distance. Tables of travel times as a function of slowness are appropriately interpolated using the τ-spline method (BULAND and CHAPMAN, 1983) to find the theoretical travel time for the reference model at any given epicentral distance. We locate over 24,000 selected events using more than 2,500,000 observations. In Figure 1 we present a simple and cumulative histogram illustrating the degree to which our preliminary location shifts the event epicenters with respect to the original ISC locations, computed predominantly with the Jeffreys-Bullen regional and teleseismic arrival times of P-wave phases (ADAMS *et al.*, 1982; VAN DER HILST and ENGDAHL, 1992). Estimates of the mean, median, rms and standard deviation for the epicenters relative to the ISC locations, are quoted in the figure caption.

2.2. Summary Rays Computation

After 1-D location, we divide the Earth's surface with a grid of 1656 roughly equal area equatorial 5° × 5° cells, 50-km thick, and compute teleseismic summary rays and residuals. A summary ray is constructed considering all individual rays connecting two cells and averaging them into one bundle, which is treated as one datum; its delay time being the average of the corresponding individual travel-time

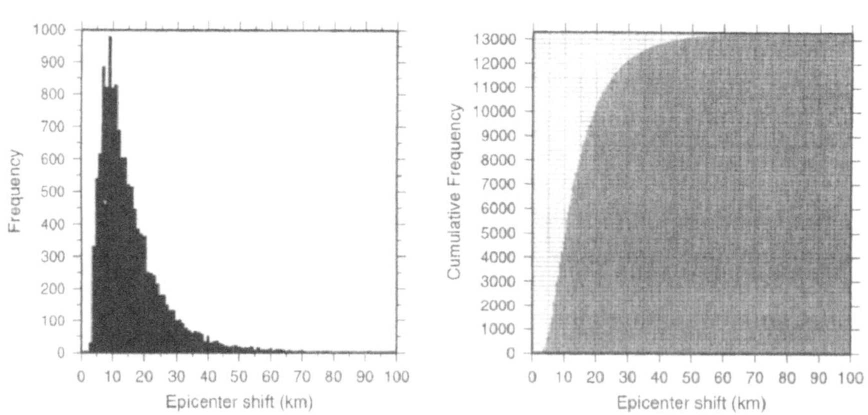

Figure 1
Histogram and cumulative plot of epicenter shifts, with respect to ISC locations, after our preliminary location with SP6 reference model. The relocation vectors are averaged over 1 km bins. The average in the epicenter relocation is 14.3 km, the median 12.5 km, the rms 16.3 km, the standard deviation 7.8 km and the maximum value about 90 km.

residuals. Since all the rays connecting the same pair of cells follow closely travelling paths in the mantle, the value of the averaged residuals is a measure of the heterogeneity encountered along the path. Summary ray formulation has been often employed in travel-time tomography (e.g., CREAGER and JORDAN, 1986; GUDMUNDSSON *et al.*, 1990; MORELLI and DZIEWONSKI, 1991; VASCO *et al.*, 1994; ROBERTSON and WOODHOUSE, 1995). Its use, attributing the same weight to each connection between source and receiver cells, allows reduction of the geographical dominance of regions with dense concentration of stations or seismic events, filtering out the effects of noise and small-scale structure which cannot be modeled, and reduction of data redundancy in tomographic inversions. For the purpose of this study, the summary ray formulation provides a simple but detailed and effective representation of the Earth's heterogeneity field.

The summary residual between cells α and β, $R_{\alpha\beta}$, is calculated by averaging individual residuals of rays from all events located in cell α ($\mathscr{E}_\alpha$ is the set of their indices) and recorded by a station in cell β ($\mathscr{S}^i_\beta$ are all stations in cell β which recorded event i), and from all events in cell β originating rays observed at a station located in cell α:

$$R_{\alpha\beta} = \frac{\sum_{i\in\mathscr{E}_\alpha}\sum_{j\in\mathscr{S}^i_\beta} r_{ij}w_{ij} + \sum_{k\in\mathscr{E}_\beta}\sum_{l\in\mathscr{S}^k_\alpha} r_{kl}w_{kl}}{\sum_{i\in\mathscr{E}_\alpha}\sum_{j\in\mathscr{S}^i_\beta} w_{ij} + \sum_{k\in\mathscr{E}_\beta}\sum_{l\in\mathscr{S}^k_\alpha} w_{kl}} \tag{2}$$

where w_{ij} is again the uniform reduction weighting function specified in (1). Obviously, $R_{\alpha\beta} = R_{\beta\alpha}$.

We obtain about 36,000 teleseismic summary rays, distributed over 872 out of the 1656 cells covering the entire Earth's surface, with multiplicity (the number of contributory rays in each bundle) reaching a few hundred in the most populated cells. Among 872 cells, 165 contain only stations, 413 only events and the remaining 294 both events and stations. Figure 2 provides an outlook to the geographical distribution of populated cells, showing the number of events/stations for each of them. About one half of the 36,000 teleseismic summary rays are only source-to-receiver (connecting an event-only cell with a station-only cell).

Some statistics relating to the summary rays and cell coverage are illustrated in the histograms of Figure 3. In Figures 3a and 3b the number of cells holding a given number of sources and receivers, respectively, is plotted. Though the distribution is very long-tailed, the most populated classes are those of cells containing less than ten sources/receivers. The minimum number of individual rays considered to form a summary ray is 1. The histogram in Figure 3c displays the number of summary rays with a given multiplicity. Summary rays with a single contributory ray are in the order of 15,000. However, we will see in Section 4 that a multiplicity of more than 5 individual rays is required to use the corresponding summary residuals as empirical corrections in our relocation scheme. The number of summary rays with 5 or more original rays is in the order of 19,000, while about

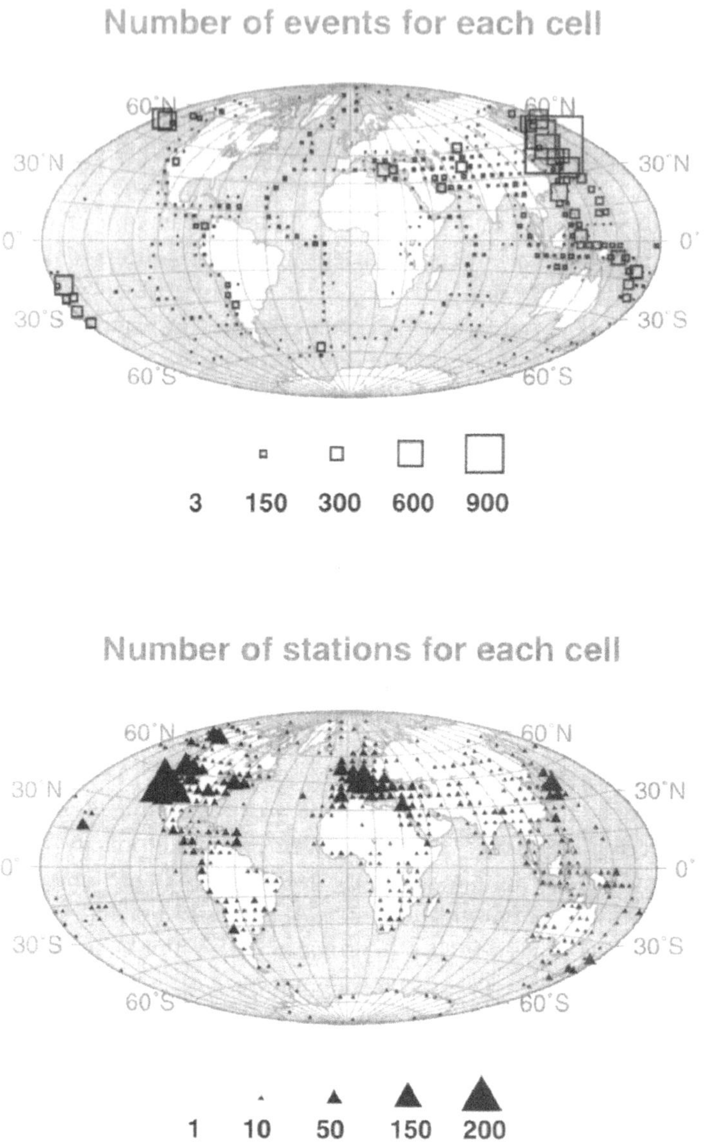

Figure 2
Geographical distribution of populated cells. The symbols size is proportional to the number of events/stations contained in each cell.

14,000 have a multiplicity of 10 or more. The last histogram, Figure 3d, illustrates the number of cells possessing a given number of summary rays: approximately 850 cells have 10 or more summary rays, the median of the distribution being approximately 150 summary rays.

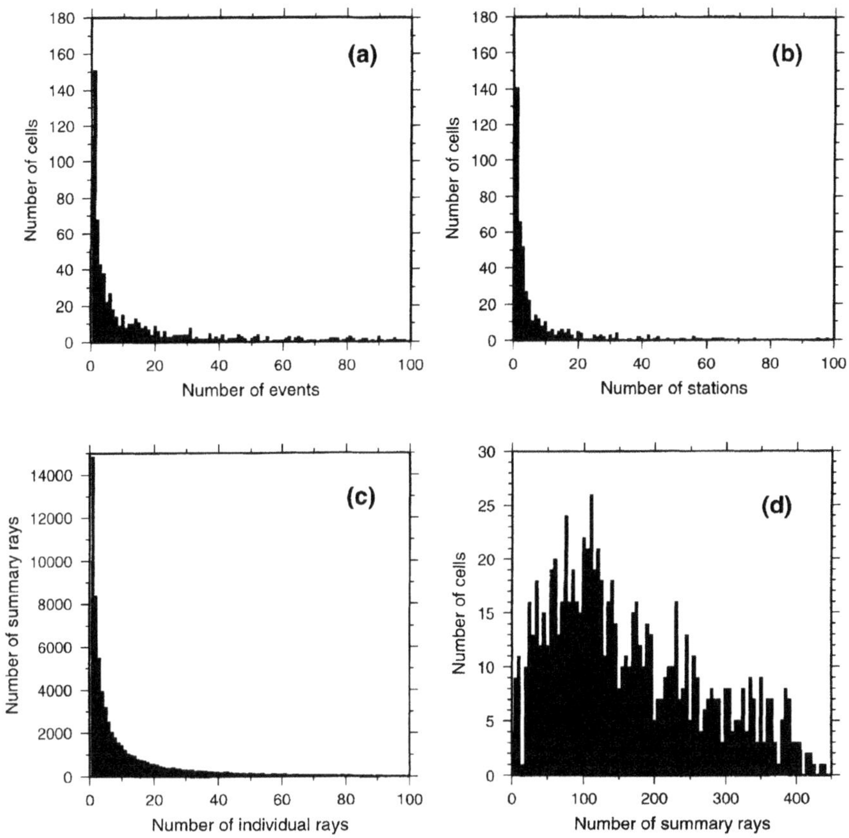

Figure 3
Histograms displaying statistics pertaining to cells and summary rays coverage: (*a*) the number of cells containing a given number of events, (*b*) the number of cells containing a given number of stations, (*c*) the number of summary rays composed by a given number of individual rays, (*d*) the number of cells having a given number of summary rays (plotted in 5 summary rays bins).

3. Data Analysis

3.1. Residual Maps

Summary residuals thus account for the delay time anomalies between each pair of cells, which often exhibit systematic features, reflecting the influence of large-scale structure heterogeneities in wave propagation. The maps in Figure 4 present examples of residuals heterogeneity as seen from six sample cells, in a representation where summary residuals are projected onto locations corresponding to the other ray tip. Each map portrays the pattern of systematic residuals that a station/event located within a particular cell, marked on the plot by a black star, would observe/produce for earthquakes/stations located far and wide. The best coverage is obviously exhibited by cells including both stations and events (*a*, *b*, *c*, *e* in Fig. 4), because their connections are both to seismically active regions and to stations. All maps display summary residuals which provide a representation of aspherical

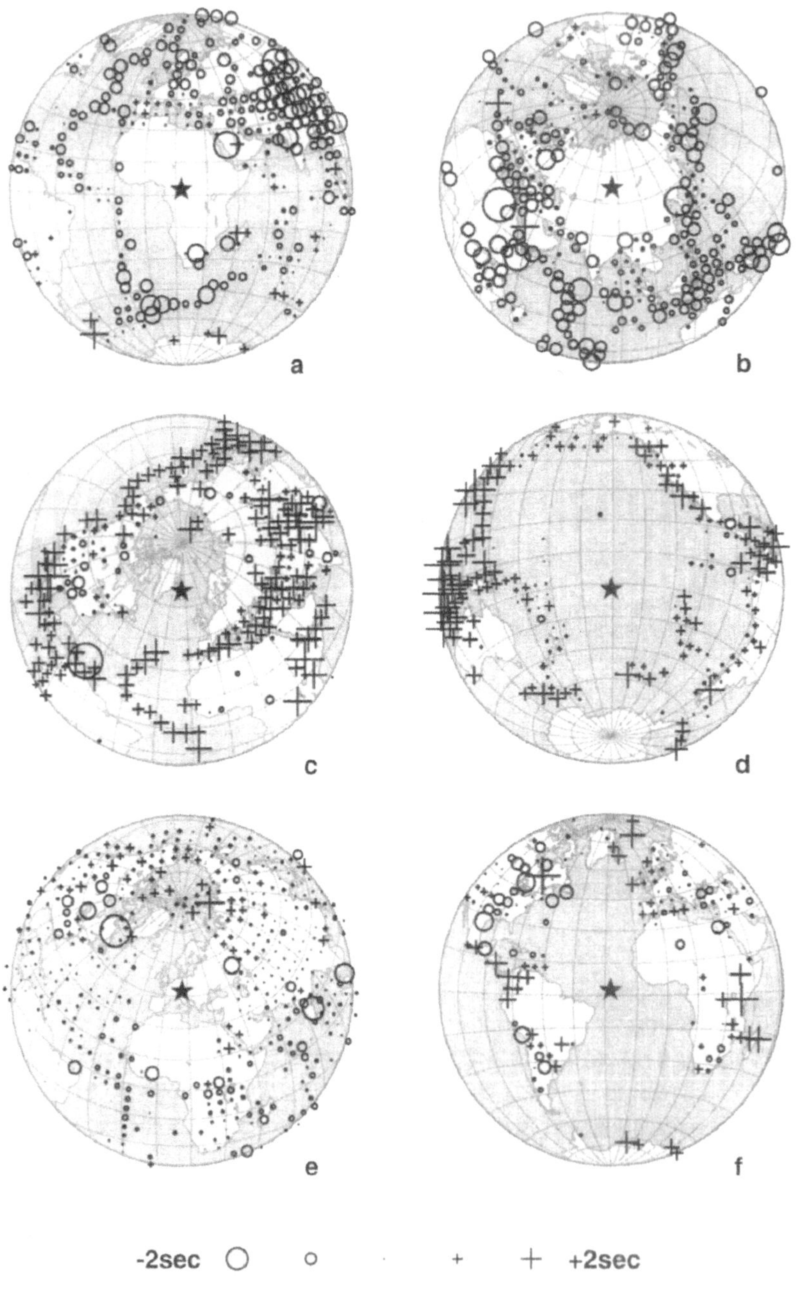

heterogeneity in *P*-wave propagation, significant in amplitude, geographically coherent and correlated to existing tectonic features, with pluses to indicate late arrivals and circles early arrivals.

The pattern and sign of residuals varies from quite homogeneous to rather heterogeneous as shown in a few examples in Figure 4. The two cells located on ancient continental shield regions (Africa and Siberia) exhibit mainly negative residuals, over the whole distance and azimuthal range, indicating that the teleseismic rays sample structures with velocity anomalies faster than the average, connected to the cold roots of continents. The two cells located in oceanic regions, corresponding to Icelandic and Tahitian hotspots, nicely show a rather uniform pattern of positive delays, due to low velocities at depth. Less uniform are the residuals displayed by the cell located in central Europe and the one situated astride the Mid-Atlantic Ridge, nonetheless showing regional consistency given by large-scale velocity anomalies sampled by the rays.

The complete set of residual anomalies is adequate to provide a measure of the Earth's laterally heterogeneous structure and can be seen as an empirical substitute for a three-dimensional velocity model, for the purpose of earthquake location.

3.2. Cell Static Corrections

Information regarding the mean structure beneath a particular cell can be obtained by averaging all summary residuals pertaining to the fan of teleseismic summary rays departing from it. This is somehow analogous to the static station correction terms (CLEARY and HALES, 1966). Station statics are usually expressed, in the form of a constant term, by the average of all travel-time residuals recorded at a station, and it has been demonstrated that they carry significant information concerning vertically-averaged upper mantle structure (SOURIAU and WOODHOUSE, 1985; ROBERTSON and WOODHOUSE, 1997). Since our cells may contain both stations and events, we introduce the concept of *cell static* correction, therefore a generalization of the classic station static correction (PIROMALLO and MORELLI, 1998).

The static term for cell α is given by averaging the $N_{\mathscr{R}_\alpha}$ summary residuals associated to it

◄

Figure 4

Teleseismic *P*-wave summary residuals projected in the distance range 30° to 90°. Maps are centered on reference cells located in (*a*) Central Africa (with 4 events and stations BNG, BCAO), (*b*) Russia (with 3 events and stations ELT, NVS, UKR), (*c*) Iceland (with 20 events and station AKU), (*d*) Tahiti (with stations TAH, PPT, PPN, AFR, TVO, PAE and no events), (*e*) Central Europe (with 13 events and stations PRU, KHC, LJU, TRI, TRS, PAD, ZLA, CTI, BHG, WET), (*f*) Mid-Atlantic Ridge (with 63 events and no stations). Circles and pluses represent early (negative) and late (positive) arrivals, respectively. All cells contain both stations and events except cell *d* with stations only and cell *f* with events only.

$$T_\alpha = \frac{\sum_{\beta \in \mathscr{R}_\alpha} R_{\alpha\beta}}{N_{\mathscr{R}_\alpha}} \quad . \tag{3}$$

They correspond to velocity perturbations below each cell and give a broader geographical coverage than studies based on the station-only approach, mainly adding information of seismogenic areas such as oceanic ridges and rifting areas. Figure 5 illustrates the map of cell statics and a map showing the number of summary rays used to compute them; each cell being again 5° × 5°, in a plot centered 0E. The cell static corrections we obtain range in value from −3.4 to +3.5 s, with 0.47 s root-mean-squares and a 0.2 s positive average. We recognize general features shared by the zero-degree statics found by different authors (DZIEWONSKI and ANDERSON, 1983; HAGER and CLAYTON, 1989; TOY, 1989; ZHOU and WANG, 1994) and the *P*-residuals patch corrections by ENGDAHL *et al.* (1998), showing that they depend little on the 1-D reference model. They are strongly correlated with surface tectonic patterns, as observed by CLEARY and HALES (1966), with shield areas (i.e., Canadian, Fennoscandian, Asian, Indian, Australian and South African Shields) showing negative delays, resulting from a faster than average velocity structure, and active continental regions (North and South American Cordillera, Mediterranean, Tibetan Plateau and southeastern Australia) exhibiting positive delays, due to slower than average velocities in the underlying mantle. Cells situated astride ocean ridges and rifting areas indicate small positive delays on average, particularly evident, for example, along the Mid-Atlantic and Indian Ridges; while Mid-Oceanic Island regions have larger positive residuals.

In Figure 6 the norm of cell statics is plotted versus the number of sources, the number of stations and the number of summary rays. In Figures 6a and 6b the circles are attributed to cell statics relative to cells holding only events or only stations, respectively, while the dots indicate values obtained for cells with both sources and receivers. We observe that cell static norms for cells containing only sources lie mostly below 0.6 s (Fig. 6a), while those pertaining to cells containing only stations have larger values, distributed up to 1.5–2 s (Fig. 6b). Instead, if we consider the hybrid cells, the corrections exhibit larger values in both cases. Moreover, a manner of decay of cell static corrections with an increasing number of stations can be recognized in Figure 6b, while a similar trend as a function of the number of sources cannot be clearly inferred from Figure 6a. The norm of cell statics versus the number of summary rays is presented in Figure 6c, merging the information of the two maps of Figure 5. An overall decrease in the norm of cell static corrections calculated for cells with a large number of summary rays is apparent.

It could be objected that cell statics for cells that do not contain receivers should be identically zero. In fact, the sum of all travel-time residuals for an event is set to zero by the origin time shift during the location process, thus absorbing all of the path effects common below the source. Therefore it is not straightforward to understand why cells containing only events exhibit static corrections slightly

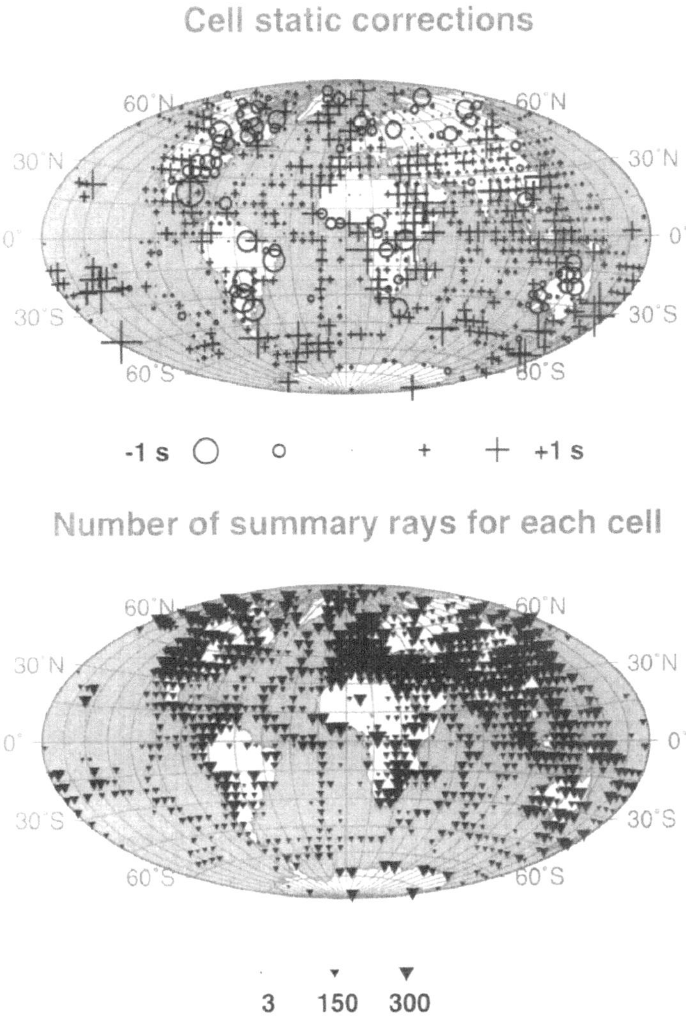

Figure 5
Cell static residuals and number of summary rays used to compute them, for roughly equal area equatorial 5° × 5° cells, in a plot centered on 0E. Cell statics are computed according to Equation (2). They are a generalization of station statics, with which they share large-scale common features, carrying, in addition, the contribution of seismogenic areas as oceanic ridges.

different from zero, as shown in Figure 6a. We recall here an explanation that has been considered in PIROMALLO and MORELLI (1998). The location procedure sets to zero the sum of residuals of both regional and teleseismic rays for each event. Conversely, in the computation of cell static corrections, teleseismic rays only are treated, and the sum of their residuals for a specific event is not necessarily null. Moreover, a cell static correction is computed by averaging residuals of summary, rather than individual, rays. The two averages would obviously coincide only in the

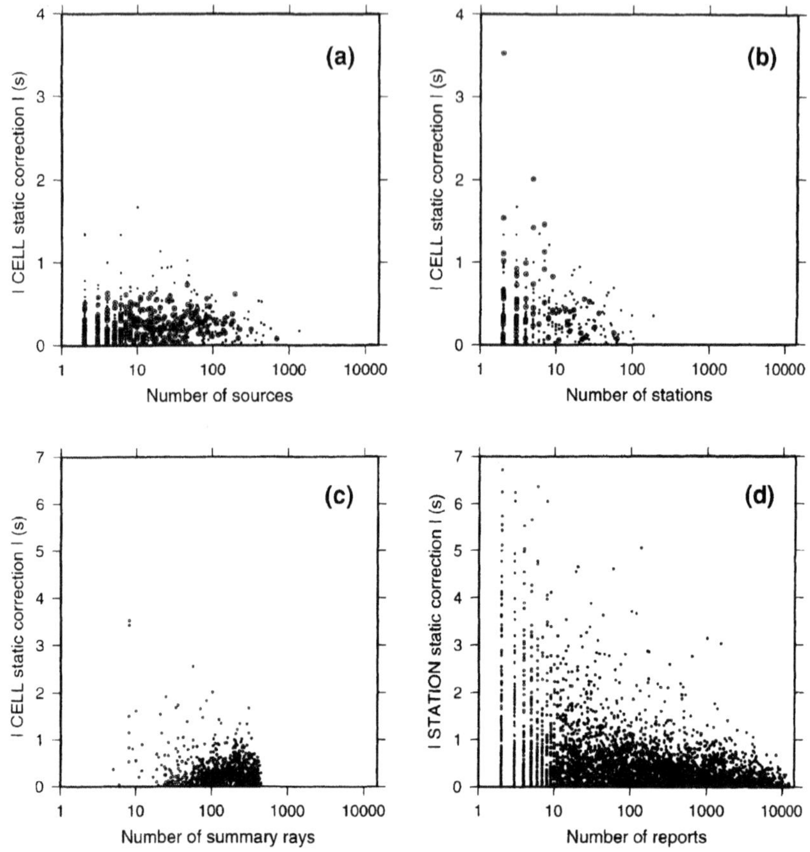

Figure 6

(*a*) Norm of cell static corrections versus number of sources (circles mark values relative to cells containing only sources). (*b*) Norm of cell static corrections versus number of receivers (circles mark values relative to cells containing only receivers). (*c*) Norm of cell static corrections versus number of summary rays. (*d*) Norm of station static corrections versus number of reports at each station. All plots are displayed using a logarithmic scale on the *x* axis. Observe that the scale on the *y* axis is not the same in the top (*a*, *b*) and bottom (*c*, *d*) panels.

case each summary residual was weighted by its multiplicity. These reasons explain why we may expect source-only cells to be non-null.

4. *Relocation with Empirical Heterogeneity Corrections*

4.1. Method

Both residual maps and cell statics show that a significant, regionally consistent, structural signal remains after conventional 1-D location. This signal provides empirical corrections to *P*-wave travel times, which are distance and azimuth dependent, thus reflecting the Earth's three-dimensional structure. The summary residuals between couples of cells are properly combined to obtain a simple Empirical Heterogeneity Correction (EHC) scheme to be applied to an iterative event relocation, following the method illustrated by PIROMALLO and MORELLI (1998). A summary residual is retained as an EHC, provided that its multiplicity is larger than five individual rays.

We do not incorporate in the EHC method any weighting factor to assign a modelling error to the correction. However, in order to evaluate the stability of a summary residual, the dependence of ray bundles on multiplicity could be addressed. The multiplicity of summary residuals or, better, the variance of summary residuals as a function of their multiplicity, could in fact provide a model for assessing the statistical reliability and relative weight of each summary ray, as suggested in PIROMALLO and MORELLI (1998). If such a weighting of each summary residual would retain the actual information contained in the data, on the other hand it would also invalidate one of the major benefits inherent to the summary ray formulation: improving uniformity in geographical coverage. This is the reason why we chose not to include any relative weighting of summary residuals in the EHC formulation.

In our relocation scheme we introduce, in addition, the station corrections calculated after the computation of empirical heterogeneity corrections. The relocation is thus performed using them both. This last step (called joint EHC hereafter) is undertaken because station statics have been seen to improve locations for both 1-D and 3-D models (SMITH and EKSTRÖM, 1996), providing a local correction below the stations. Since there is an inevitable trade-off between station corrections and 3-D structure, our choice of computing station corrections *after* summary corrections implies that we try to map into station corrections only the remaining signal. In Figure 7 we plot a histogram and cumulative histogram of station static terms after the removal of EHC corrections, showing that the majority of zero-degree terms exhibit delays smaller than 1 s. Residuals of such a value, as shown in Figure 6d where the norm of station statics is plotted versus the number of reports, mostly pertain to stations with a large (≥100) number of recordings. A clear scatter in the norm of station statics can be observed for stations with a few recordings (Fig. 6d). Since these values are likely to be rather unstable, we require at least five reports for each station correction to be used in the joint EHC. Figure 8 is a blow-up of global station statics centered on the Italian peninsula and surrounding regions, an area approximately covering nine meshes of the $5° \times 5°$ grid. The regional-scale pattern of zero-degree statics is rather heterogeneous, both in sign and amplitude, while over a smaller scale coherent features appear, related to known tectonic structures, in good agreement with other studies (see for example BABUSKA *et al.*, 1984). Summary residuals average the signal over each cell, with the possible consequence of cancelling these shorter wavelength variations in some areas. Hence the use of station corrections may partly remedy this inconvenience.

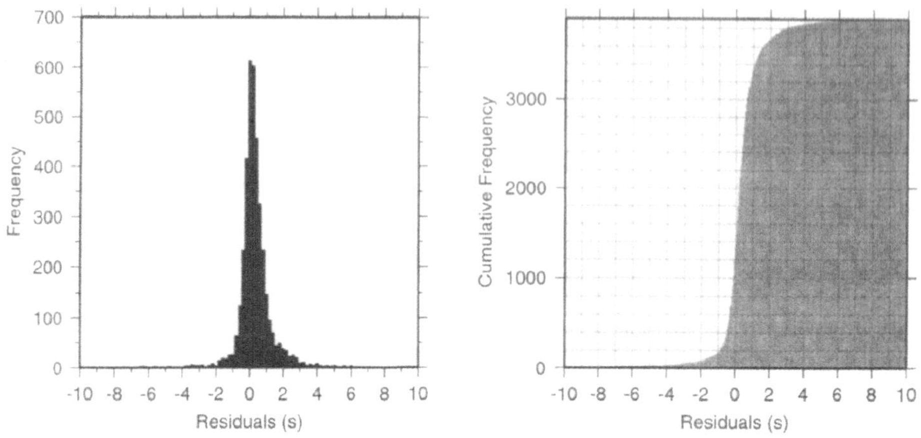

Figure 7
Histogram and cumulative histogram of station static terms after the removal of EHC corrections.

4.2. Results

The EHC relocation tends to displace epicenters, in order to compensate for the bias due to delays originated by previously unaccounted velocity anomalies. To understand how the EHC method works in practice consider the following trivial example. In general a one-dimensional location scheme would bias, for instance, a hypocenter towards a positive velocity anomaly, to shorten the ray in fact speeded up by faster material. This mislocation would cause some time delay to be left, as a consequence of the masked structural signal. By correcting for this travel-time residual anomaly, the EHC scheme will move the hypocenter closer to its actual location, i.e., away from the fast anomaly.

Epicenter relocation vectors (≤40 km) resulting from the relocation performed with the joint EHC procedure are illustrated in Figure 9 for three selected seismic regions. We plot in Figure 9 relocation vectors for shallow, well-constrained events with more than 150 teleseismic reporting stations, in order to reduce the clutter in the figure and to investigate whether these epicentral shifts are systematic. Although the pattern of relocation vectors may sometimes be mixed and scattered, rather systematic trends prevail in many areas, determined by the joint effect of the anomalies amplitude and distribution, and the location of stations and events. See for example in Figure 9c the coherent shifts for earthquakes all along the western coast of Central and South America, roughly perpendicular to the slab strike. In Figures 9a and 9b systematic orientations are to be sought on a smaller scale.

Over the whole set of events considered, the standard 1-D location yields a rms weighted residual of 0.90 s, reduced to 0.78 s by the subsequent EHC relocation and further to 0.75 s by the joint EHC relocation. The computed rms length of relocation

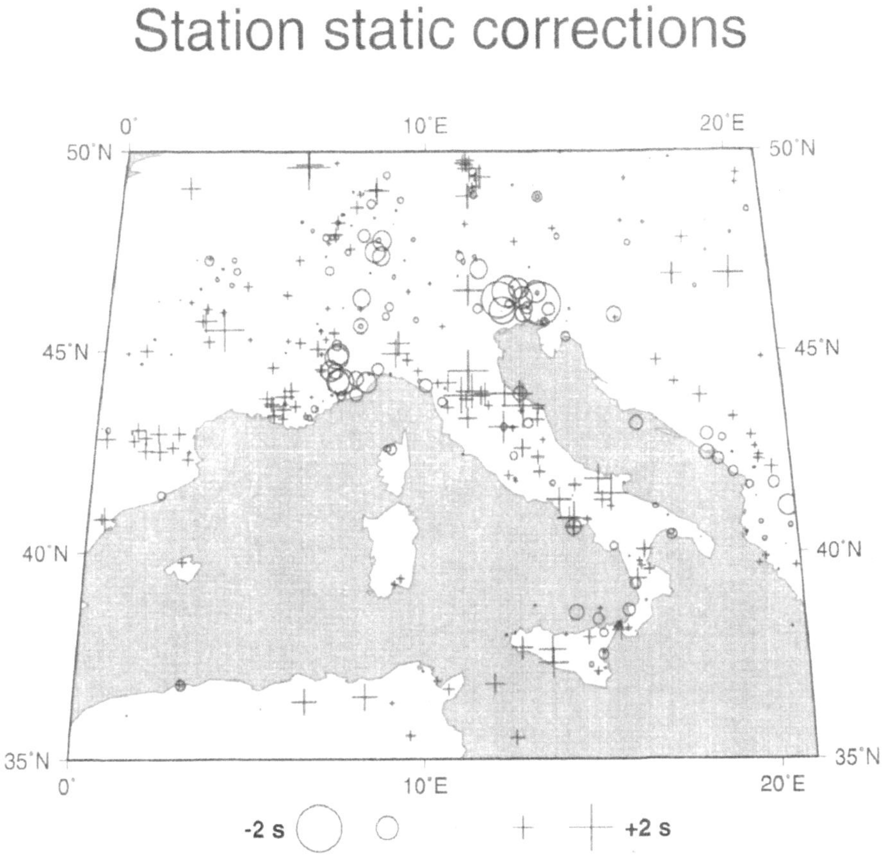

Figure 8
Blow-up of global station statics centered on the Italian peninsula. The illustrated area covers approximately nine meshes of the 5° × 5° grid. Smaller-scale coherent features appear, related to known tectonic structures.

vectors is 10.2 km and the mean 8.9 km, while the rms change in origin time is 0.4 s, the mean being −0.1 s.

5. *Testing EHC Scheme against Calibration Events*

5.1. *Choice of Test Events*

To test the efficiency of the EHC with respect to a simple 1-D location method, we compare, for a subset of seismic events and explosions included in the global data set, the locations obtained with both techniques to those obtained by independent data. We use, for this purpose, the data set of 25 explosions tabulated by SMITH and EKSTRÖM (1996), whose locations and origin times are well-known by non-seismic

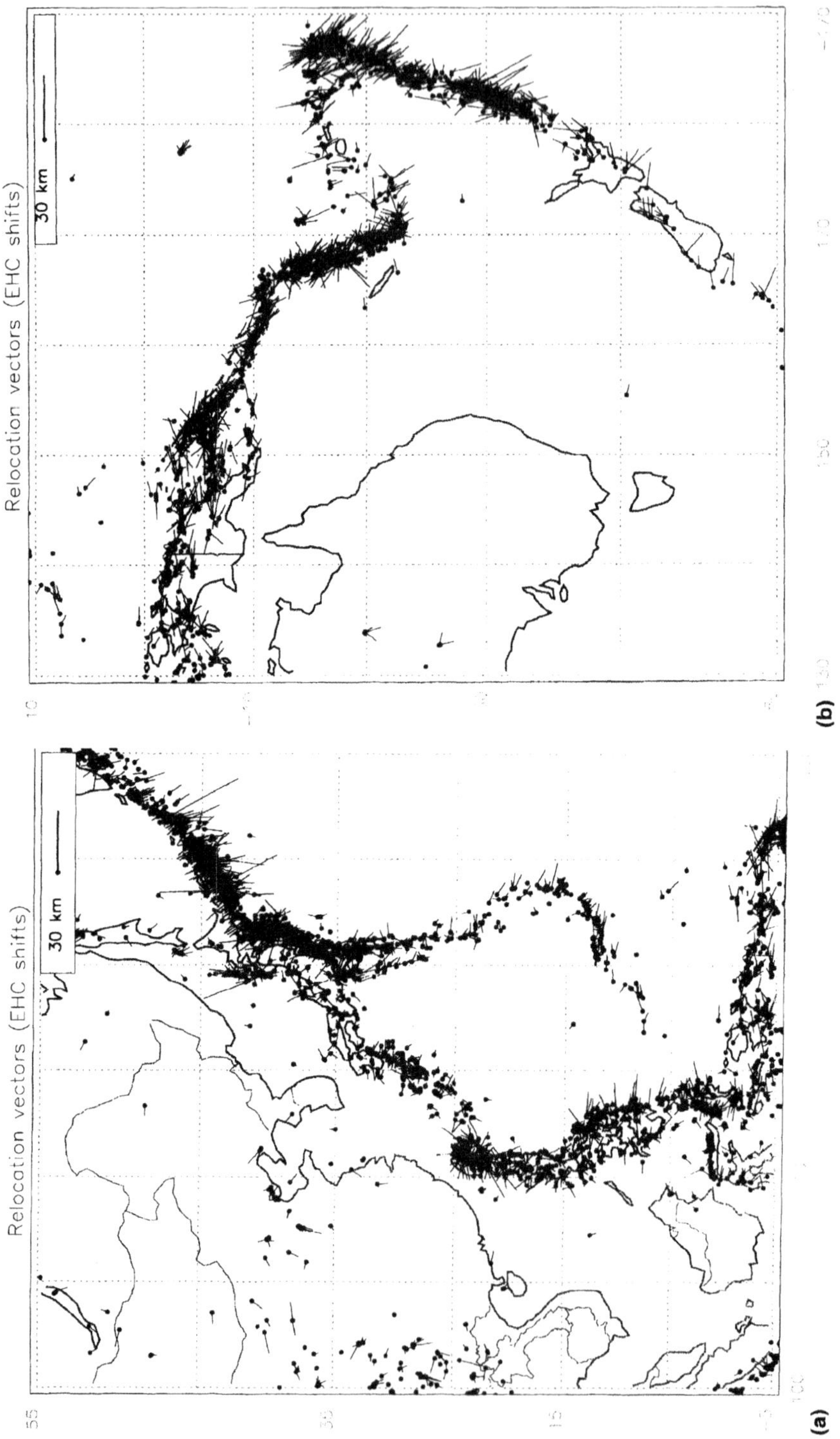
Relocation vectors (EHC shifts)
30 km
Relocation vectors (EHC shifts)
30 km
(b)
(a)

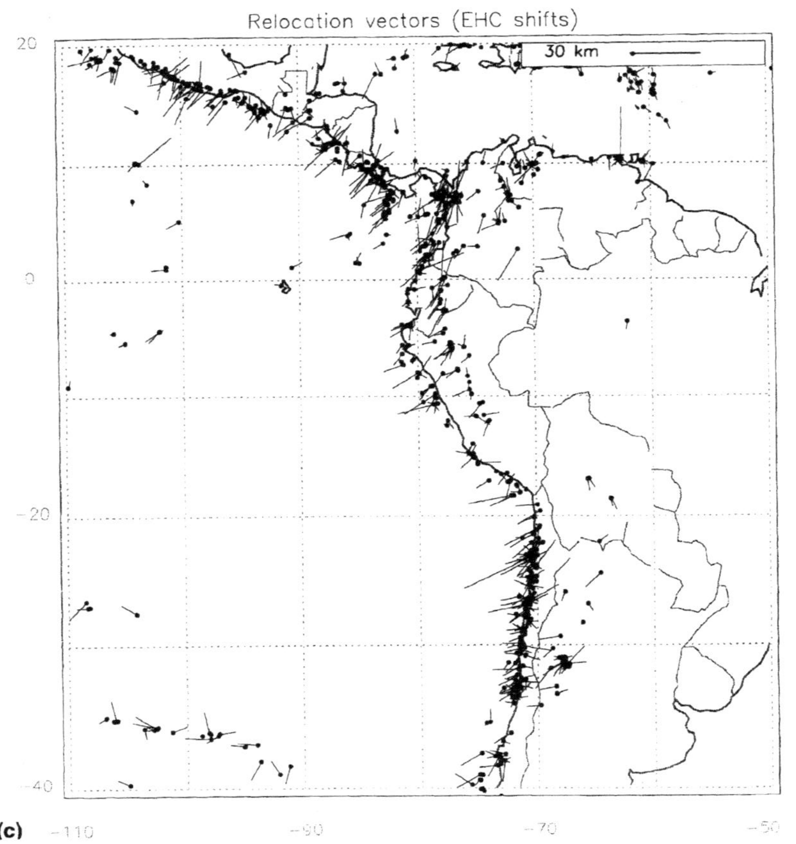

(c)

Figure 9

Epicentral shifts for shallow events, with ≥150 teleseismic arrivals, resulting from the relocation performed with the joint EHC procedure, for three selected seismic regions: (*a*) northwestern Pacific, (*b*) southwestern Pacific, (*c*) South America. The dots mark the 1-D locations, the segments show the direction of the EHC relocation shift and are proportional to the displacement. A 30-km shift is shown for scale.

means, and the data set of 83 earthquake test events compiled by KENNETT and ENGDHAL (1991), whose hypocenters have been determined by using local network data. In order to cull from the ISC Bulletin as many of these test events as possible, we release the condition on the largest azimuthal gap and on the minimum number of reports described in Section 2.1. However, since the condition of shallow depth is retained, 14 earthquake are discarded; their hypocentral depth being greater than 50 km.

5.2. Test Results

Figure 10 illustrates the summary residual maps for six cells, the first three containing the sites of explosion test events (Kazakhstan, Aleutians and Nevada) and the last three of earthquake test events (Greece, Japan and Hawaii). All but the cell covering Kazakhstan, also contain other stations and events. The pattern of residuals displayed by these maps arises from a combination of regional deviations from the average 1-D model and the station/event coverage. Each individual travel time from the test site to the recording site will be corrected by the corresponding summary residual.

The rms differences between the known locations and those predicted by the 1-D model and the EHC model are summarized in Tables 1 and 2 for explosions and earthquakes, respectively. The smaller is the rms difference, the closer is the calculated location to the actual one. The joint EHC location, with both empirical and static corrections, yields a further, though smaller, improvement (see Tables 1 and 2), showing that most of the lateral heterogeneity is modeled by simple EHC. The rms improvement is 10% for explosions (17% if the joint EHC location is applied) and 9% for earthquakes (12% by joint EHC location). Although the distance from the known location found using the EHC is smaller than the mislocation of the SP6 model, we find that the origin times are not well fit, especially for earthquakes. A reason for this is suggested by SMITH and EKSTRÖM (1996) and is related to the error in origin time that is possibly induced by errors in depth when dealing with teleseismic relocations.

The hypocentral coordinates of the explosions and earthquakes that are shifted during the relocation procedure are quoted in Table 3 and Table 4, respectively,

▶

Figure 10
Summary residual maps for six cells, the first three centered on the sites of explosion test events: (*a*) Kazakhstan (with 214 events), (*b*) Aleutians (with 426 events and stations AMC, SHY, AME, AEB, ASB, ASC, ANA, SSI, LSI, SMY, RAT, ACF, AND) and (*c*) Nevada (with 147 events and stations EUR, TNP, BCN, GSC, CLC, MSU, ARUT, KNB); the last three centered on the sites of earthquake test events: (*d*) Greece (with 289 events and stations ATH, ATU, PAT, VLS, VAM, JAN, PAIG, VGL, VSK, VFI, VSI and more), (*e*) Japan (with 370 events and stations MRK, OFU, MIZ, MAT, MAJO, FKS and many more) and (*f*) Hawaii (with 19 events and stations HVO, KKH, HPO, KHU, PAH, MKH and more).

specifying for each event also the largest open azimuth, the number of reports and the mislocation due to 1-D and joint EHC relocation. The mislocation due to the joint EHC model versus that due to 1-D model is plotted in Figure 11 for each

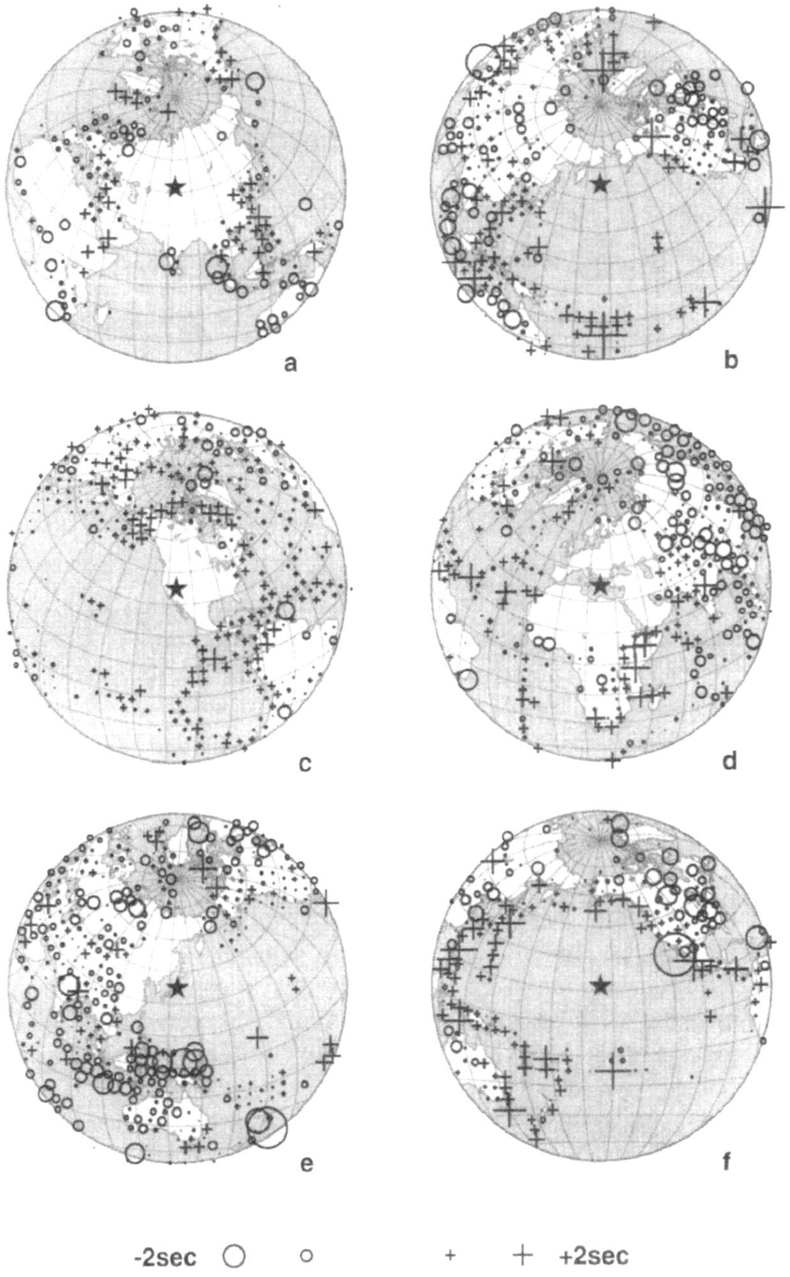

Table 1

Root-mean-square's difference between known hypocentral parameters of explosions and those found by relocation with a one-dimensional model (SP6), with empirical heterogeneity corrections (EHC) and with a joint relocation (EHC + static corrections).

Model	Rms misfit to known location (km)
SP6	9.95
SP6 + EHC	8.90
SP6 + EHC + static corr.	8.26

Table 2

Root-mean-square's difference between known hypocentral parameters of earthquakes and those found by relocation with a one-dimensional model (SP6), with empirical heterogeneity corrections (EHC) and with a joint relocation (EHC + static corrections).

Model	Rms misfit to known location (km)
SP6	16.39
SP6 + EHC	14.87
SP6 + EHC + static corr.	14.37

explosion and in Figure 12 for each earthquake, showing that in both cases the location errors are considerably smaller using EHC procedure.

The EHC procedure, compared to the simple 1-D location, reduces by more than 5% the mislocation vector length of 14 explosions out of 23 (Boxcar is most improved). The locations of two explosions do not significantly move, and the remaining seven worsen (up to 35% increase in the mislocation vector length for a site in Kazakhstan). Explosions are coded with different symbols for different sites in Figure 11. The big outlier is the Cannikin, whose starting 1-D location already has a large shift with respect to the actual site. Nonetheless, the EHC allows partial remedy to this large mislocation, reducing the vector length by 28%. Locations for the test site of Nevada (pluses), constrained by a large number of well-distributed observations, are improved, while those for Kazakhstan (stars) degrade, and those for Eurasia (squares and triangles) become only slightly better. Location errors for earthquakes are larger than those for explosions, and more scattered.

The majority of locations (36 out of 59) is more than 5% better than those found using SP6 model, and a few of them are largely improved (for events in Greece and Japan, for example, as can be easily deduced from Table 4). However, 16 event locations are shifted away from the actual position and 7 are only slightly changed. We have found no clear dependence of the mislocation vector length on the largest

Table 3

Hypocentral coordinates of the 23 explosions used and shifted during the relocation procedure. For each explosion the table specifies the date, origin time, geographical coordinates and depth, obtained by independent data, reported by SMITH *and* EKSTRÖM *(1996). The largest open azimuth, the number of reports and the mislocation vector length due to 1-D and joint EHC relocation computed in this study are quoted.*

Date			Time			Lat. (deg.)	Long. (deg.)	Dep. (km)	Gap (deg.)	N	SP6 (km)	EHC (km)	
1965	7	15	14	16	8.10	37.197	-74.352	0.0	96	44	18.63	11.20	Chase
1967	5	20	15	0	0.20	37.130	-116.064	0.0	73	66	10.53	7.61	Commodore
1968	1	19	18	15	0.10	38.634	-116.215	0.0	72	80	10.06	5.81	Faultless
1968	4	26	15	0	0.10	37.295	-116.456	0.0	80	84	11.41	4.91	Boxcar
1969	9	10	21	0	0.10	39.406	-107.948	0.0	160	35	7.92	4.91	Rulison
1969	11	30	3	32	59.70	49.924	78.956	0.0	32	97	5.38	7.28	
1971	4	25	3	32	59.90	49.769	78.034	0.0	32	100	10.06	9.82	
1971	11	6	22	0	0.06	51.472	179.107	0.0	52	129	26.80	19.27	Cannikin
1972	8	16	3	16	59.80	49.765	78.059	0.0	61	51	9.05	10.98	
1972	11	2	1	27	0.20	49.927	78.758	0.0	32	118	10.30	10.98	
1973	5	17	16	0	0.00	39.793	-108.366	0.0	186	42	11.82	16.58	Rio Blanco
1977	8	10	22	0	0.10	50.956	110.983	0.0	106	45	5.81	6.94	
1983	7	10	4	4	59.94	51.367	53.327	0.0	50	66	9.05	9.57	
1983	7	10	4	9	59.85	51.380	53.340	0.0	61	61	9.57	8.50	
1983	9	24	5	5	0.03	46.788	48.297	0.0	90	37	4.39	3.11	
1983	9	24	5	10	0.08	46.767	48.311	0.0	112	34	4.39	2.20	
1983	9	24	5	15	0.14	46.749	48.303	0.0	90	35	3.80	4.39	
1983	9	24	5	19	59.93	46.754	48.289	0.0	90	38	3.80	4.91	
1983	9	24	5	25	0.00	46.766	48.274	0.0	90	37	4.39	3.11	
1984	7	21	2	59	59.81	51.358	53.319	0.0	57	68	6.94	6.21	
1984	7	21	3	4	59.71	51.390	53.351	0.0	56	69	8.22	7.28	
1984	7	21	3	9	59.84	51.371	53.337	0.0	54	69	6.59	6.21	
1988	7	7	15	5	30.07	37.252	-116.377	0.0	109	76	12.03	9.05	Alamo

azimuthal gap, which in only a few cases exceeds 90 degrees for the test events considered here, or the number of reports, which is quite large for most events and always greater than 30 observations. Moreover, test events are mostly located in cells for which both sources and stations were used to construct summary rays. The few events that are positioned in cells where only sources contribute to summary residuals, do not suggest any evident correlation between a possible worsening of the location and the use of earthquakes to derive corrections.

Figure 13 shows the mislocation vectors for explosions, whose sites are indicated by a star on the map and earthquakes, whose locations are marked by a dot. The dot is located at the true epicenter and the segment points in the direction found by joint EHC relocation, and is linearly proportional to the mislocation distance. In some areas, where explosions or earthquakes are clustered in space, they exhibit a strong consistency among mislocation vector directions. This can be explained by the fact that these events are likely recorded by approximately the same set of stations and are thus subject to the same set of corrections.

Table 4

Hypocentral coordinates of the 59 earthquakes used and shifted during the relocation procedure. For each explosion the table specifies the date, origin time, geographical coordinates and depth, obtained by independent data, reported by KENNETT *and* ENGDHAL *(1991). The largest open azimuth, the number of reports and the mislocation vector length due to 1-D and joint EHC relocation computed in this study are quoted.*

Date			Time			Lat. (deg.)	Long. (deg.)	Dep. (km)	Gap (deg.)	N	SP6 (km)	EHC (km)	
1968	10	14	2	58	50.30	−31.600	117.000	5.0	32	65	12.42	4.39	Meckering (Australia)
1973	4	26	20	26	30.60	19.903	−155.130	48.0	59	112	10.06	9.57	Kalapana (Hawaii, USA)
1975	2	4	11	36	5.93	40.567	122.764	6.6	37	135	10.53	12.42	Haicheng (China)
1977	12	25	16	18	51.36	38.967	70.601	2.9	94	67	22.71	20.60	Garm (USSR)
1979	8	6	17	5	22.39	37.114	−121.496	7.0	97	83	14.57	15.68	Coyote Lake (CA, USA)
1980	7	27	18	52	21.52	38.195	−83.937	16.6	144	57	30.74	27.86	Sharpsburg (KY, USA)
1981	2	14	6	9	27.18	46.351	−122.238	8.5	172	48	13.54	14.23	Elk Lake (WA, USA)
1981	2	25	2	35	52.20	38.140	23.080	6.0	34	106	15.68	13.54	Greece
1982	1	18	19	27	24.50	39.840	24.470	15.0	41	110	21.29	5.81	Greece
1982	3	21	2	32	6.00	42.156	142.531	27.4	35	169	16.28	11.62	Urakawa-Oki (Japan)
1982	3	21	10	22	35.25	42.228	142.600	24.3	85	135	8.78	5.38	Urakawa-Oki (Japan)
1983	1	5	2	1	1.00	54.730	162.920	20.0	55	130	11.41	10.30	Kamchatka (USSR)
1983	1	17	12	41	30.70	37.960	20.230	6.0	33	117	25.80	19.76	Greece
1983	3	10	0	27	48.90	43.820	147.420	40.0	75	152	7.28	6.94	Kuril Islands (USSR)
1983	3	14	12	12	44.40	39.300	54.600	10.0	67	83	12.23	9.05	Turkmen (USSR)
1983	3	17	7	11	26.70	51.550	142.410	20.0	84	107	5.38	6.21	Sakhalin Island (USSR)
1983	5	2	23	42	38.15	36.232	−120.310	10.0	55	122	15.68	8.50	Coalinga (CA, USA)
1983	7	22	2	39	54.07	36.240	−120.409	7.3	71	78	19.39	13.36	Coalinga (CA, USA)
1983	8	8	3	47	58.60	35.536	139.046	18.1	83	123	12.99	12.42	Kanto-Tokai (Japan)
1983	10	7	10	18	46.32	43.946	−74.252	7.8	80	60	12.80	10.53	Goodnow (NY, USA)
1983	10	28	14	6	6.79	43.968	−113.899	16.0	55	114	10.30	9.57	Borah Peak (ID, USA)
1983	10	30	4	12	26.90	40.310	42.100	15.0	50	123	6.94	7.28	Turkey
1983	10	30	16	51	55.94	35.420	133.920	10.6	94	104	17.43	16.14	Tottori (Japan)
1983	11	16	16	12	58.57	19.419	−155.453	12.5	47	133	10.53	5.81	Kaoiki (Hawaii, USA)
1984	2	22	5	44	37.30	39.340	54.070	15.0	67	87	11.62	10.30	Turkmen (USSR)
1984	3	19	20	28	39.00	40.380	63.360	15.0	23	149	3.80	4.39	Uzbek (USSR)
1984	4	24	21	15	18.74	37.311	−121.679	8.8	54	102	9.05	12.80	Morgan Hill (CA, USA)
1984	6	24	13	29	40.20	−43.590	170.640	9.0	66	48	12.42	11.20	Macaulay (New Zealand)

1984	9	13	23	48	49.10	35.807	137.554	1.1	40	139	15.99	14.40	Kanto-Tokai (Japan)
1984	10	25	14	38	28.30	40.090	21.610	4.0	59	66	12.80	12.23	Greece
1984	10	26	20	22	19.70	39.250	71.260	15.0	63	136	12.61	9.32	Tajik (USSR)
1984	11	1	18	43	39.40	55.230	163.720	20.0	46	165	8.22	8.78	Kamchatka (USSR)
1984	12	28	10	37	50.50	56.290	163.490	13.0	47	173	10.30	15.21	Kamchatka (USSR)
1985	3	16	14	53	59.90	17.066	−62.379	15.8	67	107	17.70	15.37	Eastern Caribbean
1985	3	17	10	41	36.72	−32.735	−71.778	23.2	57	73	40.43	31.82	Chile
1985	7	13	18	54	14.89	25.771	−45.053	6.4	59	82	10.30	11.62	North Atlantic Ridge
1985	8	23	12	41	56.80	39.370	75.440	20.0	32	161	22.18	15.53	Xinjiang (China)
1985	9	10	1	26	4.80	60.500	168.860	15.0	108	111	9.57	7.28	Siberia (USSR)
1985	10	13	15	59	52.00	40.300	69.800	10.0	39	137	3.11	3.80	Tajik (USSR)
1985	10	29	13	13	40.20	36.780	54.800	20.0	49	139	17.57	15.37	Iran
1986	1	26	19	20	51.08	36.809	−121.268	6.0	114	75	18.37	21.74	Hollister (CA, USA)
1986	3	31	11	55	39.94	37.468	−121.694	6.4	71	84	33.23	19.02	Mt. Lewis (CA, USA)
1986	4	26	7	35	14.88	32.212	76.281	13.4	78	107	15.05	14.57	India
1986	5	2	10	30	2.90	55.070	163.710	20.0	46	157	12.61	12.99	Kamchatka (USSR)
1986	5	13	8	44	0.80	41.450	43.700	10.0	64	109	15.21	14.40	Turkey-USSR
1986	5	20	5	25	49.58	24.082	121.592	15.8	39	127	10.53	15.53	Taiwan
1986	7	8	9	20	44.44	33.999	−116.610	11.7	73	97	15.37	17.70	Palm Springs (CA, USA)
1986	7	21	14	42	26.03	37.533	−118.446	9.7	71	103	18.24	10.98	Chalfont Valley (CA, USA)
1986	10	10	17	49	24.71	13.650	−89.200	10.9	151	59	34.65	24.15	El Salvador
1986	11	22	0	41	44.50	34.616	139.446	2.1	57	132	20.13	22.18	Kanto Tokai (Japan)
1987	3	21	12	8	57.86	8.725	−83.582	35.1	150	54	50.17	50.41	Costa Rica
1987	5	25	11	31	54.70	63.909	−19.779	11.3	56	146	16.28	14.57	Vatnafjoll (Iceland)
1987	7	31	23	56	57.83	40.418	−124.406	17.7	67	95	16.28	14.40	Eureka (CA, USA)
1987	10	1	14	42	19.47	34.049	−118.081	14.4	56	96	12.61	6.21	Whittier (CA, USA)
1987	11	24	13	15	56.54	33.015	−115.849	1.9	73	92	9.82	17.70	Superstition Hill (CA, USA)
1987	12	17	2	8	17.30	35.372	140.519	47.3	31	161	26.71	22.50	Kamto-Tokai (Japan)
1988	1	22	0	35	59.50	−19.817	133.878	6.5	49	79	9.57	7.92	Tennant Creek (Australia)
1988	3	25	17	20	54.35	−31.410	−67.977	24.4	93	47	12.03	6.94	Argentina
1988	11	25	23	46	4.50	48.117	−71.183	28.9	39	147	5.38	9.05	Saguenay (Canada)
1989	1	23	14	6	28.91	61.933	4.432	26.2	93	69	5.38	3.80	Norway

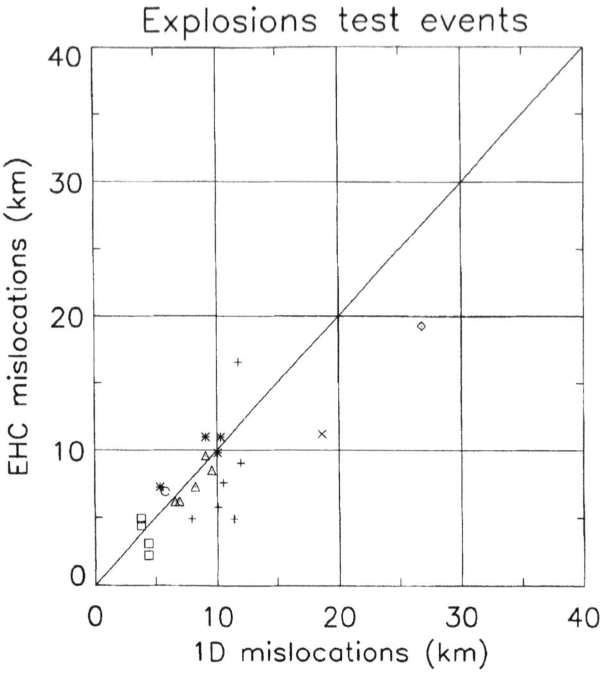

Figure 11
Explosion test events: mislocation due to joint EHC relocation versus that due to 1-D standard location for each explosion. Explosions are coded with different symbols for different sites: pluses for Nevada, squares and triangles for Eurasia, stars for Kazakhstan.

6. *Discussion and Conclusions*

As discussed by PIROMALLO and MORELLI (1998), EHCs in fact may not be corrections for the full travel-time effect of 3-D heterogeneity, because signal due to the initial mislocation cannot be entirely suppressed. They likely also include contributions from all other possible sources of systematic variation of travel time which could be hypothesized. Nevertheless, the EHC approach is very efficient to apply as compared to the construction of a 3-D Earth model. In fact, to use summary ray residuals to compute a 3-D perturbation to the reference *P*-velocity distribution in the Earth, and then adopt this 3-D model to calculate geographically-varying travel times appear to be unnecessary, time consuming and potentially unstable steps. Our technique, dealing directly with summary residuals, avoids the relative uncertainty in locating anomalies and detecting their actual amplitude, possibly arising from modelling errors typical of tomographic techniques.

The improvement in location of global teleseismic events shown in this study is smaller in comparison to that obtained for regional events by PIROMALLO and MORELLI (1998), which was estimated by means of synthetic tests to be about twice as

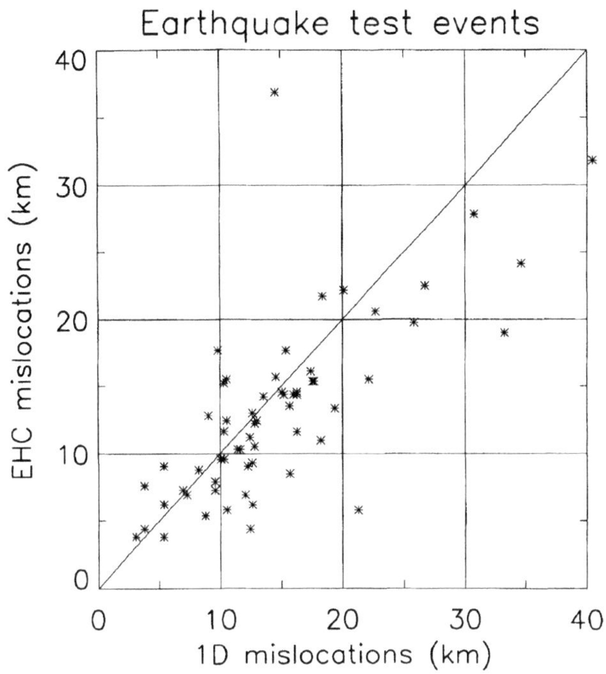

Figure 12
Earthquake test events: mislocation due to joint EHC relocation versus that due to 1-D standard location for each earthquake.

large. This can in part be explained by the fact that the use of a finer parameterization (smaller cells) and of regional phases, allowed a more accurate reproduction of the 3-D variation of travel-time heterogeneities on a regional scale. The cell size determines the heterogeneity scale length to which summary rays are sensitive (GUDMUNDSSON *et al.*, 1990), and a more detailed, smoother, representation could be in principle obtained, using a finer parameterization, also on a global scale. However, one should be aware that smaller cells can conveniently be employed only where data density is higher, because cell size trades off with the population of ray bundles.

The rms length of joint EHC relocation vectors, with respect to locations obtained with SP6 spherically symmetric model, is about 10 km. The joint EHC relocation yields an rms of weighted residual of 0.75 s (0.90 s for the standard 1-D location). The pattern of relocation vectors for the whole set of events considered reveals both rather systematic trends in some areas (for instance, it appears to be strongly related to slab geometry) and scattered features in other regions.

The effectiveness of the EHC technique, as compared to simple 1-D location, is checked in a test performed on calibration events of known locations. The rms length of the mislocation vectors for the test events, compared to their original mislocation in the reference 1-D model, is reduced in the joint EHC relocation by 17% for

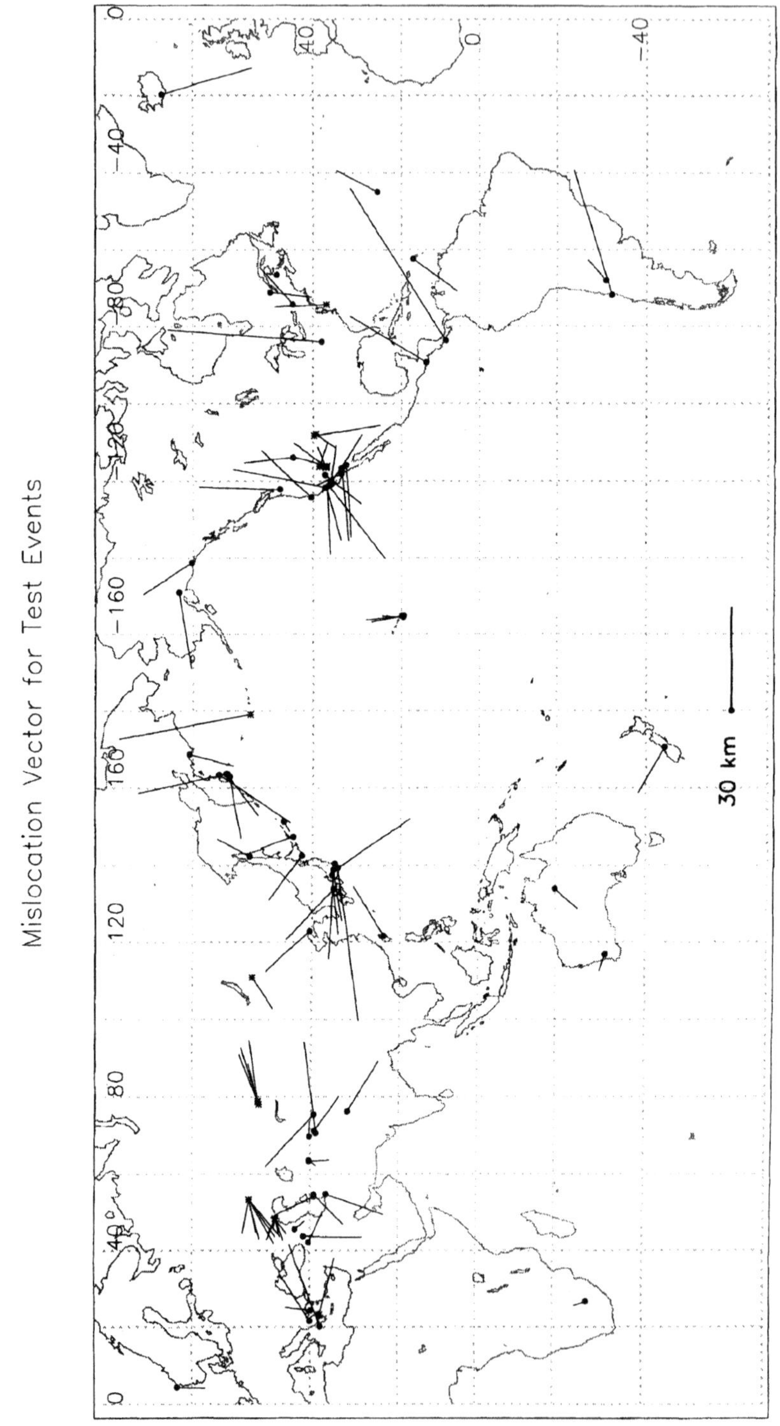

Figure 13
Mislocation vectors for explosion (marked by star) and earthquake (marked by a dot) test events. The star/dot is the actual event location and the segment points in the direction found by joint EHC relocation, and is linearly proportional to the mislocation distance.

explosions and by 12% for earthquakes. The rms improvement in mislocation vectors that we find in relocating explosions is smaller than that achieved by SMITH and EKSTRÖM (1996) (about a 40% reduction) using the 3-D global model S&P12/WM13 (SU and DZIEWONSKI, 1993), and of the same order of that obtained by BIJWAARD *et al.* (1998) in the joint inversion for 3-D slowness anomalies, station corrections and cluster hypocentral parameters (about an 18% reduction), using the reference Earth model ak135 of KENNETT *et al.* (1995). The rms reduction for mislocation vectors of earthquake test events is instead larger than that reported by SMITH and EKSTRÖM (1996) (roughly a 9%). The quantitative comparison between rms mislocation shifts obtained using approximately the same data set but different location procedures may have limited meaning; however, our technique demonstrates the ability to provide an effective alternative to the use of 3-D structural models, reaching the same value of effectiveness in improving event locations.

The application of our method can be also viewed as a preliminary step for preprocessing travel-time data to be used for studies of seismic imaging, since it allows limitation of the structural signal mapped into biased earthquake locations, thus satisfying one of the linearization requirements.

Acknowledgments

We would like to thank K. L. McLaughlin for careful review and helpful comments that enhanced this paper. We are grateful to E. Boschi for encouraging this study.

REFERENCES

ADAMS, R. D., HUGHES, A. A., and MCGREGOR, D. M. (1982), *Analysis Procedures at the International Seismological Centre*, Phys. Earth Planet. Inter. *30*, 85–93.

BABUSKA, V., PLOMEROVA, J., and SILENY, J. (1984), *Spatial Variation of P Residuals and Deep Structure of the European Lithosphere*, Geophys. J. R. Astron. Soc. *79*, 363–383.

BIJWAARD, H., SPAKMAN, W., and ENGDAHL, E. R. (1998), *Closing the Gap between Regional and Global Travel-time Tomography*, J. Geophys. Res. *103*, 30,055–30,078.

BOSCHI, L., and DZIEWONSKI, A. M. (1999), *High and Low Resolution Earth Models; The Implications of Different Approaches to Global Tomography*, J. Geophys. Res. *104*, 25567–25594.

BULAND, R., and CHAPMAN, C. H. (1983), *The Computation of Seismic Travel Times*, Bull. Seismol. Soc. Am. *73*, 1271–1302.

CLEARY, J., and HALES, A. L. (1966), *An Analysis of the Travel Times of P Waves to North American Stations, in the Distance Range 30° to 100°*, Bull. Seismol. Soc. Am. *56*, 467–489.

CREAGER, K. C., and JORDAN, T. H. (1986), *Aspherical Structure of the Core-mantle Boundary from PKP Travel Times*, Geophys. Res. Lett. *13*, 1497–1500.

DOUGLAS, A. (1967), *Joint Epicentre Determination*, Nature *215*, 47–48.

DZIEWONSKI, A. M. (1984), *Mapping the Lower Mantle: Determination of Lateral Heterogeneity in P-velocity up to Degree and Order* 6, J. Geophys. Res. *89*, 5929–5952.

DZIEWONSKI, A. M., and ANDERSON, D. L. (1983), *Travel Times and Station Corrections for P Waves at Teleseismic Distances*, J. Geophys. Res. *88*, 3295–3314.

DZIEWONSKI, A. M., and GILBERT, F. (1976), *The Effect of Small Aspherical Perturbations on Travel Times and a Re-examination of the Corrections for Ellipticity*, Geophys. J. R. Astron. Soc. *44*, 7–16.

ENGDAHL, E. R. (1998), *Development of an Archive of Seismic Ground-truth Events Globally in Support of Monitoring under the CTBT*, Paper presented at 20th Annual Seismic Research Symposium on Monitoring a CTBT, Santa Fe, 21–23 Sep., 1998.

ENGDAHL, E. R., VAN DER HILST, R. D., and BULAND, R. (1998), *Global Teleseismic Earthquake Relocation with Improved Travel Times and Procedures for Depth Determination*, Bull. Seismol. Soc. Am. *88*, 722–743.

FIRBAS, P., FUCHS, K., and MOONEY, W. D. *Calibration of seismograph network may meet Test-Ban Treaty's monitoring needs, Upper Mantle Heterogeneities from Active and Passive Seismology*, Proceedings of the NATO ARW Moscow April 1997, 199–213 (ed. K. Fuchs) NATO-ASI Series (Kluwer Acad., Norwell, Mass., 1997).

FIRBAS, P., PESHKOV, A. B., and RYABOY, V. (1998), *From IASPE191 Global Model to a 3-D Model for CTBT Monitoring*, *EOS Trans. AGU 79*(35).

FITCH, T. J. (1975), *Compressional Velocity in Source Regions of Deep Earthquakes: an Application of the Master Event Technique*, Earth Planet. Sci. Lett. *26*, 156–166.

FREEDMAN, H. W. (1967), *A Statistical Discussion of P Residuals from Explosions*, Bull. Seismol. Soc. Am. *57*, 545–561.

GUDMUNDSSON, O., DAVIES, J. H., and CLAYTON, R. W. (1990), *Stochastic Analysis of Global Travel-time Data: Mantle Heterogeneity and Random Errors in the ISC Data*, Geophys. J. Int. *102*, 25–43.

HAGER, B. H., and CLAYTON, R. W. *Constraints on the structure of mantle convection using seismic observations, flow models, and the geoid.* In *Mantle Convection* (ed. W.R. Peltier) (Gordon and Breach, New York, 1989) pp. 657–763.

HERRIN, E., and TAGGART, J. (1968), *Regional Variations in P Travel Times*, Bull. Seismol. Soc. Am. *58*, 1325–1337.

JEFFREYS, H. (1932), *An Alternative to the Rejection of Observations*, Proc. R. Soc. Lond. A *187*, 78-87.

JEFFREYS, H. *Theory of Probability* (Clarendon Press, Oxford, 1939).

JEFFREYS, H., and BULLEN, K. E., *Seismological Tables* (British Association for the Advancement of Science, London, 1940).

KENNETT, B. L. N., and ENGDAHL, E. R. (1991), *Travel Times for Global Earthquake Location and Phase Identification*, Geophys. J. Int. *105*, 429–465.

KENNETT, B. L. N., ENGDAHL, E. R., and BULAND, R. (1995), *Constraints on Seismic Velocities in the Earth from Travel Times*, Geophys. J. Int. *122*, 108–124.

LILWALL, R. C., and DOUGLAS, A. (1970), *Estimation of P-wave Travel Times Using the Joint Epicentre Method*, Geophys. J. R. Astron. Soc. *19*, 165–181.

MORELLI, A., and DZIEWONSKI, A. *Joint determination of lateral heterogeneity and earthquake location.* In *Glacial Isostasy, Sea-level and Mantle Rheology* (eds. R. Sabadini *et al.*) (Kluwer Academic Publishers, The Netherlands, 1991) pp. 515–534.

MORELLI, A., and DZIEWONSKI, A. (1993), *Body Wave Travel Times and a Spherically Symmetric P- and S-wave Velocity Model*, Geophysical J. Int. *112*, 178–194.

PIROMALLO, C., and MORELLI, A. (1998), *P-wave Propagation Heterogeneity and Earthquake Location in the Mediterranean Region*, Geophysical J. Int. *135*, 232–254.

ROBERTSON, G. S., and WOODHOUSE, J. H. (1995), *Evidence for Proportionality of P and S Heterogeneity in the Lower Mantle*, Geophys. J. Int. *123*, 85–116.

ROBERTSON, G. S., and WOODHOUSE, J. H. (1997), *Comparison of P- and S-station Corrections and their Relationship to Upper Mantle Structure*, J. Geophys. Res. *102*, 27,355–27,366.

SMITH, G. P., and EKSTRÖM, G. (1996), *Improving Teleseismic Event Locations Using a Three-Dimensional Earth Model*, Bull. Seismol. Soc. Am. *86*, 788–796.

SPAKMAN, W., and NOLET, G. *Imaging algorithms, accuracy and resolution in delay time tomography*, In *Mathematical Geophysics* (eds. Vlaar N.J., Nolet G., Wortel M.J.R. and Cloething S.A.P.L.) (D. Reidel Publishing Company, 1988), pp. 155–187.

SPAKMAN, W., VAN DER LEE, S. and VAN DER HILST, R. D. (1993), *Travel-time Tomography of the European–Mediterranean Mantle Down to 1400 km*, Phys. Earth Planet. Interiors *79*, 3–74.

SOURIAU, A., and WOODHOUSE, J. H. (1985), *A worldwide Comparison of Predicted S-wave Delays from a Three-dimensional Upper Mantle Model with P-wave Station Corrections*, Phys. Earth Planet. Interiors, *39*, 75–88.

SU, W., and DZIEWONSKI, A. M. (1993), *Joint 3-D Inversion for P- and S-velocity in the Mantle*, EOS *74*, 557.

TOY, K., *Tomographic Analysis od ISC Travel Time Data for Earth Structure*, Ph.D. Thesis, Univ. of Calif., San Diego, 1989.

VAN DER HILST, R. D., and ENGDAHL, E. R. (1992), *Step-wise Relocation of ISC Earthquake Hypocenters for Linearized Tomographic Imaging of Slab Structure*, Phys. Earth Planet. Inter. *75*, 39–53.

VAN DER HILST, R. D., ENGDAHL E. R., and SPAKMAN, W. (1993), *Tomographic Inversion of P and pP Data for Aspherical Mantle Structure below the Northwest Pacific Region*, Geophys. J. Int. *115*, 264–302.

VASCO, D. W., PULLIAM, J. R., JOHNSON L. R., and EARLE, P. S. (1994), *Robust Inversion of IASPE191 Travel Time Residuals for Mantle P- and S-velocity Structure, Earthquake Mislocations, and Station Corrections*, J. Geophys. Res. *99*, 13,727–13,755.

ZHOU, H., and WANG, H. (1994), *A Revisit to P Wave Travel-time Statics at Teleseismic Stations*, J. Geophys. Res. *99*, 17,849–17,862.

(Received July 15, 1999, revised February 4, 2000, accepted February 9, 2000)

Pure appl. geophys. 158 (2001) 349–374
0033–4553/01/020349–25 $ 1.50 + 0.20/0

Pure and Applied Geophysics

Robust Bearing Estimation for Three-component Stations

JOHN P. CLAASSEN[1]

Abstract—A robust bearing estimation process for 3-component stations has been developed and explored. The method, called SEEC for Search, Estimate, Evaluate and Correct, intelligently exploits the inherent information in the arrival at every step of the process to achieve near-optimal results. In particular, the approach uses a consistent framework to define the optimal time-frequency windows on which to make estimates, to make the bearing estimates themselves, to construct metrics helpful in choosing the better estimates or admitting that the bearing is immeasurable, and finally to apply bias corrections when calibration information is available to yield a single final estimate. The algorithm was applied to a small but challenging set of events in a seismically active region. It demonstrated remarkable utility by providing better estimates and insights than previously available. Various monitoring implications are noted from these findings.

Key words: Bearings, estimation, polarization, corrections, location, association.

Introduction

As the monitoring thresholds of global and regional networks are lowered, bearing estimates become more important to the processes which associate sparse detections and which locate events. Current methods of estimating bearings from observations by three-component (3-c) stations lack both accuracy and precision (ALEWINE, 1992; KOCH and KRADOLFER, 1997). Methods are required which will develop all the precision inherently available in the arrival, determine the measurability of the arrival, provide better estimates of the bias induced by the medium, permit estimates at lower SNRs, provide an indication of the precision of the estimate, and develop insights into the effects of the medium on the estimates.

Various approaches have been advocated to estimate the polarization properties and bearings associated with seismic arrivals. The theoretical basis for defining the polarization properties of signals appears to have arisen from the seminal work of BORN and WOLF (1975) where analytical (complex-valued) signals and notions of coherency were introduced for narrowband signals. It was perhaps this work that

[1] Monitoring Technologies Department, Sandia National Laboratories, Albuquerque, NM 87185-0975, U.S.A., E-mail: jsclaass@swcp.com

motivated SAMSON (1983) to define the degree to which signals are polarized in higher dimensional spaces. Later efforts by VIDALE (1986) reminded us how important it is to define polarization in terms of analytical signals and that the general vibrational modes associated with three-component observations are elliptically polarized. MAGOTRA *et al.* (1987) showed how the horizontal components of an entire wavetrain could be used to estimate backazimuths. Here the polarity of the z component relative to the horizontal signal components was used to remove the azimuthal ambiguity. JURKEVICS (1988), limiting the analysis of P arrivals to real-valued (as opposed to complex-valued) signals, demonstrated how to construct an observational covariance matrix by averaging the covariance matrix over 3-c elements of an array to realize bearing estimates having more precision. Although Jurkevics advocated the use of polarization properties to isolate intervals on which to make bearing estimates, no viable rationale to isolate the appropriate estimation interval was offered. Indeed, Jurkevics suggested using signal frequency bands having the best SNR relative to the background noise as a method for defining a measurement interval. However, this approach fails in practice because it does not account for signal-induced noise. In a more recent work WAŁCK and CHAEL (1991) evaluated various combinations of frequency bands and time-windows in the context of four bearing estimation techniques (MAGOTRA *et al.*, 1987; FLINN, 1965; CHRISTOFFERSSON *et al.*, 1988; ROBERTS *et al.*, 1989) to demonstrate the bearing estimation capability of the three-component RSTN. They demonstrated that bearings for events in eastern North America could be estimated with an accuracy of 6 degrees rms if the estimates from station RSSD, residing on a "pathological" geology, are eliminated. In a notable work by LILLY and PARK (1995) complex-valued orthogonal Slepian wavelets were used to elegantly estimate the polarization properties of seismic signals.

The 3-c bearing estimation method disclosed in this work differs from prior efforts in that it makes the following assumptions: 1) not all frequencies or time intervals will be equally effective in producing measurable bearings, 2) only those time-frequency windows which have good polarization properties will result in precise estimates, 3) the estimates themselves may be frequency-dependent, 4) those estimates having the largest effective degrees of freedom (a confidence measure) will experience the greatest variance reduction and are the best candidates to combine into a refined estimate which may include bias corrections, and 5) the bearing estimates associated with an event from a given epicentral region will, in general, require correction for frequency-dependent lateral refraction or diffraction. A bearing estimation method, called SEEC for Search, Estimate, Evaluate, and Correct, was developed on these underlying assumptions. Earlier disclosures of this technique can be found in CLAASSEN (1998a, b).

In the following section the SEEC methodology is disclosed. To evaluate the bearing estimator the method was applied to a small but challenging set of regional events observed in an active tectonic region of the western United States by a station located near Pinedale, Wyoming. In subsequent sections the results of this evaluation are described and discussed. To assist in the interpretation of the results, various geological features in and about the station are described. In a final section conclusions regarding the effectiveness of the SEEC method are offered and several monitoring implications are highlighted.

The Basis for Implementing the SEEC Method

The block diagram of Figure 1 provides an overview of the SEEC approach. Three-component signals are spectrally decomposed by a quadrature filter bank. The resulting quadrature time series are then searched over short time intervals for suitable polarization properties. SEEC produces bearing estimates on the favorable time intervals of adequate duration and also constructs metrics helpful in evaluating and selecting estimates or discarding all estimates when none is possible. Bias corrections are applied to the accepted estimates and a single final estimate is then constructed from the bias-corrected estimates in a near-optimal way.

This section details the basis for implementing the SEEC method. It addresses the procedure by which to design the quadrature filter bank. Several useful polarization measures are introduced. A pseudo-SNR estimator is constructed from these measures. The manner in which the SNR estimator is used to search the 3-c records for the more favorably polarized regions in time and frequency is described. Two related methods of estimating the bearings on contiguous well-polarized time-frequency intervals and their relative utility are then presented. Finally, a method of

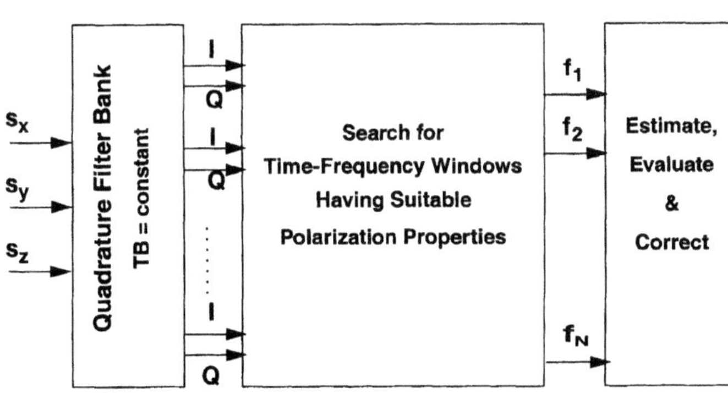

Figure 1
A block diagram of the SEEC bearing estimation process is shown here. The quadrature filter bank decomposes the three-component signals into overlapping spectral bands. A search process finds those time-spectral intervals having favorable polarization properties. Bearing estimates are made on those intervals. The estimates having experienced sufficient variance reduction are included in a frequency-dependent bias correction process to achieve near-optimal results.

evaluating and selecting the estimates having better precision and a means of correcting them for the bias introduced by the earth are addressed.

The Design of the Quadrature Filter Bank

To search spectral domains for those time intervals having the best polarization properties, the 3-c signals were filtered by a quadrature digital filter bank whose individual filters are based on orthogonal Slepian wavelets (LILLY and PARK, 1995) having a constant time-bandwidth product (TB). The wavelets were weighted (tapered) with a Hanning window to reduce spectral leakage into the filter sidebands. Experience with actual signals had demonstrated that the reduction of leakage from adjacent sidebands is essential to prevent signals in adjacent bands from influencing the bearing estimate in the band of interest. The constant TB property of the filters permits one to logarithmically space the filter center frequencies for efficient spectral coverage. The rationale for defining the individual filters in the bank using the notation of LILLY and PARK (1995) is described below.

If the entire frequency interval of interest is defined by $[f_1, f_2]$ and N is the desired number of filters to cover that bandwidth, the frequency separation for the filters in logarithmic units can be specified by

$$d_f = \log\left(\frac{f_2}{f_1}\right)/(N-1). \tag{1}$$

The filter center frequencies are, therefore, given by

$$f_c(n) = 10^{\log[f_1+(n-1)d_f]} \tag{2}$$

where $n = 1, 2, \ldots, N$. If p is defined as the target $TB/2$ product and p_o is defined as the target center frequency-time product for the filter, then the required half bandwidth for the n-th filter is

$$f_w(n) = pf_c(n)/p_o \tag{3}$$

and the required filter length in samples is

$$M_w(n) = nint[pf_s/f_w(n)] \tag{4}$$

where *nint* is the nearest integer function and f_s is the sampling frequency. If $M_w(n)$ is even, its order is increased by 1 to simplify the alignment of the signals emerging from digital filters having different lengths. The actual resulting wavelet parameters are given by

$$p_c(n) = f_c(n)M_w(n)/f_s \tag{5}$$

and

$$p_w(n) = f_w(n)M_w(n)/f_s \tag{6}$$

$p_c(n)$ and $p_w(n)$ will closely approximate p and p_o, respectively, for every n, but will be constrained by the integral property of $M_w(n)$.

As shown by Lilly and Park, the parameters $M_w(n), p_c(n)$, and $p_w(n)$ are used to define a Toeplitz matrix whose eigenvectors define a family of orthogonal wavelets centered at $f_c(n)$. Adjacent pairs of eigenvectors form quadrature filters having almost identical bandpass characteristics. Although the dominant eigenvector is the solution to the posed bandwidth maximization problem, the subsequent eigenvector pairs can be used to expand the frequency band to the target width. In the filtering operation, quadrature filter pairs contribute statistically *orthogonal* complex-valued (analytic) signals centered on $f_c(n)$. The quadrature signals after filtering with tapered filters become statistically *independent* and have a slightly larger bandwidth.

A provision was included in the SEEC process to partially whiten the signals with a low-order prediction method thereby reducing leakage of dominant spectral content into the desired passband when a strong narrowband signal is processed. This technique was found to be essential in extracting good estimates at the lowest frequencies, which are sensitive to estimation error resulting from leakage of unwanted frequencies as well as from intense background noise.

Setting the filter parameters to $p = 5$ and $p_o = 2$ appeared to work best for signals sampled at 50 Hz. Although these parameters should produce three quadrature pairs for every center frequency to realize the specified bandwidth (see LILLY and PARK, 1995), the quadrature pairs were limited to two. This choice, after Hanning weighting, not only provided the proper frequency overlap between filter bands but also provided suitable spectral resolution to define a bearing unique to that center frequency (recall the spectral leakage problem). The band of interest ranged from 0.5 Hz to 15 Hz and was covered by 12 sets of quadrature filters. The spectral characteristic of this filter bank is shown in Figure 2. The Hanning taper reduced filter sidelobes to about −50 dB relative to the central passband compared to the −25 dB that occurred without tapering.

Searching for the Polarized Regions in the Waveforms

The frequency band centered on frequency f_c having half bandwidth f_w is spanned by $M > 0$ quadrature filter pairs (M is not related to M_w). The argument n is dropped at this point for convenience. Define a vector of $2k + 1$ sequential quadrature (complex-valued) signal samples centered on the sample j from the i-th sensor axis and m-th quadrature filter as

$$\mathbf{s}_{im}(f_c, j) = [s_{im}(j-k) \quad s_{im}(j-k+1) \ldots s_{im}(j) \ldots s_{im}(j+k-1) \quad s_{im}(j+k)]^T \qquad (7)$$

where $i \in \{x, y, z\}, m = 1, 2, \ldots, \mathrm{M}$, and T is the transpose operator. Here bold symbols will denote vectors and matrices. Vectors of samples from the M quadrature filters are used to prepare a rectangular observation matrix of the form

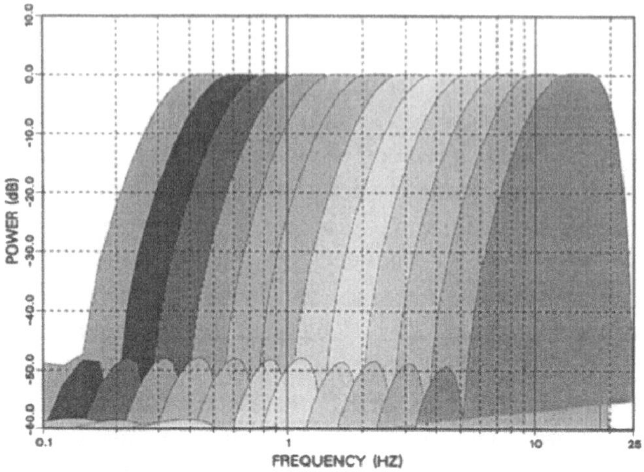

Figure 2
The bandpass characteristics of the filters in the quadrature filter bank are illustrated here. The individual filters, characterized by a constant TB product, have center frequencies which are logarithmically spaced to efficiently cover the band of interest. Frequency decomposition of the incoming signal is important because, as shown in this work, bearings can be strongly frequency-dependent.

$$\mathbf{R}(f_c, j) = \begin{bmatrix} \mathbf{s}_{x1}(f_c, j) & \mathbf{s}_{y1}(f_c, j) & \mathbf{s}_{z1}(f_c, j) \\ \vdots & \vdots & \vdots \\ \mathbf{s}_{xM}(f_c, j) & \mathbf{s}_{yM}(f_c, j) & \mathbf{s}_{zM}(f_c, j) \end{bmatrix}. \tag{8}$$

Each vector row contributes additional degrees of freedom to the observation matrix through additional frequency content. The singular values λ_ℓ and right singular vectors $\mathbf{e}_\ell$ of $\mathbf{R}$, where $\ell = 1, 2$, or 3, are used to form polarization measures which are used to search intervals of time having suitable polarization properties. The measures found to be useful in defining polarized signal segments are defined below.

These include the degree of polarization

$$\mathrm{DOP}(f_c, j) = \frac{\frac{1}{\sqrt{2}}\sqrt{(\lambda_1 - \lambda_2)^2 + (\lambda_2 - \lambda_3)^2 + (\lambda_3 - \lambda_1)^2}}{\lambda_1 + \lambda_2 + \lambda_3} \tag{9}$$

the degree of linear polarization

$$\mathrm{DOL}(f_c, j) = \frac{\mathrm{Real}^2(e_{1x}) + \mathrm{Real}^2(e_{1y}) + \mathrm{Real}^2(e_{1z})}{||\mathbf{e}_1||^2} \tag{10}$$

where $\mathbf{e}_1$ is the dominant singular vector, and the degree of dyadicy

$$\mathrm{DOD}(f_c, j) = \frac{\lambda_1}{\lambda_1 + \lambda_2 + \lambda_3}. \tag{11}$$

The definition of the degree of linear polarization (DOL) is thought to be unique to this work. It is premised on the fact that linear polarization must be defined by real-valued singular vectors. The DOL polarization measures how close the dominant eigenvector is to being real-valued. This definition of "linearity" is important to characterizing a propagation environment in which multipath, mode conversion, and scattering introduce noise which may be correlated with the linearly polarized transmission.

From the above definitions various pseudo-signal-to-noise ratios (SNR) can be constructed to include

$$\mathrm{SNR}_1(f_c, j) = \frac{\mathrm{DOD}(f_c, j)}{1 - \mathrm{DOD}(f_c, j)} \tag{12}$$

and by analogy

$$\mathrm{SNR}_2(f_c, j) = \frac{\mathrm{DOP}(f_c, j)}{1 - \mathrm{DOP}(f_c, j)}. \tag{13}$$

These definitions of SNR actually account for noise components which may be uncorrelated or correlated with the signal. The pseudo-SNRs are based on coherency-like notions but are not strictly rigorous. This is readily understood from Equation (12) for the case of noncoherent noise and a stationary polarized signal. Under these assumptions, $\lambda_1 = \sigma_s^2 + \sigma_n^2$ and $\lambda_2 = \lambda_3 = \sigma_n^2$ where σ_s^2 is the signal variance and σ_n^2 is the noise variance. In this case SNR_1 becomes

$$\mathrm{SNR}_1(f_c, j) = \frac{\sigma_s^2 + \sigma_n^2}{2\sigma_n^2}. \tag{14}$$

When the SNR becomes large, this pseudo-SNR approaches half the actual SNR.

When estimating the polarization properties the lengths of the signal vectors contributing to **R** were restricted to a fraction of the filter length, typically 1/4 or 1/2. At each center frequency this selection assures that the polarization properties of the signal can be identified with sufficient temporal resolution while retaining one or two cycle(s) of the filtered waveform to fill **R** (see Equation (8)).

To illustrate the screening capability of two of the above polarization measures, DOP and DOL were computed and ploted for a regional *P* arrival. These properties were portrayed by gray-tone encoding the intensity of the measure in a two-dimensional setting, where time runs horizontally and frequency vertically. In interpreting these polarization spectrograms, the reader is cautioned that a polarization response can occur before the actual onset of the signal since the quadrature filters are non-causal.

The intensity plot of Figure 3 illustrates the degree to which polarization DOP is localized in time and frequency for a *P* arrival. Since the actual type of polarization is not indicated by this measure, DOP must be assisted by the test for linear polarization. The plot for Figure 4 indicates that the DOL delineates the linearly

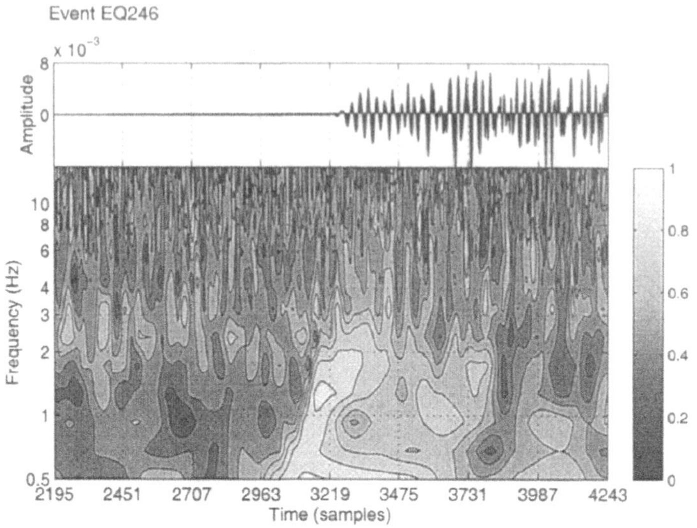

Figure 3
This gray-scale encoded contour plot illustrates how the degree of polarization (DOP) can localize time and frequency intervals having highly polarized content. These estimates were formed from the complex-valued signals produced by the filter bank illustrated in Figure 2. When the DOP approaches one, the signal is regarded as highly polarized. On the other hand when the DOP approaches zero, the signal is regarded as unpolarized. The vertical component of the signal contributing to the DOP estimates appears in the trace above and has been synchronized with the DOP estimates. Be aware that a polarization response can occur before the onset of the signal because the filters are non-causal.

polarized regions clearly. It is not surprising to see that both the background noise and signal frequently exhibit linear polarization content over short time intervals. However, the waveform energy must be concentrated in the principal component if it is to be truly linearly polarized. The DOD is capable of demonstrating this concentration. Since an empirical investigation indicated that DOP is highly correlated with DOD, DOP may be as a substitute for DOD. It was also observed that the DOP resolves polarization in time and frequency better than does the DOD.

In view of the properties noted above, the interval selection process was actually based on a pseudo-SNR defined in the following way

$$\mathrm{SNR}_3(f_c,j) = \frac{\sqrt{\mathrm{DOP}(f_c,j)\mathrm{DOD}(f_c,j)}\mathrm{DOL}_{xy}(f_c,j)}{1.0 - \sqrt{\mathrm{DOP}(f_c,j)\mathrm{DOD}(f_c,j)}\mathrm{DOL}_{xy}(f_c,j)}. \tag{15}$$

The use of both DOP and DOD through a geometric mean assures that both measures influence the interval selection process with DOP tending to increase the resolution. In addition DOL was reformulated as

$$\mathrm{DOL}_{xy}(f_c,j) = \frac{\mathrm{Real}^2(e_{1x}) + \mathrm{Real}^2(e_{1y})}{|e_{1x}|^2 + |e_{1y}|^2}. \tag{16}$$

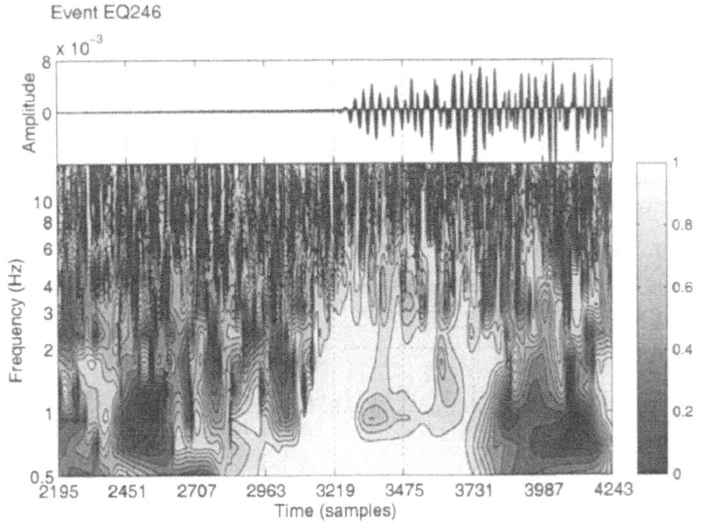

Figure 4
This gray-scale encoded contour plot illustrates how the degree of linear polarization (DOL) as defined in this work provides additional information for locating polarized content for the same event illustrated in Figure 3. Whenever DOL approaches one within a time-frequency interval, the signal or noise is potentially linearly polarized. Rigorously, those intervals are only linearly polarized when the degree of dyadicy nears one. This polarization measure together with others serve in selecting favorable intervals on which to make bearing estimates.

This form of DOL was motivated by the fact that the SVD algorithm conveniently returned a real-valued z component for the principal eigenvector (the z component served as the phase reference for the eigenvector). Hence this form for DOL slightly heightens the sensitivity of SNR_3 to linear polarized components by imposing real valuedness on the two horizontal components which really convey the bearing information.

Estimating the Bearings within the Same Mathematical Framework

Bearing estimates can be formed in a number of ways as discussed by WAŁCK and CHAEL (1991). In this work the estimates of the bearing and apparent incident angles are based on the right principal singular vector of a rectangular observation matrix (Equation (8)) associated with a selected window simply because the method is compatible with the formulation for the polarization search parameters. Contiguous points are regarded as candidates for a bearing estimate if the psuedo-$SNR_3(f_c, j)$, now denoted as $\mathbf{r}(f_c, j)$, exceeds one for every j and the duration of the concatenated subintervals is longer than two cycles. Suppose the points $j \in \{J - L, J - L + 1, \ldots, J, \ldots, J + L - 1, J + L\}$ satisfy this SNR criteria. On such an interval, two approaches, both based on the principal component method, were taken to estimate the angles.

The first approach makes an estimate on the subinterval of $2k + 1$ points for each j. Let the collection of resulting bearing and incident angle estimates be given by $[\phi(j), \theta(j)], j \in \{J - L, J - L + 1, \ldots, J, \ldots, J + L - 1, J + L\}$. To assure physically realizable incident angles, when $\theta(j) < 90°$ for any j, set $\theta(j) = 180° - \theta(j)$ and $\phi(j) = \phi(j) + 180°$. This action is mathematically justified by the fact that if **e** is a singular vector, then $-\mathbf{e}$ will also be an eigenvector. Since the pseudo-SNR will vary throughout the interval, it is appropriate to estimate the bearing on the basis of a weighted mean given by

$$\bar{\Phi}(f_c, J) = \left(\sum_{J-L}^{J+L} r(f_c, j)\phi(j)\right) \Big/ \left(\sum_{J-L}^{J+L} r(f_c, j)\right). \tag{17}$$

A variance given by

$$\operatorname{var}[\bar{\Phi}(f_c, J)] = \left(\sum_{J-L}^{J+L} r(f_c, j)\phi(j)\phi(j)\right) \Big/ \left(\sum_{J-L}^{J+L} r(f_c, j)\right) - \bar{\Phi}(f_c, J)^2 \tag{18}$$

and an average psuedo-SNR given by

$$r_{\bar{\Phi}} = \left(\sum_{J-L}^{J+L} r(f_c, j)\right) \Big/ (2L + 1) \tag{19}$$

may also be associated with $\bar{\Phi}$. Admittedly the estimates of the means and variances are biased because the estimates are based on correlated (overlapping) samples. However, if the variance is small, the mean established in this manner has been found to be acceptable in practice. Furthermore, if the variance is large, the estimate should be discarded anyway.

The second method of estimating the bearing simply constructs an observation matrix over the entire interval centered on J and uses its right principal singular vector and eigenvalues to construct a single estimate (Φ, Θ) and its pseudo-SNR given by Equation (15) with $j = J$, here denoted as r_Φ.

The underlying polarization parameters are dictated by the entire interval as opposed to the subintervals. The observation matrix, however, is modified in a special way to emphasize the individual rows having a better pseudo-SNR, namely,

$$\mathbf{R}(f_c, f_w) = \mathbf{W} \begin{bmatrix} s_{x1}(f_c, (J-L)) & s_{y1}(f_c, J-L) & s_{z1}(f_c, J-L) \\ \cdot & \cdot & \cdot \\ s_{xM}(f_c, J-L) & s_{yM}(f_c, J-L) & s_{zM}(f_c, J-L) \\ \vdots & \vdots & \vdots \\ s_{x1}(f_c, (J+L)) & s_{y1}(f_c, J+L) & s_{z1}(f_c, J+L) \\ \cdot & \cdot & \cdot \\ s_{xM}(f_c, J+L) & s_{yM}(f_c, J+L) & s_{zM}(f_c, J+L) \end{bmatrix} \tag{20}$$

where $\mathbf{W}$ is a $(2L+1)M$ by $(2L+1)M$ weighting matrix given by

$$\mathbf{W} = \begin{bmatrix} r(f_c, J-L)\mathbf{I} & & & & 0 \\ & r(f_c, J-L+1)\mathbf{I} & & & \\ & & r(f_c, J)\mathbf{I} & & \\ & & & r(f_c, J+L-1)\mathbf{I} & \\ 0 & & & & r(f_c, J+L)\mathbf{I} \end{bmatrix} \tag{21}$$

and $\mathbf{I}$ is a M by M identity matrix and the individual $r(f_c, j)$ are the pseudo-SNRs associated with the individual $2k+1$ long subintervals surrounding each $j \in \{J-L, J-L+1, \ldots, J, \ldots, J+L-1, J+L\}$. Here, the row elements in Equation (20) are not column vectors.

In practice, estimates of either type having shallow apparent elevation (incident) angles were discarded operationally because the bearings were typically absurd. The use and choice of the two bearing estimators are treated in the following subsection.

Evaluating and Correcting the Estimates

SEEC yields a bearing for each time-frequency window considered measurable by the criterion described above. To select the better estimates from the entire set of bearings, an information theoretic criterion commonly associated with estimation theory is applied to each estimate. The criterion simply states that the variance of an estimate is reduced by 1.0/(SNR*TB) where SNR and TB characterize the signals contributing to the estimate. For an example of this principle, see Equation (11) in an article by SHUMWAY and KIM (1998). As a consequence, estimates whose SNR*TB exceeds a given threshold are accepted as precision estimates. The others are discarded. In this work SNR*TB is defined as the *effective* degrees of freedom (DOF) associated with an estimate. Experience with events from the same locale has shown that the estimates have utility when DOF > 10.

Since two estimates are potentially available for each interval satisfying the DOF criterion, a method was constructed to select between the two estimates. Since the standard deviation is a measure of consistency of the bearings (singular vectors) throughout the interval, estimates of either form are only accepted if $\text{var}^{1/2}[\bar{\Phi}(f_c, J)] < 15$ degrees. Furthermore, $\bar{\Phi}$ was accepted over Φ only if $r_{\Phi}\text{TB} > 10$ and $r_{\bar{\Phi}}\text{TB}$ exceeded $r_{\Phi}\text{TB}$ by 5. This choice indicates a strong preference for estimates over the entire interval as opposed the average estimate over many subintervals. This presumes that better estimates are possible if the rectangular observation matrix has a large number of effective degrees of freedom. Of course the polarization coherency must be sustained throughout the entire interval to build an estimate on self-consistent information. $\text{var}[\bar{\Phi}(f_c, J)]$ is an indicator as to whether this consistency really exists. This selection process is applied to all time-frequency intervals satisfying the DOF requirement and may result in several estimates or possibly none.

When multiple estimates satisfying the acceptance criterion are available for an event, the effects of frequency-dependent bearing biases may have to be addressed. Only then can all the inherent accuracy and precision associated with the accepted bearings be achieved. The biases for an epicentral region of the earth may be determined by using well-located reference events within that region. Suppose that the frequency-dependent estimates $\hat{\Phi}_i(f_c)$ and their degrees of freedom $\mathrm{DOF}_i(f_c)$ as observed by a particular station are available for reference events $i = \{1, 2, \ldots, I\}$ within the epicentral region. The frequency-dependent biases are compared with the true (great-circle) bearings Φ_i to establish the bearing bias in the following near-optimal manner

$$\mathrm{bias}(f_c) = \frac{\sum_i [\hat{\Phi}_i(f_c) - \Phi_i] \cdot \mathrm{DOF}_i(f_c)}{\sum_i \mathrm{DOF}_i(f_c)}. \tag{22}$$

Once the regional bearing biases are established as a function of frequency for a particular station, estimates $\hat{\Phi}(f_c)$ for a subsequent event with effective degrees of freedom $\mathrm{DOF}(f_c)$ arising from the same epicentral region may be corrected for bias in the following near-optimal manner

$$\hat{\Phi} = \frac{\sum_{f_c} [\hat{\Phi}(f_c) + \mathrm{bias}(f_c)] \cdot \mathrm{DOF}(f_c)}{\sum_{f_c} \mathrm{DOF}(f_c)} \tag{23}$$

to achieve a single bearing with all the inherently available accuracy and precision. Conceivably the summation could also extend over different time intervals as well as frequency intervals.

Application of SEEC to Event Data

The polarization-based bearing estimator was applied to a reasonably large set of regional P arrivals arising from events located in the western US and observed by a high-frequency station located near Pinedale, Wyoming. The locations of the events analyzed as part of the entire effort appear on the map of Figure 5. The event truth for the events appearing in this article is listed in Table 1. The arrivals were extracted from the S-3 seismometer channels, which had been electronically altered to provide an acceleration output having a nominal 0.5 to 80 Hz passband which was sampled at 200 Hz. The bearing estimates were restricted to the central frequencies associated with the filter bank (0.5 to 15 Hz) after having decimated the data to a 50 Hz sample

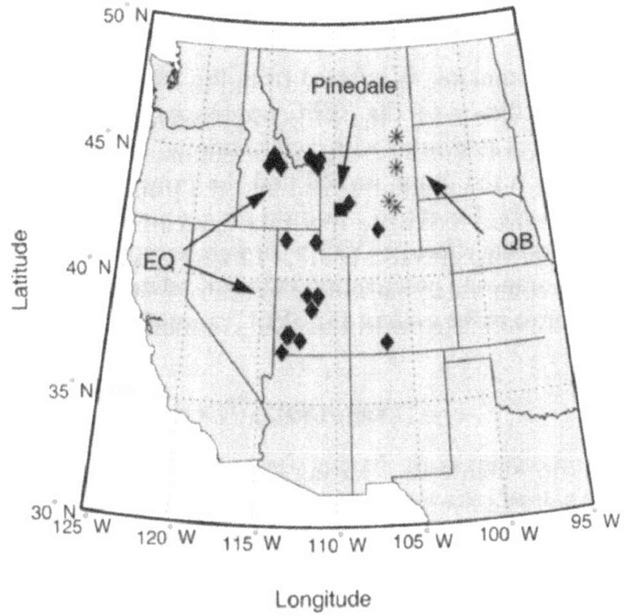

Figure 5
The location of events contributing to an evaluation of the SEEC algorithm are shown above. Earthquakes (EQ) are designated by shaded diamonds and quarry blasts (QB) by asterisks. Not all events shown here are addressed within this paper but are located on the map to illustrate that the SEEC algorithm was applied to a reasonably large set of events sampling different paths. The location of the Pinedale station, whose observations contributed to the evaluation, is identified by a shaded square.

Table 1

The ground truth for events appearing within this article

ID	Latitude	Longitude	State	Date	Time	Mag.	Depth	Range
EQ246	37.090	−113.472	UT	09/02/92	10:26:20.9	5.8	15	6.43
EQ355	37.567	−112.322	UT	12/19/91	20:26:35.7	3.6	7	5.62
EQ181a	37.736	−113.171	UT	06/29/92	01:59:06.0	4.1	1.4	5.75
EQ181b	37.733	−113.173	UT	06/29/91	22:20:45.8	3.8	3.1	5.71
EQ081	37.816	−112.995	UT	03/22/91	14:59:59.2	3.2	2.0	5.61
EQ221	38.788	−111.587	UT	08/09/91	08:49:20.0	3.5		4.27
EQ176	38.783	−111.554	UT	06/24/92	07:31:20.2	4.4		4.27
EQ193	38.322	−111.123	UT	07/11/92	13:23:07.8	4.4		4.27
EQ161	39.302	−111.159	UT	06/09/92	23:30:18.6	3.4	6.1	3.7
EQ074	39.352	−111.171	UT	03/13/91	20:33:14.5	3.2	9.0	3.7
EQ303	39.340	−111.161	UT	10/29/92	06:57:43.1	2.9	1.5	3.6
EQ326	39.332	−111.155	UT	11/19/92	06:46:57.5	3.0	5.7	3.6
QB288	44.352	−105.391	WY	10/14/92	17:31:52.8			3.4
QB151	42.780	−105.610	WY	05/31/91	21:44:53.3			2.9
EQ031	43.020	−108.947	WY	01/31/92	12:01:47.9	3.4	5 (g)	0.5
EQ133	44.533	−111.312	MT	05/13/91	14:13:05.8	3.4	5 (g)	2.1
EQ134	44.438	−111.217	MT	05/14/91	01:49:43.7	3.1	5 (g)	2.1

rate. Contiguous samples were included in a measurable interval if $r(f_c, j) > 1$ for every j in the interval.

The bearing error estimates, as inferred from the NEIC locations, are presented here to illustrate the efficacy of the SEEC approach. The SEEC errors are also compared with the errors associated with a conventional broadband estimator whose solutions were based on the initial arrival and the principal eigenvector of a real-valued covariance matrix. Except to illustrate the features of the SEEC estimator, only those estimates having effective DOF > 10 were accepted and plotted. For most of the error plots, a composite polarization indicator related to SNR_3 appears above the bearing error plot to make evident the SEEC rationale. The composite indicator is defined by

$$i_c = (\mathrm{DOP} \cdot \mathrm{DOD})^{1/2}\mathrm{DOL} \tag{24}$$

and nears unity on subintervals having favorable polarization properties. The bearing error lines extend over the measurable time-intervals as defined by the SNR criterion. The frequency at which the estimate was made is symbol encoded in accord with a key appearing in the attached legend. The key also provides the effective DOF and the actual bearing estimates (PHI). The following events were selected to illustrate the characteristics of SEEC and the influence of the earth's propagation environment on the bearing estimates.

Event QB288

This arrival was produced by a quarry shot in eastern Wyoming at a range of 3.4 degrees. The broadband signal-to-noise ratio for this arrival is poor as implied by the trace of Figure 6a. However, SEEC was able to identify spectral components and time intervals having usable polarized states at the middle and lower frequencies as illustrated in polarization spectrogram (Fig. 6a). The bearing error graph of Figure 6b illustrates that nearly correct estimates can be produced while the broadband approach based on a real covariance matrix fails. Here acceptable estimates primarily appear at the onset of the arrival.

Event EQ246

The estimator was applied to an arrival emanating from a magnitude 5.8 earthquake in southern Utah at a range of 6.4 degrees. The arrival produced strong, linearly polarized content at frequencies less than 3 Hz as illustrated in Figure 7a. The higher frequency selectivity of the constant TB filters at the lower frequencies enabled SEEC to isolate polarized components at frequencies as low as 0.5 Hz, a frequency at which this event would not have been detected.

An interesting bearing dispersion characteristic emerged for this arrival as indicated by the graph of Figure 7b. The bearings move with frequency to the left of

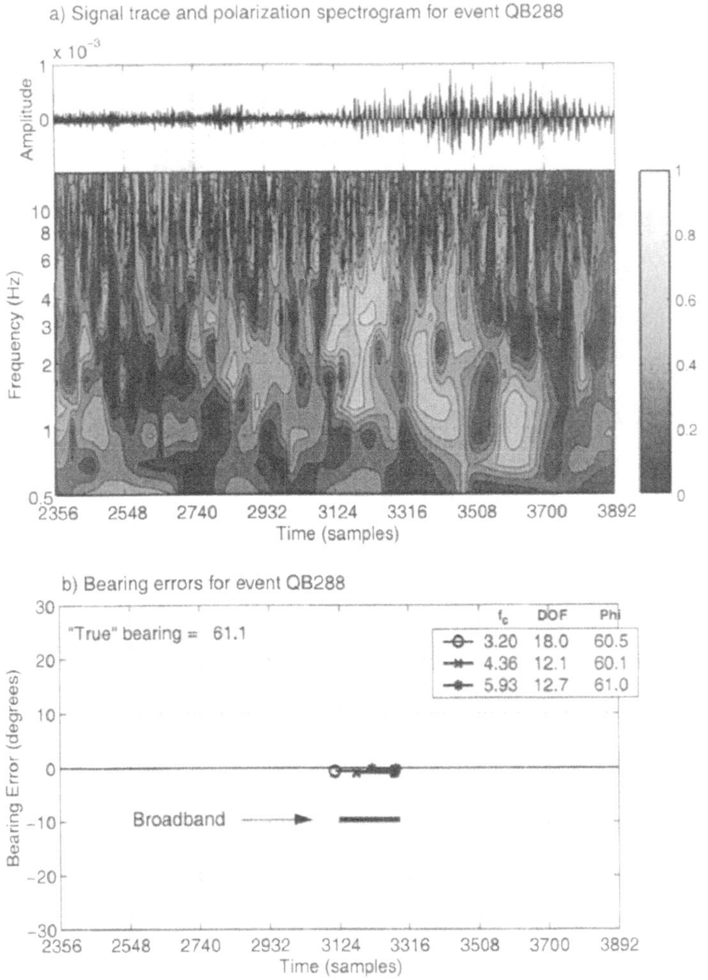

Figure 6

(a) Appearing in the upper graph is a gray-scaled encoded composite polarization indicator which identifies several disjoint windows in time and frequency having weak linear polarized content for event QB288. The indicator approaches one when linearly polarized content is present. The signal from the vertical seismometer appears above the polarization measure. (b) The lower graph illustrates the bearing errors associated with frequency-time intervals carefully defined by the polarization indicator. Error measures are extended over the estimation interval and are only illustrated for estimates having effective degrees of freedom (DOF) exceeding 10. The solid bar denotes the bearing error associated with an estimate based on the entire available spectrum as opposed to the frequency intervals defined by the quadrature filter bank. This result demonstrates that the SEEC approach is effective in defining useful intervals on which to make accurate bearing estimates.

the event as observed at the recording station. The most accurate bearing estimates occurred at the lowest frequencies. Perhaps frequency-dependent lateral refraction or diffraction causes bearing errors as large as 24 degrees to the left of the Pinedale

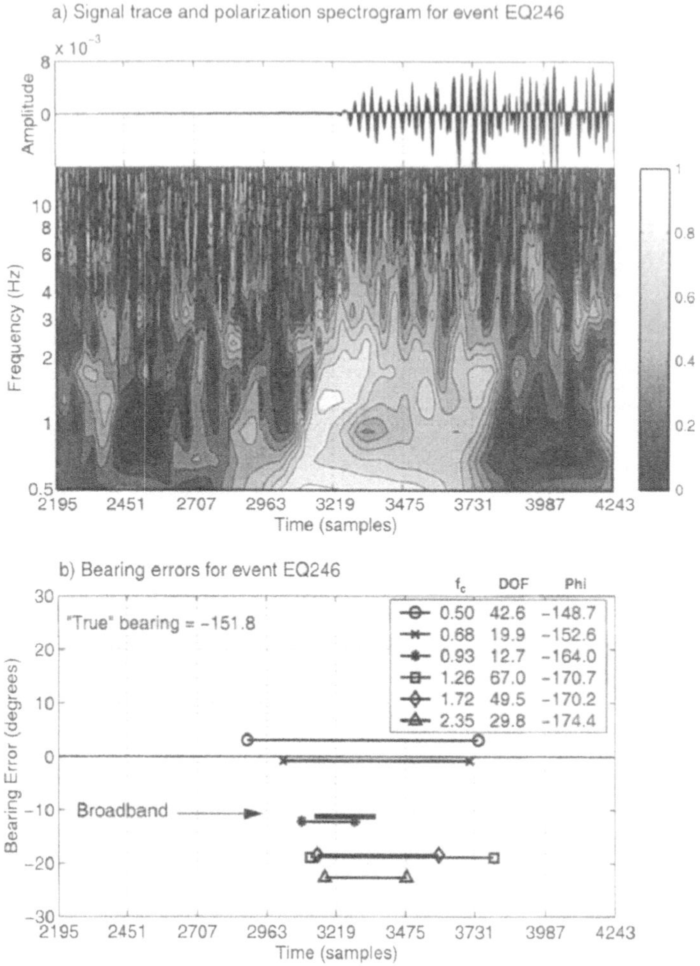

Figure 7

(a) The composite polarization indicator finds significant polarized content for event EQ246 at frequencies less than 3.0 Hz. The selectivity of the filter bank was effective in isolating polarized components at the lowest frequencies over long time intervals. The corresponding signal from the vertical seismometer appears above the polarization spectrum. (b) The lower graph illustrates the bearing errors associated with the estimation intervals having DOF greater than 10. Bearing dispersion to the left of the great circle path is evident in the error characteristic. The lower frequencies clearly experienced less frequency-dependent refraction or diffraction along this Utah to Wyoming path as one would expect.

station. Other events in the vicinity of this one exhibited similar bearing dispersion characteristics. In fact this bearing characteristics was evident for events along the Wasatch range and throughout northern Utah. Comparison of the conventional broadband estimate with the SEEC estimates appears to indicate that the broadband estimate is a frequency-weighted average of the SEEC bearings.

Event EQ031

The graph of Figure 8a illustrates the strength of the SEEC approach in finding useful polarized content in time and frequency. Here linearly polarized content was identified in both the first and second *P* arrivals over broad frequency intervals. This wavetrain was produced by a local earthquake that occurred just 1/2 degree east of the Pinedale station.

As evident in Figure 8b, frequency-dependent bearings can also occur at short ranges. The ground truth suggested a reference bearing in the NE quadrant; whereas, the low frequency estimate suggests that the event arose from the SE quadrant. The errors have been referenced to an assumed bearing of 107 degrees because a slight error in location for a nearby event causes large errors in bearing. The arrivals appear to experience bearing dispersion to the left of the station, although one cannot state this with complete confidence because the effective DOF are limited. The second *P* arrival, however, appears to confirm this possibility, because bearing estimates for the second arrival almost replicate the estimates for the first arrival. This possibly suggests that these two arrivals experienced similar crustal effects.

Events EQ133 and EQ134

Events EQ133 and EQ134 are two nearly colocated earthquakes in the Hebgen Lake area west of Yellowstone Park. The nearly identical waveforms (except for amplitude) provide additional confidence that these two events are colocated. The bearing estimates for these two events are shown in Figures 9a and 9b, respectively. The bearings for the two events cluster to the right of the great circle paths with mean biases of 25.9 and 25.4 degrees, respectively. The bearing estimates for these events suggest that the bearings may be slightly dispersed in frequency to the right as well, although orderliness with frequency is not uniquely evident. Comparisons of the estimates for the two events suggest that the SEEC method can reasonably replicate bearings with a precision somewhere between 1.2 to 1.8 degrees even though the DOF are limited. Furthermore, a comparison of the mean unweighted bearings for the two events gives an indication of the potential increase in precision possible when the individual estimates are combined in a suitable scheme. The conventional broadband estimates, whose errors are located by the arrow heads, fall among the SEEC estimates.

It is interesting to note that even though the amplitude for the initial arrival for event EQ133 was twice that of EQ134 and the estimation TB products are similar, the available DOF did not rise proportionately with the square of the event amplitude. This illustrates that the arrivals contain noise components which are correlated to the signal amplitude. As a consequence, the SNR remains nearly constant.

Event QB151

The error estimate of Figure 10b shows that the best bearing may not always be available at the onset of the arrival. This arrival arose from a quarry blast in eastern

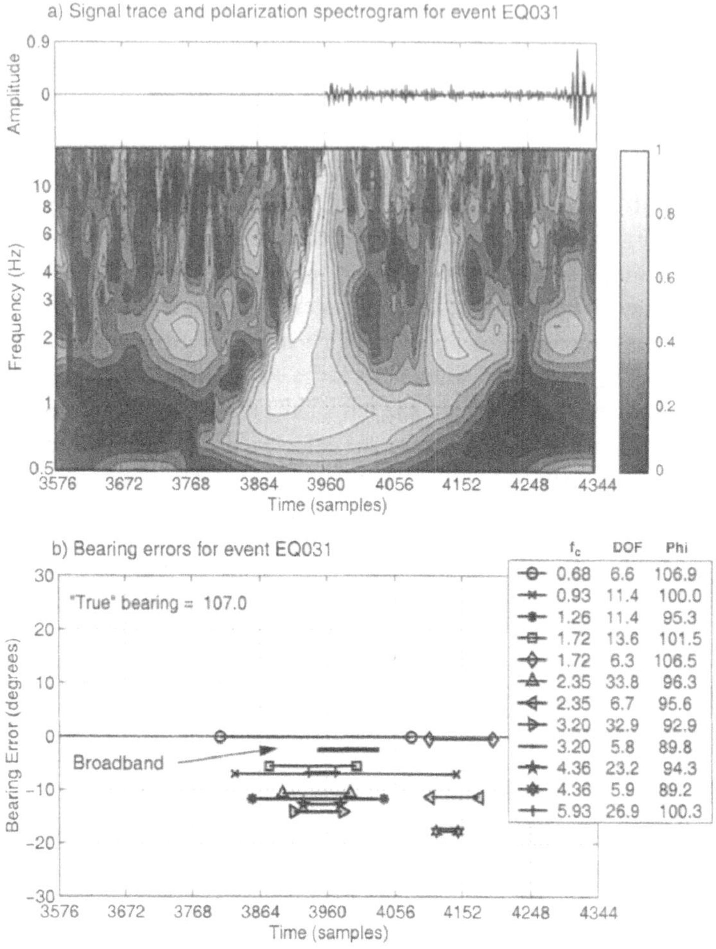

Figure 8

(a) The composite polarization indicator finds linear polarized content within two time intervals for an arrival associated with a local earthquake, EQ031. The corresponding arrivals are evident in the vertical seismometer trace above the polarization spectrum. (b) Bearing errors appear in the lower graph. The acceptance threshold (DOF) was lowered to illustrate the bearing errors for the weak second arrival. Both arrivals appear to confirm that the bearing dispersion occurs to the left of the presumed line of sight for both the first and second *P* arrival. This result illustrates that frequency-dependent bearings can also be observed at short ranges and in this case are induced by shallow structures.

Wyoming but was overridden by the wavetrain of a preceding event. The polarization spectrogram shown in Figure 10a demonstrates that the searching method can find frequency-time intervals that yield measurable estimates in the presence of interference. It is interesting to note that estimates on the initial arrival, whose errors were comparable to the broadband estimate, were also available from the SEEC process but were discarded on the basis of too few DOF.

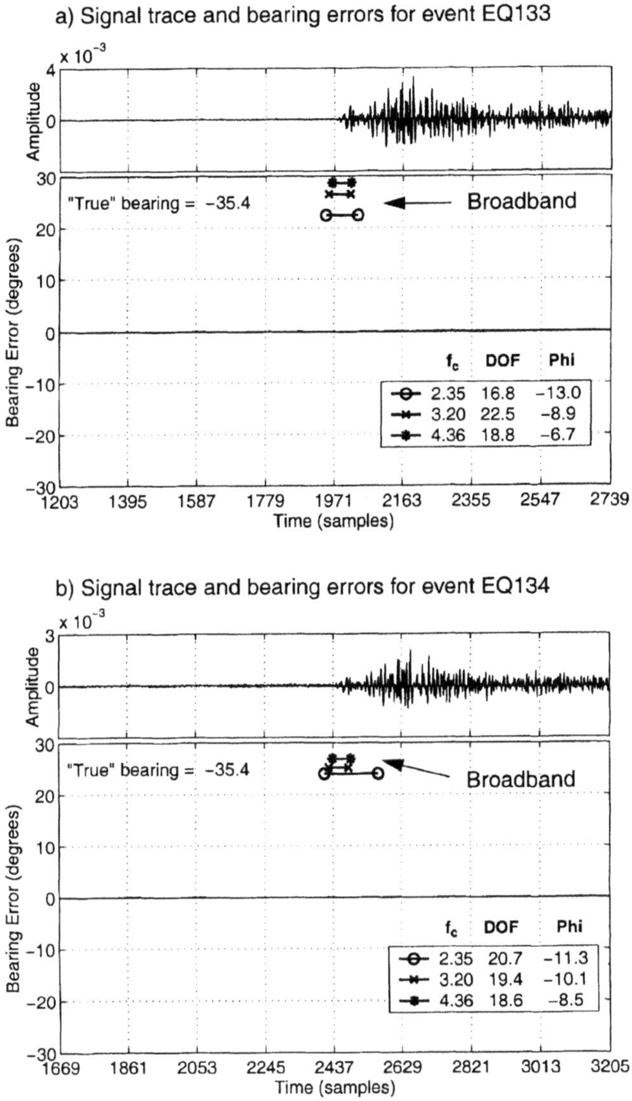

Figure 9

(a) An earthquake (EQ133) near Hebgen Lake yields a relatively tight but heavily biased bearing error pattern when observed at the Pinedale station. (b) A nearly colocated earthquake (EQ134) yields a similar bearing pattern. A comparision of the error estimates for these two events provides a quantitative indication of the relative precision (~ 1.5 degrees) with which 3-c estimates can be made with the SEEC process. It is interesting to note that even though the available power in event EQ134 was less than 1/4 than that in event EQ133, the effective degrees of freedom are comparable for those frequency windows having approximately the same time span. This implies that the noise is correlated with the signal.

The Advantages in Correcting for Bias over Frequency

The bearing biases were determined for four earthquakes in southern Utah using the method embodied in Equation (22). The estimated biases were then applied to the same event data to remove the biases in accord with Equation (23). The actual statistics were based on the "leave-one-out method" wherein the corrected estimates for an event used biases computed from the remaining events. The results appear in

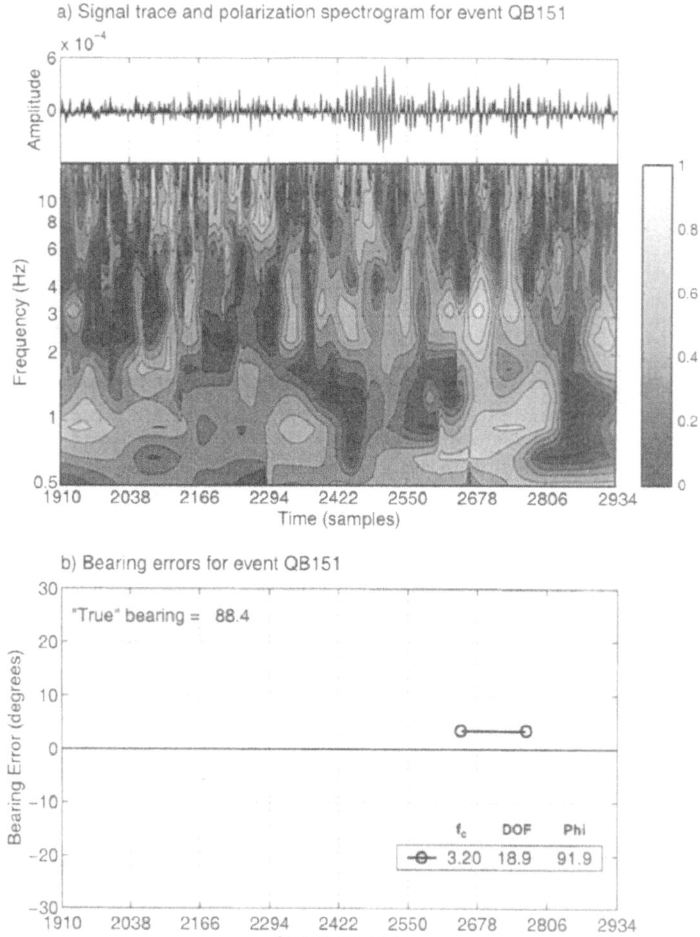

Figure 10

(a) As observed in the trace above, this arrival experienced interference by a preceding arrival. Despite the poor SNR, the interval search process found several polarized regions in the coda of the primary arrival for event QB151. (b) The strongest polarized interval in the P coda resulted in a single usable estimate having a sufficient number of DOF. A large error is associated with the broadband estimate at the onset of the event. The arrival time was inferrred from the NEIC origin time and the P-wave travel-time characteristic.

Table 2. The rms accuracy for this small collection of events is 1.6 degrees. This compares favorably with the rms error 7.6 degrees associated with the individual estimates before creating a weighted average.

This procedure was repeated for 7 events in central Utah. The results are shown in Table 3. SEEC yielded an accuracy of 1.5 degrees rms for the weighted average technique. The accuracy of the collection of individual estimates was 5.9 degrees rms. These results demonstrate that individual low accuracy estimates at different frequencies for a common event can be corrected for bias and optimally combined to realize improved accuracy. The lack of a dramatic improvement is thought to be attributed to the small sample size and the inaccuracies in the ground truth.

Discussion of the Approach and the Results

In the SEEC method quadrature filters spectrally decompose the signals into overlapping bands. The use of wavelets having a constant TB characteristic to define the filters is attractive because the bandwidth is suitably matched to signal and noise environments where narrower filtering is required at the lower frequencies to improve the signal-to-noise ratio. The retention of the phase sensitivity across multiple channels through the use of quadrature filters is important to isolating the coherent signal components in a multipath and scattering propagation environment. Multiple quadrature filters at a given center frequency effectively widen the bandwidth about

Table 2

The bearing errors after correction for regional biases are shown here for events occurring in southern Utah. The DOF *reflects the accumulative precision from estimates contributing at different frequencies and time intervals for a given event*

Event	DOF	Actual (deg.)	Estimate (deg.)	Error (deg.)
EQ246	175.51	−150.80	−149.01	1.79
EQ081	74.78	−150.90	−151.21	−0.31
EQ181a	43.45	−149.80	−148.51	1.29
EQ181b	130.75	−150.00	−152.19	−2.19

Table 3

The bearing errors after correction for regional biases are shown here for events occurring in the middle of Utah. The accumulated effective DOF *associated with each event is also indicated*

Event	DOF	Actual (deg.)	Estimate (deg.)	Error (deg.)
EQ176	243.29	−158.52	−159.89	−1.37
EQ221	65.34	−158.17	−155.52	2.65
EQ074	105.65	−159.82	−158.09	1.73
EQ161	184.34	−160.21	−162.01	−1.80
EQ303	53.65	−160.00	−159.40	0.60
EQ193	187.41	−160.53	−160.62	−0.09
EQ326	150.87	−160.10	−160.40	0.30

the center frequency while providing additional independent rows in the observation matrix to enhance the accuracy of the estimate.

A composite polarization search parameter is effective in identifying time-frequency intervals most likely to produce bearing estimates having good precision. A unique and effective linear polarization measure was included in the search parameter. This approach is thought to avoid the limitations associated with fixed temporal and frequency windows or, for that matter, with accepting the detection window as the defining window, either of which may miss important polarized content.

An information theoretic variance reduction parameter associated with the selected intervals is effective in evaluating and selecting the better bearing estimates from the set of defining intervals. This variance reduction parameter was specifically found to be useful in calibrating subregions for bearing biases and for correcting subsequent observations for biases in a near-optimal setting. The bias correction uses the select bearing estimates and their corrections in a weighted average scheme to embed as much of the inherent accuracy and precision available into a single final estimate. The entries of Tables 2 and 3 show the potential effectiveness of this method.

Bearing biases have been noted in many of the observations at the Pinedale station. Of the biased observations, some reflected a strong orderly bearing dispersion with frequency. Bearing dispersion was evident for events arising throughout Utah and from a local event just east of the station. This frequency dependence may be attributable to a wavelength effect where the longer wavelengths are not as sensitive to thin or small structures as are the shorter wavelengths. Biases have been typically attributed to dipping receiver structures at some stations, an example of which was documented in a recent work by BEAR *et al.* (1999) for the three-component array at Pinon Flats, California. They assert that the azimuthally varying particle-motion anomalies are frequency-dependent with some anomalies larger than 40 degrees. However, systematic trends with frequency are not strongly evident in their results.

The results here suggest that more than a simple dipping structure at the receiver accounts for the observed bearings at the Pinedale station. Possibly source, path and receiver structures are all influencing the observations depending on the epicentral region. Some of the complexity in the observations may be attributed to the complexity of the crustal structure in the vicinity of the Pinedale station. As inferred from the work of SMITHSON *et al.* (1979) the Pinedale station is situated on the southwestern edge of the Wind River thrust fault which dips to the northeast beneath the Wind River Mountains and runs parallel to the mountain range. The Wind River thrust fault overrides the Green River sedimentary basin to the southwest of the mountain range. This thrust fault apparently gave rise to the Pacific Creek anticline which is a faulted gentle fold in the basin in front of the Wind River fault. Synclines and anticlines, to include the Pinedale anticline, also surround the station (see for example, LAW and SPENCER (1989)). These folded and faulted zones could give rise to the observed bearing characteristic.

Another possible explanation for the bearing biases are the horizontal velocity gradients in the upper mantle of the western US. Estimates of these gradients at a depth of 100 km are illustrated by the color intensity map of Figure 11. The color-encoded scale covers velocity changes of ±5% about 8.23 km/sec. These velocity gradients undoubtedly extrapolate to the uppermost mantle (45–60 km) and lower crust and therefore influence regional *P* arrivals. These gradients may account for refractions as great as 25 degrees as inferred from Snell's law. The bearings associated with the events in Colorado could be influenced by the velocity contrast across the Archean-Proterozoic suture zone along the Wyoming-Colorado border where island arcs are thought to have collided with the North American plate (VERNON *et al.*, 1998). Some evidence for this velocity contrast is evident in the gradient map.

Conclusions and Monitoring Implications

These results have shown that it is essential to search the temporal-frequency windows for regions having suitable polarization properties before attempting a bearing estimate. Limiting bearing estimates to frequencies having the best detection SNR may miss useful bearing information. Furthermore, on some occasions the useful time-interval may not occur at the onset of the *P* arrival, as a result of interference or an emerging arrival.

SEEC is premised on the assumption that the estimates may be frequency-dependent. The method is therefore capable of measuring bearing dispersion, thereby providing bearing biases as a function of frequency. In view of the frequency dependence, it is important to avoid contaminating a bearing estimate at a given frequency through leakage into the sidebands of a filter. Filters having low sidelobe characteristics, as a consequence, are essential to realizing unbiased estimates. When bandpass filtering is inadequate alone as in the case of strong, narrow band signals, it is helpful to partially whiten the signals of interest before bandpass filtering. For measurable events SEEC can provide additional insight at differing time-frequency windows to make appropriate judgements for constructing a single final estimate in comparison to conventional approaches which directly produce a single outcome without regard to the effects of the underlying frequency content.

In addition, bearings should be estimated on as many time and frequency intervals as possible to realize the inherently available precision. To assist in achieving accuracy, near-optimal methods were identified to measure the frequency-dependent biases for a given locale and to correct subsequent estimates from that locale for frequency-dependent biases. The near-optimally combined and corrected bearing estimates at various time-frequency intervals were shown to improve the accuracy over that associated with making bias corrections at the individual frequencies. An information theoretic information criterion was found to be helpful

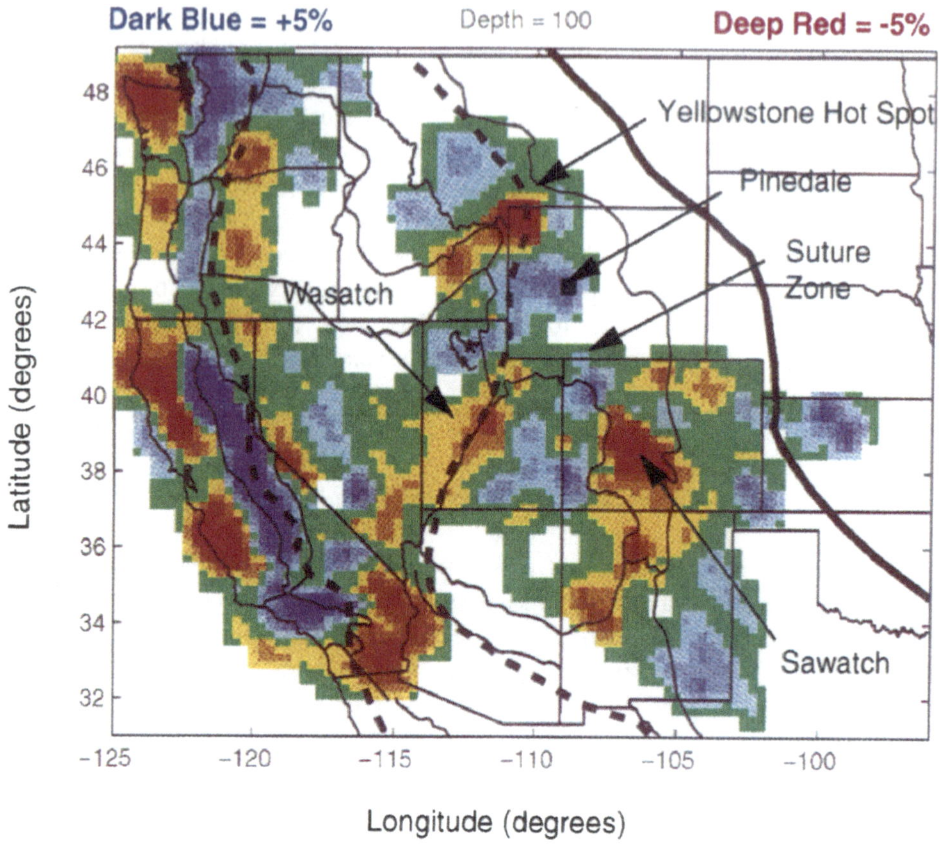

Figure 11
The lateral *P*-velocity gradients at a depth of 100 km across the western United States are illustrated through a color-encoded map. Various tectonic features in relationship to the velocity gradients are also labelled. The percent deviations in velocity are referenced to a mean velocity of 8.23 km/sec. A simple examination, based on Snell's law, suggests that rays could bend laterally by as much as 25 degrees (*the graph was provided by the courtesy of Ken Dueker, University of Colorado*).

in evaluating and selecting the higher precision bearings before making the bias inferences or corrections.

There must be an awareness that bearing estimates can be frequency-dependent in an operational setting. To do otherwise is to accept poor precision in the estimates for events arising from the same locale simply because the estimates will be effected by the differing frequency content from event to event. Further, making bearing corrections without accounting for the frequency content in the estimate may produce inaccurate corrections.

There must also be an awareness that not all arrivals will yield useful bearing estimates. The SEEC approach provides insight to the measurability of an event.

When an arrival is measurable, a careful search in frequency and time for polarized content is essential to identify useful estimation intervals. Estimates at a low frequency, when available, will often yield a near bias-free estimate. The lowest frequency estimate is probably the most useful bearing in associating detections when building events. When locating events, bearings will make a better contribution when the estimates are appropriately selected and have their respective biases removed to form a weighted average of the corrected bearings.

In the future the utility of SEEC method should be further established on larger data sets which include observations in stable as well as tectonic regions. This effort should "tune" the parameters in the algorithm for optimal performance. The resulting algorithm should then be compared with operational bearing estimation methods. If the SEEC approach demonstrates to be an effective approach, methods of integrating the technique in an operational context must also be considered. Other efforts should also consider extending SEEC to other types of polarized arrivals, to codas, and to arrays. It is anticipated that the SEEC algorithm will eventually appear in MatSeis, a publicly available seismic analysis tool (HARRIS and YOUNG, 1996).

Acknowledgments

The author is grateful to Timothy McDonald, Mark Harris, and Greg Elbring of Sandia National Laboratories for providing helpful reviews of this work. The waveform and event truth used in this work were contributed by D. B. Carr of Sandia National Laboratories. Special thanks go to Mark Harris and Patrick Vaughan for assisting in the preparation of the graphics. Sandia is a multiprogram laboratory operated by Sandia Corporation, a Lockheed Martin Company, for the United States Department of Energy under Contract DE-AC-94AL85000.

REFERENCES

ALEWINE, R. W. (1992), *The role of arrays in global monitoring systems*, Proc. of the GERESS Symposium, Waldkirchen, 2224 pp.

BEAR, L. K., PAVLIS, G. L., and BOKELMANN, G. H. R. (1999), *Multi-wavelet Analysis of Three-component Seismic Arrays: Application of Measure Effective Anisotropy at Pinon Flats, California*, Bull. Seismol. Soc. of Am. *89*, 693–705.

BORN, M., and WOLF, E., *Principles of Optics* (Pergamon Press 1975).

CHRISTOFFERSSON, A., HUSEBYE, E. S., and INGATE, S. F. (1988), *Wavefield Decomposition Using ML-probabilities in Modeling Single-site 3-component Records*, Geophysics J. *93*, 197–213.

CLAASSEN, J. P. (1998a), *In Search of a Robust 3-component Bearing Estimator*, 93rd Annual Meeting of the Seismol. Soc. of Am., Boulder, CO.

CLAASSEN, J. P. (1998b), *Robust and Intelligent Bearing Estimation*, Proc. of 20th Annual Seismic Research Symposium on the Monitoring a CTBT, Santa Fe, NM, 495–504.

FLINN, E. A. (1965), *Signal Analysis Using Rectilinearity and Direction of Particle Motion*, Proc. IEEE, *53*, 1884–1876.

HARRIS, M., and YOUNG, C. (1996), *MatSeis: A Seismic GUI and Toolbox for MATLAB*, Seismol. Res. Lett. *67*(2), 267–269.

JURKEVICS, A. (1988), *Polarization Analysis of Three-component Array Data*, Bull. Seismol. Soc. Am. *78*, 1725–1743.

KOCH, K., and KRADOLFER, U. (1997), *Investigation of Azimuth Residuals Observed at Stations of the GSETT-3 Alpha Network*, Bull. Seismol. Soc. Am. *87*, 1576–1597.

LAW, B. E., and SPENCER, C. W. (1989), *Geology of Tight Gas Reservoirs in the Pindeale Anticline Area, Wyoming*, U.S. Geological Survey Bulletin 1886.

LILLY, J. M., and PARK, J. (1995), *Multiwavelet Spectral and Polarization Analysis of Seismic Records*, Geophys. J. Internat. *122*, 1001–1021.

MAGOTRA, N., AHMED, N., and CHAEL, E. P. (1987), *Seismic Event Detection and Source Location Using Single Station (three component) Data*, Bull. Seismol. Soc. Am. *77*, 958–971.

ROBERTS, R. G., CHRISTOFFERSSON, A., and CASSIDY, F. (1989), *Real-time Event Detection, Phase Identification, and Source Location Estimation Using Single-station Three-component Seismic Data*, Geophys. J. *97*, 471–480.

SAMSON, J. C. (1983), *The Spectral Matrix, Eigenvalues, and Principal Components in the Analysis of Multi-channel Geophysical Data*, Annals Geophysicae *1*, 115–119.

SHUMWAY, R. H., and KIM, S. E. (1998), *Detection and Estimation on Small Infrasonic Arrays*, Proc. of the 20th Annual Seismic Research Symposium on Monitoring a CTBT, pp. 628–637.

SMITHSON, S. B., BREWER, J. A., KAUFMAN, S., OLIVER, J. E., and HURUCH, C. A. (1979), *Structure of the Laramide Wind River Uplift, Wyoming, from the COCORP Deep Reflection Data and from Gravity Data*, J. of Geophys. Res. *84*, B11, 5955–5972.

VERNON, F., DUEKER, K., and PAVLIS, G. (1998), *Shakedown at the Gates of the Lodore*, IRIS Newsletter *17*, 1–4.

VIDALE, J. E. (1986), *Complex Polarization Analysis of Particle Motion*, Bull. Seismol. Soc. Am. *76*, 1393–1405.

WAŁCK, M. C., and CHAEL, E. P. (1991), *Optimal Backazimuth Estimation for Three-component Recordings of Regional Seismic Events*, Bull. Seismol. Soc. Am. *81*, 643–666.

(Received June 30, 1999, revised December 12, 1999, accepted February 9, 2000)

Pure appl. geophys. 158 (2001) 375–396
0033-4553/01/020375-22 $ 1.50 + 0.20/0

Pure and Applied Geophysics

Slowness Corrections — One Way to Improve IDC Products

JOHANNES SCHWEITZER[1]

Abstract—The first step to identify and locate a seismic event is the association of observed onsets with common seismic sources. This is especially important in the context of monitoring the Comprehensive Nuclear-Test-Ban Treaty (CTBT) at the International Data Center (IDC) being developed in Vienna, Austria. Well-defined slowness measurements are very useful for associating seismic phases to presumed seismic events.

Shortly after installation of the first seismic arrays, systematic discrepancies between measured and theoretically predicted slowness values were observed, and therefore slowness measurements of seismic stations should be calibrated. The observed slownesses measured with small aperture arrays, some of which will be included in the International Monitoring System (IMS) now being implemented for verifying compliance with the CTBT, show large scatter and deviations from theoretically expected values. However, in this study a method is presented, by which mean slowness corrections can be derived, which show relatively stable patterns specific to each array.

The correction of measured slowness values of these arrays clearly improved the single array location capabilities. Applying slowness corrections with seismic phases observed by ARCES, FINES, GERES, and NORES, and associated to seismic events in the bulletins of the prototype International Data Center (pIDC) in Arlington, VA, also clearly demonstrates the advantages of these corrections. For arrays with large slowness deviations that are due to the influence of a dipping layer, the corrections were modeled with a sine function depending on the measured azimuth. In addition, the measured values can be weighted with the corresponding uncertainties known from the process of deriving the mean corrections.

Key words: Array seismology, slowness corrections, array calibration.

1. Introduction

One advantage of the data processing at the prototype International Data Center (pIDC) in Arlington, VA is the extensive measurement of all relevant parameters of the detected phases. This data extraction will be very similar at the International Data Center (IDC) in Vienna, Austria, now being established for monitoring the Comprehensive Nuclear-Test-Ban Treaty (CTBT).

The first step to identify and locate a seismic event is the association of observed onsets with common seismic sources. During this process the challenge is to build only real events and to verify all detected phases to the fullest extent possible.

[1] NORSAR, P. O. Box 51, N-2027 Kjeller, Norway. E-mail: johannes@norsar.no

However, in practice many detected and reported phases remain unexplained. These are mostly false alarms, phases from small local and regional events, or later phases of already defined events, and they include observations of smaller events by very sensitive seismic stations at teleseismic distances, which are not reported by local or regional authorities. The latter case is especially important for monitoring the CTBT.

In this study, the phrase "slowness" is used for the total slowness vector $\vec{s}$, derived from the ray parameter r (i.e., the inverse of the horizontal apparent velocity v) and the backazimuth Θ of a plane wave observed with an array or at a 3C station. The ray parameter is defined as

$$r = \frac{1}{v} \cdot \frac{\pi \cdot 6371}{180} \text{ in} \left[\frac{s}{\text{deg}}\right],$$

and the slowness vector is defined in polar coordinates as $\vec{s} = (r, \Theta)$ or in Cartesian coordinates as $\vec{s} = (r \cdot \sin\Theta, r \cdot \cos\Theta)$.

Well-defined slowness measurements are very useful for associating seismic onsets with events. Slowness measurements are also helpful for locating events defined by a small number of observations, and they are used by the pIDC during all steps to produce its final bulletins. Therefore, the estimation of possibly systematic slowness deviations and statistical information about the scatter of individually measured slowness values are part of the generally needed calibration of all seismic stations participating in the International Monitoring System (IMS) to obtain the best possible input for the event location procedures (HARJES *et al.*, 1994).

Slowness measurements can be made either by measuring for a detected phase the polarization of the particle motion at 3C stations and then calculating backazimuth and apparent velocity, or by observing directly the propagation of the wavefront with seismic arrays. Many studies of slowness deviations and their causes can be found for large- and medium-sized arrays (e.g., OTSUKA, 1966a, b; GREENFIELD and SHEPPARD, 1969; JOHNSON, 1969; ENGDAHL and FELIX, 1971; BERTEUSSEN, 1974, 1976; NOPONEN, 1974; VERMEULEN and DOORNBOS, 1977; AKI *et al.*, 1977; WALCK and MINSTER, 1982; FABER *et al.*, 1986; KRÜGER and WEBER, 1992).

In this work a method to derive slowness corrections for small aperture seismic arrays will be presented and it will be shown how such corrections can improve the data interpretation at the pIDC. Small aperture arrays have, as expected from their array-transfer functions, a lower resolution in measuring the slowness of a plane wave. Therefore, deviations of observed slownesses from theoretically estimated ones have mostly been investigated in the context of estimation uncertainties (e.g., BAME *et al.*, 1990; HARRIS, 1990; SUTEAU-HENSON, 1990; CARR, 1993) and not as systematic effects due to lateral heterogeneities or dipping layers in the earth. However, observations with the GERES array (SCHWEITZER, 1992; HARJES *et al.*, 1994) revealed that at small aperture arrays, despite the scatter, systematic slowness deviations can be observed and consequently a first set of slowness calibrations for European small aperture arrays (the Apatity array, ARCES, FINES, GERES,

NORES, and SPITS) was derived (SCHWEITZER, 1994). More recently, similar studies were published for many of the IMS stations (LIN and ROECKER, 1996; BONDÁR, 1997; KOCH and KRADOLFER, 1997, 1999; TIBULEAC and HERRIN, 1997). As far as the above-mentioned European arrays were investigated in these studies, the authors confirmed the former results of SCHWEITZER (1994, 1995) and SCHWEITZER and KVÆRNA (1995).

In the following, a methodology for estimating mean slowness corrections will be briefly described. The results of applying this method will be shown for the small aperture arrays with the best documented slowness corrections (ARCES, FINES, GERES, and NORES). All automatically detected and analyzed phases from these arrays were investigated to search for systematic patterns in slowness deviations. It is known that these automatically measured slowness parameters (apparent velocity or ray parameter and azimuth) usually exhibit a larger scatter than manually analyzed onsets. Since standard slowness measurements at the pIDC (and IDC) are only done automatically, it is important to take this into account by deriving slowness corrections from automatic measurements also. SCHWEITZER (1994) demonstrated for a selected subset of slowness corrections for GERES that manually and automatically derived slowness corrections differ in their scatter, however not in their major trends.

As result of this effort, mean slowness corrections were calculated and are now available to automatically correct the observed slowness measurements over a large range of backazimuth and ray-parameter values. In addition, standard deviations for the mean slowness corrections were estimated. All these results should be used to increase the stability and quality of both the phase associations and the event locations based on calibrated slowness observations.

2. *The Data Sources*

At NORSAR, small aperture arrays have been used for more than one decade to locate seismic events. Automatically estimated slowness observations for small aperture arrays have been available since January 1, 1989. To obtain the slowness bias from theoretically expected values a list of reference events was used. Therefore, such a list was compiled for the time period January 1, 1989 to August 5, 1995. The main sources for this list were the bulletins of ISC and PDE. But because these bulletins are not complete for all observable smaller events in Europe, the following local and regional catalogues from Europe were added: the bulletins of the Seismological Institutes in Helsinki and Bergen for Scandinavia, a list of confirmed quarry blasts from the Kola peninsula, a local Bulletin of the Vogtland/Western Bohemian region of earthquake swarms for Central Europe, a list of precisely located events from the Polish mining areas, and a list of confirmed quarry blasts in Bavaria/Germany and in the Czech Republic (for details see Table 1). All these event lists were merged together and multiple entries were carefully eliminated. Table 1 also

Table 1

Contributions of the different bulletins to the list of reference events. (1) P. Wiejacz and J. Niewiadomski (both Polish Academy of Sciences, Warszawa) several pers. communications in 1991 and 1992, (2) Compiled at the Institute of Geophysics at Ruhr-University Bochum, and supported by several pers. communications with J. Zednik (Czech Academy of Sciences, Praha) in 1992, (3) Mykkeltveit (1992), (4) Neunhöfer (1998).

Bulletin	Time Period	Number of Events
ISC	Jan 1, 1989 – Jul 31, 1992	120,393
PDE monthly	Mar 1, 1990 – Dec 31, 1994	46,163
PDE weekly	Jan 1, 1995 – Aug 5, 1995	6,472
Helsinki	Jan 1, 1989 – Jun 30, 1994	3,428
Bergen	Jan 1, 1989 – Mar 31, 1994	1,319
Polish mines (1)	Jun 27, 1990 – Jun 12, 1992	418
Bavarian and Czech quarries (2)	Jan 9, 1991 – Oct 14, 1992	285
Kola mines (3)	Jun 15, 1991 – Oct 23, 1992	195
Vogtland (4)	Jan 31, 1991 – Nov 16, 1992	74
Total	Jan 1, 1989 – Aug 5, 1995	178,747

supplies information regarding the amount of contributions from each source to the final list of 178,747 reference events.

For these events, the epicentral distances, the corresponding backazimuth values, and the theoretical onset times of all interesting seismic phases, as well as the predicted slownesses, had to be calculated and compared with automatic estimates from detections at the arrays investigated. The theoretical ray parameters and onset times were calculated using the tau-spline interpolation software of Buland and Chapman (1983) and the IASPEI91 travel-time tables (Kennett *et al.*, 1991). The IASPEI91 model was chosen because it is used as the standard model at the pIDC. However, other radially symmetric velocity models would not change the observed backazimuth deviations but only add changes to the theoretical ray parameters. If other global or regional horizontally layered velocity models should be used, the corresponding slowness corrections could be easily recalculated.

The availability of slowness observations is not the same for all the arrays and mostly reflects the successive extension of the European array network (see Table 3). For a sufficient coverage of the whole interesting slowness space with observations, a long observing period is needed (minimum several years). This is not only because the distribution of seismic events (natural or man-made) is changing with time but also because gaps due to station problems must be filled. Therefore the work presented here should be frequently updated for all arrays of interest.

Association of Observed Onsets with Theoretically Estimated Onsets

To obtain reliable slowness corrections, the association criteria of observations to events must be carefully defined. In this study the following procedure was chosen:

a) For each event, backazimuth and distance were calculated for each array which observed the event.

b) To obtain an optimum coverage of the slowness space most of the theoretically possible phases as included in the IASPEI91 tables (KENNETT and ENGDAHL, 1991) were compared with the detected onsets. Using distance, depth, and origin time of the event, the absolute onset times of first and later onsets were calculated for all arrays considered. To reduce the number of erroneous associations, some restrictions related to epicentral distance and event magnitude were introduced, some secondary onsets were only associated if an earlier phase of the same event was also associable, and additionally, all theoretical phases to be considered for comparison with observed ones had to be separated in time by at least three seconds. Slowness corrections were not only calculated for *P*-type onsets but also for local/regional *S* phases. The latter are usually observed and analyzed at small aperture arrays and are important for all identification and location procedures at local/regional distances.

Table 2

*Phases for which ray parameter and azimuth values were used in this study to estimate the slowness corrections. For core phases all branches were used (*i.e. *ab, bc, and df). "mag" means this phase was used only if the event magnitude was larger than 4.0; "*" means this phase was used only if another earlier onset from the same event was observed. In addition the table contains the number of observed phases found for each array and used in this study.*

Phase	Restrictions	ARCES	FINES	GERES	NORES
Pg, Sg	$\Delta < 10°$	78, 236	221, 344	576, 740	137, 508
SgSg	$4° \leq \Delta < 10°$; *	1	–	1	9
Pb	$\Delta < 15°$	487	119	1014	1070
Sb	$\Delta < 15°$	760	419	1074	1398
PbPb, SbSb	$4° \leq \Delta < 15°$; *	51, 242	13, 207	35, 35	43, 367
Pn	–	2376	1885	3100	2961
Sn	–	1617	1109	1386	1371
PnPn	$4° \leq \Delta < 18°$	651	259	1085	691
SnSn	$4° \leq \Delta < 18°$; *	181	89	208	173
P	–	14950	9663	7307	12816
pP	–	4850	2093	1533	3176
sP	–	1756	867	1114	1174
PP	$50° \leq \Delta \leq 100°$; mag	370	159	319	509
PcP, ScP	$10° \leq \Delta$; mag; *	1424, 129	444, 37	523, 81	648, 24
Pdiff	$110° \leq \Delta$	673	150	350	277
sPdiff	$110° \leq \Delta$; if magnitude < 4.0 then *	191	32	95	77
PKP, SKP	–	4110, 263	1778, 178	2832, 143	3065, 189
pPKP, sPKP	–	79, 39	101, 51	345, 96	240, 120
PKiKP	$\Delta \geq 80°$	980	361	714	446
pPKiKP, sPKiKP	$\Delta \geq 80°$; if magnitude < 4.0 then *	138, 81	69, 81	128, 163	58, 59
PKKP	$\Delta \geq 30°$; mag; *	326	76	122	83
P′P′	mag; *	99	63	129	36

Table 3

Number of associated and used slowness observations for the investigated arrays and the analyzed time periods.

Array	Analyzed Time Period	Number of Detections		Mean Slowness Deviation [s/deg]
		Associated	Used	
ARCES	Jan 01, 1989 – Aug 05, 1995	52,028	37,138	1.84
FINES	Nov 23, 1989 – Aug 05, 1995	29,699	20,868	2.15
GERES	Oct 17, 1990 – Aug 05, 1995	32,801	25,248	1.96
NORES	Jan 28, 1989 – Aug 05, 1995	41,934	31,725	2.06

The list of phases used with the restrictions applied is given in Table 2. In addition, Table 3 illustrates how often the different phases contributed to the total number of slowness observations for each array investigated, respectively.

c) This list of onsets for every event was then compared with the parameters automatically extracted for all detected onsets for each array. Data from the GERES array were automatically analyzed at NORSAR and at the Ruhr-University in Bochum, Germany, where a specially tuned data processing method had been developed (e.g., HARJES *et al.*, 1993). Therefore, for all GERES observations after November 9, 1990, the results from the automatic data processing in Bochum were used for this study.

To define a theoretically expected phase as observed, the residual between detected and theoretically estimated onset time must be less than 10 seconds, and the length of the slowness correction vector must be less than 10 s/deg. In the case of more than one detection in the time interval of ± 10 seconds around the theoretical onset, the onset with the smallest travel-time residual and an acceptable slowness deviation was associated. Sometimes, onsets of two or more events had approximately the same arrival time at a seismic station and the list of associations had to be checked for such situations. In such cases all the associations were discarded from further use.

Additional restrictions for the selection of slowness observations used in this study led to a smaller but more stable set of observed slowness deviations (for details see SCHWEITZER, 1994). Figure 1 shows for NORES all 31,594 slowness values used and gives an impression of the coverage of the slowness space and the scatter of the

▶

Figure 1
The upper plot shows all 31,725 NORES slowness observations used in this study, plotted as symbols at the endpoints of the slowness vectors. At the bottom, all corresponding theoretical slowness vectors are shown. Because of the velocity structure in the uppermost layers of the earth, the theoretical ray parameter of local and regional phases is nearly constant for a wide distance range. The six outermost circles correspond with Sg, Sb, Sn, Pg, Pb, and Pn onsets.

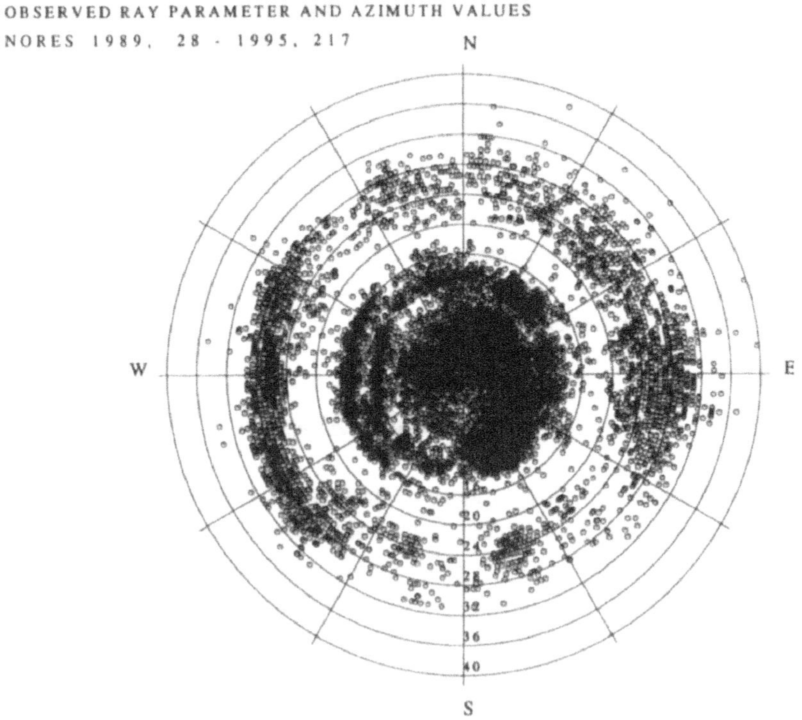
OBSERVED RAY PARAMETER AND AZIMUTH VALUES
NORES 1989, 28 - 1995, 217
N
W
E
S
20
24
28
32
36
40

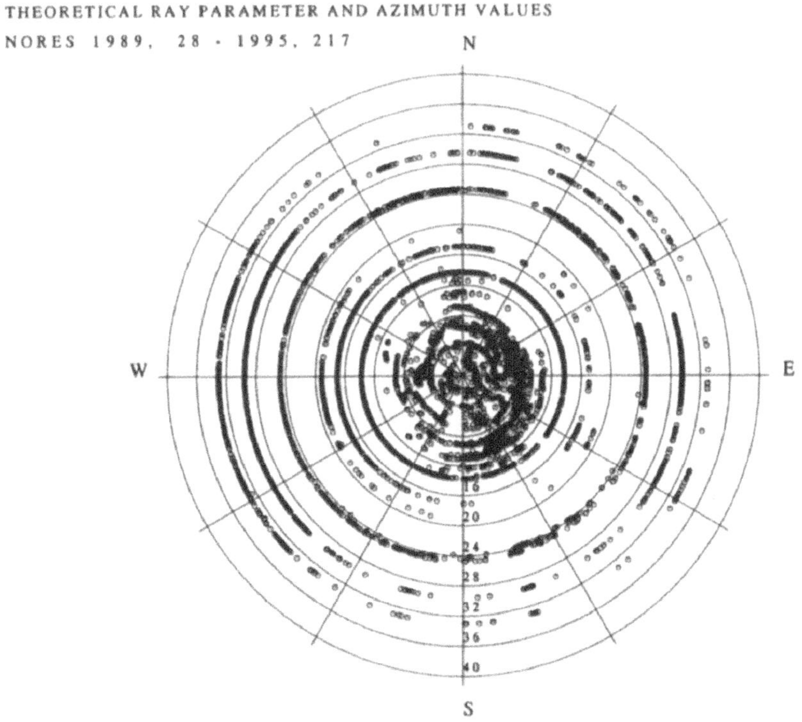
THEORETICAL RAY PARAMETER AND AZIMUTH VALUES
NORES 1989, 28 - 1995, 217
N
W
E
S
16
20
24
28
32
36
40

data. In this study, the slowness vectors are plotted with symbols at their endpoints, using polar-coordinates. On the top all observed slowness values are seen, and at the bottom the corresponding theoretical values are plotted. All *P*-type onsets have theoretical ray parameters smaller than 20 s/deg and the local and regional *S*-type phases have theoretical ray parameters between 20 and 40 s/deg. The separation between a *P* and an *S* phase with small aperture arrays is usually based on this ray parameter difference. Because of the velocity structure in the uppermost layers of the earth, the theoretical ray parameters of local and regional phases are nearly constant for a wide distance range. Therefore, the six outermost circles correspond with Sg, Sb, Sn, Pg, Pb, and Pn onsets.

3. Mean Slowness Corrections

Owing to the relatively large scatter in single observations of the small aperture arrays, it became necessary to calculate mean slowness corrections. For this purpose, slowness space was divided into bins. To divide slowness space into approximately equally sized areas, the size of the bins was changed for different ray parameters between 1 s/deg times 30 deg and 2 s/deg times 5 deg, in total, we got 1849 bins. In this study only slowness corrections which are based on at least three observations in one bin are shown (compare Fig. 1 and Fig. 5 which display single observations and mean slowness correction vectors for NORES). No explicit relation can be given between observation and theory because of the scatter in the observed data, therefore two different types of slowness corrections can be calculated, depending on which slowness, the observed or the theoretical one, was used as reference.

One type of correction vectors $\overrightarrow{soc}$ points from one bin of observations $\overrightarrow{so}$ (see in Fig. 1 on the top) to the most likely "true" theoretical slowness $\overrightarrow{st}$ (see Fig. 1 at the bottom). The mean correction vector of the *j*-th bin is than defined as mean value of the *n* slowness deviations in this bin:

$$\overrightarrow{soc}_j = \frac{\sum_{i=1}^{n} (\overrightarrow{so_i} - \overrightarrow{st_i})}{n}.$$

The other type of correction vectors $\overrightarrow{stc}$ points from one bin of the theoretical slowness vectors $\overrightarrow{st}$ (see Fig. 1 at the bottom) to the most likely corresponding

▶

Figure 2
The upper plot shows 799 ARCES slowness correction vectors using the observations as reference. At the bottom the 451 corresponding slowness corrections relative to the theoretical values (predicted corrections) are seen. The symbol size corresponds to the number of single observations per bin. The maximum number of observations per slowness correction in one bin is 859 (top) and 3199 (bottom). For further details see text.

CORRECTION VECTORS FOR ARCES

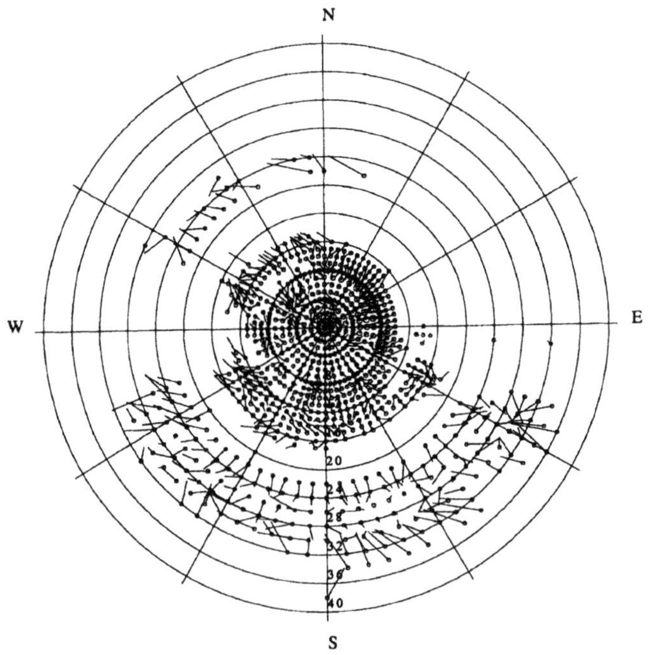

PREDICTED SLOWNESS DEVIATIONS FOR ARCES

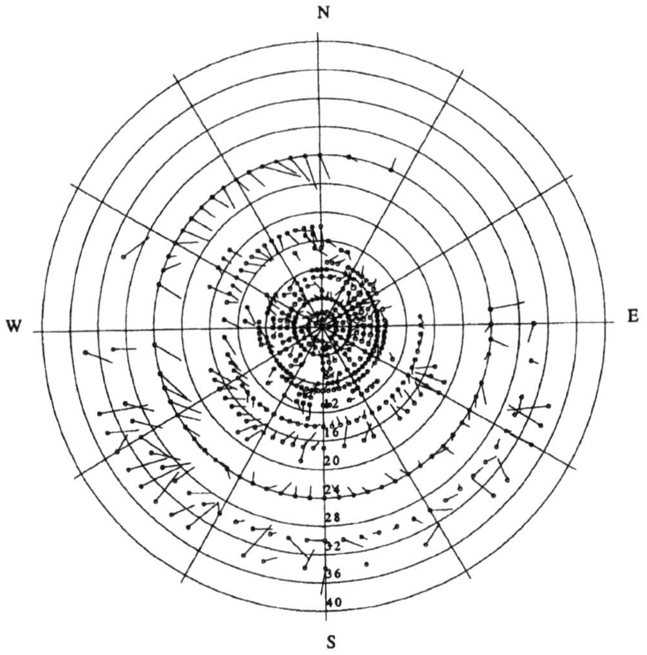

observed slowness $\vec{so}$ (see e.g. Fig. 1 on the top). The mean correction vector of the k-th bin is than defined as mean value of the m slowness deviations in this bin:

$$\overrightarrow{stc_k} = \frac{\sum_{i=1}^{m} (\overrightarrow{st_i} - \overrightarrow{so_i})}{m}.$$

Figures 2–5 present the slowness corrections for ARCES, FINES, GERES, and NORES. For each figure, the top part shows the correction vectors $\overrightarrow{soc}$ and the bottom part the correction vectors $\overrightarrow{stc}$. The symbol size corresponds to the number of single slowness corrections which were observed per bin, and the vectors plotted require addition to the reference value to gain the corrected backazimuth and ray parameter.

Besides the mean slowness correction vectors, the corresponding standard deviations were also calculated for each bin. These standard deviations can be used in association or location programs to weight single observations. Table 3 gives mean values of the length of the single slowness correction vectors.

The two types of slowness corrections are useful for different applications. The first one ($\overrightarrow{soc}$) can be used to correct systematic slowness deviations of observations and to reduce the scatter in these data. The latter ($\overrightarrow{stc}$) predicts for theoretical slownesses known from the reference model systematic slowness deviations, although it cannot predict the scatter caused by low slowness resolution or noise.

4. *Testing the Mean Slowness Corrections of Small Aperture Arrays*

4.1. *Single Array Locations*

For the four arrays investigated, ARCES, FINES, GERES, and NORES, the data base of slowness-correction vectors was sufficiently dense that these corrections could be applied for locating local and regional seismic events. In this way, the correction vectors could be used to improve the single array locations for these arrays.

The single array location procedure RONAPP (MYKKELTVEIT and BUNGUM, 1984) uses the TTAZLOC algorithm (BRATT and BACHE, 1988) and locates events with travel time and backazimuth information as input data. In these algorithms apparent velocities of the detected onsets are only used to identify the different seismic phases. Because the apparent velocities of local and regional P and S phases

▶

Figure 3

As Figure 2 but for FINES. The number of slowness corrections is 881 (relative to observation) and 451 (relative to the theory). The maximum number of observations per slowness correction in one bin is 482 (top) and 1403 (bottom).

CORRECTION VECTORS FOR FINES

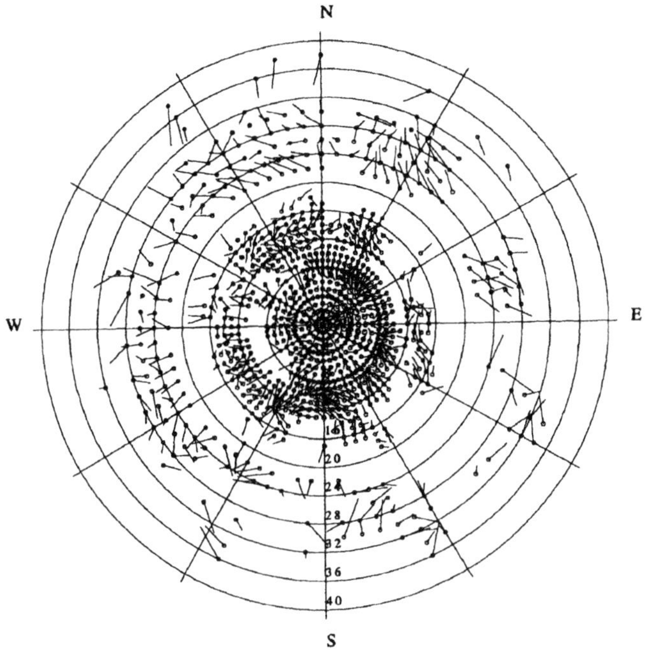

PREDICTED SLOWNESS DEVIATIONS FOR FINES

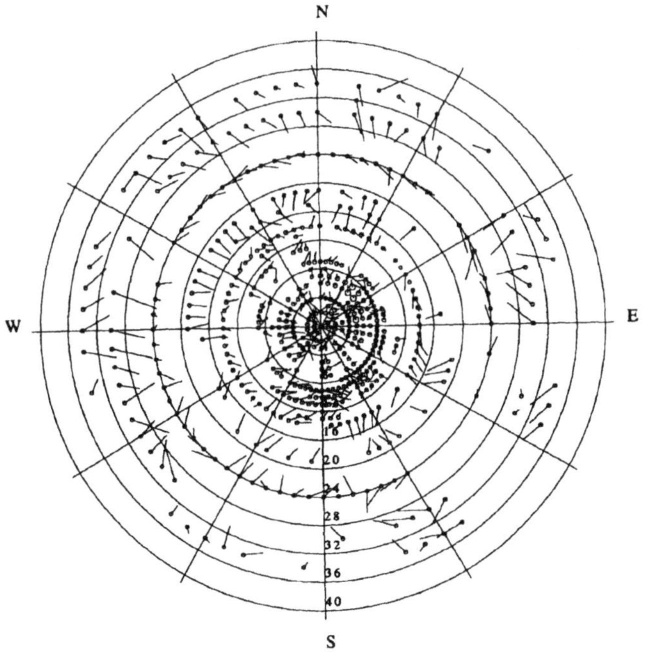

are nearly constant with respect to the epicentral distance, the distance of an event is determined by inverting the travel-time difference between P and S phase onsets and not by inverting the ray parameter. The uncertainties of the estimated parameters (onset time, backazimuth, and apparent velocity/ray parameter) were calculated from the signal-to-noise ratio (SNR) and the coherence of the signals. Therefore, correcting automatically estimated slowness observations with the mean slowness corrections presented here mainly influences the location algorithm in changing the backazimuth of the observed phases, and the application of slowness corrections is independent from the velocity model used as a reference to derive the slowness corrections.

To assess the mean slowness corrections for the four arrays cited above, the whole data set of detected phases for 1994 was reprocessed. Figure 6 shows all 25,612 events defined and located by the four small aperture arrays ARCES, FINES, GERES, and NORES during the original single-array data processing. The map clearly shows the concentration of the seismicity at known source regions (natural or man-made). In addition, we can see a scattered distribution of events located at larger distances from the arrays. The two circles of events around NORES and ARCES are an artifact of the real-time algorithm used at that time for these two arrays.

Figure 7 shows the 24,946 relocated events after correcting the automatically estimated slowness values (as apparent velocity and backazimuth) with the mean slowness corrections. For phases which fall in bins where no corrections were available, the original values remained unchanged. The reduction of the number of defined and located seismic events by about 2.5% is caused mostly by removing events distant from the arrays. These discarded events were mostly defined due to misassociations and errors in the processing rules. Notice the reduced number of artificial events scattered in the background in comparison to Figure 6.

Because most of the events located by the regional arrays are due to man-made activity, this large number of relocated events cannot be compared with independent bulletins. Therefore the evaluation of the results can only be done in a more qualitative way. It is clearly seen that the concentration of events around known source regions in Europe is considerably higher and the azimuthal scatter in particular is smaller. This clearly shows the positive effect of correcting the observed slowness with the mean slowness corrections.

▶

Figure 4

As Figure 2 but for GERES. The number of slowness corrections is 855 (relative to observation) and 528 (relative to the theory). The maximum number of observations per slowness correction in one bin is 698 (top) and 1278 (bottom).

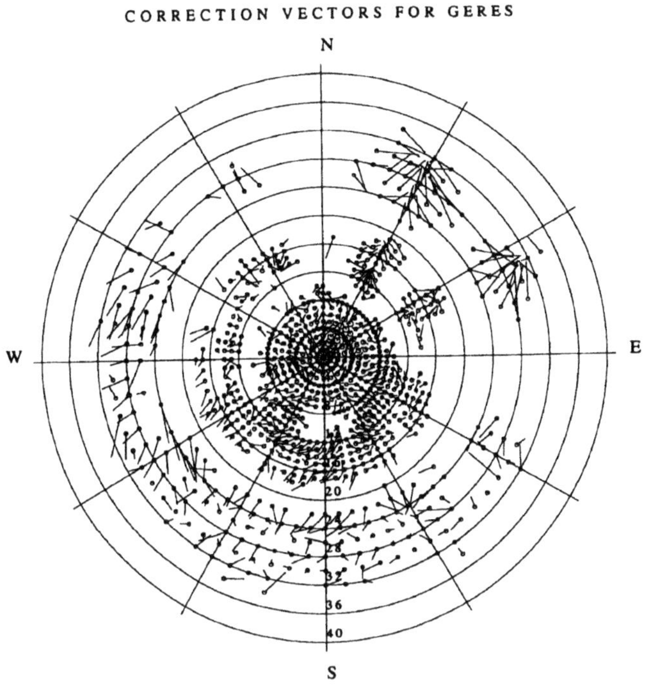
CORRECTION VECTORS FOR GERES
N
W
E
S
20
28
32
36
40

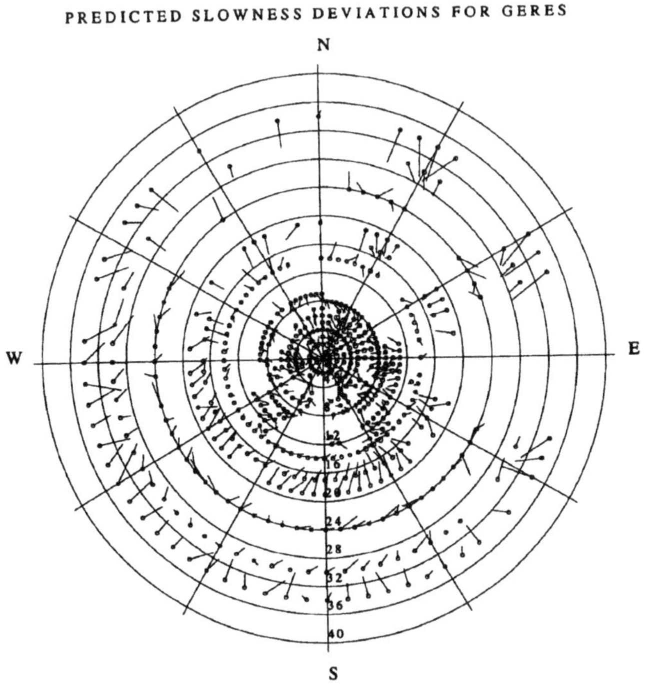
PREDICTED SLOWNESS DEVIATIONS FOR GERES
N
W
E
S
12
16
20
24
28
32
36
40

4.2. Slowness Residuals in the REBs

The GSETT-3 Reviewed Event Bulletins (REBs) contain numerous (automatically) estimated ray parameter and backazimuth values observed for the small aperture arrays ARCES, FINES, GERES, and NORES. In addition, the REBs also list the estimated ray parameter and backazimuth residuals with respect to the location in the REBs and to the used velocity model. Although these slowness values are not always used as defining parameters in the final location of the seismic events, the measured ray parameter and azimuth of a detected onset play an important role during the identification and association process at the pIDC.

All REB-events located with at least 10 defining phases during the first 10 months of the GSETT-3 experiment (January 1 – October 31, 1995) were investigated for onsets at the four small aperture arrays. These events were assumed to have a location precision that allowed an evaluation of the slowness corrections. The listed ray parameter and backazimuth residuals were transformed into slowness-error vectors. Whenever the length of such a vector was smaller than 6 s/deg and the travel-time residual of the onset was smaller than 6 s, this onset was defined as a valid association, and the slowness vector appearing in the REBs was corrected with the mean slowness correction derived. For this, only the slowness correction vectors were used, as plotted on top of Figures 2–5. When the observed slowness fell in a bin without a known correction, no correction was applied. However, a more sophisticated interpolation scheme should be used if the application of mean slowness corrections is implemented at the IDC. Figure 8 shows the results of this investigation for each array. The dotted line delineates the distribution of the vector lengths between theoretical and observed slowness without any correction, and the solid line shows the distribution of the vector lengths after applying the slowness corrections. The two distributions are normalized relative to their maxima. The corrected slowness values clearly show, in addition, smaller errors and should therefore be used in the data processing at the IDC (see Table 4).

5. Observation of a Dipping Layer Structure at SPITS

For the small aperture array on Spitsbergen (SPITS), relatively large slowness deviations are observed. Because of the small aperture of only 1 km of this 9 element array, single slowness estimates show in addition a relatively large scatter. For including this array in the Generalized Beam Forming (GBF) location algorithm at

►

Figure 5

As Figure 2 but for NORES. The number of slowness corrections is 963 (relative to observation) and 529 (relative to the theory). The maximum number of observations per slowness correction in one bin is 728 (top) and 2169 (bottom).

CORRECTION VECTORS FOR NORES

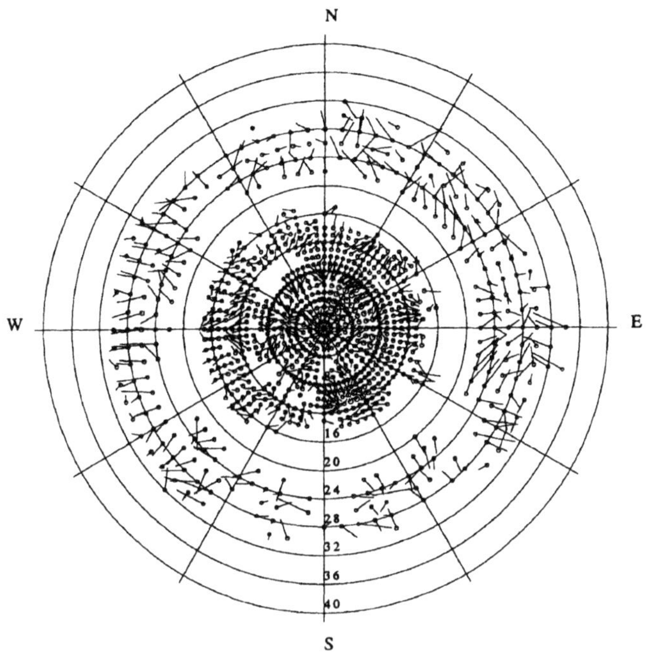

PREDICTED SLOWNESS DEVIATIONS FOR NORES

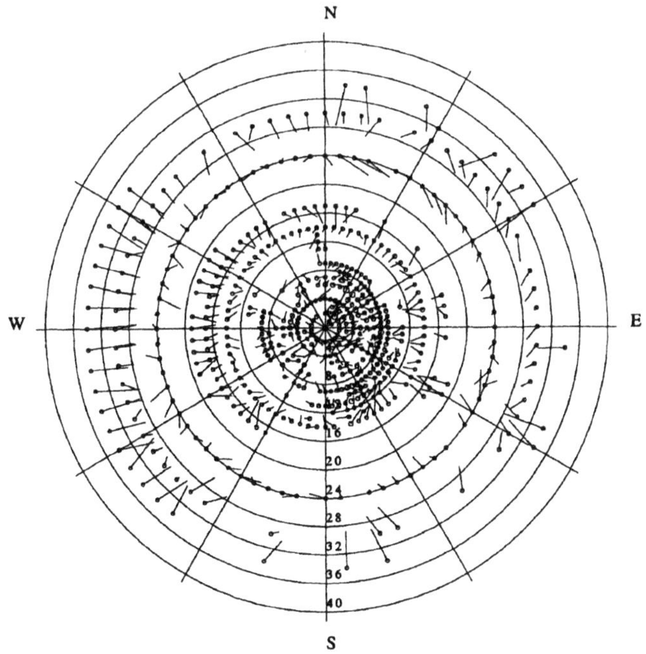

ARCES, FINES, GERES, NORES - 1994, ORIGINAL

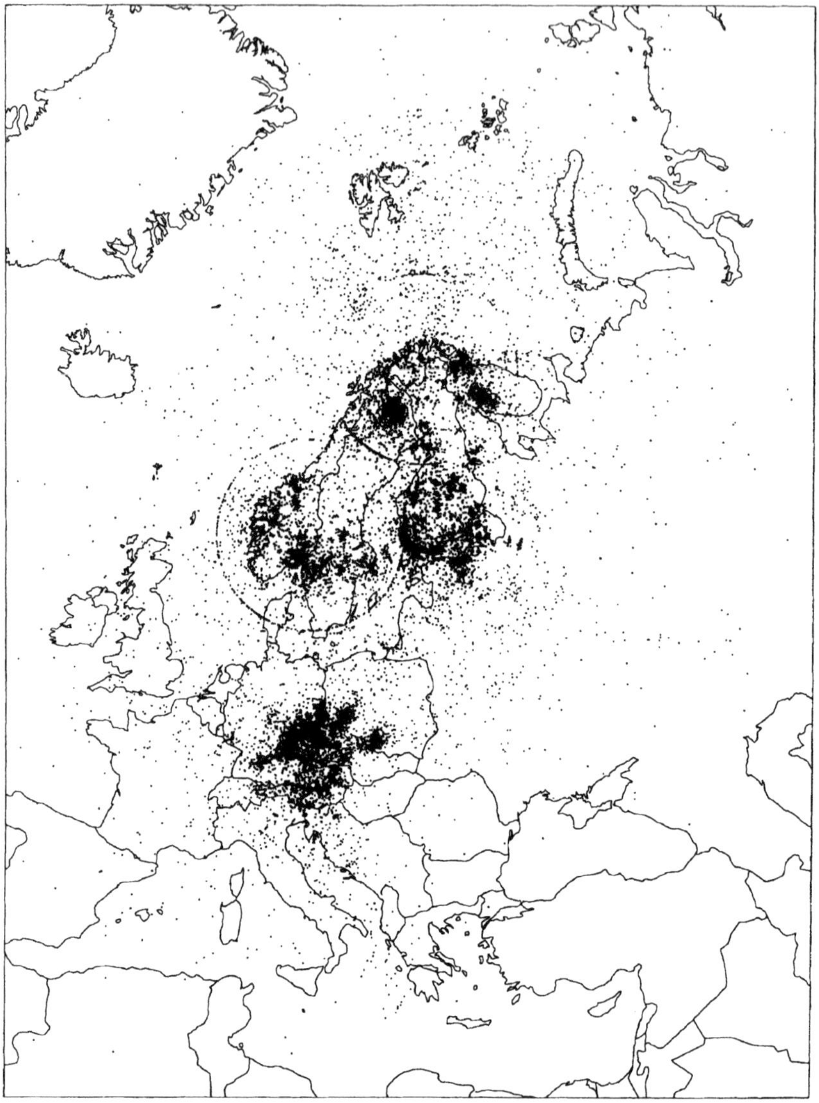

Figure 6
All 25,612 events automatically located in near real time during 1994 by ARCES, FINES, GERES, and NORES using the originally estimated apparent velocities and azimuth values. The displayed locations are single array solutions and the map includes solutions from several arrays for particular events.

NORSAR (RINGDAL and KVÆRNA, 1989; KVÆRNA *et al.*, 1999), these variations were accommodated with a simple approach:

Figure 9 shows Pn apparent velocity estimates from a set of manually located events at local/regional distances from this array. The corresponding apparent

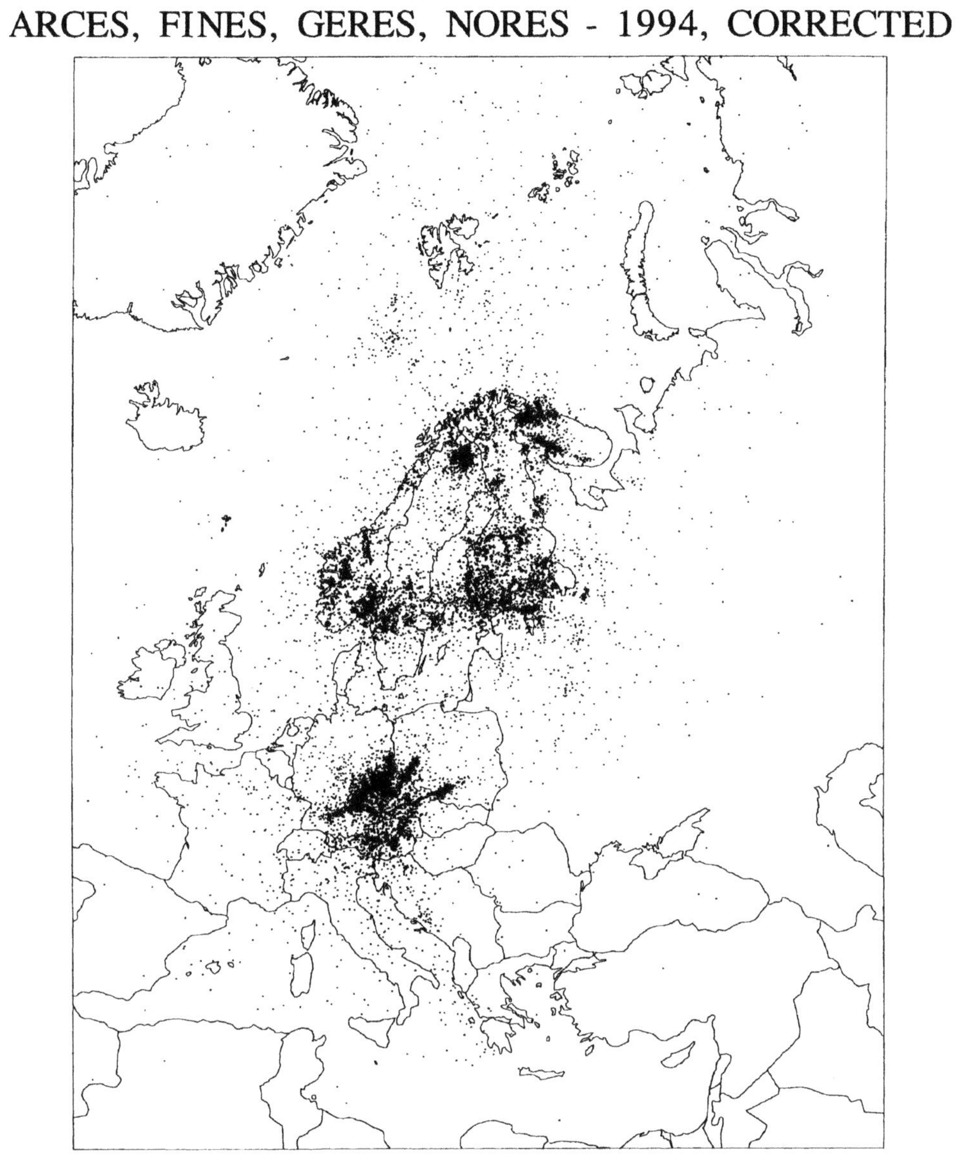

Figure 7
All 24,946 events for 1994 relocated after correcting the apparent velocities and azimuth values with the mean slowness corrections of ARCES, FINES, GERES, and NORES. The shown locations are single array solutions and the map includes solutions from several arrays for particular events.

velocities and azimuth values were automatically estimated and are influenced by the actual noise conditions, the relatively low slowness resolution of the SPITS array, and maybe the different frequency contents of the signals. For instance, CLAASSEN (1998) showed an example in which systematic changes of slowness deviations can be

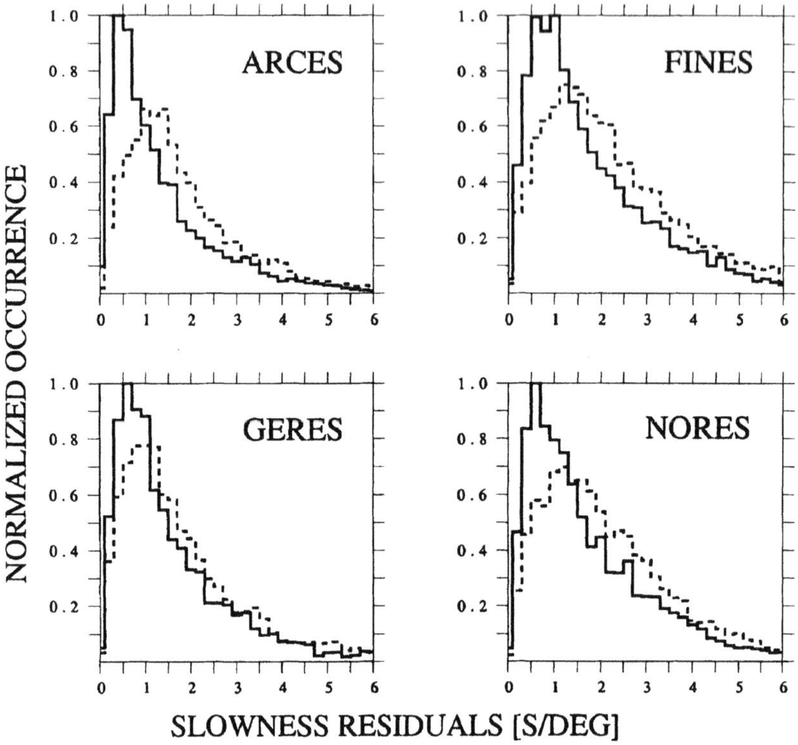

Figure 8
Distributions of slowness residuals in the REBs, i.e. the vector lengths between the theoretical and the observed slowness vectors, for each of the small aperture arrays investigated. The dotted lines depict the original residuals and the solid lines show the remaining residuals after applying the mean slowness corrections. All distributions were normalized for each array separately.

Table 4

Some statistical parameters of observed and corrected slowness deviations.

Array	Number of Observations	Observed Slowness Deviations		Slowness Deviations after the Correction	
		mean [s/deg]	median [s/deg]	mean [s/deg]	median [s/deg]
ARCES	7,183	1.767	1.476	1.350	0.963
FINES	7,746	2.164	1.897	1.830	1.407
GERES	5,142	1.732	1.404	1.547	1.135
NORES	5,308	2.071	1.812	1.783	1.351

observed with respect to signal frequency. However, distinctly higher velocities to the northwest, and predominantly low velocities in the southeast can be seen for all investigated phases (Pn, Pg, Sn, and Sg/Lg). After obtaining geologic information on the crust below the SPITS array (e.g., SIGMOND, 1992), parts of this effect can be explained by sedimentary layers in the uppermost crust dipping to the

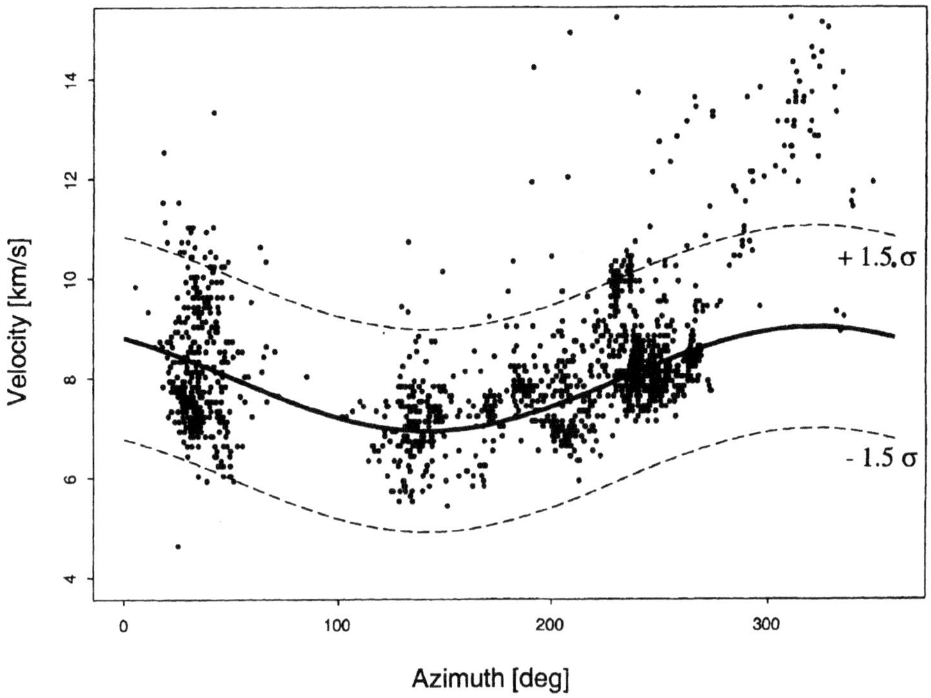

Figure 9
Plot of Pn apparent velocity versus azimuth observed at the SPITS array. After binning the Pn data in distance and azimuth, a sine function was fit to the velocities. For details see text. The dotted lines show the $\pm 1.5\sigma$ range.

northwest as demonstrated by NIAZI (1966) or ZENGENI (1970) for other dipping layer structures. After binning the Pn data in distance (Δ) and backazimuth (Φ), a function of the form

$$v = \text{const} + f_1 \cdot \Delta + f_2 \cdot \sin(\Phi + \Psi)$$

was fitted to the observed Pn apparent velocities v. The estimated values for the modeled factors (const, f_1, f_2) and azimuth of the dipping layer Ψ can be found in Table 5. A value of $\pm 1.5\sigma$ was used as a guideline for the scatter in the data (see Fig. 9). The phases Pg, Sn, and Sg/Lg show similar patterns and they were also modeled with such a function (see Table 5). The differences in Ψ for the different phases are small and can be explained with differences in backazimuth coverage and scatter of the observed data. Different slowness deviations for different frequencies (i.e., wavelengths) can be used as an indication of the geometry of disturbing structures. The fact that the effect is very similar for all investigated phases (Pn, Pg, Sn, and Sg/Lg), usually automatically processed in different dominant frequency bands, can be interpreted that the dipping structure below the SPITS array is thick enough to influence all signals around and above 1 Hz very similarly. The high

Table 5

Parameters to model the apparent velocities at SPITS with a sine-like function (for details see text).

Phase	Constant	f_1	f_2	Ψ
Pn	8.052	−2.368e-4	1.040	308
Pg	7.077	1.637e-3	1.353	328
Sn	4.723	−3.793e-4	0.817	300
Sg/Lg	4.252	1.975e-3	0.863	302

apparent velocities for observations with a backazimuth of about 300° can only be explained by additional lateral heterogeneities on the ray paths of these observations.

6. *Conclusion*

Although a large scatter can be observed for single slowness estimations, mean slowness corrections could be defined and estimated with their standard deviations for the four small aperture arrays ARCES, FINES, GERES, and NORES. All corrections were calculated with respect to the radially symmetric earth model IASPE191, which is also in use at the pIDC as the standard model.

In section 4.1 it was demonstrated that a reduction of the scatter for single observations and a correction for mean slowness errors is especially needed for single array location routines. That these slowness corrections can now be regularly applied to correct automatically estimated slowness and azimuth values is discussed in section 4.2. For teleseismic distances, mean slowness corrections are useful if the ray parameter and backazimuth fit better in theoretically defined limits for the association of first and later arrivals. Correcting the ray parameter and backazimuth of detected onsets will also aid reduction of the huge number of unassociated detections, which is significantly needed for a successful monitoring of the CTBT. In section 5 it was shown that large slowness deviations due to the influence of dipping layer structures can be simply modeled with an azimuth-dependent sine function.

In addition, the measured and corrected slowness values should be weighted in all applications with the corresponding uncertainties known from the mean corrections. This will reduce the influence of scatter in the observations due to the different resolution limits in estimating the slowness of seismic phases at the different arrays or stations.

If enough data are available, a set of specific corrections for source–receiver pair combinations can be estimated and later used to obtain better defined predicted slowness values (see SCHWEITZER and KVÆRNA, 1995). This can also reduce the influence of slowness errors in the GBF event defining and location method.

To invert the mean slowness-correction vectors with respect to the earth's structure could be the topic of future studies.

Acknowledgments

Parts of this work were accomplished during my employment with the GERESS project at the Ruhr-University Bochum, Germany and this research was partly sponsored by the Nuclear Treaty Programs Office of the U.S. Department of Defense and monitored by AFTAC, Patrick AFB, FL32925, under contract no. F08650-96-C-0001. I thank Lyla Taylor for her assistance in preparing Figure 9 and Table 5. The critical comments of Ken Muirhead and an anonymous reviewer are acknowledged. This is NORSAR contribution No. 674.

References

AKI, K., CHRISTOFFERSON, A., and HUSEBYE, E. S. (1977), *Determination of the Three-dimensional Seismic Structure of the Lithosphere*, J. Geophys. Res. *82*, 277–296.

BAME, D. A., WALCK, M. C., and HIEBERT-DODD, K. L. (1990), *Azimuth Estimation Capabilities of the NORESS Regional Seismic Array*, Bull. Seismol. Soc. Am. *80*, 1999–2015.

BERTEUSSEN, K. A. (1974), *NORSAR Location Calibrations and Time Delay Corrections*, NTNF/NORSAR Scientific Report *2-73/74*, 45 pp.

BERTEUSSEN, K. A. (1976), *The Origin of Slowness and Azimuth Anomalies at Large Arrays*, Bull. Seismol. Soc. Am. *66*, 719–741.

BONDÁR, I. (1997), *Slowness-azimuth Station Corrections (SASC) for Teleseismic Phases at the IMS Network*, Center of Monitoring Research Technical Report *CMR-97/25*.

BRATT, S., and BACHE, T. C. (1988), *Locating Events with a Sparse Network of Regional Arrays*, Bull. Seismol. Soc. Am. *78*, 780–798.

BULAND, R., and CHAPMAN, C. H. (1983), *The Computation of Seismic Travel Times*, Bull. Seismol. Soc. Am. *73*, 1271–1302.

CARR, D. B. (1993), *Azimuth Estimation Capabilities of the ARCESS Regional Seismic Array*, Bull. Seismol. Soc. Am. *83*, 1213–1231.

CLAASSEN, J. P. (1998), *Robust and intelligent bearing estimation*. In *20th Annual Seismic Research Symposium on Monitoring a Comprehensive Test-Ban-Treaty (CTBT)*, September 21–23, 1998, Proceedings, 479–488.

GREENFIELD, R. J., and SHEPPARD, R. M (1969), *The Moho Depth Variations under the LASA and their Effect on dT/dΔ Measurements*, Bull. Seismol. Soc. Am. *59*, 409–420.

ENGDAHL, E. R., and FELIX, C. P. (1971), *Nature of Travel-time Anomalies at Lasa*, J. Geophys. Res. *76*, 2706–2715.

FABER, S., PLOMEROVÁ, J., and BABUŠKA, V. (1986), *Deep-seated Lateral Velocity Variations beneath the GRF-array Inferred from Mislocation Patterns and P Residuals*, J. Geophys. *60*, 139–148.

HARJES, H.-P., JOST, M. L., SCHWEITZER, J., and GESTERMANN, N. (1993), *Automatic Seismogram Analysis at GERESS*, Computers and Geosciences *19*, 157–166.

HARJES, H.-P., JOST, M. L., and SCHWEITZER, J. (1994), *Preliminary Calibration of Candidate Alpha Stations in the GSETT-3 Network*, Annali di Geofisica *37*, 382–396.

HARRIS, D. B. (1990), *Comparison of the Direction Estimation Performance of High-frequency Seismic Arrays and Three-component Stations*, Bull. Seismol. Soc. Am. *80*, 1951–1968.

JOHNSON, L. R. (1969), *Array Measurements of P Velocities in the Lower Mantle*, Bull. Seismol. Soc. Am. *59*, 973–1008.

KENNETT, B. L. N., and ENGDAHL, E. R. (1991), *Travel Times for Global Earthquake Location and Phase Identification*, Geophys. J. Int. *105*, 429–466.

KOCH, M., and KRADOLFER, U. (1997), *Investigation of Azimuth Residuals Observed at Stations of the GSETT-3 Alpha Network*, Bull. Seismol. Soc. Am. *87*, 1576–1597.

KOCH, M., and KRADOLFER, U. (1999), *Determination of Mislocation Vectors to Evaluate Bias at GSETT-3 Primary Stations*, J. Seism. *3*, 139–151.

KRÜGER, F., and WEBER, M. (1992), *The Effect of Low Velocity Sediments on the Mislocation Vectors of the GRF Array*, Geophys. J. Int. *108*, 387–393.

KVÆRNA, T., SCHWEITZER, J., TAYLOR, L., and RINGDAL, F. (1999), *Monitoring of the European Arctic Using Regional Generalized Beamforming*, NORSAR Scientific Report *2-98/99*, 78–94.

LIN, C.-H., and ROECKER, S. W. (1996), *P-wave Backazimuth Anomalies by a Small-aperture Seismic Array at Pinyon Flat, Southern California: Implications for Structure and Source Location*, Bull. Seismol. Soc. Am. *86*, 470–476.

MYKKELTVEIT, S. (1992), *Mining Explosions in the Khibiny Massif (Kola Peninsula of Russia) Recorded at the Apatity Three-component Station*, Phillips Laboratory, Hanscom AFB, MA, USA Report *PL-TR-92-2253*.

MYKKELTVEIT, S., and BUNGUM, H. (1984), *Processing of Regional Seismic Events Using Data from Small-aperture Arrays*, Bull. Seismol. Soc. Am. *74*, 2313–2333.

NIAZI M. (1966), *Corrections to Apparent Azimuths and Travel-time Residuals for a Dipping Mohorovicic Discontinuity*, Bull. Seismol. Soc. Am. *56*, 491–509.

NOPONEN, I. (1974), *Seismic Ray Direction Anomalies Caused by Deep Structure in Fennoscandia*, Bull. Seismol. Soc. Am. 64, 1931–1941.

NEUNHÖFER, H. (1998), *Das Bulletin der lokalen Erdbeben im Vogtland 1962–1997 (Bulletin of the Vogtland/Western Bohemia earthquakes)*, Deutsche Geophysikalische Gesellschaft (DGG) Mitteilungen *4/1998*, 2–7.

OTSUKA, M. (1966a), *Azimuth and Slowness Anomalies of Seismic Waves Measured on the Central California Seismographic Array. Part I. Observations*, Bull. Seismol. Soc. Am. *56*, 223–239.

OTSUKA, M. (1966b), *Azimuth and Slowness Anomalies of Seismic Waves Measured on the Central California Seismographic Array. Part II. Interpretation*, Bull. Seismol. Soc. Am. *56*, 655–675.

RINGDAL, F., and KVÆRNA, T. (1989), *A Multichannel Processing Approach to Real Time Network Detection, Phase Association and Threshold Monitoring*, Bull. Seismol. Soc. Am. *79*, 1927–1940.

SCHWEITZER, J. (1992), *The GERESS contribution to the GSETT-2 experiment 1991*. In *Advanced Waveform Research Methods for GERESS Recordings*. DARPA Annual Report No. AFOSR-90-0189 Scientific Report No. 2, *PL-TR-92-2142*, ADA253686, pp. 25–49.

SCHWEITZER, J. (1994), *Mislocation Vectors for Small Aperture Arrays — A First Step Towards Calibrating GSETT-3 Stations*, NORSAR Scientific Report *1-94/95*, 104–118.

SCHWEITZER, J. (1995), *An Assessment of the Estimated Mean Mislocation Vectors for Small Aperture Arrays*, NORSAR Scientific Report *1-95/96*, 158–163.

SCHWEITZER, J., and KVÆRNA, T. (1995), *Mapping of Azimuth Anomalies from Array Observations*, NORSAR Scientific Report 2-94/95, 158–163.

SIGMOND, E. M. O. (1992), *Bedrock Map, Norway and Adjacent Ocean Areas*, Norges Geologiske Undersøkelse 1986–1992.

SUTEAU-HENSON, A. (1990), *Estimating Azimuth and Slowness from Three-component and Array Stations*, Bull. Seismol. Soc. Am. *80*, 1987–1998.

TIBULEAC, I., and HERRIN, E. (1997), *Calibration Studies at TXAR*, Seism. Res. Lett. *68*, 353–365.

VERMEULEN J. M., and DOORNBÖS, D. J. (1977), *Mantle Heterogeneity and Mislocation Patterns for Seismic Networks*, J. Geophys. *43*, 545–559.

WALCK, M. C., and MINSTER, J. B. (1982), *Relative Array Analysis of Upper Mantle Lateral Velocity Variations in Southern California*, J. Geophys. Res. *87*, 1757–1772.

ZENGENI, T. G. (1970), *A Note on Azimuthal Correction for dT/dD for a Single Dipping Plane Interface*, Bull. Seismol. Soc. Am. *60*, 299–306.

(Received July 15, 1999, revised January 20, 2000, accepted February 9, 2000)

Pure appl. geophys. 158 (2000) 397–419
0033–4553/01/020397–23 $ 1.50 + 0.20/0

Pure and Applied Geophysics

Definition of Exclusion Zones Using Seismic Data

Y. BARTAL,[1] M. VILLAGRAN,[1,2] Y. BEN HORIN,[1] G. LEONARD[3] and M. JOSWIG[4]

Abstract—In verifying compliance with the Comprehensive Nuclear-Test-Ban Treaty (CTBT), there is a motivation to be effective, efficient and economical and to prevent abuse of the right to conduct an On-site Inspection (OSI) in the territory of a challenged State Party. In particular, it is in the interest of a State Party to avoid irrelevant search in specific areas. In this study we propose several techniques to determine 'exclusion zones', which are defined as areas where an event could not have possibly occurred. All techniques are based on simple ideas of arrival time differences between seismic stations and thus are less prone to modeling errors compared to standard event location methods. The techniques proposed are: angular sector exclusion based on a tripartite micro array, half-space exclusion based on a station pair, and closed area exclusion based on circumferential networks.

Key words: CTBT, nuclear explosions, event location, seismic array, on-site inspection.

Introduction

In verifying compliance with the Comprehensive Nuclear-Test-Ban Treaty (CTBT), there is a motivation to be effective, efficient and economical and to prevent abuse of the right for one State Party to call for an On-site Inspection (OSI) on the territory of another challenged State Party. This can be achieved by several means (BARTAL *et al.*, 1998), related to accurate and precise location and identification of seismic events:

1. Demonstrating that the triggered seismic event is not a nuclear explosion (different characteristics).
2. Relocating the event such that its epicenter is outside the state's borders.
3. Relocating the event such that its epicenter is away from sensitive installations.
4. Relocating the event such that its depth is not feasible for an explosion.
5. Reducing the potential search area by minimizing the 2-D error region. This error region is usually defined as the 90% confidence error ellipse of the epicenter.

[1] National Data Center, Israel
[2] Institute of Solid Earth Physics, University of Bergen, Norway
[3] IAEC, Israel
[4] Tel Aviv University, Israel

In points 2–4 above the uncertainty associated with the epicenter and depth estimates should be included (point 5 already deals with the uncertainty). The conventional way to estimate an event's hypocenter is to find the hypocenter's parameters that minimize the discrepancy between measured and calculated travel times (azimuth and slowness may also be used). The error in the hypocenter determination is composed of measurement and modeling errors. The measurement error is relatively easy to estimate, however the modeling error is mostly unknown and usually has a bias factor. Moreover, there might be an error associated with local minima and with nonconvergent solutions.

In case there is significant evidence to suspect that a nuclear test has occurred, one or more State Parties could call for an OSI over an area of up to 1000 squared kilometers within the territory of another State Party. In that case, it is in the interest of the Inspected State Party (ISP) to avoid irrelevant search in general and in particular to avoid a search in specifically sensitive areas. Excluding such zones will reduce the target area. It is also in the interest of the CTBT Organization (CTBTO) to improve the cost effectiveness of the search by reducing the search area and thus reducing the manpower, equipment and time needed for the search.

In this study we propose several techniques to determine 'exclusion zones', defined as areas where an event could not have possibly occurred. All techniques are based on simple ideas of arrival time differences between seismic stations and thus are less prone to modeling errors compared to standard event location methods.

The techniques proposed are:

— Angular sector exclusion based on a tripartite micro array. Here we calculate the backazimuth to the event including uncertainty. This defines an angular sector where the event's epicenter must be, and this in turn defines the complementary angular exclusion zone where the epicenter cannot be (provided any azimuth bias is properly calibrated). It should be noted that the common practice for single station backazimuth calculation is 3-component analysis. We demonstrate that the proposed technique is superior.

— Half-space exclusion based on a station pair. Here we use a pair of stations close to one another and we assume that the event must be closer to the one station which records it first than to the other station. This defines a half space where the epicenter must be and the complementary half-space exclusion zone where the epicenter cannot be. Note that this is only true in a velocity structure that is laterally homogeneous.

— Closed area exclusion based on circumferential networks. Here we surround the closed area with a network of at least three stations. If the stations are far enough from the closed area then for an event inside the area, the arrival time difference between any two stations cannot be larger than a set maximum. This maximum is based on an assumed apparent velocity and the maximum distance difference between a point inside the closed area and any two stations in the network. If it can be shown for a specific event that the time difference between some two stations is larger than a set threshold, then the closed area can be declared as an exclusion zone.

It should be emphasized that our main concern in these 'exclusion zones' techniques is for epicentral distances of a few tens of km. For a distant event, a potential search area over an error ellipse of 1000 km^2 as required by the CTBT is already far enough from the zone we would like to exclude.

Tripartite Micro Array

The azimuth calculation method using tripartite arrays, based on arrival time differences, is a well-known technique and has been used by many workers since the early fifties (ASADA and SUZUKI, 1950).

As an example, WARD and GREGERSEN (1973) have used tripartite arrays (1–2 km aperture) and compared their capabilities from an azimuth determination point of view, to those of a network composed of 20 stations. They have found that array-based azimuth evaluations agreed fairly well with network estimations.

Recently CRISTOFFERSSON and ROBERTS (1996) discussed the performance of a tripartite array as compared to a three-component single station. They concluded that an array would outperform a single station.

A tripartite micro array with 1.5 km aperture experimentally installed in northern Israel, demonstrated that it can balance the performance of 8 existing local stations (RABINOWITZ and JOSWIG, 1995).

The same micro array equipment was installed in southern Israel for the purpose of this study from July 1998 until February 1999 as a feasibility experiment for testing exclusion zones techniques.

Figure 1 shows one of the three identical arrays' stations. The digitizer and seismometer were sheltered inside the concrete structure to which the solar panels and satellite dish were fastened.

Experimental Setup

The equipment installed was the same as was previously installed in Northern Israel. It included three sites; each hosting a three-component short period (1 Hz) seismometer system with a dynamic range of 120 dB, a sampling rate of 62.5 Hz and a 16-bit A/D conversion (LE-3D Seismometers and Mars88 units from Lennartz Electronic). The three sites formed a triangle of ~250 m aperture, located about 20 km southwest to the Dead Sea. The array site locations are shown in Table 1. The time for a seismic wavefront to cross the array was about 50 ms (apparent velocity of about 5 km/s). The stations used solar energy for power supply and satellite communications to the Israeli National Data Center (NDC). This setup enabled easy reconfiguration of the array although this option was not utilized. A more complete description appears in LEONARD and TIKOCHINSKI (1997).

Figure 1
One station showing the concrete box, solar panels and satellite dish.

The uncertainty in signal time picks was assumed to be half a sampling interval, or 8 ms for signals with strong signal-to-noise ratio. This time pick uncertainty could be reduced to 2 ms by increasing the sampling rate to 250 Hz however we did not exploit this option.

Seismic Events Used

A set of 11 quarry explosions and 4 earthquakes have been chosen within the period July–October 1998. Table 2 lists the events and Figure 2 shows their spatial distribution. A combination of phase readings from Israel Seismic Network (ISN) and several stations connected directly to the NDC (three broadbands and the

Table 1

Stations locations

Station	Latitude (deg., km)	Longitude (deg., km)	Height (m)
DIM1	31.023611 (Y = 48.127)	35.170811 (X = 166.461)	420
DIM2	31.023447 (Y = 48.119)	35.169327 (X = 166.197)	420
DIM3	31.021326 (Y = 47.898)	35.021326 (X = 166.339)	420

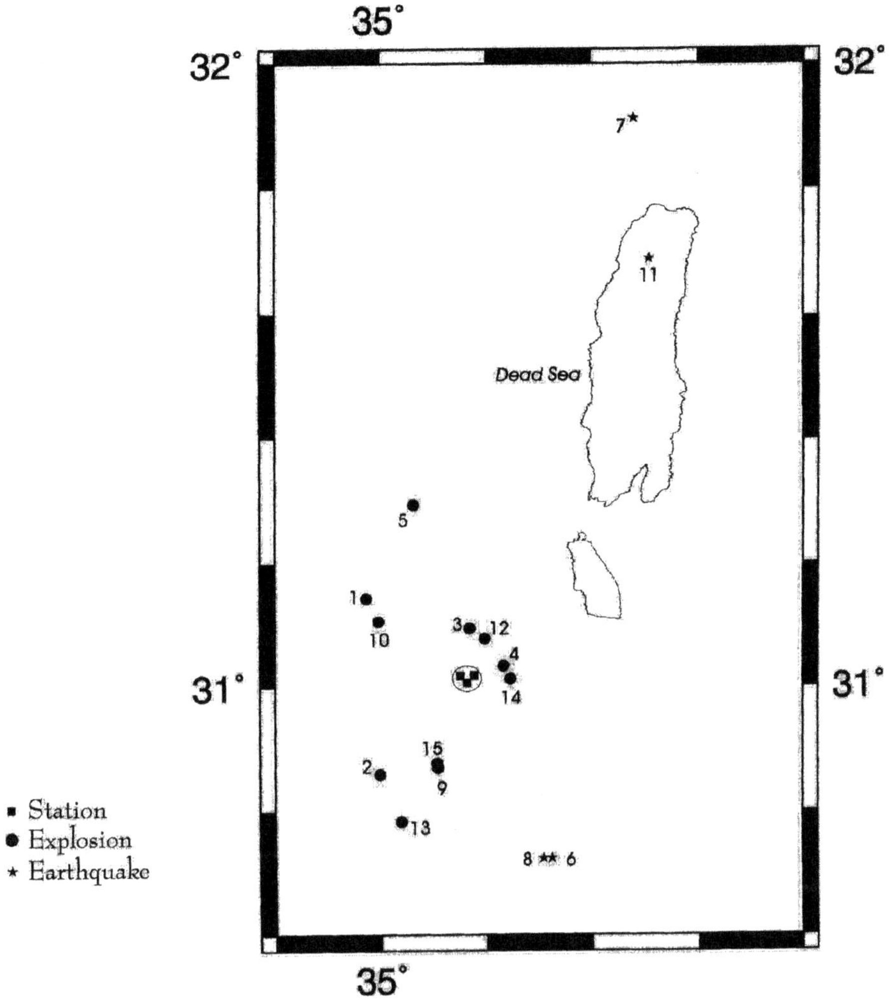

Figure 2
Spatial distribution of epicenters used for the experiment.

tripartite array) were used for location of the events. These epicentral locations are used to define the location backazimuth, which was considered precise enough to be used as a reference. Magnitude determination was done using only the stations connected directly to the NDC.

Azimuth Calculation Using Arrival Time Differences

This technique is based on arrival time differences between the three stations. A detailed formulation appears in Appendix 1. It should be emphasized that one does not need to pick the onset times, which may not be clear enough. All that is needed is to choose the corresponding times of the same wave pattern at all three stations. This can be the first maximum, first minimum or any other common feature. What the formulation needs is actually the time differences, therefore the absolute times are also not necessary. In that respect the uncertainty is less than what would have been if absolute accuracy was needed. The time picks were made manually by human judgement however cross-correlation techniques could be used for greater accuracy. It should be noted that due to the small aperture, travel-time differences are small and quite sensitive to errors. From Appendix 1 one can see that for a laterally homogeneous velocity structure, the azimuthal error dependence on the specific station layout is scaled by the slowness. This means that the error in degree is doubled for a teleseismic or regional phase with double the apparent velocity. The results will thus naturally degrade for remote events due to smaller phase slowness.

The sampling rate we used was set to a relatively low value due to bandwidth limits in the data transmission and the preference for the complete, 3-C recording instead of just sampling the vertical component with two- or even four-fold resolution. One way to escape this constraint is an exact interpolation. Once the Nyquist condition is sufficiently fulfilled due to low-pass filtering before the Analog to Digital Converter, the signal contains only frequencies up to Nyquist frequency. In that case the complete signal can be reconstructed from its samples so that the Fourier representation and its retransformation into the time domain will allow for arbitrary exact interpolation between the sampled values (KORN and KORN, 1968). One can then decrease the sampling interval at least down to 25% (i.e., 2 ms).

Data Analysis

The data have been analyzed as follows:

1. Noise spectra using Acceleration Power Spectral Density in dB relative to 1 m/sec^2 (Fig. 3).

2. Event location and magnitude calculation (Fig. 4). For the magnitude determination we used amplitude measurements on synthesized Wood Anderson seismograms for the M_L calculation, using the HUTTON and BOORE (1987) distance-attenuation for Southern California into Richter's magnitude scale definition for both earthquakes and explosions. All three components were used for magnitude determinations.

3. Signal-to-Noise Ratio (SNR) calculation for the vertical component of each station for each event. We used the spectral ratio of 2 seconds after over 2 seconds before the *P-wave* onset where the data is 8 pole Butterworth bandpass-filtered within 0.8–10.0 Hz (Fig. 5).

▶

Figure 3
Acceleration Power Spectral Density (in dB relative to 1 m/s^2) of the noise at DIM tripartite array.

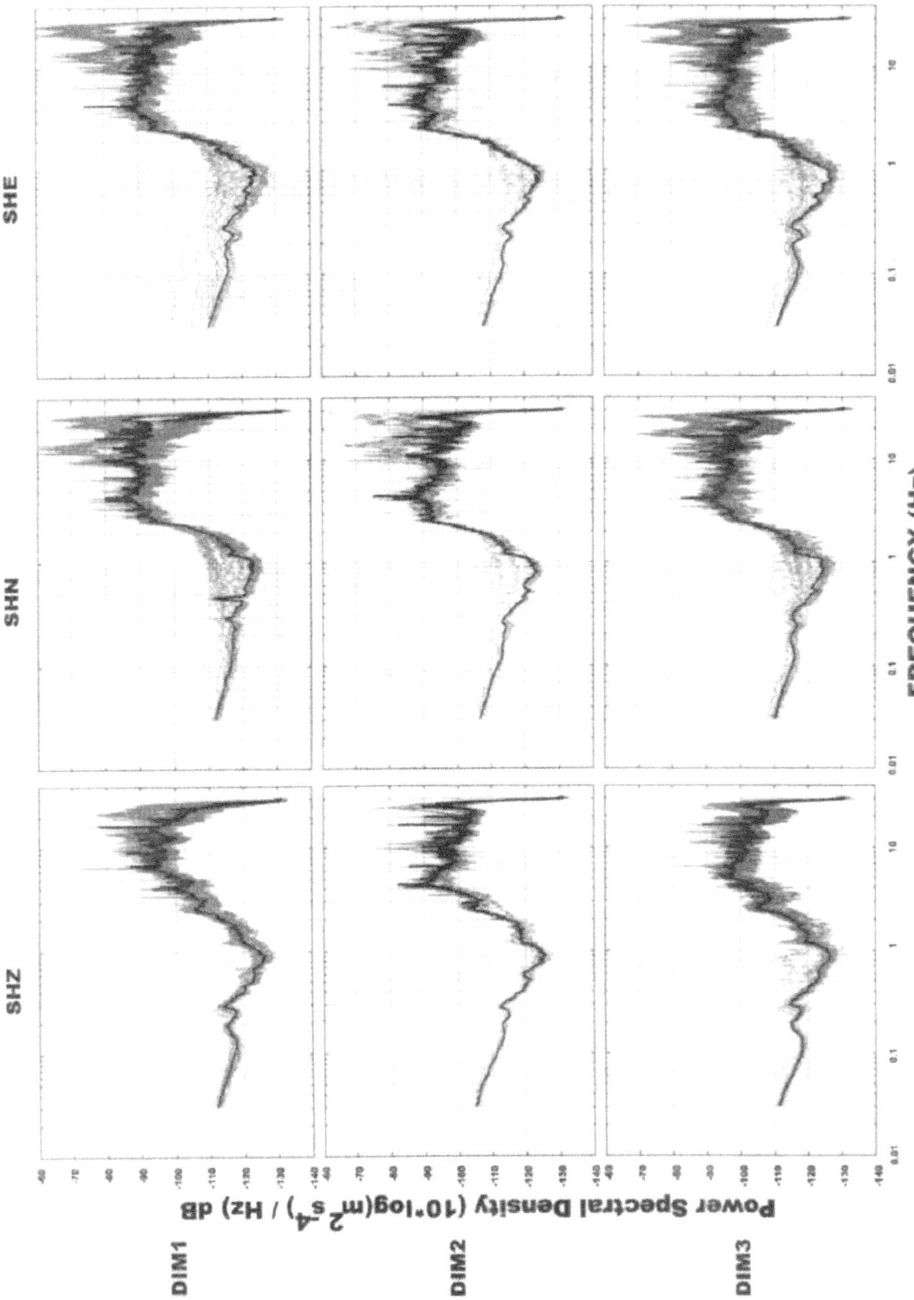
SHZ
SHN
SHE
DIM1
DIM2
DIM3
Power Spectral Density (10*log(m^2s^-4) / Hz) dB
FREQUENCY (Hz)

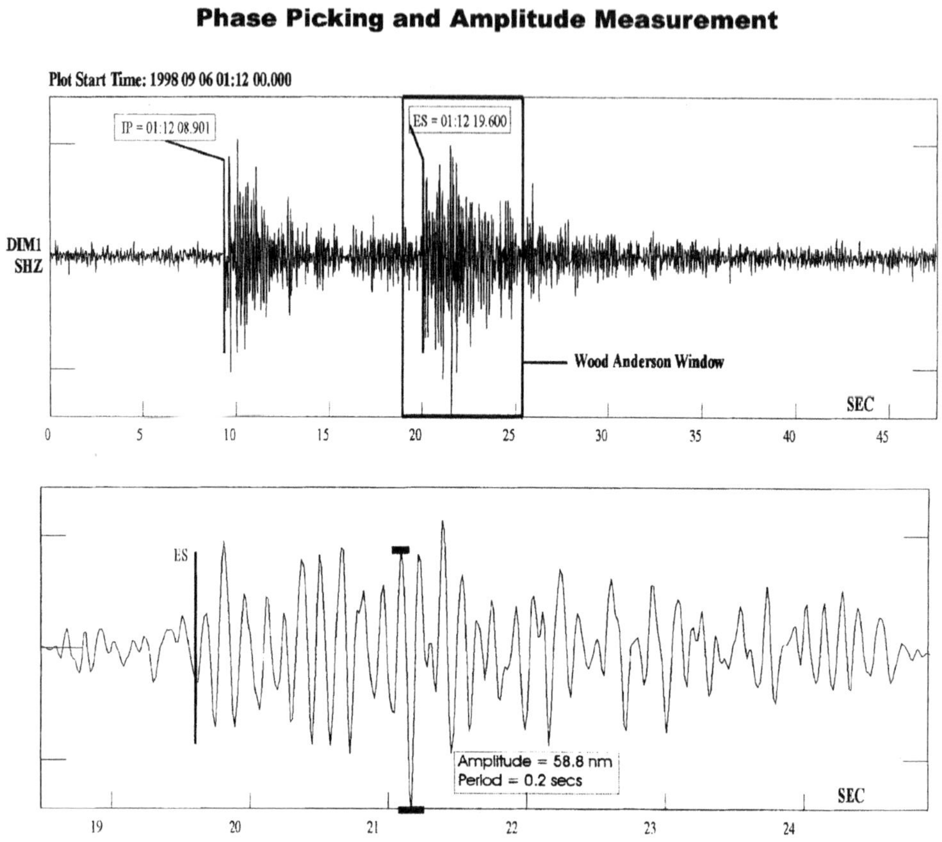

Stat Chan	Amp (nm)	*ML*
DIM1 SHZ	58.8	2.1
DIM1 SHN	104.7	2.3
DIM1 SHE	105.1	2.3
DIM2 SHZ	57.9	2.1
DIM2 SHN	80.2	2.2
DIM2 SHE	82.0	2.2
DIM3 SHZ	41.1	1.9
DIM3 SHN	56.0	2.1
DIM3 SHE	45.3	2.0

Hypocentral Distance 81.5 km

Location (Hypocenter Program):
Event Type: **Earthquake**
Latitude: 31.687 N
Longitude: 35.504 E
Origin Time: 1998 09 06 0111 55.00
Number of stations used for location: 26
RMS = 0.3, Average NDC-*ML* = 2.1 (GII 1.9)

Figure 4
An example of phase picking and magnitude calculation.

4. Calculation of backazimuth and slowness values including uncertainties according to the formulation in Appendix 1. We set a window around the *P-wave* onset of the vertical components of about 1.3 seconds displaying the raw seismograms in counts as a function of sample number to enable better accuracy. Wave-motion picks were then selected which seemed to be correlated in the three stations of the tripartite array. An error of half a sample (8 msec) was assumed to be reasonable for these calculations (Fig. 6).

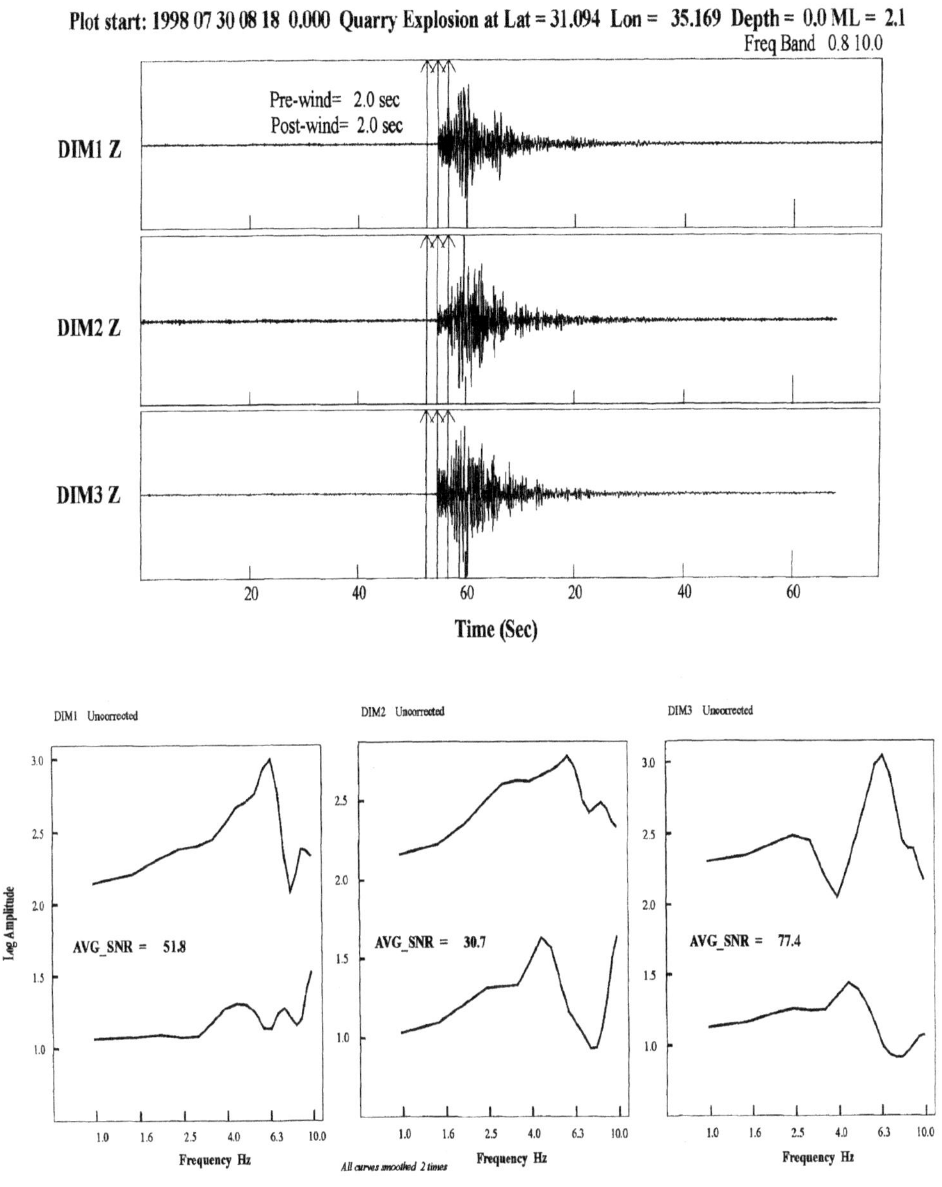

Figure 5
An example of the signal-to-noise ratio SNR calculation.

5. For the purpose of comparing results with the technique described in item 4, we used three-component analysis for the calculation of backazimuth. This is accomplished by using auto- and cross-correlations of the three orthogonal components of a single station as explained later (ROBERTS *et al.*, 1989).

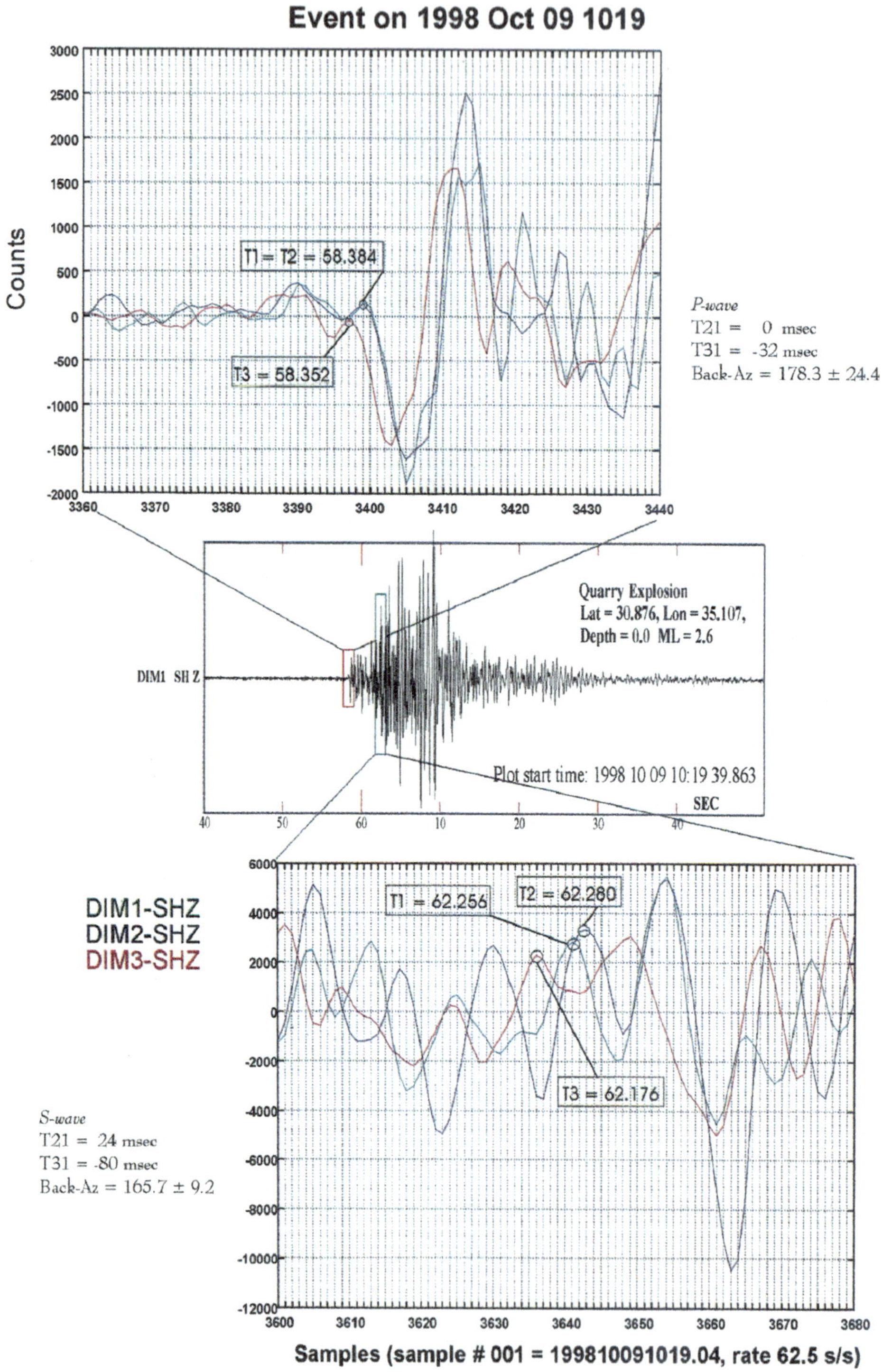
Event on 1998 Oct 09 1019
Counts
T1 = T2 = 58.384
T3 = 58.352
P-wave
T21 = 0 msec
T31 = -32 msec
Back-Az = 178.3 ± 24.4
Quarry Explosion
Lat = 30.876, Lon = 35.107,
Depth = 0.0 ML = 2.6
DIM1 SH Z
Plot start time: 1998 10 09 10:19 39.863
SEC
DIM1-SHZ
DIM2-SHZ
DIM3-SHZ
T1 = 62.256
T2 = 62.280
T3 = 62.176
S-wave
T21 = 24 msec
T31 = -80 msec
Back-Az = 165.7 ± 9.2
Samples (sample # 001 = 199810091019.04, rate 62.5 s/s)

Results

The Acceleration Power Spectral Density representation (in dB relative to 1 m/sec^2) illustrates that the area of the tripartite array location is characterized by a high level of noise (above −130 dB) for the 0.03–20.0 Hz frequency band for all channels at all sites (Fig. 3). This fact, in addition to a relatively low sensitivity setting of the equipment, made it difficult to obtain better micro-seismicity recording. As seen in the results of SNR in Table 2, even at close distances, events with magnitudes below $M_L = 2$ have a high level of noise that makes it difficult to identify correlated wave picks around the *P-wave* onset.

The combination of phase readings of NDC and ISN gives good event hypocentral results (RMS below 0.6) for all events. This enabled us to use the location backazimuth as a reference for the experiment. Magnitude results using NDC stations are in agreement with ISN magnitude values (which are duration magnitudes) within ±0.5 magnitude unit. All events are below magnitude of 2.7. Figure 4 presents an example of magnitude calculation and phase picking in one of the stations of the tripartite array.

During the course of the measurements for backazimuth calculation, several important features that made the precision difficult appeared, and certain assumptions were made to compensate for them:

a) Considering the shape of the seismograms, some waveform picks were chosen between sampling points (eye interpolation) assuming the waveform peak lies there.

b) When low SNR mixed with lack of defining peaks in the waveform made it impossible to choose the picks from the *P* phase, waveform picks are chosen from the *S* phase, expecting higher slowness (events 5 and 13).

c) When backazimuth results are wrong (nominal tripartite backazimuth differed from the location backazimuth by 20 degrees or more which we chose as a reasonable limit) and SNR is good, a waveform pick was attempted from the *S* phase and results are compared (failed to find picks from *S* phase for event 7, succeeded for event 11). It should be noted that in a real-life situation when the event location is not known, we will not know that we need to try *S* phase. What we do demonstrate here is that statistically this method works, just as the 90% location error ellipse covers 90% of the events, however for a single event it might be quite far from the ground-truth location. For this study in which the locations were quite accurate, it was interesting to show whether the accuracy could still be improved.

◀

Figure 6
An example of backazimuth calculation using both the *P wave* and *S wave* around onset times at the tripartite array.

Table 2

Results of analysis

No.	Date & Time	Lat. N	Long. E	Dep. km	ΔT (ms)		N Sta	RMS	SNR			Baz (deg.)	3-comp. Baz			Loc. Baz (deg.)	Dist. (km)	Slow (s/km)	Vel. (km/s)	M_L		Ev ID
					t_{21}	t_{31}			Dim1	Dim2	Dim3		Dim1	Dim2	Dim3					NDC	GII	
1	19980724085404.9	31.143	34.973	0.0F	−48	8	15	0.4	2.6	2.9	3.2	**305.8 ± 9.4**	80	338	106	**305.9 ± 5.8**	22.72	0.23 ± 0.07	4.36 ± 1.33	2.0	2.3	E
2	19980726085753.8	30.858	34.999	0.0F	−24	−40	14	0.3	5.6	3.3	4.6	**214.1 ± 18.0**	242	126	**137**	**221.1 ± 5.5**	24.25	0.16 ± 0.06	6.45 ± 2.64	1.7	1.9	E
3	19980730081852.2	31.094	35.169	0.0F	16	48	25	0.5	51.8	30.7	77.4	**17.0 ± 20.8**	**20**	**37**	**242**	**2.0 ± 13.8**	7.90	0.19 ± 0.09	5.31 ± 2.42	2.1	2.6	E
	Ground truth for #3	31.104	35.188	0.0												**12.9**	9.29					
4	19980803105315.9	31.034	35.235	0.0F	64	32	17	0.4	92.5	112.6	140.5	**87.5 ± 8.4**	**85**	**102**	**92**	**79.2 ± 15.6**	6.71	0.24 ± 0.06	4.13 ± 1.06	2.2	2.3	E
5A	19980803110552.4	31.293	35.064	0.0F	−40	72	14	0.5	3.3	1.8	7.6	**337.8 ± 9.2**	243	**304**	201	**342.0 ± 5.9**	31.48	0.43 ± 0.09	2.31 ± 0.47	2.0	1.7	E
6	19980814170745.7	30.724	35.329	11.8	40	−16	19	0.6	1.3	1.2	1.6	**134.4 ± 18.1**	129	N	207	**154.7 ± 7.4**	36.60	0.22 ± 0.09	4.57 ± 1.96	1.6	1.5	Q
7	19980817014315.9	31.909	35.482	14.3	24	32	32	0.3	13.5	19.0	9.8	**43.5 ± 30.1**	**57**	**10**	**87**	**16.8 ± 2.2**	102.72	0.13 ± 0.09	7.81 ± 5.74	2.6	2.7	Q
8	19980824132330.1	30.723	35.308	7.2	32	−48	14	0.6	7.7	4.9	6.3	**155.0 ± 13.1**	76	162	131	**157.7 ± 7.4**	35.90	0.31 ± 0.09	3.25 ± 0.96	2.0	2.0	Q
9	19980901093639.9	30.869	35.109	0.0F	−12	−52	16	0.2	10.3	9.8	14.8	**198.4 ± 16.3**	**227**	**193**	**228**	**197.6 ± 9.3**	18.01	0.24 ± 0.09	4.13 ± 1.48	2.7	2.5	E
10	19980904082409.3	31.106	34.996	0.0F	−56	−8	21	0.5	9.7	12.9	33.6	**290.3 ± 16.5**	**254**	**WD**	**272**	**299.6 ± 5.6**	18.60	0.23 ± 0.08	4.37 ± 1.58	2.3	2.8	E
11	19980906011155.0	31.689	35.507	16.0	24	32	26	0.3	19.4	14.3	9.3	**43.5 ± 30.1**	**15**	**35**	1	**23.5 ± 3.1**	80.54	0.13 ± 0.09	7.81 ± 5.74	2.1	1.9	Q
11A					40	72						**31.3 ± 14.4**						0.28 ± 0.09	3.60 ± 1.20			
12	19980906093115.4	31.077	35.198	0.0F	32	56	13	0.3	86.0	38.2	39.9	**32.3 ± 18.4**	327	27	41	**26.9 ± 30.9**	6.63	0.22 ± 0.09	4.62 ± 1.98	2.2	-	E
13A	19980907103825.7	30.783	35.040	0.0F	8	−96	12	0.5	3.2	2.5	2.3	**174.4 ± 8.5**	N	N	N	**204.3 ± 7.2**	29.18	0.44 ± 0.08	2.25 ± 0.39	1.6	2.0	E
14	19980908112850.7	31.013	35.248	0.0F	72	24	13	0.6	82.5	50.1	35.7	**98.5 ± 14.5**	**35**	**82**	**125**	**98.7 ± 16.3**	7.91	0.28 ± 0.07	3.61 ± 0.93	2.1	2.6	E
15	19981009101954.1	30.876	35.107	0.0F	0	−32	7	0.4	15.7	25.5	22.4	**178.3 ± 24.4**	**188**	**159**	**214**	**199.0 ± 19.7**	17.22	0.14 ± 0.07	7.04 ± 3.61	2.6	-	E
15A					24	−80						**165.7 ± 9.2**						0.42 ± 0.08	2.40 ± 0.48			

- 'Baz' is the backazimuth based on the time differences technique, '3-comp Baz' on three-component analysis and 'Loc Baz' on the network solution.
- Event 3 was a planned calibration event whose location is exact (and thus backazimuth too).
- **N** in the 3-comp. column means high noise level.
- **WD** in the 3-comp. column means waveform disturbed.
- Bold values in the 3-comp. column indicate that the coherence value CO is greater or equal to 0.1.
- An 'A' in the first column means the backazimuth was calculated using *S* phase instead of *P* phase

Table 2 shows that for 14 events out of 15 (93.33%) the tripartite backazimuth agreed with the reference value within the uncertainties, despite the low SNR for some of them (below 3.0 in at least one component). The exception is event 13 for which there was a gap of 15 degrees between the closest backazimuth values (174.4 + 8.5 compared to 204.3 − 7.2). Values of slowness ranged between 0.13–0.31 s/km (3.23–7.69 km/s) for the *P-wave* picks and between 0.22–0.44 s/km (2.27–4.55 km/s) for the *S-wave* picks. Out of the 15 events, 4 events had an epicentral distance below 10 km (all with good results), 9 events – between 10 and 50 km (only one with bad results) and 2 events – beyond 50 km (all with good results, although relatively more biased than for the shorter distances).

For some cases the SNR is too low and the combination with an emergent arrival of waveforms phases made the selection of correlated picks not only difficult but unreliable. For events 5 and 13 a wave point was chosen within the *S-wave* train. For event 5 the procedure was successful and a very good backazimuth value resulted. For event 13 we chose what seemed to be reasonable picks, nonetheless the result of backazimuth does not fit the reference value (including uncertainties).

Events 7 and 11 are the special cases of this experiment, both being earthquakes over 80 km distant from the tripartite array. Both have very good SNR in the three stations although the backazimuth is very biased from the reference value obtained using the network location (20 degrees and more, nonetheless still acceptable within the uncertainties). Since the location backazimuth had a low uncertainty, it was expected to yield better results, therefore we tried the same method as for low SNR events, namely seeking for time picks in the *S-wave* train.

Event 7 is probably the most exceptional case of all. The SNR is good (above 9.0 for the three stations) however once into the time series after the *P-wave* arrival, it is impossible to find a correlated wave segment. Even an attempt to find a point within the *S-wave* train failed. The epicentral distance to the event exceeds 100 km and the source is 14 km deep north of the Dead Sea. It may be the case that since the waves through the laterally heterogeneous crust of the Jordan Valley rift find the fastest way to propagate to a certain direction/point (Fermat's principle), their arrival at the array of stations is at a different direction than the straight source to station line.

For event 11 the backazimuth calculated using the *S-wave* train is better than the result obtained using the *P-wave* train. As for event 7, there may be an extra factor perturbing the waveform's arrival at the observation point. The source of this event is approximately 80 km distance from the array and 16 km deep.

As an extra exercise for the purpose of checking results, results for one event with good backazimuth values using the *P-wave* train, were compared to results using the *S-wave* train (event 15). The comparison shows good agreement between the two (see Fig. 6).

Three-component (3-C) Analysis

The three-component analysis utilizes all three components of ground motion as recorded by a single 3-C seismic station designed to estimate the azimuth of the incident seismic wave. We calculated these azimuth values employing the SEISAN software package (HAVSKOV, 1997), which uses the approach by ROBERTS *et al.* (1989). This formulation appears in Appendix 2.

Results

For the three-component analysis, extensive work had to be done and the results are shown in Table 2. There are two groups of data: the very good SNR data and the medium to low SNR data. For the first group, possessing a very clear view of the first motion facilitates selection of a narrow time window around the onset and the results of backazimuth calculation are mostly acceptable (events 3, 4, 7, 9, 10, 11, 14, 15).

For the second group, the analysis became very complicated. The fact that the very first motion of the onset is unclear, makes it obligatory to select a wider time window than for the first group. This led to random results and it was impossible to repeat the results without a certain order. For this group we repeated the calculation using 0.5, 1.0 and 2.0 seconds time window around the approximate position of the first onset. This was done three times using raw-unfiltered data and filtered data for the 1–5 Hz and 5–10 Hz frequency bands. The azimuth chosen was the one that had the best coherence value even if this does not match the coherence value requirement (CO parameter is greater or equal to 0.1) suggested by ROBERTS *et al.* (1989). The results of this case are in general not good (events 1, 2, 5, 6, 8, 12, 13).

Events 6 and 13 had one or more channels which were too noisy to attempt a three-component analysis. For event 10 (good SNR) one of the stations showed disturbed records on the *P-wave* onset (could be by a pedestrian or car passing very near it), however the other two agreed with the results of the location backazimuth.

Events 3 and 12 experienced a special problem that prevented us from obtaining similar results on the backazimuth at the three stations. For some reason, there was a similarity between the vertical components but not between the horizontal components.

The results presented demonstrate that the 3-C technique is much less reliable than the tripartite procedure for determining an event's backazimuth. Furthermore, for the 3-C analysis one has to set the time window of the seismogram over which the analysis is performed. This window is set based on the personal judgement of the analyst, which might cause variability in the results.

Epicentral Bounding Using the Arrival Order Technique

The Arrival Order Technique (AOT), developed by SAMBRIDGE and KENNETH (1989) was used for bounding the epicentral location of all 15 events recorded by the tripartite array (Table 2). In our case, this simple method utilizes the order of the arrival times at the three stations in order to delimit the area where the epicenter could be geometrical. The basic assumption is that the velocity structure is laterally homogeneous and thus stations with the earlier time picks are closer to the epicenter. Therefore, each pair of stations (D1-D2, D2-D3 and D1-D3) provides a geometrical constraint on the epicenter position. For example, if $t_1 < t_2$, the epicenter must lie in the half-space defined by the perpendicular bisector of the line joining D1 and D2, which is closer to D1. The three pairs define three constraints which define an angular sector where the epicenter could be.

Figure 7 displays the three mentioned bisectors: D1-D2, D2-D3 and D1-D3. In such a way, three constraints are obtained in our case, and together they define the area in which the epicenter is assumed to be. Also shown in the figure are the error ellipses of the events.

The results are summarized in Table 3, where the epicenter location is tested for its fulfillment of the three mentioned constraints. An event with three fulfilled constraints obviously indicates an agreement between the epicenter location and the AOT.

As shown in the table, there exists a full agreement between the epicenter location and the AOT for 14 of the 15 events (93.33%). For event 13 one of three constraints was violated. This event was also the only exception in comparing backazimuth values as mentioned. For event 15 one constraint is undecided (bisector D1-D2). If we examine the number of constraints, then 44 of 45 constraints are fulfilled (97.78%).

The hypocenter location is calculated based on *P* phases only, using HYPOCENTER program version 3.2 (LIENERT, 1994), which is the main location program of the SEISAN software. The calculation of the error ellipse was performed as stated in FLINN (1965). A similar method is routinely used by the International Data Centre (IDC) at the CTBT Organization (CTBTO). The total standard deviation error used (which includes both modeling and measurement errors) was 0.37 seconds (BARTAL *et al.*, 1998).

Circumferential Networks

It is proposed here to define closed area exclusion based on circumferential networks. Here we surround the closed area with a network of at least three stations (see Fig. 8).

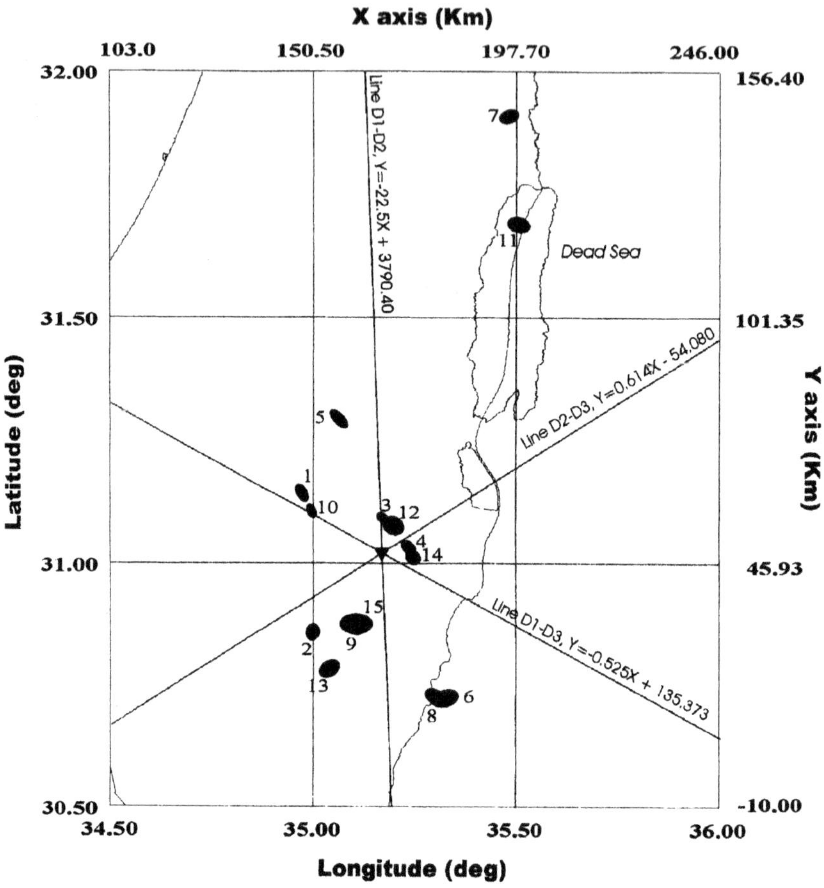

Figure 7
Illustration to the Arrival Order Technique (AOT). Following Table 3 (arrival time order column) it is possible to constrain the events' epicenters.

If the stations are far enough from the closed area, then for an event inside the area, the arrival time difference between any two stations cannot be larger than a set maximum. This maximum is based on an assumed maximum apparent velocity and the maximum distance difference between a point inside the closed area and any two stations in the network. If we find any pair of stations in the network whose arrival time difference exceeds that maximum, consequently the epicenter could not be inside the closed area.

Assume that the distance between any two stations in Figure 8 is R and the circle's radius is r. Suppose also that the maximum apparent velocity is V. Then the maximum time difference between any two stations for an event inside the circle is less than $2r/V$. The minimum time difference between any two stations for an event

Table 3

*Arrival Order Technique for delimiting an epicentral location area for the events of Table 2. 'Yes' indicates that the epicenter location was inside the constrained zone. '**No**' indicates that it was outside. '**Yes**' (highlighted) indicates that the epicenter was inside the constrained zone, although some of the error ellipse was outside. 'smaj', 'smin' and 'azim' are the semi-major, semi-minor and azimuth parameters of the 90% error ellipse based on the network solution*

Event #	D1-D2	D2-D3	D1-D3	Arrival time order	smaj	smin	azim	Comments
1	Yes	Yes	Yes	2,1,3	2.3	1.2	149	
2	Yes	Yes	Yes	3,2,1	2.0	1.6	47	
3	**Yes**	Yes	Yes	1,2,3	1.5	0.9	114	Error ellipse intersects D1-D2
4	Yes	Yes	Yes	1,3,2	2.1	1.2	122	
5	Yes	Yes	Yes	2,1,3	2.9	1.3	128	
6	Yes	Yes	Yes	3,1,2	3.4	1.6	83	
7	Yes	Yes	Yes	1,2,3	2.7	1.3	79	
8	Yes	Yes	Yes	3,1,2	3.5	1.6	114	
9	Yes	Yes	Yes	3,1,2	2.2	1.5	90	
10	Yes	Yes	**Yes**	2,3,1	1.8	1.0	140	Error ellipse intersects D1-D3
11	Yes	Yes	Yes	1,2,3	3.1	1.5	98	
12	Yes	Yes	Yes	1,2,3	2.9	1.9	103	
13	**No**	Yes	Yes	3,1,2	2.9	1.7	73	Could be wrong time picks
14	Yes	Yes	Yes	1,3,2	2.1	1.3	110	
15	Yes	Yes	Yes	3,(1,2)	4.6	2.0	91	

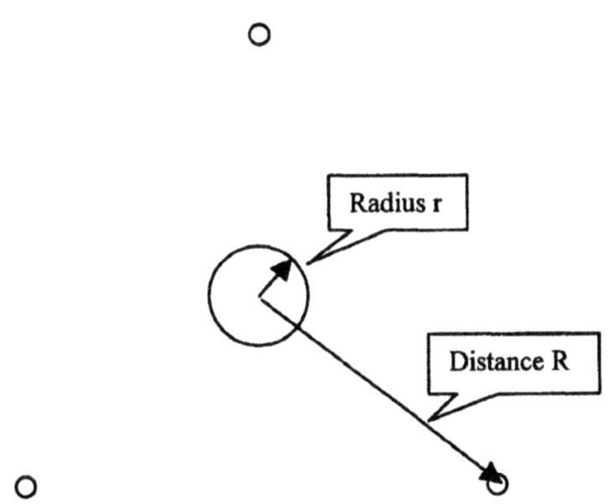

Figure 8
A closed area to be excluded (center) and a circumferential network of three stations.

outside the triangle area in Figure 8 is $R \cdot \cos 30/V$. Thus to make sure that an event outside the triangle cannot be mislocated inside the circle, we need the following condition

$$2r/V < R \cdot \cos 30/V$$

or

$$R > 2.3\,r.$$

Still larger values than that will insure the validity of the argument. However the larger R is, the larger also is the "grey" area of the triangle excluding the circle. Also the assumption of a plane wavefront and homogeneous crustal medium may not hold.

This technique of circumferential network has not been tested however its testing is planned in the near future together with more testings of the other two techniques described.

Summary

It is the interest of Signatory States Parties to the CTBT to avoid abuse of the treaty and especially to avoid unjustified OSI in sensitive areas. In that respect it is desirable to find techniques, in addition to the standard event location, which can be used to define exclusion zones. These zones are defined such that an event whose epicenter is outside the exclusion zone cannot be located inside it. Using exclusion zones, one can also reduce the search area as defined by the standard location techniques. Reduction of the search area is in the interest of the ISP as well as the CTBTO in order to reduce the intrusiveness and increase the cost effectiveness of an OSI.

In this study we described three techniques for defining exclusion zones:

— Angular sector exclusion based on a tripartite micro-array.

— Half-space exclusion based on a station pair.

— Closed area exclusion based on circumferential networks.

The first two techniques have been tested and the results presented reveal their feasibility. The third technique was described but not tested yet. Its testing is planned together with more testing of the first two techniques soon.

The main advantage of all three proposed techniques is their simplicity, being based on the arrival times at the stations. There is no seismological model involved and thus no modeling errors except for the basic assumption of laterally homogeneous velocity structure. The only uncertainty stems from time picks errors which are relatively easy to estimate.

Simple enough to be understood from a political point of view, these techniques may be relatively easy in their application and presentation to the CTBTO Executive Council in the context of deciding where an on-site inspection should and should not be conducted.

The basic methods presented in this study are not perfect and are intended to complement more formal and thorough methods, which are not perfect either. One should employ all available data and analysis methods in order to arrive at the best decision concerning a possible OSI.

Appendix 1

Azimuth Calculation Using a Tripartite Array

Let the tripartite array elements be denoted by indices $i = 1, 2, 3$ and the apparent velocity by v. Then

$$v = \frac{\Delta y'}{\Delta t}.$$

Here $\Delta y'$ denotes the difference in y coordinates of any pair of elements in a rotated axis system whose y axis is directed towards the event. This difference is expressed in terms of the difference in x and y coordinates in the original axis systems by

$$\Delta y' = \Delta x \sin\alpha + \Delta y \cos\alpha$$

so that

$$v = \frac{\Delta x \sin\alpha + \Delta y \cos\alpha}{\Delta t}.$$

Now we can write this equation for two pairs of elements

$$v = \frac{x_{12} \sin\alpha + y_{12} \cos\alpha}{t_{12}} = \frac{x_{13} \sin\alpha + y_{13} \cos\alpha}{t_{13}}.$$

Here we denote $x_{ij} = \Delta x_{ij}$, $y_{ij} = \Delta y_{ij}$, $t_{ij} = \Delta t_{ij}$

$$\Rightarrow ctg\,\alpha = \frac{-x_{13}t_{12} + x_{12}t_{13}}{y_{13}t_{12} - y_{12}t_{13}} = \frac{-x_{13} + x_{12}z}{y_{13} - y_{12}z}; \quad z \equiv \frac{t_{13}}{t_{12}}$$

from which the backazimuth angle is calculated by the inverse function.

$$\text{if}\,(\alpha < 0)\,\text{then}\,\alpha \rightarrow \alpha + 360.$$

The velocity v and slowness s are expressed by:

$$v = \frac{x_{12}y_{13} - x_{13}y_{12}}{\sqrt{(y_{13}t_{12} - y_{12}t_{13})^2 + (-x_{13}t_{12} + x_{12}t_{13})^2}} = \frac{1}{s}.$$

Uncertainty Calculation

If we define $\gamma = \Pi/2 - \alpha$ then

$$|d\gamma| = |d\alpha|\, tg\gamma = ctg\,\alpha$$

$$d\gamma = \frac{d(tg\gamma)}{(tg\gamma)'} = \frac{1}{1 + tg^2\gamma} \cdot \frac{d(tg\gamma)}{dz} \cdot \sum_{i=1}^{3} \frac{\partial z}{\partial t_i} dt_i$$

$$\frac{d(tg\gamma)}{dz} = \frac{(y_{13} - y_{12}z) \cdot x_{12} - (-x_{13} + x_{12}z) \cdot (-y_{12})}{(y_{13} - y_{12}z)^2}$$

$$= \frac{y_{13}x_{12} - x_{13}y_{12} - (y_{12}x_{12} - x_{12}y_{12}) \cdot z}{(y_{13} - y_{12}z)^2} = \frac{y_{13}x_{12} - x_{13}y_{12}}{(y_{13} - y_{12}z)^2}$$

$$\Rightarrow \frac{1}{1 + tg^2\gamma} \cdot \frac{d(tg\gamma)}{dz} = \frac{y_{13}x_{12} - x_{13}y_{12}}{(y_{13} - y_{12}z)^2 + (-x_{13} + x_{12}z)^2}$$

$$\frac{\partial z}{\partial t_1} = \frac{\partial}{\partial t_1}\left(\frac{t_1 - t_3}{t_1 - t_2}\right) = \frac{(t_1 - t_2) \cdot 1 - (t_1 - t_3) \cdot 1}{(t_1 - t_2)^2} = \frac{t_3 - t_2}{(t_1 - t_2)^2} \equiv \frac{t_{32}}{t_{12}^2}$$

$$\frac{\partial z}{\partial t_2} = \frac{\partial}{\partial t_2}\left(\frac{t_1 - t_3}{t_1 - t_2}\right) = \frac{(t_1 - t_2) \cdot 0 - (t_1 - t_3) \cdot (-1)}{(t_1 - t_2)^2} = \frac{t_1 - t_3}{(t_1 - t_2)^2} \equiv \frac{t_{13}}{t_{12}^2}$$

$$\frac{\partial z}{\partial t_3} = \frac{\partial}{\partial t_3}\left(\frac{t_1 - t_3}{t_1 - t_2}\right) = \frac{(t_1 - t_2) \cdot (-1) - (t_1 - t_3) \cdot 0}{(t_1 - t_2)^2} = \frac{(t_1 - t_2)}{(t_1 - t_2)^2} \equiv \frac{-t_{12}}{t_{12}^2}$$

$$\Rightarrow d\gamma = \frac{(y_{13}x_{12} - x_{13}y_{12}) \cdot (t_{32}dt_1 + t_{13}dt_2 + t_{12}dt_3)}{(y_{13}t_{12} - y_{12}t_{13})^2 + (-x_{13}t_{12} + x_{12}t_{13})^2}$$

$$\Rightarrow |d\alpha| = |d\gamma| \le \frac{|y_{13}x_{12} - x_{13}y_{12}| \cdot (|t_{32}dt_1| + |t_{13}dt_2| + |t_{12}dt_3|)}{(y_{13}t_{12} - y_{12}t_{13})^2 + (-x_{13}t_{12} + x_{12}t_{13})^2}$$

$$= \frac{|t_{32}dt_1| + |t_{13}dt_2| + |t_{12}dt_3|}{|y_{13}x_{12} - x_{13}y_{12}| \cdot s^2}.$$

It can be shown that

$$s_x = \frac{y_{13}t_{12} - y_{12}t_{13}}{x_{12}y_{13} - y_{12}x_{13}}$$

$$s_y = \frac{-x_{13}t_{12} + x_{12}t_{13}}{x_{12}y_{13} - y_{12}x_{13}}$$

so that

$$s = (s_x^2 + s_y^2)^{1/2} \Rightarrow ds = 1/2(s_x^2 + s_y^2)^{-1/2} \cdot (2s_x ds_x + 2s_y ds_y) = \frac{s_x ds_x + s_y ds_y}{s}$$

$$ds_x = \sum_{i=1}^{3} \frac{\partial s_x}{\partial t_i} dt_i = \frac{(y_{13} - y_{12}) \cdot dt_1 + (-y_{13}) \cdot dt_2 + y_{12} \cdot dt_3}{x_{12}y_{13} - y_{12}x_{13}}$$

$$ds_y = \sum_{i=1}^{3} \frac{\partial s_y}{\partial t_i} dt_i = \frac{(-x_{13} + x_{12}) \cdot dt_1 + x_{13} \cdot dt_2 + (-x_{12}) \cdot dt_3}{x_{12}y_{13} - y_{12}x_{13}}$$

$$\Rightarrow ds \leq |ds| \leq \left|\frac{s_x/s}{x_{12}y_{13} - y_{12}x_{13}}\right| \cdot \Big(|(y_{13} - y_{12}) \cdot dt_1| + |(-y_{13}) \cdot dt_2| + |y_{12} \cdot dt_3|\Big)$$
$$+ \left|\frac{s_y/s}{x_{12}y_{13} - y_{12}x_{13}}\right| \cdot \Big(|(-x_{13} + x_{12}) \cdot dt_1| + |(x_{13}) \cdot dt_2| + |(-x_{12}) \cdot dt_3|\Big)$$

$$v = \frac{1}{s} \Rightarrow |dv| = \left|\frac{-1}{s^2} ds\right| = \left|\frac{ds}{s^2}\right|.$$

Appendix 2

Azimuth Calculation Using Three-components Analysis

The seismic signal recorded by the three-component sensor is defined by the following expression

$$y(t) = \begin{pmatrix} y_n(t) \\ y_e(t) \\ y_z(t) \end{pmatrix}.$$

Here the subscripts n, e and z refer, respectively, to the north, east and vertical components. We should mention that **y**(t) is assumed to describe a *P-wave* signal arriving at some angle to the vertical, and thus appearing on both vertical and horizontal components.

The backazimuth (from the station to the source) *az* is thus given by the following expression

$$\tan(az) = \frac{-\langle y_e y_z \rangle}{-\langle y_n y_z \rangle}.$$

Here $\langle f \rangle$ represents the time average of f over a user-defined time window T. The three components of **y**(t) should be "cleaned" from their DC part, representing the DC noise level. The DC contributions are evaluated by averaging every component over T.

One can write the z component as a linear combination of both horizontal components

$$y_z(t) = Ay_n(t) + By_e(t) + N(t).$$

Here $N(t)$ is a noise term and

$$A = -R\cos(az)$$
$$B = -R\sin(az),$$

R being defined by

$$R = \frac{\langle y_z y_z \rangle}{\sqrt{(\langle y_n y_z \rangle^2 + \langle y_e y_z \rangle^2)}}.$$

The predicted coherence is defined by ROBERTS *et al.* (1989) to be

$$\mathrm{CO} = 1 - \frac{\langle (y_z - Ay_n - By_e)^2 \rangle}{\langle y_z y_z \rangle}.$$

It is clear that in an ideal case, i.e. $y(t)$ is a noise-free signal, the coherence coefficient CO will be 1. In a real case, it will be less than 1. ROBERTS *et al.* (1989) used the three-component analysis as an event detector. For this purpose, one should choose a threshold value for CO above which an event is assumed to be detected. In SEISAN, CO is utilized as a quality indicator of the three-component analysis for a given event: 3-C estimation is assumed to be reliable only if CO is greater or equal to 0.1.

REFERENCES

ASADA, T., and SUZUKI, S. (1950), *On Microearthquakes Observed after the Imaichi Earthquake, December 26, 1949*, Bull. Earthquake Res. Inst., Tokyo Univ. Vol. 28, pp. 415–421.

BARTAL, Y., SOMER, Z., LEONARD, G., BEN HORIN, Y., and TIKOCHINSKI, G. (1998), *Optimal Seismic Networks in Israel in the Context of the Comprehensive Test-Ban-Treaty (CTBT) and the Cooperating National Facilities (CNF)*, Report SNRC 2818, Soreq Nuclear Research Center, Yavne, Israel.

CRISTOFFERSSON, A., and ROBERTS, R. *Trinity or Verity?, Monitoring a Comprehensive Test-Ban-Treaty* (eds. Husebye and Dainty) (NATO ASI Series, Kluwer Academic Publishers, 1996).

FLINN, E. A. (1965), *Confidence Regions and Error Determinations for Seismic Event Location*, Rev. Geophys. *3*, 157–185.

HAVSKOV, J. (1997), *The SEISAN Earthquake Analysis Software for the IBM PC and SUN*, Version 6.0, Institute of Solid Earth Physics, University of Bergen, Norway.

HUTTON, L. K., and BOORE, D. M. (1987), *The M_L Scale in Southern California*, Bull. Seismol. Soc. Am. *77*, 2074–2094.

KORN, G. A., and KORN, T. M. (1968), *Mathematical Handbook for Scientists and Engineers*, 2nd edition, (McGraw-Hill, 1968) pp. 18.11-2.

LIENERT, B. R. (1994), *HYPOCENTER 3.2, A Computer Program for Locating Earthquakes Locally, Regionaly and Globaly*, Hawaii Institute of Geophysics and Planetology.

LEONARD, G., and TIKOCHINSKI, G. (1997), *A Proposal for Installation of Seismic Stations in KAMAG Area*, CTBT project (in Hebrew).

RABINOWITZ, N., and JOSWIG, M. (1995), *Detection, Localization and Identification of Local Seismic Events by a Small Array in Israel*, Scientific Progress Report, GIF I-0914-123.08/91.

ROBERTS, R. G., CHRISTOFFERSSON, A., and CASSIDY, F. (1989), *Real-time Event Detection, Phase Identification and Source Location Estimation Using Single Station Three-component Seismic Data*, Geophys. J. *97*, 471–480.

SAMBRIDGE, M. S., and KENNETT, B. L. N. (1986), *A Novel Method of Hypocentre Location*, Geophys. J. R. astr. Soc. *87*, 679–697.

WARD, PETER L., and GREGERSEN, SOREN (1973), *Comparison of Earthquake Locations Determined with Data from a Network of Stations and Small Tripartite Arrays of Kilauea Volcano, Hawaii*, Bull. Seismol. Soc. Am. *63*, 679–711.

(Received July 30, 1999, revised November 7, 1999, accepted February 9, 2000)

Notes to Authors

PAGEOPH welcomes original contributions in English (and occasionally in French and German) on all aspects of geophysics. All manuscripts should be submitted to the Regular Issues Editor-in-Chief, in triplicate, formatted with double spacing and wide margins. For further details see the following paragraphs.

Format of Manuscripts

Length and Page Charges: A paper should not exceed 16 printed pages including tables and figures. For articles exceeding 16 printed pages the authors will be charged sFr. 80.00 for each additional page. No page charges, except those for color prints, are required for contributors to special issues.

Title Page: This should include the the complete title, full names and addresses of all authors. In addition, corresponding authors should provide their fax number and e-mail address if they are available.

Abbreviated Title: It is necessary to indicate an abbreviated title, which will be used as a running head (no more than 50 characters including spaces).

Abstract: The abstract should be in English, and in the language of the text, if different. It has to be of no more than 10 sentences and should be concise and self-contained.

Keywords: Up to 6 keywords should be listed, suitable for incorporation into information-retrieval systems.

Text: The text must include a citation for each item listed under References; the approximate position of each figure should be indicated in the text. The metric system should be used throughout the text, figures, and tables.

Tables: Tables are to be presented on separate pages, with a brief title for each.

Figures: Figure captions and legends are to be typed on a separate page or pages as the last element in the manuscript. Make sure that line thickness and lettering allow an adequate size reduction. Heliographic or photocopies are not suitable for reproduction. Highquality, glossy, photographic prints must be submitted. Color prints are permitted but authors will be charged for them.

References: They are to be listed in alphabetical order in the following style:
Journal article: Haurwitz, B., and Cowley, A.D. (1973), The Diurnal and Semidurnal Barometric Oscillations, Global Distribution and Annual Variation, Pure Appl. Geophys. 102, 193-222.

Whole book: Bath, M., Introduction to Seismology (Birkhäuser, Basel 1973).

Article in a book: Haurwitz, B., and Cowley, A.D., Barometric oscillations, In Introduction to Seismology (ed. Bath. M.) (Birkhäuser, Basel 1973) pp. 193-222.

Submission of Manuscripts

Manuscripts must be submitted in triplicate, formatted with double line spacing and wide margins. Copies of the figure should be attached at the end of the manuscript. Original, high-quality, glossy figures may be submitted later. All manuscripts pages, including references, tables, and captions, should be numbered consecutively, starting with the title page as page one

All manuscripts should be submitted to Regular Issues Editor-in-Chief
Brian Mitchell
Department of Earth & Atmospheric Sciences
Saint Louis University
3507 Laclede Avenue
St. Louis, MO 63103, USA
e-mail: mitchell@eas.slu.edu

Delivering manuscripts in diskette form may substantially facilitate the publication process provided certain points are taken into consideration:

Texts on diskette should be delivered in either DOS or Macintosh format. They should be saved and delivered in two separate versions:
- with standard text format as offered by your word processing program, and
- in addition in Rich Text Format (RTF) or, as a last resort, as an ASCII file.

Numerous word processing programs offer these options when saving the text.
The final hard copy of the manuscript should be submitted together with the diskette.
The electronic and printed version must be absolutely identical.
All pictorial and graphic illustrations should be delivered as hard copy originals and must be 200% of the final printed size.
Digital drawings and graphs should be submitted in Encapsulated PostScript (EPS) or Tag Image File Format (TIFF) form. Do not fail to include a hard copy for ready viewing. Back-up copies of the diskettes must be kept. Diskette must be adequately protected for transport.

Galley Proofs
Unless indicated otherwise, galley proofs will be sent to the first-named author directly from Birkhäuser Verlag AG and should be returned with the least possible delay. Textual alterations made in the galley proof stage will be charged to the author. One copy of the corrected proof is to be returned immediately to
Editorial Office
Attn. Mrs. Renate D'Arcangelo
Harvard University
233 Pierce Hall, 29 Oxford Street
Cambridge, MA 02138, USA

The editorial office assumes no responsibility for delayed proofs, errors in the original manuscript, or major alterations in proofs for any reason.

Reprints
The authors will receive 50 reprints of each article without charge. Additional reprints may be ordered in lots of 50 when the final corrected page proofs are returned. Orders submitted thereafter are subject to considerably higher rates.

GPSR Compliance

The European Union's (EU) General Product Safety Regulation (GPSR) is a set of rules that requires consumer products to be safe and our obligations to ensure this.

If you have any concerns about our products, you can contact us on ProductSafety@springernature.com

In case Publisher is established outside the EU, the EU authorized representative is:

Springer Nature Customer Service Center GmbH
Europaplatz 3
69115 Heidelberg, Germany

Batch number: 10726322

Printed by Printforce, the Netherlands